Günter G. Hoffmann

Infrared and Raman Spectroscopy

Also of interest

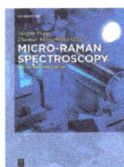

Micro-Raman Spectroscopy.
Theory and Application
Edited by Jürgen Popp, Thomas Mayerhöfer, 2020
ISBN 978-3-11-051479-7, e-ISBN (PDF) 978-3-11-051531-2

Polymer Surface Characterization
Edited by Luigia Sabbatini, Elvira De Giglio, 2022
ISBN 978-3-11-070104-3, e-ISBN (PDF) 978-3-11-070114-2

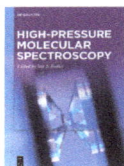

High-pressure Molecular Spectroscopy
Edited by Ian S. Butler, 2022
ISBN 978-3-11-066528-4, e-ISBN (PDF) 978-3-11-066861-2

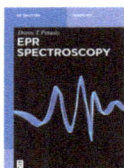

EPR Spectroscopy
Doros T. Petasis, 2022
ISBN 978-3-11-041753-1, e-ISBN (PDF) 978-3-11-042357-0

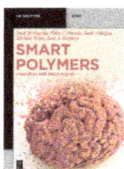

Smart Polymers.
Principles and Applications
José Miguel García, Félix Clemente García, José Antonio Reglero Ruiz,
Saúl Vallejos, Miriam Trigo-López, 2022
ISBN 978-1-5015-2240-6, e-ISBN (PDF) 978-1-5015-2246-8

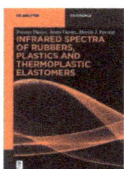

Infrared Spectra of Rubbers, Plastics and Thermoplastic Elastomers
Yvonne Davies, Jason Davies, Martin J. Forrest, 2019
ISBN 978-3-11-064408-1, e-ISBN (PDF) 978-3-11-064575-0

Günter G. Hoffmann

Infrared and Raman Spectroscopy

Principles and Applications

DE GRUYTER

Author
Prof. em. Dr. Günter G. Hoffmann
Hoffmann Datentechnik ㅂㅂㅂ
Wachstr. 29
46045 Oberhausen
Germany
hoffmann-oberhausen@t-online.de

ISBN 978-3-11-071754-9
e-ISBN (PDF) 978-3-11-071755-6
e-ISBN (EPUB) 978-3-11-071770-9

Library of Congress Control Number: 2023934180

Bibliographic information published by the Deutsche Nationalbibliothek
The Deutsche Nationalbibliothek lists this publication in the Deutsche Nationalbibliografie;
detailed bibliographic data are available on the Internet at http://dnb.dnb.de.

© 2023 Walter de Gruyter GmbH, Berlin/Boston
Cover image: Günter G. Hoffmann
Typesetting: Integra Software Services Pvt. Ltd.
Printing and binding: CPI books GmbH, Leck

www.degruyter.com

To Heike, Marcel, Vanessa, David, and Charlotte

I wish to thank Joel Stein and Walter de Gruyter GmbH for the opportunity to write this book and have it published.

I also would like to thank Ute Skambraks from DeGruyter and Suruthi Manogarane from Integra-PDY IN for their effort in the production of the book.

The disposition of drawings and Raman data by Dr. Armin Gembus, Dr. Jürgen Savatski, Michael Müller, and Bruker Optik GmbH is gratefully acknowledged.

Special thanks are due to Prof. Dr. H. W. Siesler, Essen, for carefully reading and commenting the manuscript.

Preface

Quite a few excellent books about vibrational spectroscopy have already been published. So why write a new one? The last years have seen the birth of new techniques and, first of all, a wealth of new applications. Therefore, a lot of new users need an introduction to these techniques and applications, but, if they are new to vibrational spectroscopy, an introduction to the parent techniques as well. A book like this will unavoidably contain some errors; hence, constructive criticism is always welcome. Please send errors, critics, comments, and remarks to VibSpec-GGH@t-online.de.

As a huge number of papers on vibrational spectroscopy have been published, the citations given in the chapters have to be taken as examples only, and, where appropriate, have been taken from the newer literature, while not disregarding older seminal works.

This book covers the literature until the beginning of 2023.

https://doi.org/10.1515/9783110717556-202

Contents

1 Introduction

Vibrational spectroscopies can detect and analyse vibrations in molecules. Mainly two different forms are used today: infrared and Raman spectroscopy.

In the most common form of infrared spectroscopy a substance is transmitted by infrared radiation of changing wavelength and the absorption plotted against wavelength. In the most common form of Raman spectroscopy a substance is irradiated by a laser. The intensity of the resulting radiation of lower wavelength is then plotted against its wavelength. In this case wavelength means the difference of its wavelength compared to the laser. Instead of wavelength most often its reciprocal wavenumber is used.

Vibrational spectroscopy is used by chemists to characterize their substances. If the spectra of substances are known, analytical chemists can use them to analyze a mixture of chemicals. Finally, samples may be analysed with spatial resolution, needing sophisticated (and expensive) instruments.

An exceptional (and extremely expensive) application of infrared spectroscopy is used on the James Webb Space Telescope (JWST). This instrument collects electromagnetic radiation with its 6.5 m diameter mirror (Figure 1.1). As expected, the instrument will not only take spectra of single points in space but also images of astronomical objects using focal plane arrays. The telescope carries instruments for the near infrared working in the 0.6–5 µm range and for the mid infrared working in the 5–28.5 µm region of the electromagnetic spectrum. The latter will be able to detect light from the first (oldest) galaxies. If a galaxy is very far from the observer, its visible and ultraviolet light is shifted by the cosmological redshift to this region. Early galaxies formed about 13.5 billion years from now, that is only a few hundred million years from the "Big Bang". The Hubble Space Telescope already has a limited NIR capability, enabling it to look at galaxies 500 million years from the "Big Bang", whereas Webb will be able to image galaxies formed only 200 million years from the beginning of the universe. Infrared detectors can see newly forming stars and faintly visible comets as well as objects in the Kuiper Belt. As an example, Figure 1.2 shows the carina nebula imaged by Hubble. With visible light, only a few stars can be detected in dust and gas. Near infrared light penetrates the dust and shows a plethora of stars.

The detectors of the JWST are described in detail by Rauscher *et al.* (2014).

Infrared radiation is also emitted from cooler objects and it is able to penetrate gas and dust clouds better than visible light. This is clearly visualized in the spectacular Figure 1.1.

The near-infrared imager and slitless spectrograph part of the FGS/NIRISS will be used to investigate the following science objectives: first light detection, exoplanet detection and characterization, and exoplanet transit spectroscopy.

https://doi.org/10.1515/9783110717556-001

Figure 1.1: The James Webb Space Telescope (artist's concept above) will be one of the primary instruments scientists use to continue the search for planets outside our solar system. Credits: NASA Goddard Space Flight Center from Greenbelt, MD, USA – James Webb Space Telescope: https://www.jwst.nasa.gov/content/webbLaunch/assets/images/mirrorAlignment/instrumentsCommOverallCompositeImage-1200px.jpg.

Figure 1.2: Carina nebula in visible light (left) and near-infrared light (right) as seen by the Hubble space telescope. Credits: NASA, ESA.

If the planet moves along the surface of its star, the starlight will be absorbed in the region of the chemicals in the planets atmosphere. Signals of vegetation (if present) should be visible too.

The focal plane of the infrared detector consists of two HgCdTe sensor chip assemblies. Each chip is a 2D array of 2,048 × 2,048 pixels, 18 µm pitch, hybridized onto a dedicated Read-Out integrated circuit. It will enable researchers to identify the "fingerprints" of molecules (Figure 1.3) like water, carbon dioxide, methane, and ammonia, which can't be identified with any other existing instruments.

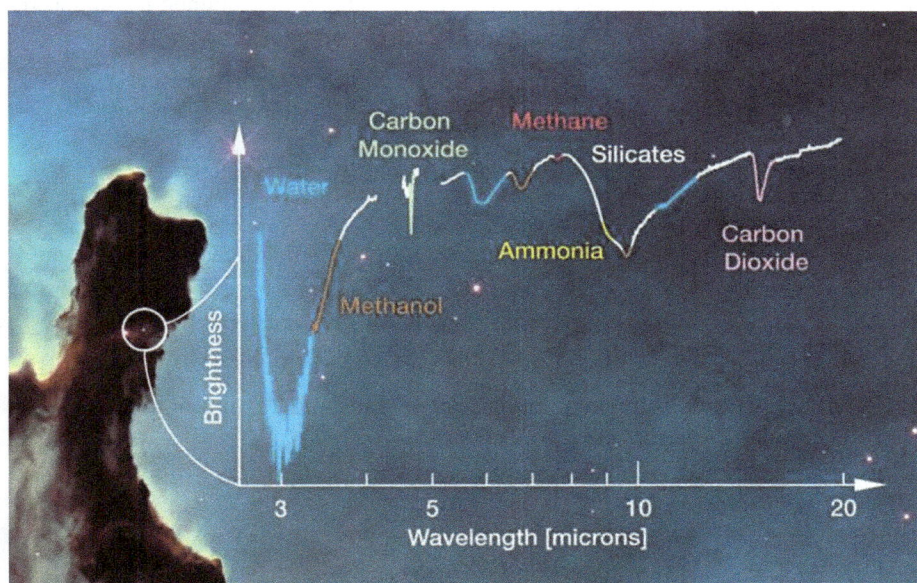

Figure 1.3: Compounds in space which can be detected by Webb. Credits: NASA, ESA, the Hubble Heritage Team, and M. McClure (Universiteit van Amsterdam) and A. Boogert (University of Hawaii).

1.1 Vibrational spectral libraries

A few libraries of vibrational spectra are available, some free, but most of them have to be bought or paid per spectrum (descriptions taken from the companies website). Some of the libraries (but not all) are described below:

1) Aldrich Raman Condensed Phase Library
This library represents a comprehensive collection of 18,454 FT-Raman spectra. It contains many common chemicals found in the *Aldrich Handbook of Fine Chemicals*.

A list of the Aldrich spectra can be accessed:

https://assets.thermofisher.com/TFS-Assets/CAD/Specification-Sheets/D02332~.pdf

2) Nicolet Library of FT-IR and Raman Spectra
The Nicolet library of Fourier transform infrared (FT-IR) and Raman spectra contains a broad choice of common chemicals that can be found in the Aldrich Handbook of Fine Chemicals. This spectral collection includes 3,119 FT-IR spectra and 3,119 matched Raman spectra, which are of interest to analytical laboratories.

https://www.thermofisher.com/order/catalog/product/833-009601

3) FT-Raman Forensic Library
This library is designed to assist forensic scientists and investigators in fast and easy identification of common street drugs. It contains a collection of 175 common drugs and related compounds frequently encountered in forensic analysis. It contains spectra of controlled substances, common contaminants, and "cutting" agents. The Raman technique provides a unique benefit in forensic science since sample analysis can be completed directly through plastic bags and glass containers.

http://www.thermo.com.cn/Resources/200802/productPDF_2080.pdf

4) Cayman Chemical™ Raman Library
This library, containing 187 spectra, includes synthetic cannabinoids, synthetic cathinones, synthetic piperazines, and tryptamines, which are the basis for newer drugs of abuse such as spice, bath salts, legal ecstasy, and novel psychoactive drugs.

http://www.thermofisher.com/order/catalog/product/834-105000

5) National Institute of Advanced Industrial Science and Technology (AIST) Spectral Database for Organic Compounds (SDBS)
From this database 3,573 Raman spectra of organic compounds are available and about 54,100 FT-IR spectra.

This is a free service!

http://sdbs.db.aist.go.jp/sdbs/cgi-bin/cre_index.cgi

6) KnowItAll Raman Spectral Library
Bio-Rad produced high-quality Raman spectral databases from their renowned Sadtler databases. The KnowItAll Raman Spectral Library offers access to over 339,000 IR spectra and 25,000 Raman spectra

https://sciencesolutions.wiley.com/solutions/technique/ir/knowitall-ir-collection/

7) FDM
FDM publishes FT-IR, ATR/FT-IR, and Raman libraries.

Separate libraries can be obtained for polymers, retail adhesives and sealants, plastic, organics, and inorganics. Together the FT-IR libraries contain 2,728 spectra taken at a spectral range of 400 to 4,000 cm^{-1} with a resolution of 4 cm^{-1}, and the Raman libraries contain 3,050 spectra measured with 780 nm excitation and a spectral range of 3,417 to 200 cm^{-1}.

http://www.fdmspectra.com/

8) S.T. Japan-USA

This database contains more than 170,000 ATR-FT-IR, FT-IR transmission, Raman, and NIR spectra. New spectra are continously added to the spectral libraries. The NIR database – 4,200 cm^{-1} to 10,000 cm^{-1} – includes 6,049 NIR spectra.

The largest FT-IR collection, the KBr database, contains 22,995 spectra measured using KBr sample preparation with a spectral range of 4,000 to 400 cm^{-1}.

The Raman collection contains 20,112 Raman spectra. The samples were measured with an excitation laser wavelength of 1,064, 785, 532, or 488 nm and are covering a spectral range from 4,000 cm^{-1} to 200 cm^{-1}.

http://www.stjapan-usa.com

9) NIST Chemistry WebBook

The NIST Chemistry WebBook shows evaluated infrared reference spectra from the Coblentz IR spectral collection.

This is a free service!

https://webbook.nist.gov/

The websites mentioned have been last accessed in February 2023. We cannot take responsibility for the content of the websites whose links are refered to in this book.

Primpke *et al.* (2018) describe reference database design for the automated analysis of microplastic samples based on FT-IR spectroscopy.

Books and Reviews

A number of excellent books and reviews about vibrational spectroscopy have been published and are used throughout this book.

Atkins, P. W. (Atkins 1983): *Molecular Quantum Mechanics*, 1983.

Bellamy, L. J. (Bellamy 1966): *Ultrarotspektrum und chemische Konstitution*, 1966.

Bentley, F. F.; Smithson, L. D.; Rozek, A. L. (Bentley, Smithson *et al.* 1968): *Infrared Spectra and Characteristic Frequencies 700–300 cm^{-1}*, 1968.

Berger, S.; and Sicker, D. (Berger and Sicker 2009): *Classics in Spectroscopy–Isolation and Structure and Elucidation of Natural Products*, 2009.

Brittain, E. F. H.; George, W. O.; and Wells, C. H. J. (Brittain, George *et al.* 1970): *Introduction to Molecular Spectroscopy*, 1970.

Califano, S. (Califano 1997): *Vibrational States*, 1976.

Chalmers, J. M.; and Griffiths, P. R.; (Eds.) (Chalmers and Griffths 2002): *Handbook of Vibrational Spectroscopy*, 2002.

Clark, R. J. H.; Hester, R. E., (Eds.): *Advances in Infrared and Raman Spectroscopy*, 1984.

Coblentz, W. W.: *Investigations of Infra-Red Spectra*, 1905.

Colthup, N. B.; Daly, L. H.; and Wiberley, S. E. (Colthup, Daly *et al.* 1990): *Introduction to Infrared and Raman Spectroscopy*, 3rd Ed., 1990.

Crocombe, R. A.; Leary, P. E.; Kammrath, B. W. (Eds): *Portable Spectroscopy and Spectrometry*, 2021.

Volume 1 (Technology and Instrumentation) (Crocombe, Leary *et al.* 2021).

Volume 2 (Applications) (Crocombe, Leary *et al.* 2021).

Cross, A. D. (Cross 1964): *Introduction to Practical Infrared Spectroscopy*, 1964.

Daimay, L.V.; Colthup, N.B.; Fateley, W.G.; Grasselli, J.G. (Lin-Vien, Colthup *et al.* 1991): *Handbook of Infrared and Raman Characteristic Frequencies of Organic Molecules*, 1991.

Das, R. S.; Agrawal, Y.K. (Das and Agrawal 2011): *Raman spectroscopy: Recent advancements, techniques and applications*, 2011.

Dollish, F. R.; Fateley W. G.; Bentley, F. F. (Dollish, Fateley *et al.* 1973): *Characteristic Raman Frequencies of Organic Compounds*, 1974.

Freeman, S. K. (Freeman 1965): *Interpretive Spectroscopy*, 1965.

Glazebrook, R.: *Dictionary of Applied Physics, Volume 4: Light, Sound, Radiology*, 1923.

Grasselli, J. G.; Bulkin, B. J. (Grasselli and Bulkin 1991): *Analytical Raman Spectroscopy*, 1991.

Griffiths, P. R.; de Haseth, J. A. (Griffiths and de Haseth 2007): *Fourier Transform Infrared Spectrometry*, 2nd Ed., 2007.

Günzler, H.; Böck, H.: *IR-Spektroskopie–Eine Einführung*, 1975.

Harmony, M. D. (Harmony 1972): *Introduction to Molecular Energies and Spectra*, 1972.

Herzberg, G. (Herzberg 1945): *Infrared and Raman Spectra*, 1945.

Hendra, P.; Jones, C.; Warner, G. (Hendra, Jones *et al.* 1991): *Fourier-Transform Raman Spectroscopy– Instrumentation and Applications*, 1991.

Hoffmann, G. G. (Hoffmann 2019): *Raman Spectroscopy–Principles and Applications in Chemistry, Physics, Materials Science, and Biology*, 2019.

Huber, K. P.; Herzberg, G. (Huber and Herzberg): *Constants of Diatomic Molecules*, 1979.

Hummel, D. O. (Hummel 1974): *Polymer Spectroscopy*, 1974.

Kawata, S.; Ohtsu (Kawata, Ohtsu *et al.* 2001): *Near-Field Optics and Surface Plasmon Polaritons*, 2001.

Kawata, S.; Ohtsu (Kawata, Ohtsu *et al.* 2002): *Nano-Optics*, 2002.

Kawata, S.; Shalaev, V. M. (Kawata and Shalaev 2007): *Tip Enhancement*, 2007.

Kohlrausch, K.W.F. (Kohlrausch 1931): *Der Smekal-Raman-Effekt*, 1931.

Kohlrausch, K.W.F. (Kohlrausch 1938): *Der Smekal-Raman-Effekt–Ergänzungsband*, 1938.

Kohlrausch, K.W.F. (Kohlrausch 1943): *Ramanspektren*, 1943.

Kuzmany, H. (Kuzmany 2009): *Solid-State Spectroscopy*, 2nd Ed., 2009.

Long, D. A. (Long 2002): *The Raman effect. A unified Treatment of the Theory of Raman Scattering by Molecules*, 2002.

Nafie (Nafie 2011), L. A.: *Vibrational Optical Activity–Principles and Applications*, 2011.

Novotny, L.; Hecht, B. (Novotny and Hecht 2012): *Principles of Nano-Optics*, 2012.

Pachler, K. G.; Matlok, F.; Gremlich, H.-U. (Pachler, Matlok *et al.* 1988): *Merck FT-IR Atlas*, 1988.

Polavarapu, P. L.: *Vibrational Spectra: Principles and Applications with Emphasis on optical Activity*, 1998.

Schlücker, S. (Schlücker 2010): *Surface Enhanced Raman Spectroscopy*, 2010.

Schrader, B. (Schrader 1996): *Raman/Infrared Atlas of Organic Compounds*, 2nd Ed., 1989.

Schrader, B. (Schrader 1995): *Infrared and Raman Spectroscopy: Methods and Applications*, 1995.

Shalaev, V. M.; Kawata, S. (Shalaev and Kawata 2007): *Nanophotonics with Surface Plasmons*, 2007.

Sicker *et al.* (Sicker, Zeller *et al.* 2019): *Natural Products - Isolation, Structure Elucidation, History*, 2019.

Siesler, H. W.; Ozaki, Y.; Kawata, S.; Heise, H. M. (Siesler, Ozaki *et al.* 2002): *Near Infrared Spectroscopy*, 2002.

Socrates, G. (Socrates 2001): *Infrared and Raman Characteristic Group Frequencies*, 3rd Ed., 2001.

Stahl, E.; Schild, W. (Stahl and Schild 1981): *Isolierung und Charakterisierung von Naturstoffen*, 1981.

Träger, F. (Ed.) (Träger 2007): *Springer Handbook of Lasers and Optics*, 2007.

Yates, J. T.; Madey, T. E.: (Yates 1987) *Vibrational Spectroscopy of Molecules on Surfaces*, 1987.

Vandenabeele, P. (Vandenabeele 2013): *Practical Raman Spectroscopy - an Introduction*, 2013.

Varsányi, G. (Varsányi 1969): *Vibrational Spectra of Benzene Derivatives*, 1969.

Volkmann, H. (Volkmann 1972): *Handbuch der Infrarot-Spektroskopie*, 1972.

Workman, J. (Workman 2000): *The Handbook of Organic Compounds, Three-Volume Set: NIR, IR, R, and UV-Vis Spectra Featuring Polymers and Surfactants*, 2000.

Workman, J.; and Weyer, L. (Workman and Weyer 2012): *Practical Guide to Interpretive Near-Infrared Spectroscopy*, 2012.

Weidlein, J.; Müller, U. und Dehnicke, K. (Weidlein, Müller *et al.* 1981): *Schwingungsfrequenzen I–Hauptgruppenelemente*, 1981.

Weidlein, J.; Müller, U. und Dehnicke, K. (Weidlein, Müller *et al.* 1982): *Schwingungsspektroskopie*, 1982.

Weidlein, J.; Müller, U. und Dehnicke, K. (Weidlein, Müller *et al.* 1986): *Schwingungsfrequenzen II–Nebengruppenelemente*, 1986.

Wilson, E. B.; Decius, J. C.; and Cross, P. C. (Wilson, Decius *et al.* 1955): *Molecular Vibrations*, 1955.

Woodward, L.A. (Woodward 1972): *Introduction to the Theory of Molecular Vibrations and Vibrational Spectroscopy*, 1972.

2 History

Wiliam Herschel (1800), better known for his astronomical works, discovered infrared radiation in 1800. Studying the visible spectrum of the sun, Herschel observed a non-visible radiation near the red end of the solar spectrum, detected by using a blackened mercury thermometer (Figure 2.2).

Vibrational spectroscopy plays an important role in chemistry, being very helpful in the structure determination of chemical compounds. Vibrations of molecules can be excited by infrared radiation. In 1882, W. de W. Abney and E. R. Festing (1882) by means of photography investigated the absorption spectra of 52 compounds to 1.2 μm, which was the limit of the sensibility of their photographic plates.

First full-scale IR spectra were published by William Weber Coblentz (Figure 2.1) not earlier than 1905; later a catalogue of the IR spectra of hundreds of organic compounds and some others were issued by Coblentz (Thomas 1991) . IR spectroscopy provides chemists with "fingerprints" of their compounds, but IR spectroscopy did not become common practice before the late 1930s, when the first commercial instruments were built by Perkin-Elmer and Beckman company.

Figure 2.1: William Weber Coblentz. Copyright © Wikimedia commons (Harris and Ewing, National Bureau of Standards, courtesy AIP Emilio Segrè Visual Archives).

The interested reader may find more information in a detailed review named "Infrared Spectroscopy – Past and Present", written by Barth and Haris (2009).

Raman spectroscopy, discovered about a quarter century later than IR spectroscopy, in principle provides the same information as its complementary technique, but with less disturbance by overtone and combination bands. Studying the same vibrations, it provides the identity and composition of the sample studied. Even external

https://doi.org/10.1515/9783110717556-002

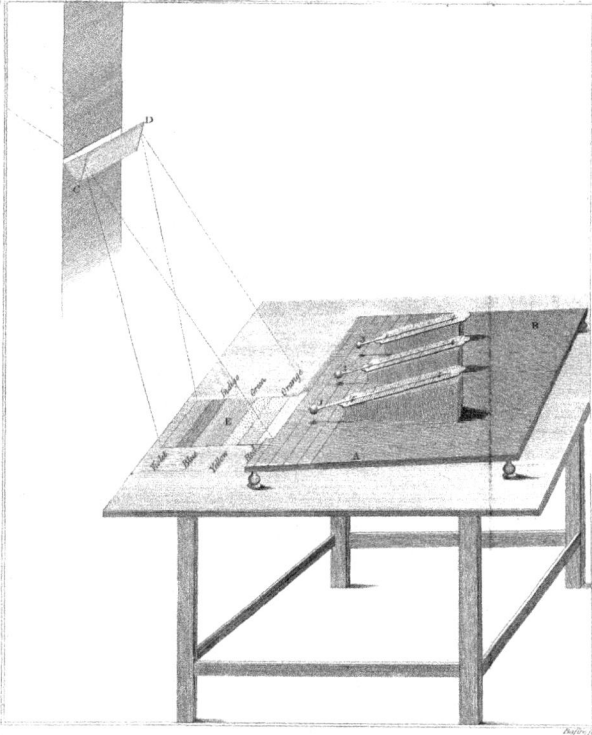

Figure 2.2: Drawing of Herschel's famous experiment dispersing sunlight. The experiment that discovered infrared radiation in 1800. A prism dispersed sunlight, the spectrum fell on a table, and a moveable stand with mounted thermometers. Thermometers 1 and 2 were exposed to the radiation, whereas thermometer 3 served as a control (Herschel 1800). Copyright ©1800 the Royal Society, London, with permission.

perturbations (strain, temperature, pressure) on molecules and their alignment can be analysed. The technique was limited to bulk material for a long time. The use of a microscope allowed for the study of samples down to 200–300 nm spatial resolution. According to Abbe's law a better resolution could not be achieved with a microscope:

$$d = 0.5\ \lambda/\mathrm{NA} \qquad \text{Abbe's law} \qquad (2.1)$$

where λ is the wavelength of radiation used and NA the numerical aperture of the microscope objective.

This situation changed when it was found that the electromagnetic field at the surface of a laser-irradiated sample can be hugely enhanced by using gold or silver as the tip of the scanning element in AFM. Now a resolution of 20 nm is routinely possible, in special cases even a resolution of less than 1 nm!

A recent article by Daniel Rabinovich (2016) brought my focus to a stamp portraying Sir Chandrasekhara Venkata Raman. I managed to buy a first-day cover of this

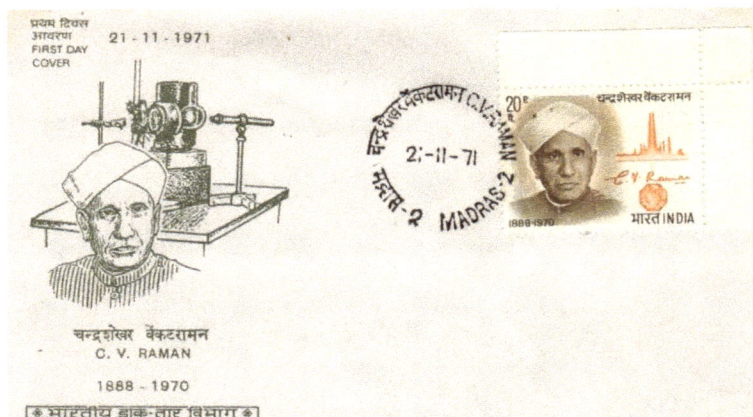

Figure 2.3: First-day cover honouring Sir Chandrasekhara Venkata Raman (collection of the author).

stamp (Figure 2.3), also showing the simple apparatus Raman used. The stamp shows Raman's portrait, his signature, the Raman spectrum of carbon tetrachloride, and a diamond.

Infrared and Raman spectroscopy are not only used to analyse bulk chemicals, but is more and more used to acquire images of samples. The size of those samples ranges from a few nanometres in molecules to thousands of light years in astronomy.

Long after the discovery of Herschel, the first systematic recording of infrared spectra started in 1905, whereas the first Raman spectrum was recorded not earlier than 1928. In a very short paper transmitted by teletype, Raman and Krishnan (1928) reported a new type of secondary radiation. This type of radiation was theoretically predicted by Adolf Smekal (1923) and experimentally proved independently one week earlier in 1928 by Grigory Landsberg and Leonid Isaakovich Mandelstam (1928).

Soon after the appearance of these seminal publications an avalanche of papers on Raman spectroscopy appeared. Kohlrausch in his 1931 book (Kohlrausch 1931) counts already 417 papers, and in his 1938 supplementary book (Kohlrausch 1938) counts 1,291 new papers.

The invention of the laser by Maiman in 1960 again multiplied the flood of Raman papers.

A few historical papers

Herschel, W. (1800). "Investigation of the Powers of the Prismatic Colours to Heat and Illuminate Objects; With Remarks, That Prove the Different Refrangibility of Radiant Heat. To Which is Added, an Inquiry into the Method of Viewing the Sun Advantageously, with Telescopes of Large Apertures and High Magnifying Powers." Philosophical Transactions of the Royal Society **90**: 255–283.

Herschel, W. (1800). "Experiments on the Refrangibility of the Invisible Rays of the Sun." Philosophical Transactions of the Royal Society **90**: 284–292.

Herschel, W. (1800). "Experiments on the Solar, and on the Terrestrial Rays that Occasion Heat; With a Comparative View of the Laws Which Occasion Them, Are Subject, in Order to Determine Whether They Are the Same, or Different. Part I and Part II." Philosophical Transactions of the Royal Society **90**: 293–326 and 437–538.

Smekal, A. (1923). „Zur Quantentheorie der Dispersion." Die Naturwissenschaften **11**(Nr. 43): 873–875.

Landsberg, G. S., L. I. Mandelstam (1928). "New phenomenon in scattering of light (preliminary report)." Journal of the Russian Physico-Chemical Society, Physics Section **60**: 335.

Landsberg, G., L. Mandelstam (1928). "Eine neue Erscheinung bei der Lichtzerstreuung in Krystallen." Naturwissenschaften **16**: 557.

Landsberg, G. S., L. I. Mandelstam (1928). "Über die Lichtzerstreuung in Kristallen." Zeitschrift für Physik **50**: 769.

3 Theory

Due to the limited space in this book, theoretical aspects of vibrational spectroscopy can only be briefly addressed. The interested reader is advised to consult the relevant literature (page 5).

3.1 Introduction

If a molecule is irradiated by infrared light (photons) of an energy equal to the energetic distance between the ground state of one of its vibrations and the corresponding first excited vibrational state, it is excited to this state or, with lesser probability, to a second excited vibrational state, thereby the photons are absorbed.

A different process may occur, when the molecule is irradiated by light with an enery higher than all of its vibrational states: if a non-absorbing substance is irradiated by monochromatic radiation, the electrons of the molecules are forced to vibrations of the same frequency. The molecules form a vibrating dipole that irradiates this frequency in all directions, thus scattering the incoming radiation. The effect is called Rayleigh scattering. A very small part of the incoming radiation is emitted at a different frequency, the difference of both frequencies corresponds to a frequency of a molecular vibration: Raman radiation.

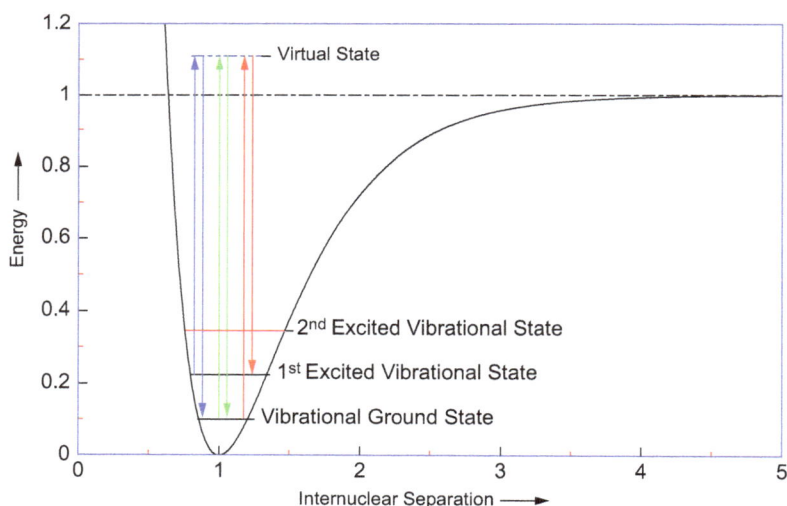

Figure 3.1: Morse potential of a diatomic molecule and energy levels symbolizing the scattered radiation. Green: Rayleigh scattering; blue: *anti*-Stokes Raman scattering, red: Stokes Raman scattering.

https://doi.org/10.1515/9783110717556-003

Normally the transition of the molecule (Figure 3.1) is from the vibrational ground state to the first excited state: the Stokes part of the Raman spectrum. But as at temperatures higher than absolute zero vibrationally excited states are also occupied according to Boltzmann's law, the transition of the molecule can also go from the first excited vibrational state to the ground state: the *anti*-Stokes part of the Raman spectrum.

The fraction of molecules in excited states can be calculated by

$$\frac{N_1}{N_0} = e^{-hc\tilde{v}_0/kT} \text{ Boltzmann's law} \tag{3.1}$$

where N_0 is the number of molecules in the vibrational ground state, N_1 is the number of molecules in the first excited state, h is Planck's constant, c is the velocity of light, $\tilde{v}_0$ is the frequency of the exciting radiation, k is Boltzmann's constant, and T is the temperature of the sample.

As an example, Figure 3.2 shows the Stokes and *anti*-Stokes part of the Raman spectrum of solid sulfur. It was taken with the radiation of a Nd:YAG laser ($\lambda_{exc} = 1,064$ nm) and an FT Raman spectrometer.

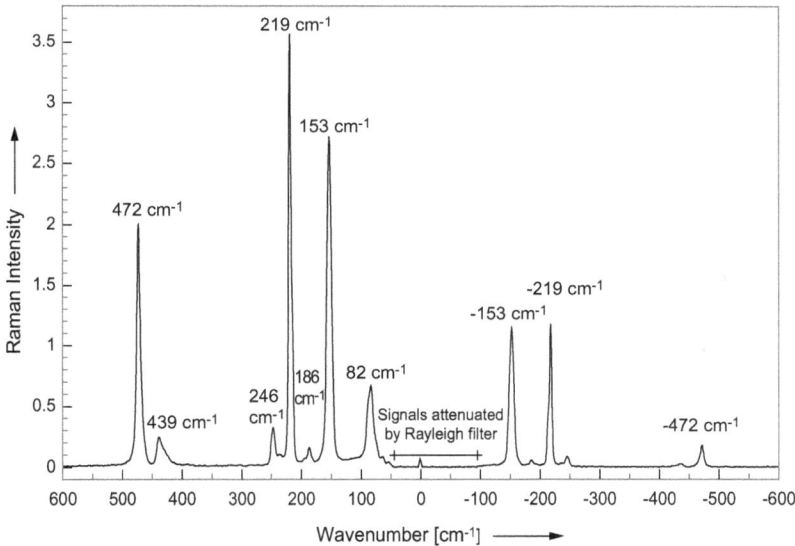

Figure 3.2: Stokes and *anti*-Stokes part of the Raman spectrum of solid sulfur ($\lambda_{exc} = 1,064$ nm, Hoffmann).

Using the following formula, the temperature of the sample can be calculated using both the Stokes and the *anti*-Stokes part of its Raman spectrum:

$$T = \frac{hc}{k}\tilde{\nu}_k \left[\ln \frac{I_k^-}{I_k^+} - \ln \left(\frac{\tilde{\nu}_0 - \tilde{\nu}_k}{\tilde{\nu}_0 + \tilde{\nu}_k} \right)^4 \right]^{-1} \tag{3.2}$$

where I_k^- and I_k^+ are the intensities of the *anti*-Stokes and Stokes scattered light, respectively, and T is the temperature of the specimen.

As an example from semiconductor industry, the Stokes and the *anti*-Stokes part of the Raman spectrum from the GaAs (110) surface has been published by Yamaguchi *et al.* (1999). The authors also calculated the temperature of the sample.

3.2 Classical theory

If a molecule consists of N atoms, then the molecule can be desribed by $3N$ cartesian coordinates x_a, y_a, z_a. $(a = 1, 2, \ldots, N)$. These coordinates can be transformed to $3N - 6$ internal coordinates q_i ($3N - 5$ for linear molecules), as there are three translational and three rotational motions.

The kinetic energy of the molecule is given by

$$2T = \sum_{a=1}^{N} m_a \left[\left(\frac{d\Delta x_a}{dt} \right)^2 + \left(\frac{d\Delta y_a}{dt} \right)^2 + \left(\frac{d\Delta z_a}{dt} \right)^2 \right] \tag{3.3}$$

We can replace the displacement coordinates Δx_1 to Δz_N by q_1 to q_{3N}:

$$q_1 = \sqrt{m_1}\Delta x_1, q_2 = \sqrt{m_1}\Delta y_1, q_3 = \sqrt{m_1}\Delta z_1, q_4 = \sqrt{m_2}\Delta x_2, \text{ etc.} \tag{3.4}$$

the mass-weighted cartesian displacement coordinates.

The kinetic energy of the molecule now reads as

$$2T = \sum_{i=1}^{3N} \dot{q}_i^2 \left(\dot{q} = \frac{dq}{dt} \right) \tag{3.5}$$

The internal mass-weighted cartesian displacement coordinates of a molecule can then be converted to normal coordinates Q and normal vibrations.

If the frequency and the phase of the motion of each coordinate of a mode are the same, this mode is called a normal mode. Its frequency is called a normal or fundamental frequency of the molecule.

The new set Q_k, $k = 1, 2, \ldots, 3N$, can be derived from the mass-weighted cartesian displacement coordinates q_i by the linear equation

$$Q_k = \sum_{i=1}^{3N} l_{ki}'' q_i \quad k = 1, 2, \cdots, 3N \tag{3.6}$$

We have to choose the coefficients l''_{ki} in a way that kinetic and potential energies are defined in the form

$$2T = \sum_{i=1}^{3N} \dot{Q}_k^2 \quad 2V = \sum_{i=1}^{3N} \lambda'_k Q_k^2 \left(\dot{Q} = \frac{dQ}{dt} \right) \tag{3.7}$$

Our sample molecule CCl_4 has $3*5 - 6 = 9$ normal vibrations (Table 3.1).

Some of the normal vibrations lead to a change in polarizability of the molecule.

The intensity of bands in the Raman spectrum of a compound is governed by the change in polarizability, a, that occurs during the vibration:

$$I_{Raman} = KI_L (\tilde{v}_0 - \tilde{v}_i)^4 \left(\frac{\partial a}{\partial Q} \right)^2 \tag{3.8}$$

where I_L is the power of the laser at the sample, $\tilde{v}_0 - \tilde{v}_i$ the wavenumber at which the band is measured, and $\partial a / \partial Q$ the change in polarizability with the normal coordinate of the vibration. This parameter is the Raman equivalent of absorptivity and is sometimes called the Raman cross section. The constant of proportionality, K, is dependent on the optical geometry, collection efficiency, detector sensitivity, and amplification.

3.3 Placzek's theory

The theoretical interpretation of the Raman effect is given in 1934 in the classical work by Placzek in the German "Handbuch der Radiologie" (Placzek 1934). The work has been translated in 1962 by the US Atomic energy commission (UCRL-256; Placzek 1934). Placzek's theory is only valid for molecules excited – but not absorbing – in the visible region ($\omega_1 \gg \omega_{v^f v^i}$).

Transitions are confined to the electronic ground state.

Following the treatment of Long (2002), the Placzek pure vibrational transition polarizability is

$$(a_{\rho\sigma})_{v^f v^i} = \langle v^f | \hat{a}_{\rho\sigma}(Q) | v^i \rangle \tag{3.9}$$

where a is the Raman scattering tensor, ρ and σ are the space-fixed axes of the molecule, f and i are the two states of the transition with their corresponding vibrational quantum numbers v^f and v^i, $\hat{a}_{\rho\sigma}$ the adiabatic polarizability, and Q are the nuclear coordinates.

3.4 *Ab initio* calculation of vibrational spectra

Theoretical vibrational spectra of compounds can be calculated with fair correlation to experimental spectra by *ab initio* methods using Hatree-Fock MO theory, but with the newer density functional theory (DFT) of Kohn and Sham (1965) results of comparable quality can be obtained, while the calculations are computationally less demanding. One has a large choice of functionals and basis sets, a compromise between precision and computational time is the use of DFT using the hybrid functional with Becke's three-parameter exchange (Becke 1988) and Lee–Yang–Parr correlation (Lee *et al.* 1988) (B3LYP) and the basis set 6–311G**. In the case of calculating Raman spectra, it is advantageous to add diffuse functions (e.g. using the basis set 6–311^{++}G**), taking into account that a transition to a virtual state is involved (e.g. Hoffmann 2003). The resulting vibrational frequencies have only to be scaled by a constant frequency scaling factor close to 1, for example, 0.9688 in the case of the mentioned method and basis set. An evaluation of scaling factors is given by Merrick *et al.* (2007) and Palafox (2018).

A good introduction to *ab initio* calculations is given by Cramer (2004) in his book on computational chemistry.

An example of the DFT calculation of vibrational spectra is shown in Figure 3.3 and Table 3.1. The Raman spectrum of isotopically pure carbon tetrachloride is calculated with the aid of the quantum chemical program Gaussian09 rev E.01 by Frisch *et al.* (2009). DFT is employed using the hybrid functional B3LYP and the aug-cc-pVTZ basis set, which means that for carbon 4s,3p,2d,1f and for chlorine 5s,4p,2d,1f functions are used. For comparison the experimental spectrum is shown in Figure 3.4. Note that with higher resolution the bands of the different isotopomers can be distinguished (inset showing the 459 cm^{-1} band).

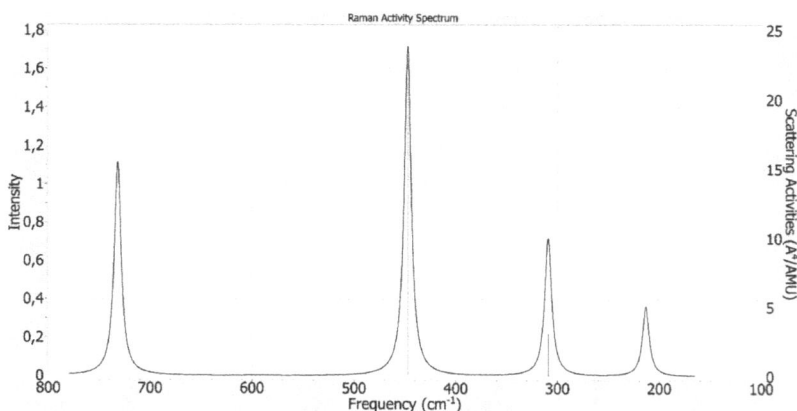

Figure 3.3: Raman spectrum of CCl$_4$, calculated with Gaussian09 (Frisch *et al.* 2009) (DFT/B3LYP/aug-cc-pVTZ which means that for carbon 4s,3p,2d,1f and for chlorine 5s,4p,2d,1f functions are used (Hoffmann 2017–2018).

Table 3.1: Calculated IR and Raman spectra of carbon tetrachloride.

Vibration number	1, 2	3, 4, 5	6	7, 8, 9
Symmetry	E	T_2	A_1	T_2
Harmonic frequencies (cm^{-1})	213.26	309.00	447.09	731.84
Reduced masses (amu)	34.969	32.461	34.969	13.021
Force constants (mdyn/Å)	0.9370	1.8262	4.1183	4.1089
IR intensities (km/mol)	0.0000	0.4344	0.0000	161.47
Raman scattering activities (Å^4/AMU)	2.2793	3.0679	22.669	5.2289
Depolarization ratios for plane-polarized incident light	0.7500	0.7500	0.0000	0.7500
Depolarization ratios for unpolarized incident light	0.8571	0.8571	0.0000	0.8571

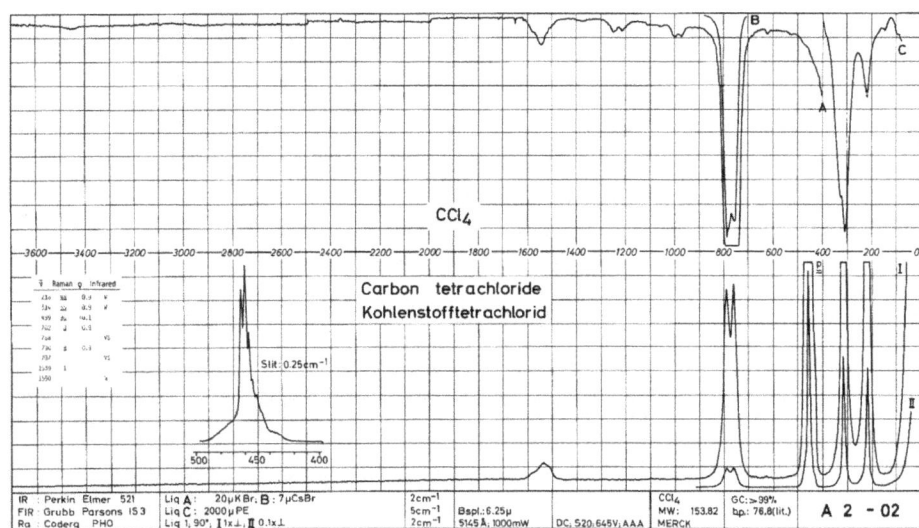

Figure 3.4: Infrared and Raman spectra of liquid CCl_4 (from Schrader (1996), Infrared Atlas of Organic Compounds, 2nd ed., page A2–02. Copyright ©1996 Wiley-VCH Verlag GmbH & Co. KGaA. Reproduced with permission).

The infrared spectrum of CCl_4 shows a broad, partly resolved band at 460 cm^{-1} (Menzies and Whiddington 1939). As chlorine is composed of two isotopes, the splitting can be attributed to the different species of the molecule. The band at 461.5 cm^{-1} belongs to $C^{35}Cl_4$ (31,6%), at 458.4 cm^{-1} to $C^{35}Cl_3^{37}Cl$ (42.2%), and at 455.1 cm^{-1} to $C^{35}Cl_2^{37}Cl_2$ (21.1%). The bands of $C^{35}Cl^{37}Cl_3$ (4.7%) and of $C^{37}Cl_4$ (0.4%) can only be detected under special conditions.

If a compound is excited by linearly polarized light, different spectra are obtained if the Raman spectrum is observed through a linear polarizer called the analyser. The analyser may have its direction of polarization either parallel or perpendicular to the polarization direction of the polarized light (see Figure 3.5). If the polarization ratio

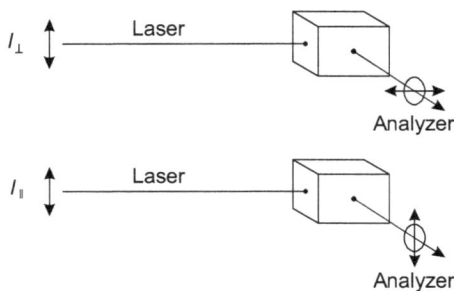

Figure 3.5: Measurement of the depolarization ratio for plane-polarized incident light.

$\rho = \frac{I_\perp}{I_\parallel}$ is between 0 and ¾ the band is called polarized; if ρ is ¾ it is called depolarized. See Table 3.1 for the calculated ρ values of CCl_4.

3.5 Group theory

Group theory is very helpful in the interpretation of vibrational spectra. The use of group theory is nicely described in the book of Cotton (1971).

To identify the symmetry group of a molecule we have to look at the molecule under study and identify any possible elements of symmetry. Table 3.2 summarizes the possible elements of symmetry and the corresponding symmetry operations. The simplest element of symmetry is the Identity E that causes no change in the molecule and is – of course – a symmetry element of every molecule. Well known to us is the mirror plane σ because we know its symmetry operation, the reflection, from our mirrors. The centre of inversion i is more complicated. Its symmetry operation is executed by moving each atom in the molecule along a straight line through the inversion centre to a point of equal distance from the inversion centre. The resulting configuration must then be indistinguishable from the original. An n-fold proper axis C_n upon an n-fold rotation about it produces an indistinguishable configuration of the molecule. A n-fold improper axis S_n operation can be thought of as an n-fold proper rotation followed by a reflection through a plane perpendicular to the rotation axis and passing through the centre of the molecule.

To start the symmetry identification of a molecule we first look for an axis of rotation. After having identified the axis of rotation with the highest n, we look for mirror planes and classify them as vertical or horizontal.

For example, for the water molecule we find a two-fold axis of rotation and a vertical mirror plane: the symmetry group of water is C_{2v} (Table 3.3).

The notation for the symmetry groups (e.g. C_{2v}) is called Schönflies notation.

On the following pages some character tables of common symmetry groups are shown together with molecules belonging to this group.

Table 3.2: Elements of symmetry and the corresponding symmetry operations.

Elements of symmetry	Name of element	Symmetry operation
E	Identity	No change
i	Centre of inversion	Inversion of all atoms through the centre
σ	Plane	Reflection through a plane
C_n	n-Fold proper axis	n-Fold rotation about an axis
S_n	n-Fold improper axis	n-Fold rotation followed by a reflection through a plane $\perp$ to the rotation axis

The following tables are chosen as the most important symmetry groups for spectroscopy of molecules. Other symmetry groups may be found in the book of Hollas (1972).

The first column contains the irreproducible presentations (symmetry species) of the group. A and B are one-dimensional species, A is symmetric, B antisymmetric with respect to the generating rotation operation. E, F, G . . . indicate two-, three-, four- . . . dimensional symmetry species. In species A and B we distinguish symmetry with respect to rotation about a C_2 axis perpendicular to the generating axis by subscripts 1 and 2. Symmetry with respect to inversion is specified by g and u (German: *gerade, ungerade*).

Rotations R, translations T, and polarizations a are shown in the last two columns, important for the prediction of the occurance of IR and Raman lines in the vibrational spectrum.

The group C_{2v} (Table 3.3) has a two-fold proper axis and two mirror planes which are perpendicular to each other. An example of a molecule belonging to this group is H_2O, and others are *ortho*- and *meta*-dichlorobenzene and difluoromethane.

Table 3.3: Character table of the symmetry group C_{2v}.

C_{2v}	E	C_2	$\sigma_v(xz)$	$\sigma'_v(yz)$	R and T	Polarization
A_1	1	1	1	1	T_z	a_{xx}, a_{yy}, a_{zz}
A_2	1	1	−1	−1	R_z	a_{xy}
B_1	1	−1	1	−1	R_y, T_x	a_{xz}
B_2	1	−1	−1	1	R_x, T_y	a_{yz}

The group C_{3v} (Table 3.4) has a three-fold proper axis and three mirror planes which are perpendicular to each other. $CHCl_3$ belongs to this group.

Table 3.4: Character table of the symmetry group C_{3v}.

C_{3v}	E	$2C_3(z)$	$3\sigma_v$	R and T	Polarization
A_1	1	1	1	T_z	$a_{xx} + a_{yy}, a_{zz}$
A_2	1	1	−1	R_z	
E	2	−1	0	$(T_x, T_y), (R_x, R_y)$	$(a_{yz}, a_{xz}), (a_{xx}, -a_{yy}, a_{xy})$

The group D_{3d} (Table 3.5) has a six-fold improper axis and and a two-fold proper axis perpendicular to it. Three mirror planes σ_d are parallel to the improper axis, containing it. The ethane molecule exists as two conformers: the eclipsed conformation belonging to symmetry group D_{3h} (Table 3.6) and the energetically more favourable staggered conformer belonging to symmetry group D_{3d}.

Table 3.5: Character table of the symmetry group D_{3d}.

$D_{3d} \equiv S_{6v}$	E	$2S_6(z)$	$2S_6^2 \equiv 2C_3$	$S_6^3 \equiv S_2 \equiv i$	$3C_2$	$3\sigma_d$	R and T	Polarization
A_{1g}	1	1	1	1	1	1		$a_{xx} + a_{yy}, a_{zz}$
A_{1u}	1	−1	1	−1	1	−1		
A_{2g}	1	1	1	1	−1	−1	R_z	
A_{2u}	1	−1	1	−1	−1	1	T_z	
E_g	2	−1	−1	2	0	0	(R_x, R_y)	$(a_{yz}, a_{xz}), (a_{xx}, -a_{yy}, a_{xy})$
E_u	2	1	−1	−2	0	0	T_x, T_y	

1,3,5-Trifluorobenzene has a three-fold axis of rotation and three two-fold axes of rotation perpendicular to it. The molecule additionally has three vertical mirror planes and one horizontal plane perpendicular to them: the molecule belongs to D_{3h}.
Another molecule belonging to D_{3h} is boron trifluoride.

Table 3.6: Character table of symmetry group D_{3h}.

D_{3h}	E	$2C_3$	$3C_2$	σ_h	$2S_3$	$3\sigma_v$	R and T	Polarization
A_1'	1	1	1	1	1	1		$a_{xx} + a_{yy}, a_{zz}$
A_2'	1	1	−1	1	1	−1	R_z	
E'	2	−1	0	2	−1	0	(T_x, T_y)	$(a_{xx}, -a_{yy}, a_{xy})$
A_1''	1	1	1	−1	−1	−1		
A_2''	1	1	−1	−1	−1	1	T_z	
E''	2	−1	0	−2	1	0	(R_x, R_y)	(a_{xz}, a_{yz})

Benzene with its two six-fold proper symmetry axes and a mirror plane in the molecular plane belongs to the point group D_{6h} (Table 3.7). This group has 12 elements of symmetry.

Table 3.7: Character table of the symmetry group D_{6h}.

D_{6h}	E	$2C_6$	$2C_3$	C_2	$3C_2'$	$3C_2''$	i	$2S_3$	$2S_6$	σ_h	$3\sigma_d$	$3\sigma_v$		Polarization
A_{1g}	1	1	1	1	1	1	1	1	1	1	1	1		α_{zz}
A_{2g}	1	1	1	1	−1	−1	1	1	1	1	−1	−1	R_z	
B_{1g}	1	−1	1	−1	1	−1	1	−1	1	−1	1	−1		
B_{2g}	1	−1	1	−1	−1	1	1	−1	1	−1	−1	1		
E_{1g}	2	1	−1	−2	0	0	2	1	−1	−2	0	0	(R_x, R_y)	$(\alpha_{xz}, \alpha_{yz})$
E_{2g}	2	−1	−1	2	0	0	2	−1	−1	2	0	0		$(\alpha_{xx\text{-}yy}, \alpha_{xy})$
A_{1u}	1	1	1	1	1	1	−1	−1	−1	−1	−1	−1		
A_{2u}	1	1	1	1	−1	−1	−1	−1	−1	−1	1	1	T_z	
B_{1u}	1	−1	1	−1	1	−1	−1	1	−1	1	−1	1		
B_{2u}	1	−1	1	−1	−1	1	−1	1	−1	1	1	−1		
E_{1u}	2	1	−1	−2	0	0	−2	−1	1	2	0	0	(T_x, T_y)	
E_{2u}	2	−1	−1	2	0	0	−2	1	1	−2	0	0		

Linear molecules like H–Cl or X–C≡C–H belong to point group $C_{\infty v}$ (Table 3.8) or, like the more symmetrical H–C≡C–H, to $D_{\infty h}$ (Table 3.12).

Table 3.8: Character table of symmetry group $C_{\infty v}$.

$C_{\infty v}$	E	$2C_\infty^{\Phi}$	$2C_\infty^{2\Phi}$	$\ldots$	$\infty\sigma_v$	R and T	Polarization
$A_1 \equiv \Sigma^+$	1	1	1	$\ldots$	1	T_z	$\alpha_{xx} + \alpha_{yy}, \alpha_{zz}$
$A_2 \equiv \Sigma^-$	1	1	1	$\ldots$	−1	R_z	
$E_1 \equiv \Pi$	2	$2\cos\Phi$	$2\cos 2\Phi$	$\ldots$	0	(T_x, T_y) (R_x, R_y)	$(\alpha_{yz}, \alpha_{xz})$
$E_2 \equiv \Delta$	2	$2\cos 2\Phi$	$2\cos 2\cdot2\Phi$	$\ldots$	0		$(\alpha_{xx} - \alpha_{yy}, \alpha_{xy})$
$E_3 \equiv \Phi$	2	$2\cos 3\Phi$	$2\cos 2\cdot3\Phi$	$\ldots$	0		
$\ldots$	$\ldots$	$\ldots$	$\ldots$	$\ldots$	$\ldots$		

The group T_d (Table 3.9) is also called the tetraeder group, and it has eight three-fold proper axes and three two-fold proper axes, all going through the centre atom of the molecule. The C_2 axes are dividing the angle between every pair of C_3 axes. Six mirror planes σ_d are parallel to the C_2 axes and containing them.

Methane and CCl_4 belong to the tetraeder group T_d.

Table 3.9: Character table of the symmetry group T_d.

T_d	E	$8C_3$	$3C_2$	$6S_4\equiv3C_2$	$6\sigma_d$	R and T	Polarization
A_1	1	1	1	1	1		$\alpha_{xx}+\alpha_{yy}+\alpha_{zz}$
A_2	1	1	1	-1	-1		
E	2	-1	2	0	0		$(2\alpha_{zz}-\alpha_{xx}-\alpha_{yy}, \alpha_{xx,}-\alpha_{yy})$
$T_1\equiv F_1$	3	0	-1	1	-1	(R_x, R_y, R_z)	
$T_2\equiv F_2$	3	0	-1	-1	1	(T_x, T_y, T_z)	$(\alpha_{xy}, \alpha_{xz}, \alpha_{yz})$

C_{70} has D_{5h} symmetry as has dicyclopentadienyl ruthenium.

Dicyclopentadienyl iron (ferrocene) has D_{5d} symmetry (Seiler and Dunitz 1979).

Buckminsterfullerene (C_{60}) is one of the rare examples of molecules with I_h symmetry (Tab. 3.11). Its structure is that of a truncated icosahedron, other objects with I_h symmetry are the pentagondodecahedron and the icosahedron (Fig. 3.6).

Table 3.10: Character tables of the symmetry groups D_{5h} and D_{5d}.

D_{5h}

D_{5h}	E	$2C_5$	$2C_5^2$	$5C_2$	σ_h	$2S_5$	$2S_5^3$	$5\sigma_v$		Polarization
A_1'	1	1	1	1	1	1	1	1		$\alpha_{xx}+\alpha_{yy}, \alpha_{zz}$
A_2'	1	1	1	1	-1	1	1	1	-1	R_z
E_1'	2	$2\cos(2\pi/5)$	$2\cos(4\pi/5)$	0	2	$2\cos(2\pi/5)$	$2\cos(4\pi/5)$	0	(T_x, T_y)	
E_2'	2	$2\cos(4\pi/5)$	$2\cos(2\pi/5)$	0	2	$2\cos(4\pi/5)$	$2\cos(2\pi/5)$	0		$(\alpha_{xx-yy}, \alpha_{xy})$
A_1''	1	1	1	1	-1	-1	-1	-1		
A_2''	1	1	1	1	-1	-1	-1	1		T_z
E_1''	2	$2\cos(2\pi/5)$	$2\cos(4\pi/5)$	0	-2	$-2\cos(2\pi/5)$	$-2\cos(4\pi/5)$	0	(R_x, R_y)	$(\alpha_{xz}, \alpha_{yz})$
E_2''	2	$2\cos(4\pi/5)$	$2\cos(2\pi/5)$	0	-2	$-2\cos(4\pi/5)$	$-2\cos(2\pi/5)$	0		

D_{5d}

D_{5d}	E	$2C_5$	$2C_5^2$	$5C_2$	i	$2S_{10}^3$	$2S_{10}$	$5\sigma_d$		Polarization
A_{1g}	1	1	1	1	1	1	1	1		$\alpha_{xx}+\alpha_{yy}, \alpha_{zz}$
A_{2g}	1	1	1	-1	1	1	1	-1	R_z	
E_{1g}	2	$2\cos(2\pi/5)$	$2\cos(4\pi/5)$	0	2	$2\cos(2\pi/5)$	$2\cos(4\pi/5)$	0	(R_x, R_y)	$(\alpha_{xz}, \alpha_{yz})$
E_{2g}	2	$2\cos(4\pi/5)$	$2\cos(2\pi/5)$	0	2	$2\cos(4\pi/5)$	$2\cos(2\pi/5)$	0		$(\alpha_{xx-yy}, \alpha_{xy})$

Table 3.10 (continued)

D_{5d}										
D_{5d}	E	$2C_5$	$2C_5^2$	$5C_2$	i	$2S_{10}^3$	$2S_{10}$	$5\sigma_d$		Polarization
A_{1u}	1	1	1	1	−1	−1	−1	−1		
A_{2u}	1	1	1	−1	−1	−1	−1	1		T_z
E_{1u}	2	$2\cos(2\pi/5)$	$2\cos(4\pi/5)$	0	−2	$-2\cos(2\pi/5)$	$-2\cos(4\pi/5)$	0		(T_x, T_y)
E_{2u}	2	$2\cos(4\pi/5)$	$2\cos(2\pi/5)$	0	−2	$-2\cos(4\pi/5)$	$-2\cos(2\pi/5)$	0		

Buckminsterfullerene (C_{60}) is one of the rare examples of molecules with I_h symmetry (Tab. 3.11). Its structure is that of a truncated icosahedron, other objects with I_h symmetry are the pentagondodecahedron and the icosahedron (Fig. 3.6).

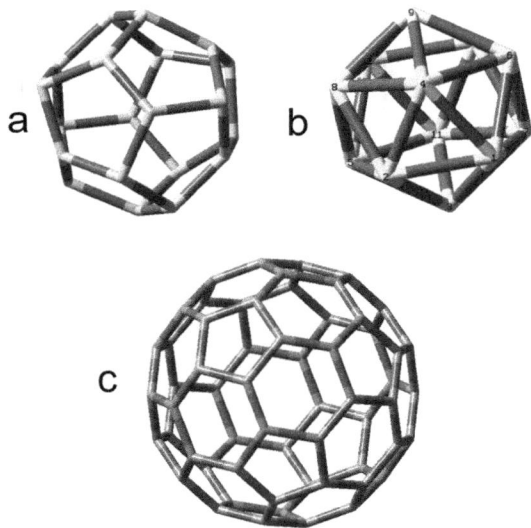

Figure 3.6: Three objects belonging to the symmetry group I_h: the pentagondodecahedron (a), the icosahedron (b), and the buckminsterfullerene structure (c).

Table 3.11: Character table of the symmetry group I_h.

I_h	E	$12C_5$	$12(C_5)^2$	$20C_3$	$15C_2$	i	$12S_{10}$	$12(S_{10})^3$	$20S_6$	15σ	T and R	Polarization
A_g	1	1	1	1	1	1	1	1	1	1		$\alpha_{xx}+\alpha_{yy}+\alpha_{zz}$
T_{1g}	3	$-2\cos(4\pi/5)$	$-2\cos(2\pi/5)$	0	-1	3	$-2\cos(2\pi/5)$	$-2\cos(4\pi/5)$	0	-1	(R_x, R_y, R_z)	
T_{2g}	3	$-2\cos(2\pi/5)$	$-2\cos(4\pi/5)$	0	-1	3	$-2\cos(4\pi/5)$	$-2\cos(2\pi/5)$	0	-1		
G_g	4	-1	-1	1	0	4	-1	-1	1	0		
H_g	5	0	0	-1	1	5	0	0	-1	1		$[2\alpha_{zz}-\alpha_{yy}-\alpha_{xx},\,\alpha_{xx}-\alpha_{yy},\,\alpha_{xy},\,\alpha_{xz},\,\alpha_{yz}]$
A_u	1	1	1	1	1	-1	-1	-1	-1	-1		
T_{1u}	3	$-2\cos(4\pi/5)$	$-2\cos(2\pi/5)$	0	-1	-3	$2\cos(2\pi/5)$	$2\cos(4\pi/5)$	0	1	(T_x, T_y, T_z)	
T_{2u}	3	$-2\cos(2\pi/5)$	$-2\cos(4\pi/5)$	0	-1	-3	$2\cos(4\pi/5)$	$2\cos(2\pi/5)$	0	1		
G_u	4	-1	-1	1	0	-4	1	1	-1	0		
H_u	5	0	0	-1	1	-5	0	0	1	-1		

Table 3.12: Character table of the symmetry group $D_{\infty h}$.

$D_{\infty h}$	E	$2\,C_\infty^\varphi$	$\dots$	$\infty\sigma_v$	i	$2\,S_\infty^\varphi$	$\dots$	∞C_2	R and T	Polarization
$A_{1g} \equiv \Sigma_g^+$	1	1	$\dots$	1	1	1	$\dots$	1		$\alpha_{xx}+\alpha_{yy},\ \alpha_{zz}$
$A_{2g} \equiv \Sigma_g^-$	1	1	$\dots$	-1	1	1	$\dots$	-1	R_z	
$E_{1g} \equiv \Pi_g$	2	$2\cos\varphi$	$\dots$	0	2	$-2\cos\varphi$	$\dots$	0	$(R_x,\ R_y)$	$(\alpha_{yz},\ \alpha_{xz})$
$E_{2g} \equiv \Delta_g$	2	$2\cos(2\varphi)$	$\dots$	0	2	$2\cos(2\varphi)$	$\dots$	0		$(\alpha_{xx},\ -\alpha_{yy},\ \alpha_{xy})$
$E_{3g} \equiv \Phi_g$	2	$2\cos(3\varphi)$	$\dots$	0	2	$-2\cos(3\varphi)$	$\dots$	0		
E_{ng}	2	$2\cos(n\varphi)$	$\dots$	0	2	$(-1)^n\,2\cos(n\varphi)$	$\dots$	0		
$\dots$	$\dots$	$\dots$	$\dots$	$\dots$	$\dots$	$\dots$	$\dots$	$\dots$		
$A_{1u} \equiv \Sigma_u^+$	1	1	$\dots$	1	-1	1	$\dots$	-1	T_z	
$A_{2u} \equiv \Sigma_u^-$	1	1	$\dots$	-1	-1	1	$\dots$	1		
$E_{1u} \equiv \Pi_u$	2	$2\cos\varphi$	$\dots$	0	-2	$2\cos\varphi$	$\dots$	0	$(T_x,\ T_y)$	
$E_{2u} \equiv \Delta_u$	2	$2\cos(2\varphi)$	$\dots$	0	-2	$-2\cos(2\varphi)$	$\dots$	0		
$E_{3u} \equiv \Phi_u$	2	$2\cos(3\varphi)$	$\dots$	0	-2	$2\cos(3\varphi)$	$\dots$	0		
E_{nu}	2	$2\cos(n\varphi)$	$\dots$	0	-2	$(-1)^{n+1}\,2\cos(n\varphi)$	$\dots$	0		
$\dots$	$\dots$	$\dots$	$\dots$	$\dots$	$\dots$	$\dots$	$\dots$	$\dots$		

4 Instruments

Vibrational spectra can be measured with two types of instruments: dispersive and interferometric, which can again be divided into scanning and non-moving spectrometers. Dispersive instruments use a prism (outdated) or a grating for the dispersion of photons of different energy, while the interferometric instruments use a Michelson interferometer for this purpose. Scanning instruments use a slit with attached detector to record the spectrum one data point after the other, while non-moving spectrometers use an array of multiple detectors (e.g. a CCD) to record the whole spectrum or interferogram at once.

4.1 Sources of radiation

Raman spectroscopy needs a source of monochromatic light, whereas IR spectroscopy needs a continuum or a tunable laser. Due to advances in light technology, the sources changed in the course of time.

4.1.1 Sunlight

Sunlight, filtered to yield nearly monochromatic light, was the first radiation source for Raman spectroscopy.

4.1.2 Lasers

The invention of the ruby laser by Maiman (1960) changed Raman spectroscopy tremendously. The acronym LASER stands for "Light Amplification by Stimulated Emission of Radiation". A very strong source of monochromatic radiation was now at hand, with a wavelength of 694.3 nm in the case of the ruby laser. A laser consists of a medium, where an excess of higher excited states ("negative temperature") is produced. This medium was a ruby rod in the first laser, where excited states of chromium atoms in a sapphire matrix are produced by a xenon flash lamp. One of the polished ends of the laser rod was covered by a 100% reflecting mirror, and the other end with a mirror of 99% reflectivity. This is the end where the laser beam is emitted.

Figure 4.1 shows a sketch of Maiman's first laser and Figure 4.2 shows a photo of its components.

Soon after the invention of the ruby laser, lasers of different wavelengths became available. The correct choice of the exciting radiation is of upmost importance. The efficiency of the Raman process is dependent on the fourth power of the wavelength, i.e. ν^4.

https://doi.org/10.1515/9783110717556-004

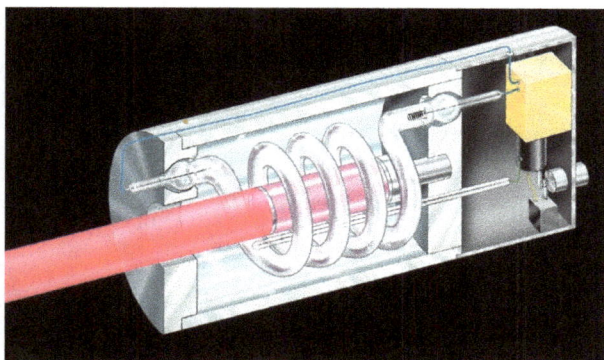

Figure 4.1: Sketch of the first ruby laser. Copyright © Wikimedia commons.

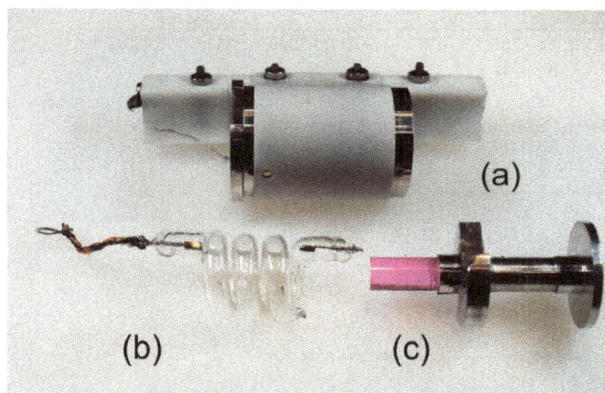

Figure 4.2: Components of Maiman's original ruby laser: (a) housing, (b) flash lamp, and (c) ruby rod with holder. Copyright ©Wikimedia commons.

Therefore, one would probably suppose that radiation in the UV is best. The problem is that most substances produce intense fluorescence spectra upon excitation by shorter wavelengths. As the excitation of fluorescence is by a factor of some 10,000 more effective than the excitation of Raman radiation, it completely obscures the Raman spectra.

Figure 4.3 shows the excitation efficiency of different laser lines normalized to the intensity at 488 nm.

Figure 4.4 shows the wavelengths of lasers used in Raman spectrometers and the corresponding regions of the Stokes Raman spectra excited by them.

The choice of lasers is not only governed by the wavelength but also by practical considerations. Diode lasers are cheap and easy to operate, but emit a rather broadband radiation. For high resolution applications, their bandwidth has to be reduced by an external cavity. For infrared spectroscopy normally a broadband source is

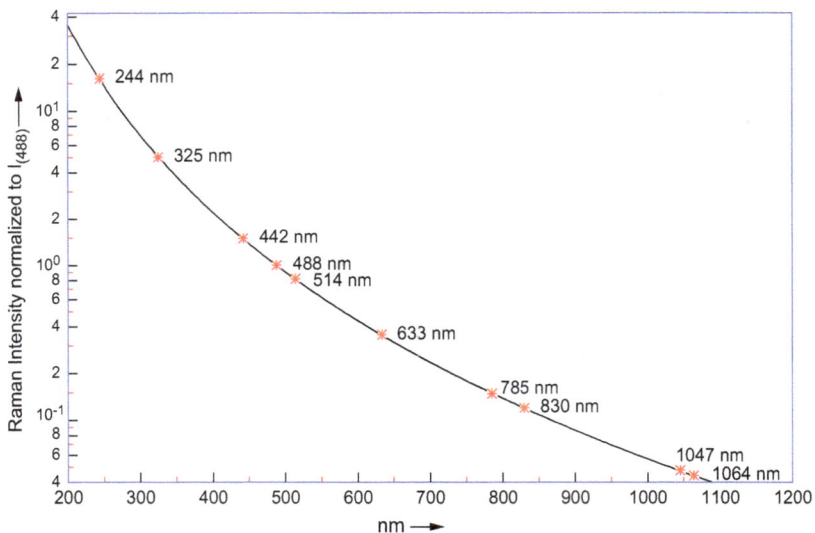

Figure 4.3: Excitation efficiency of different laser lines normalized to the intensity at 488 nm.

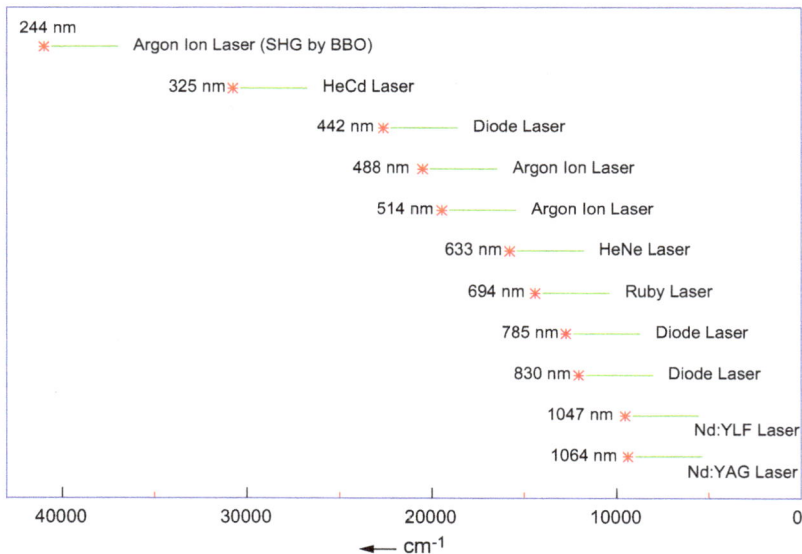

Figure 4.4: Wavelengths of Raman lasers (red stars) and regions of the excited Stokes Raman spectra (green lines).

used. But now a tunable diode laser is available: the quantum cascade laser (QCL). A QCL can emit light in the mid and far infrared.

Tuschel (2016) publishes a guide for selecting an excitation wavelength for Raman spectroscopy and used deep ultraviolet resonance Raman excitation to detect explosives (Tuschel *et al.* 2010).

4.1.3 Nernst sticks and globars

For IR spectral measurements, a polychromatic light source is needed. In former times Nernst sticks were used. These sticks are made of a mixture of 85% zirconium oxide, 7% yttrium oxide, and 3% erbium oxide and are heated by an electric current to about 1,500 K. They are quite efficient, but as their electric conductance rises with temperature and is very low at room temperature, they have to be heated to start them. During usage their operating current has to be limited by a resistor. This is very inconvenient, and it is the reason why they have been replaced by globars (Stewart and Richmond 1957). The globar is made from silicium carbide and also operates at 1,500 K. It is conductive at room temperature and needs no current limiter. For simpler (and therefore cheaper) instruments and in the NIR a spiral of a nickel-chrome wire (1,100 K) can be used.

Table 4.1: Line spectrum of mercury (data taken from Rumble; Reader and Corliss 2017).

Spectral range		Wavelength		
Ultraviolet	UV-C	184.95 nm		Main emission line
		248.3 nm		Main emission line
		253.65 nm		
		280.4 nm		
	UV-B	296.73 nm		Main emission line
		312.56 nm		
	UV-A	334.15 nm		Main emission line
		365.01 nm	i-line	Main emission line
Visible light	Violet	404,66 nm	h-line	Double line
	Violet	407.78 nm		
	Blue	435.83 nm	g-line	Main emission line
	Cyan	491.60 nm		
	Green	546.07 nm	e-line	
	Orange	576.96 nm		Main emission line
	Orange	579.07 nm		Double line
	Red	614.95 nm		
Near infrared	IR-A	1,013.97 nm	t-line	
	IR-B	1,529.88 nm		
		1,970.09 nm		
		2,325.4 nm		

4.1.4 Mercury-vapour lamp

The mercury arc, with its very intense lines, is a good radiation source for Raman, but is not very effective, as many of its lines have to be filtered out to produce monochromatic light, thereby losing a lot of power. Its lines are summarized in Table 4.1 and its main visible lines are shown in Figure 4.5.

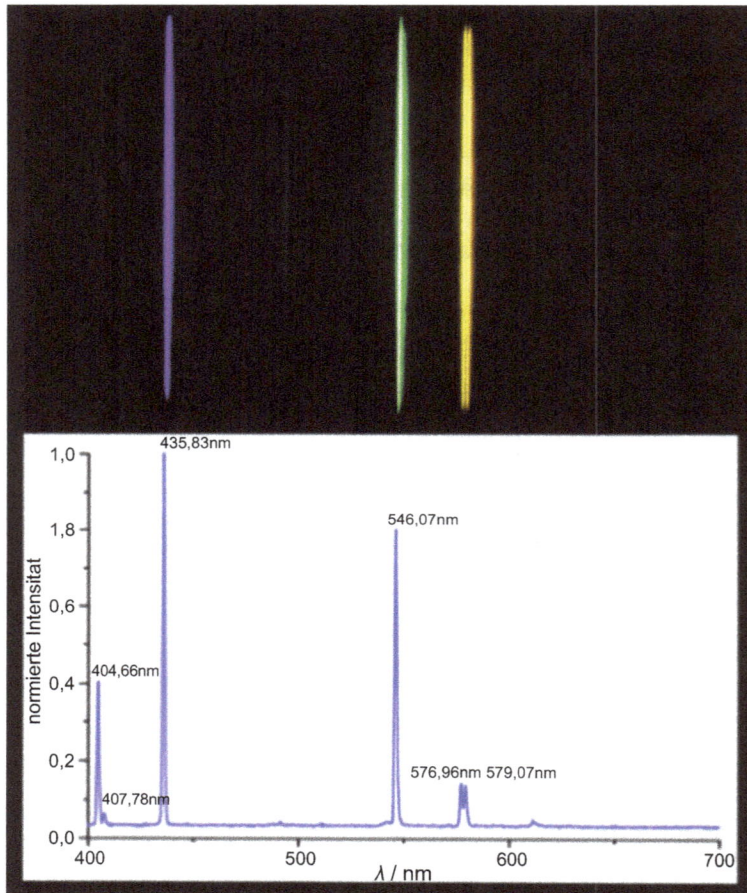

Figure 4.5: Visible spectrum of a mercury-vapour lamp. Bottom spectrum: main lines. Top insert shows also weaker lines (Copyright © Wikimedia commons).

4.2 Mono- and polychromators

As a mono- or polychromator some manufacturers still use the old-fashioned Czerny-Turner design (Figure 4.6). Here the Raman radiation that enters the device through a slit is imaged as a parallel beam by a concave mirror on a grating. The light reflected from the grating is then focused by a second concave mirror with identical properties on a slit (in the case of a monochromator) or on a detector array (in the case of a polychromator).

If one needs to measure Raman lines very close to the exciting laser line, normal filters are not efficient enough and special filters or a triple monochromator design is used (Horiba Scientific).

An account on the development and justification of FT-Raman spectroscopy is given by Hirschfeld and Chase (1986).

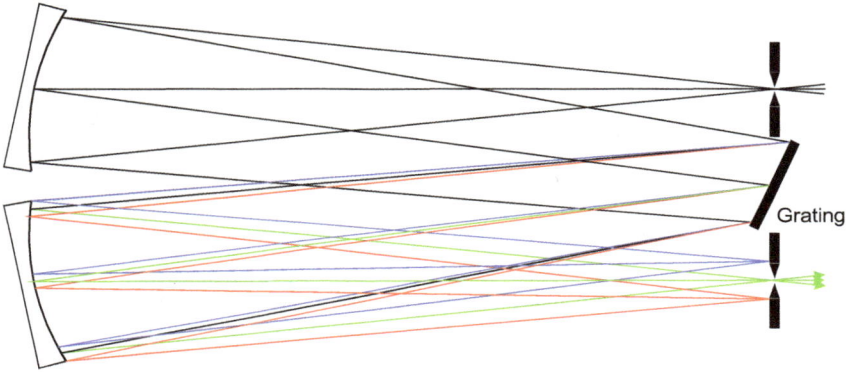

Figure 4.6: Design of the Czerny-Turner monochromator.

4.3 Filters

Filters have to be used in Raman spectrometers for two purposes: First they are needed to eliminate unwanted components from the light of the radiation source (e.g. plasma lines from gas lasers).

To remove the extremely strong Rayleigh line from the Raman spectrum, either a notch filter or an edge (low pass) filter is used. The former is used if the Stokes as well as the *anti*-Stokes part of the Raman spectrum has to be studied; the latter is cheaper and allows for measurements closer to the Rayleigh line.

4.4 Detectors

The first detector for infrared radiation was a blackened mercury thermometer. Nowadays a lot of sensitive detectors are available, some can be used at ambient; most of them have to be cooled. Rogalski (2012) reviews the history and performance (Figure 4.7) of infrared detectors. Cooled detectors for IR are the same as for Raman, noncooled detectors mainly used for IR spectroscopy are the pyroelectric TGS (triglycine

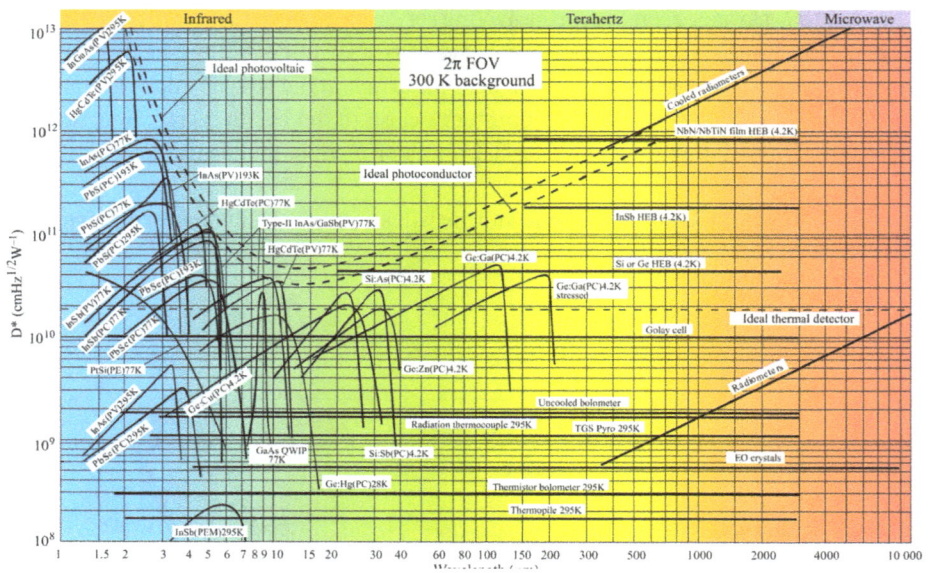

Figure 4.7: Comparison of the D^* of various available detectors when operated at the indicated temperature. Chopping frequency is 1,000 Hz for all detectors except the thermopile (10 Hz), thermocouple (10 Hz), thermistor bolometer (10 Hz), and Golay cell (10 Hz). Each detector is assumed to view a hemispherical surrounding at a temperature of 300 K. Theoretical curves for the background-limited D^* (dashed lines) for ideal photovoltaic and photoconductive detectors and thermal detectors are also shown. PC, photoconductive detector; PV, photovoltaic detector; PEM, photoelectromagnetic detector; HEB, hot electron bolometer (Rogalski 2012). Copyright ©2012 by the authors, licensed by Creative Commons (CC 3.0).

sulphate) detector or, more recently developed, the DLATGS (deuterated L-alanine-doped triglycine sulfate) detector with increased sensitivity.

Tan and Mohseni (2018) discuss emerging technologies for high-performance infrared detectors.

To detect the first Raman spectrum, a photographic plate was used. It has been employed by Raman and Krishnan to take the famous spectrum of carbon tetrachloride in 1929.

Goushcha and Tabbert (2007) provide a book chapter on optical detectors.

In place of the photographic plate nowadays an array detector is used. This can either be a diode array detector or a charge-coupled device (CCD). These detectors combine the advantages of the photographic plate (measurement of a wide spectral range and integration of the intensity of radiation) with those of photoelectric registration.

Raman detectors are fabricated from semiconductors. Figure 4.8 shows the usable wavelength range of some common semiconductor Raman detectors.

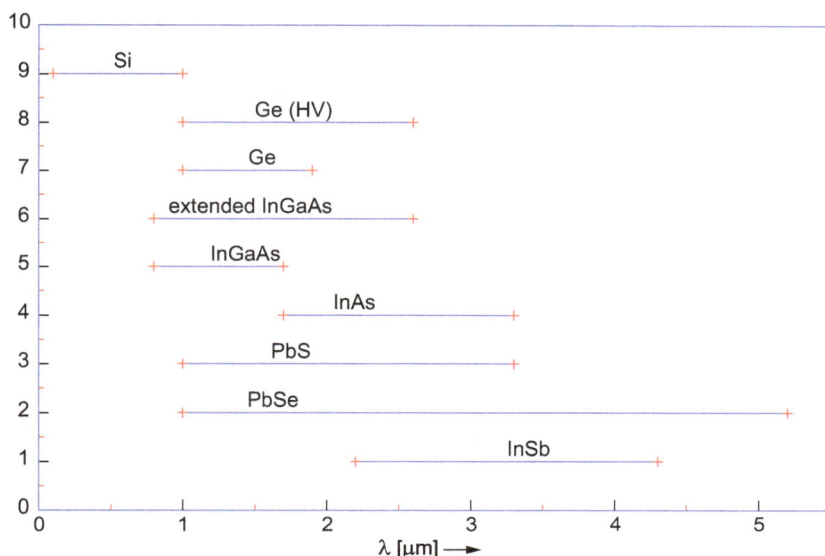

Figure 4.8: Usable wavelengths of semiconductor Raman detectors.

4.4.1 Photomultipliers

A photomultiplier is a very sensitive device that under some circumstances can even count single photons. Photons hit electrons out of the cathode material. The photons are then multiplied by an array of electrodes. The detectivity and wavelength range of a photomultiplier is dependent on its cathode material (e.g. Ko and Lee 2015; Yamamoto *et al.* 2012). Photomultipliers were very common in dispersive Raman spectrometers with double monochromators.

4.4.2 Semiconductor detectors

Silicon detectors are limited to the wavelength range between 0.1 nm (UV) and 1,000 nm (NIR). An alternative is indium antimonide, but the sensitivity is relatively low.

Germanium detectors are very sensitive if used in the photoresistive mode in the liquid nitrogen cooled state.

InGaAs was one of the first Raman detectors to be used. When cooled to 77 K, the spectral range becomes somewhat limited, with a long wavelength cut-off at 3,000 cm^{-1} Stokes shift. This eliminates observation of aromatic C–H stretching modes as well as O–H and N–H stretching modes. However, cooling does produce a better NEP for the detector. Chase (1987) shows the effect of element temperature on the long-wavelength cut-off of an InGaAs detector. Clearly, if long-wavelength response is needed the element must be no colder than dry ice temperatures. However, the optimum sensitivity is achieved at 77 K.

Chase compares newer and earlier versions of germanium and InGaAs detectors in his chapter "Fourier transform near-infrared Raman Spectroscopy" in Chalmers and Griffiths' (2002) *Handbook of Vibrational Spectroscopy*.

Most commercially available InGaAs detectors designed to run at 77 K have Dewar hold times of greater than 8 h, thus daylong operation is feasible. Overnight runs are difficult due to loss of cooling fluid. InGaAs detectors can be obtained in both down-looking and side-looking configurations for easy use in any detector module. The usual InGaAs element is a 1 mm^2 element. In several commercial FT-Raman spectrometers, this becomes the limiting aperture for the optical system.

The current version of germanium detectors must be cooled for proper noise performance and are usually operated at 77 K. Earlier versions of germanium detectors were operated as simple low-bias diodes. The noise performance was significantly inferior to InGaAs. The newer version of a germanium detector operates with high-purity germanium elements biased near 100 V. The spectral coverage goes out to 3,400 cm^{-1} Stokes shift and the sensitivity is excellent.

4.4.3 Array sensors

The detector mainly used in dispersive Raman spectrometers is the CCD. As all silicon devices, it is limited down to 1,000 nm. The quantum efficiency is quite high: back-thinned, back-illuminated CCDs reach a peak quantum efficiency of more than 90%.

A semiconductor detector is only sensitive to light with a photon energy larger than the bandgap or, put another way, with a wavelength shorter than the cut-off wavelength associated with the bandgap. This "long wavelength cutoff" extends to 3.75 μm for InAs, 0.55 μm for GaP, InP at 0.96 μm, and GaAs at 0.87 μm. Complex designs of Al$_x$Ga$_{1-x}$As-GaAs devices can be sensitive to infrared radiation.

Linear image sensors made from InGaAs are photodiode arrays sensitive to NIR wavelengths from 0.9 to 2.6 μm. These arrays have to be bonded to silicon CMOS chips, as circuitry like transistors cannot be fabricated on the InGaAs semiconductor. Front-illuminated devices are wire-bonded while back-illuminated InGaAs devices are bump-bonded to the circuitry used to bias, address, and read out the array's pixels.

Kiefer (2017) describes the simultaneous acquisition of the polarized and depolarized Raman signal with a single detector.

The impact of array detectors on Raman spectroscopy has been reviewed by Denson, Pommier, and Bonner Denton (2007).

Infrared semiconductor detectors include direct hybrid arrays of InSb and HgCdTe photodiodes that operate from 0.6 to 5 μm and of Si:As impurity band conduction detectors from 5 to 28 μm; a number of approaches to photoconductive detector arrays in the far-infrared; and bolometer arrays read out by transistors or superconducting devices in the far-infrared through millimetre-wave spectral range (Rieke 2007).

Bolometers are detailed by Rieke (2007), and Bulayev (2015) describes CMOS image sensors.

The topic of image sensor selection is treated by Ghassemi (2017).

4.5 Sample arrangements

Initially a 90° (angle between laser beam and axis of detection) arrangement was used in Raman spectroscopy, but for intensity-optimized Raman spectra a 180° arrangement (as drawn in Figure 4.9) is most favourable. A laser beam coming from below hits a small totally reflecting prism on the optical axis of the lens arrangement. The beam, that optionally can be focused, hits the sample in the focus of a high-aperture aspherical lens. The lens feeds the Raman radiation via a Rayleigh line removing filter into a spectrometer.

Schrader (1987) has developed a spherical cuvette made from chromium-free sapphire. This cuvette features a three-fold increase of collected Raman radiation compared

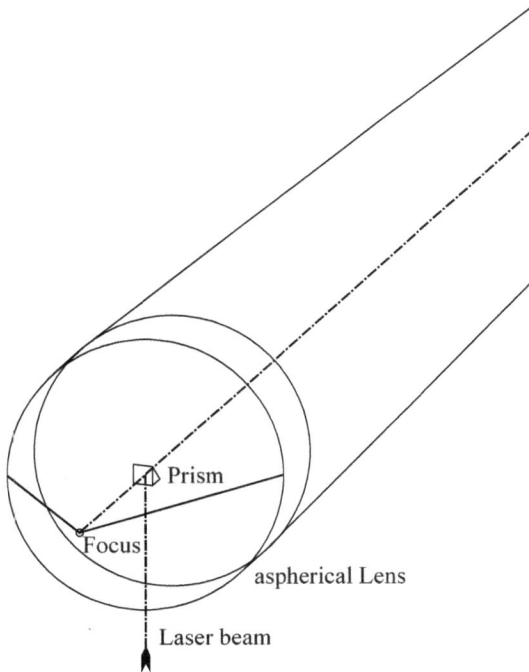

Figure 4.9: 180° backscattering arrangement.

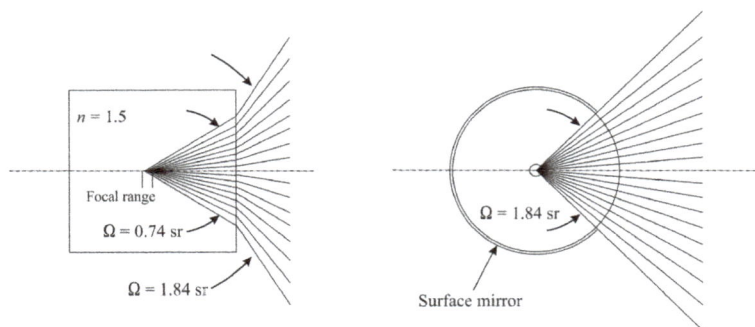

Figure 4.10: Comparison of a rectangular (left) and a spherical (right) cuvette for Raman spectroscopy. The effective solid angle is by a factor of 2.5 larger for the spherical cuvette (redrawn after Schrader (1987)).

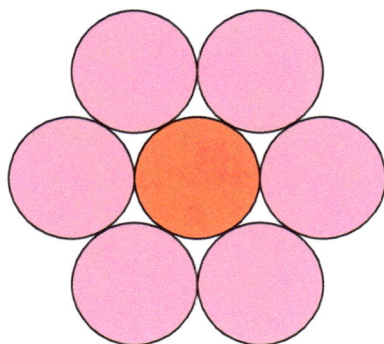

Figure 4.11: Cross section of a fibre optical probe for process control. Red: fibre for excitation by laser radiation. Pink: fibres for the collection of Raman radiation.

to a cuvette with rectangular shape. Figure 4.10 shows a comparison of a rectangular and a spherical cuvette for Raman spectroscopy.

Raman spectra can be measured from a remote sample with the aid of a fibre optical probe. The cross section of one possible arrangement of such a fibre bundle is shown in Figure 4.11: a centre fibre (red) is delivering the exciting laser radiation to the sample, while the six surrounding fibres (pink) collect the Raman radiation. This arrangement can be used advantageously to monitor the progress of a reaction in a chemical reactor.

4.6 Types of instruments

As already mentioned, Raman spectrometers can be dispersive or interferometric, with scanning or fixed components. As an example of a Raman spectrometer with no

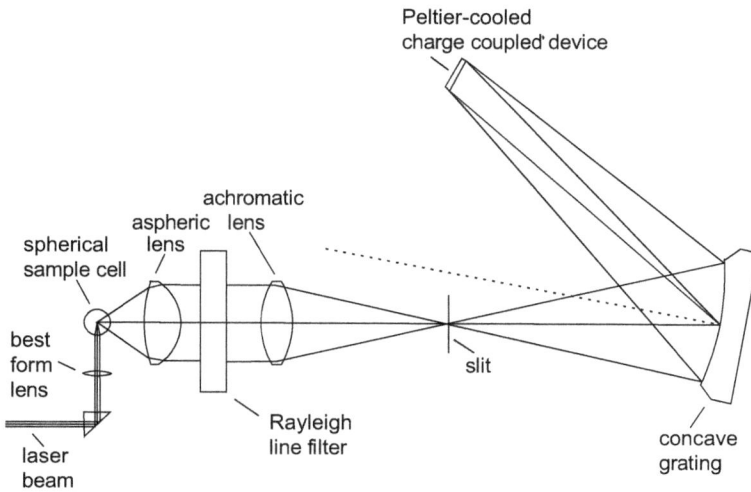

Figure 4.12: Layout of a compact Raman spectrometer using a holographic concave grating and CCD detector designed by Hoffmann *et al.* (1992).

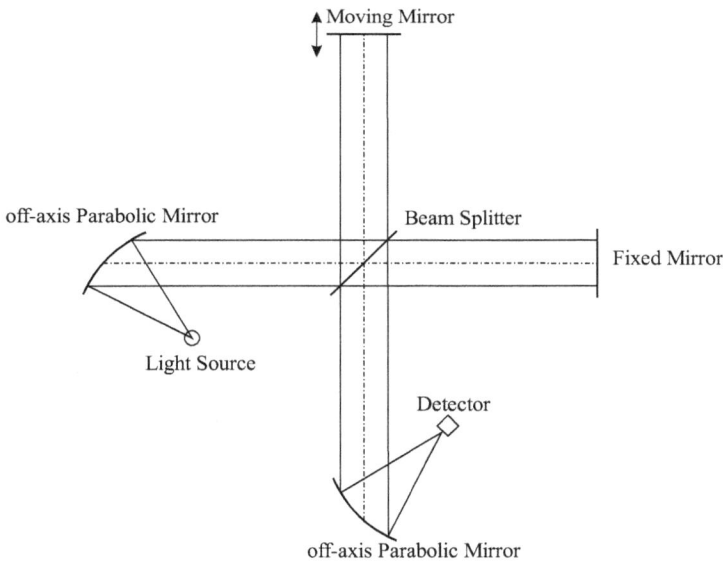

Figure 4.13: FT spectrometer using a Michelson interferometer. In the case of a FT-IR spectrometer the light source is a globar, in the case of a FT-Raman spectrometer the source is the irradiated part of a sample followed by a Rayleigh line removing filter.

moving parts, Figure 4.12 shows the layout of a compact Raman spectrometer using a holographic concave grating and CCD detector designed by Hoffmann *et al.* (1992).

The design of an FT spectrometer using a Michelson interferometer with a moving mirror is shown in Figure 4.13. In the case of an FT-IR spectrometer the light source is a Globar, in the case of a FT-Raman spectrometer the source is the laser-irradiated part of a sample followed by a Rayleigh line removing filter. This design has been used to build an FT spectrometer for the measurement of circular dichroism in the IR, where it is important that the spectrometer contains the least possible number of optical elements that can change the state of polarization (Hoffmann *et al.* 1987).

4.7 State of the art in commercial spectrometers

Commercial Raman and IR spectrometers use a Michelson interferometer or mono/polychromators for the dispersion of wavelengths. They mainly use a room-temperature InGaAs detector or the liquid nitrogen-cooled germanium diode detector for Raman, room temperature DTGS or liquid nitrogen-cooled MCT for infrared. Most instruments come with a spectral library to allow for comparison of reference spectra with the measured spectrum.

The following descriptions and technical data are based on the manufacturers' webpage data, and no guarantee on correctness and completeness of the information can be given.

The lists only provide examples, because of the huge number of commercially available instruments they cannot be complete. To make things worse, most of the spectrometers can be modified so that this book cannot provide enough space to cover all modifications.

4.7.1 Infrared spectrometers

4.7.1.1 Instruments for routine analysis

IR-spectrometers for routine and process control are affordable, robust, and reliable instruments that are optimized for their intended use, but cannot be flexibly adapted to other special uses. An example is the FT-IR gas analyser shown in Figure 4.14. Table 4.2 gives an overview of some routine instruments.

4.7.1.2 Instruments for research

IR spectrometers for research are more expensive than their routine counterparts. They are sophisticated instruments, and a lot of attachments, options, and modifications are available (see, e.g. Figure 4.15). Among them are also specialized instruments, for example instruments for the measurement of vibrational circular dichroism (VCD, see Section 18.1). Table 4.3 gives an overview of the market.

Table 4.2: IR spectrometers for routine and process control.

Supplier	Model	Measuring range cm^{-1}	Resolution cm^{-1}	Radiation source	Detector	Remarks
Thermo Fisher Scientific	MAX-iR™		1–32	VCSEL diode silicon carbide (SiC) IR source	DTGS standard InAs or MCT StarBoost	FTIR Gas Analyzer
Mettler Toledo	ReactIR	800–4,000	4		TE cooled MCT	Reaction monitoring
PerkinElmer Inc.	Spectrum 3 FT-IR	30–11,000				Tri-Range
PerkinElmer Inc.	Spectrum TWO FT-IR	350–8,300	0.5			

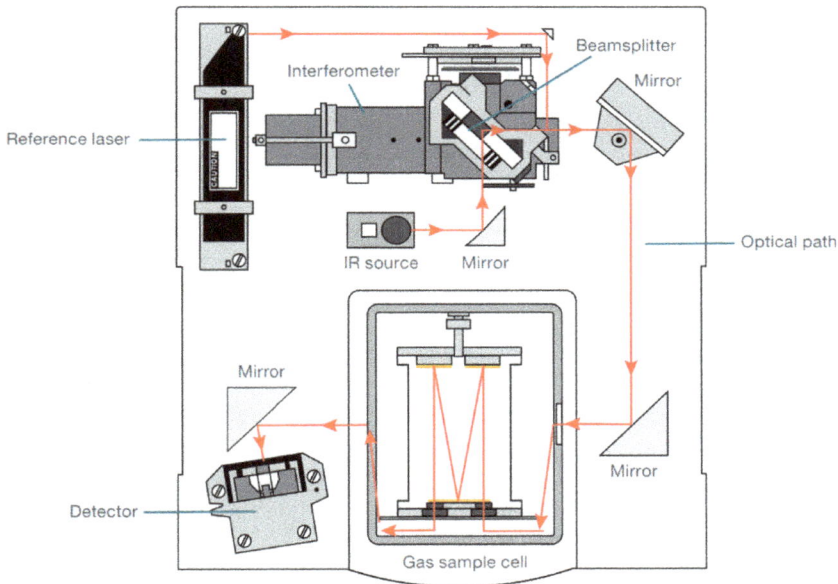

Figure 4.14: Thermo Fisher Scientific's FT-IR gas analyser (www.thermofisher.com).

For some applications of IR spectroscopy, a sample has to be analysed with spatial resolution. For this purpose, two types of specialized instruments are available (Table 4.4): the IR microscopes, giving a resolution in the micrometre range, and the nano-IR spectrometers, working in the nanometre range with the help of an atomic force microscope (AFM, see Section 17.2).

Table 4.3: IR spectrometers for research.

Supplier	Model	Measuring range cm^{-1}	Resolution cm^{-1}	Radiation source	Detector	Remarks
Bruker Optics GmbH & Co. KG	Vertex 70v	50–6,000	< 0.4 optional 0.16	cooled MIR Tungsten Hg-Arc	DTGS liquid He cooled bolometer	Vacuum Instrument
Bruker Optics GmbH & Co. KG	Vertex 80	10–50,000	< 0.2 Opt. < 0.06	cooled MIR Tungsten Hg-Arc D		Vacuum or Purgeable
Bruker Optics GmbH & Co. KG	IFS125 HR	5 to > 50,000	< 0.001	4 sources	6 detectors	atmospheric research
Shimadzu Deutschland GmbH	IRAffinity-1S	350–7,800	0.5	Ceramics	DLATGS	
Shimadzu Deutschland GmbH	IRSpirit	350–7,800	0.9	Ceramics	DLATGS	High humidity environment
Shimadzu Deutschland GmbH	IRTracer-100	350–7,800 650–5,000	0.25	Ceramics	DLATGS MCT	
Thermo Fisher Scientific	Nicolet™ iS™ 50	15–27,000	0.09	Polaris™ Infrared source		
BioTools	ChiralIR-2X™	850–4,000				Vibrational Circular Dichroism (VCD)
Bruker Optics GmbH & Co. KG	PMA 50	Polarization Modulation Accessory				VCD measurements

4.7.1.3 NIR instruments

Near-infrared (NIR) spectroscopy is an important and extremely useful method of analysis for some research areas and applications ranging from materials science via chemistry to life sciences. An (incomplete) list of NIR spectrometers is given in Table 4.5. Some instruments are hand-held.

Table 4.4: IR microscopes (IR-M) and nano-IR spectrometers (N-IR).

Supplier	Type	Model	Measuring range cm^{-1}	Resolution cm^{-1}	Spatial resolution	Detector	Radiation source
Bruker Optics GmbH & Co. KG	N-IR	Anasys nanoIR3	250–2,875	Standard 0.4	<10 nm	CCD QCL 2048 Pixel	
Shimadzu Deutschland GmbH		AIM-9000	650–5,000 400–4,600			MCT TGS	
Shimadzu Deutschland GmbH	IR-M Combination with Raman microscope	AIRsight	400–4,600 700–5,000			TGS T2SL	
JASCO Deutschland GmbH	IR-M	IRT 7000 series	400–15,000	4	Step 0.1	MCT DLATGS InSb InGaAs	
NT-MDT	N-IR	NTEGRA SPECTRA II	10-. . . Detector dependent	0.1	1 nm	TE cooled CCD, APD, PMT, FLIM	UV to IR
PerkinElmer Inc.	IR-M	Multiscope Spotlight 200i	600–7,800		10 μ	MCT LiTaO$_3$	
Thermo Fisher	IR-M	Nicolet IN5	650–7,800 450–7,600	2	1 μm	MCT-A (LN$_2$) DTGA (RT)	

Figure 4.15: The VERTEX 70v can optionally be equipped with optical components to cover the spectral range from 10 cm^{-1} in the far IR/THz through the mid- and near-IR up to the VIS/UV spectral range at 28,000 cm^{-1} (www.bruker.com/optics).

Table 4.5: NIR spectrometer.

Supplier	Model	Measuring range	Resolution	Radiation source	Detector
StellarNet Inc	RED-Wave NIR2.2-1024	900–2,300 nm	7 nm	SL1 Tungsten Halogen	InGaAs Std. 512 pixel Opt. 1024 pixel
StellarNet Inc	RED-Wave NIR2.2-1024X	1500–2,200 nm	2.8 nm	SL1 Tungsten Halogen	InGaAs Std. 512 pixel Opt. 1024 pixel
StellarNet Inc	Green wave	350–1,150 nm	0.1 nm	fiber optic cable	CCD 2048 pixel Photo Diode
PerkinElmer Inc.	Spectrum Two N	2,000–14,700 cm^{-1} 3,800–14,700 cm^{-1}	1 cm^{-1}	Ceramics	LiTaO$_3$ Std. InGaAs Opt.
Deutsche METROHM GmbH & Co. KG	DS2500 Analyzer	400–2,500 nm	8.75 nm		

4.7.2 Raman spectrometers

In a Raman spectrometer market overview, providers present numerous Raman spectrometers with essential technical data. WEKA BUSINESS MEDIEN (MEDIEN 2021).

4.7.2.1 Instruments for routine analysis and process control

As for IR measurement, Raman spectrometers for routine and process control are affordable, robust, and reliable instruments that are optimized for their intended use, but cannot be flexibly adapted to other special uses. For example, the Mettler Toledo ReactRam, with its numerous attachments, is optimized for the control of chemical reactions. An overview is given in Table 4.6.

Table 4.6: Raman spectrometers for routine and process control.

Supplier	Model	Measuring range cm^{-1}	Resolution cm^{-1}	Radiation source, power and excitation wavelength λ_{exc}	Detector
analyticon instruments gmbh	HyperFlux PRO Plus	200–3,300	< 5	Laser, 495 mW, 785 nm	Si-based CCD, cooled

Table 4.6 (continued)

Supplier	Model	Measuring range cm^{-1}	Resolution cm^{-1}	Radiation source, power and excitation wavelength λ_{exc}	Detector
Anton Paar Germany GmbH	Cora 5000	200–3,500 (532 nm) 100–2,300 (785 and 1064 nm)	9–12 (532 nm) 6–9 (785 nm) 12–17 (1064 nm)	Laser 532 nm, 785 nm und 1064 nm	2048 px CCD 256 px InGaAs
JASCO Deutschland GmbH	NRS-4500	Std. 100–8,000 Opt. 50–8,000	Std. 2 Opt. 0.7	Std. 532, Opt. 244; 266; 325; 355; 442; 488; 514,5; 633; 785; 1064	Air Cooled CCD
Mettler Toledo	ReactRam	150–3,400		Laser, 400 mW, 785 nm	Deep Cooled CCD
HORIBA Jobin Yvon GmbH	MakroRAM	100–3,400	8	Laser, 325, 405, 473, 532, 638, 785 nm	Syncerity Back illuminated NIR CCD

4.7.2.2 Raman instruments for research

Raman spectrometers for research are more expensive than their routine counterparts. They are sophisticated instruments, and a lot of attachments, options, and modifications are available. Among them are also specialized instruments, for example instruments for the measurement of Raman optical activity (ROA, see Section 18.2). Table 4.7 gives an overview of the market.

Some IR instruments can be converted to a Raman spectrometer. For example, scientists already using Bruker's VERTEX FT-IR research spectrometer can attach the FT-Raman module RAM II (Figure 4.16) to their instrument. The optics of this attachment still allow for the alternative measurement of IR and Raman of the sample.

For those scientists or OEMs, wanting to build their own specialized Raman instrument, building blocks are available from different suppliers. An example is the CiCi-Raman-785 spectrometer from Horiba Scientific (Figure 4.17).

4.7.2.3 Raman imaging spectrometers and microscopes

For some samples, it is advantageous to study them with high spatial resolution. There are three methods to solve this problem:

First: A confocal microscope is scanning the sample

Second: An image of the sample is projected on a 2D detector (CCD, etc.)

Third: The sample is scanned with an AFM or STM.

Excitation LASERs (1064 nm / 785 nm)

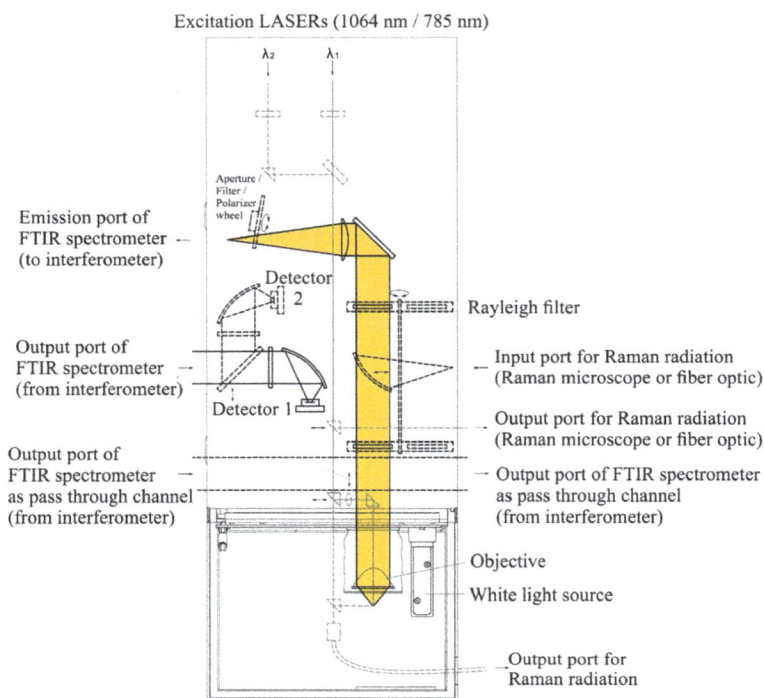

Figure 4.16: Optical path of Raman accessory RAM II (courtesy of Bruker Optik GmbH, Ettlingen).

Table 4.7: Raman spectrometers for research.

Supplier	Model	Measuring range cm^{-1}	Resolution cm^{-1}	Radiation source, power and excitation wavelength λ_{exc}	Detector
Bruker Optics GmbH & Co. KG	MultiRAM	50–3,500 (Ge) 50–3,600 (InGaAs) 50–4,200 (Si-Aval.)	< 0.5	Nd:YAG Laser air-cooled (500–2000 mW, 785 nm) Diode Laser (500 mW)	Ge (LN$_2$ cooled) InGaAs Si-Avalanche
JASCO Deutschland GmbH	NRS-5600	10–8,000	1 Opt. 0.4	Std. 50 mW 532 nm Opt. deep UV to 1064nm	CCD Std. 1024 × 255 Opt. 2048 × 512 EM-CCD and InGaAs
Teledyne Princeton Instruments	TPIR-785	80–3,650	5	Laser 785 nm	CCD 1340 x 400

Table 4.7 (continued)

Supplier	Model	Measuring range cm^{-1}	Resolution cm^{-1}	Radiation source, power and excitation wavelength λ_{exc}	Detector
Thermo Fisher Scientific	DXR3	50–3,500	5, Opt. 2	455, 532, 633, 785 nm	TE cooled EMCCD
BioTools	ChiralRaman-2X™	100–2,000		532 nm	Raman optical Activity (ROA)

Figure 4.17: The CiCi-Raman-785 spectrometer from Horiba Scientific features an aberration-corrected concave holographic grating configured with a deep-cooled Syncerity CCD camera, a new VIS-NIR detector, and a round-to-slit fibre converter for collection from a Raman probe. It offers 90% quantum efficiency at 650 nm and 70% at 900 nm, with 2,048 pixels and 1 mm collection height (www.horiba-scientific.com).

Instruments of the first and second types are listed in Table 4.8. They offer spatial resolutions in the micrometre range.

An example of the first type of instrument is the confocal Raman microscope WITec alpha 300R (Figure 4.18).

Another example of this method is the Bruker Senterra II (Figure 4.19). It is a compact Raman microscope with multiple laser input. During the Raman measurement, the visible observation using the binocular is blocked, but for visual observations the Raman path can be blocked and the path for visible observations opened. Along with the change of excitation wavelength, the filter has to be changed. A linearly movable filter changer serves this purpose. The Raman light is then fed into a spectrograph using a grating as the dispersing element and a CCD as the detector. Three different gratings are mounted on a grating turret in order to change the spectral resolution by simply turning the turret.

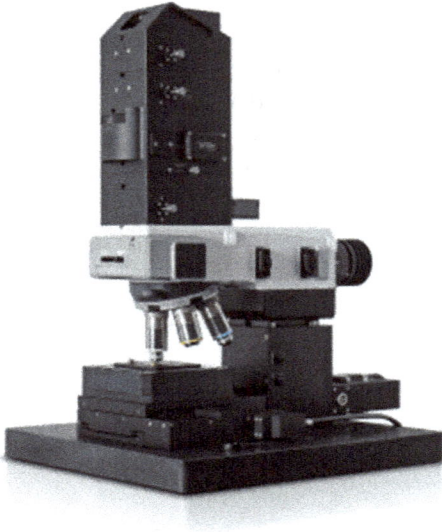

Figure 4.18: The confocal Raman microscope WITec alpha 300R (www.witec.com).

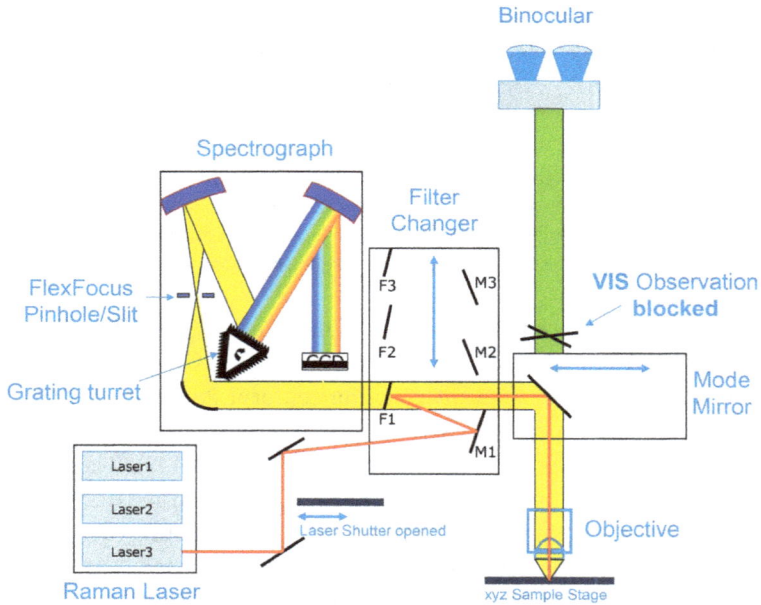

Figure 4.19: Optical path of the Raman microscope Senterra II (courtesy of Bruker Optik GmbH, Ettlingen).

Table 4.8: Raman imaging spectrometers and microscopes.

Supplier	Model	Measuring range cm^{-1}	Resolution spectral spatial cm^{-1} μm		Radiation source, power and excitation wavelength (λ_{exc})	Detector pixel
Hamamatsu Photonics Deutschland GmbH / Edinburgh Instruments	RM5	50–4,000	1.4		Laser 532 nm, 638 nm, 785 nm	High sensitivity CCD 1650 x 200
Hamamatsu Photonics Deutschland GmbH / Edinburgh Instruments	RMS 1000	5–30,000	< 0.1		532 nm, 638 nm, 785 nm	High sensitivity CCD 1650 x 200
JASCO Deutschland GmbH	NRS 7000 series	10–8,000	1 Opt. 0.4	Step 0,1	Std. 50mW, 532 nm Opt. deep UV to 1064nm	CCD std. 1024 × 255 opt 2048 × 512 EM-CCD, InGaAs
WITec Wissenschaftliche Instrumente und Technologie GmbH	alpha300 R	<10–10,000	< 0.2	0.2	633 nm	
Renishaw GmbH	inVia	(200–2,200 nm) 5–4,000 Opt. –30,000	0.3	0.25	229–1064 nm	CCD 1024 × 256
Shimadzu Deutschland GmbH	AIRsight	150–4,000 150–3,200	3		532 nm 785 nm	Combination with IR microscope
Thermo Fisher Scientific	DXR™3	50–3,500	5 Opt. 2	1	455, 532, 633, 785 nm	TE cooled EMCCD
HORIBA Jobin Yvon GmbH	LabRAM Soleil nano	30–. . . 300–1,600 nm			325, 405, 473, 532, 638, 785 nm	
HORIBA Jobin Yvon GmbH	MacroRAM	100–3,400	8		785	Back-illuminated NIR CCD

The third method gives the highest spatial resolution down to 1 nm. Instruments, using tip-enhanced Raman spectroscopy (TERS, see Section 14.1), are listed in Table 4.9.

Table 4.9: Tip-enhanced Raman spectrometers.

Supplier	Model	Measuring range cm^{-1}	Resolution spectral spatial	Radiation source, power and excitation wavelength λ_{exc}	Detector
Bruker Optics GmbH & Co. KG	Innova-IRIS	Specifications see Renishaw inVia			
HORIBA Scientific	LabRAM Nano	50–4,000			
WITec Wissenschaftliche Instrumente und Technologie GmbH	TERS module for the alpha300	Specifications see alpha300 R			
NT-MDT	NTEGRA SPECTRA II	10–. . .	0.1 cm^{-1} < 30 nm	UV to IR	TE cooled CCD, APD, PMT, FLIM

4.7.2.4 Hand-held Raman Instruments

Obliged to the US "homeland security" program, nowadays a large choice of hand-held Raman and IR spectrometers is available, as Vagnini *et al.* (2017) report.

Common to all hand-held Raman spectrometers (Table 4.10) is their ability to provide results within seconds, thus minimizing costly and time-consuming laboratory work, and that they can be used safely by inexperienced users.

Manufacturers in Pharma- and Biotechnologie have to monitor the active agent during the complete process of production to ensure the product's quality and the safety of the user. Hand-held Raman spectrometers allow for reliable identification of active agents. A large number of chemicals can be tested through closed packages, thus reducing the risk of contamination and exposition. The Bruker BRAVO (Figure 4.20) is an example of such a hand-held Raman spectrometer.

Berlizov *et al.* (2016) assess the hand-held Raman spectrometer FirstDefender for nuclear safeguards applications. As a result of Berlizov's work, the instrument's capability was extended to identify a variety of industrial yellow cakes (mixture of uranium compounds) of different origin, composition, and purity.

Table 4.10: Hand-held Raman spectrometers.

Supplier	Model	Measuring range cm^{-1}	Resolution cm^{-1}	Radiation source, power and excitation wavelength λ_{exc}	Detector
analyticon instruments gmbh	TruScan RM	250–2,875	8–10.5	Laser, 250 mW, 785 nm	2048 Pixel, CCD
Anton Paar Germany GmbH	Cora 100	400–2,300	10	Laser, 300 mW, 785 nm	Linear CCD-Array
BioTools	RamTest	100–4,000	4–6	Laser, 30 mW, 532 nm laser	CCD
Bruker Optics GmbH & Co. KG	BRAVO	300–3,200	10–12	Laser, 852 und 785 nm	CCD
Deutsche METROHM GmbH & Co. KG	TacticID 1064 ST	176–2,500		Laser, 420 mW, 1064 nm	NIR InGaAs array
B&W Tek	NanoRam	176–2,900	9	Laser, 300 mW, 785 nm	TE-cooled Linear CCD
HORIBA Jobin Yvon GmbH	AnywhereRaman™	150–3,150	5–6	Laser, 785 nm	Back-illuminated linear CCD

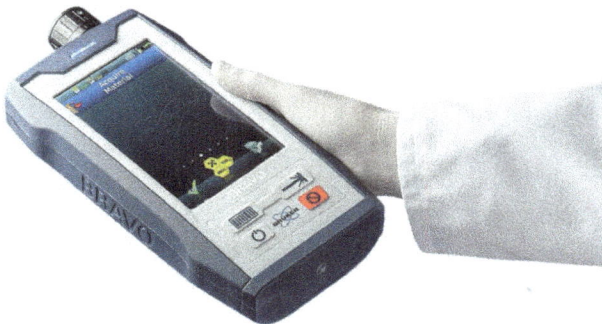

Figure 4.20: Hand-held Raman spectrometer BRAVO (courtesy of Bruker Optik GmbH, Ettlingen).

4.8 Wavenumber calibration of spectrometers

The Commission on Molecular Structure and Spectroscopy (Spectroscopy 1960) publishes tables for the calibration of infrared spectrometers in the range of 4,300–600 cm^{-1} and from 600 to 1 cm^{-1} (Cole *et al.* 1960).

As an example of calibration spectra, the infrared spectrum of hydrogen chloride (HCl) in the gas phase is shown in Figure 4.21. The splitting of the bands, caused by the two isotopes of chlorine in a ratio of 1–3, is clearly visible.

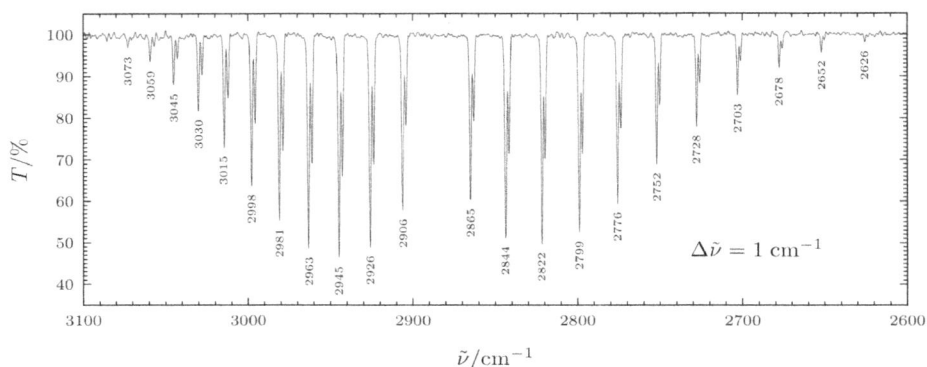

Figure 4.21: Infrared spectrum of hydrogen chloride (HCl) in the gas phase. Splitting of bands is caused by the two isotopes of chlorine (^{35}Cl and ^{37}Cl) (by mrt, CC BY-SA 2.5, https://commons.wikimedia.org/w/index.php?curid=1232895).

A very common compound for the calibration of vibrational spectrometers is indene (see Fig. 5.66). The vibrational spectra of the bicyclic compound have been examined by Klots (1995) in the vapour and the liquid state using infrared and Raman spectroscopy. Wavenumbers of the liquid spectra are measured to 0.1 cm^{-1} to re-examine the potential of indene as a calibration molecule. An assignment for its normal modes of vibration is made considering vapour band shapes, polarization ratios, and calculated frequencies from a scaling procedure of the indene AM1 force field.

Okajima and Hamaguchi (2015) use the pure rotational spectrum of N_2 for the accurate intensity calibration for low wavenumber (−150 to 150 cm^{-1}) Raman spectroscopy.

System parameters are applied by Liu and Yu (2013) for accurate wavelength calibration of grating spectrometers.

5 Vibrational spectroscopy of organic compounds

The vibrational spectrum of a compound should show 3N-6 bands, the so-called normal vibrations. With the exception of a few simple chiral molecules we will not find all of them in the Raman spectrum. This can be contributed to three main reasons:
a) Many vibrations have similar frequencies so that they occur as a single broad band
b) The intensities of some bands are to low
c) Some bands of molecules will be forbidden by symmetry

Overtones and combination bands are not as prominent in the Raman as in the infrared (IR) spectrum, but sometimes they are of observable intensity.
   The following abbreviations are used in this chapter:

IR	infrared spectrum
Ra	Raman spectrum
p	polarized
vs	very strong
s	strong
m	medium
w	weak
vw	very weak
sh	shoulder
br	broad
sr	sharp
v	variable
–	not observed

For IR intensities, see *Handbook of Infrared and Raman Characteristic Frequencies of Organic Molecules* by Daimay *et al.* (1991) and references cited therein.

5.1 Aliphatic carbon compounds

5.1.1 Alkanes

These simple organic compounds only show C–C and C–H vibrations. Hence, they can easily be identified by C–H stretching vibration bands at 2,970–2,840 cm^{-1} (medium to strong intensity in the Raman spectrum), deformational bands near 1,470 cm^{-1}, the symmetrical CH_3 deformational band near 1,380 cm^{-1} (split in compounds containing the isopropyl and *tert*-butyl group), and the CH_2 rocking band at 720 cm^{-1}.
   n-Alkane carbon–carbon stretching vibrations occur as weak bands at 1,150–1,135 cm^{-1} and 1,060–1,056 cm^{-1}. For comparison, polyethylene shows bands at 1,133 and 1,061 cm^{-1}.

https://doi.org/10.1515/9783110717556-005

Wu, Sasaki, and Shimizu (1995) publish a high-pressure Raman study of dense methane (CH_4). The deuterated compound CD_4 is studied as well.

Intensive normal co-ordinate calculations on the extended n-paraffins C_2H_6, through n-$C_{14}H_{30}$ and polyethylene have been carried out by Schachtschneider and Snyder (1963) using a perturbation method.

5.1.1.1 Acyclic aliphatic compounds

In alkanes we find three different chemical functionalities: methyl (CH_3), methylene (CH_2), and methine (CH). In the following paragraphs we will take a closer look at the possible vibrations of these groups. "Antisymmetric" in this context means "non-totally symmetric".

The vibrations of the CH_2 group can be classified according to their symmetry. They can be descibed as stretching (Figure 5.1), rocking, scissoring, twisting, and wagging (Figure 5.2).

The three-atom functionality CH_2 can perform two stretching vibrations: the symmetric stretching with both hydrogens moving in-phase and the antisymmetric C–H stretching with the two hydrogens moving out-of-phase. These vibrations have signals near 2,939 cm^{-1} (ν_{as}, Ra vs, IR vs) and near 2,850 cm^{-1} (ν_s, Ra vs, IR vs).

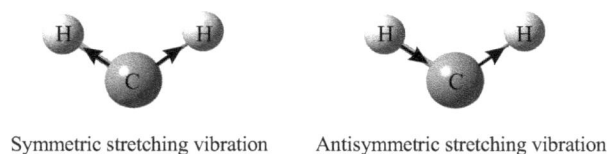

Symmetric stretching vibration   Antisymmetric stretching vibration

Figure 5.1: Stretching vibrational modes for the methylene group.

The angle H–C–H is changed during the scissoring vibration. The scissoring vibration δ_s is found at 1,470 cm^{-1} (Ra m, IR m), whereas we find the rocking vibration at 720 cm^{-1} (Ra –, IR m), the twisting vibration at 1,300 cm^{-1} (Ra s, IR –), and the wagging vibration at 1,305 cm^{-1} (Ra v, IR v).

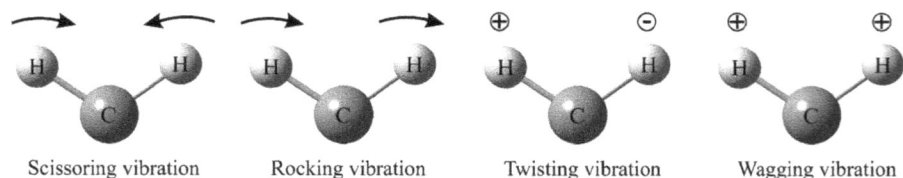

Scissoring vibration   Rocking vibration   Twisting vibration   Wagging vibration

Figure 5.2: Bending vibrational modes for CH_2.

For the methyl moiety we can observe three stretching vibrations (Figure 5.3). We find them as a medium intensity Raman band at 2,960 cm^{-1} (ν_{as}, IR s) and a medium to strong band at 2,870 cm^{-1} (v_s, degenerate, IR s). The deformation vibrations, which can be symmetric or antisymmetric, are shown in Figure 5.4. δ_s is found at 1,380 cm^{-1} (visualized for *n*-heptane in Figure 5.5) and δ_{as} at 1,470 cm^{-1}. Their Raman bands are of weak to medium intensity (IR w) and their frequency will be raised by electro-negative substituents. As an example, the Raman spectrum of *n*-heptane is shown in Figure 5.6. This spectrum shows neither the 1,380 nor the 720 cm^{-1} Raman band, as those are very weak in this compound.

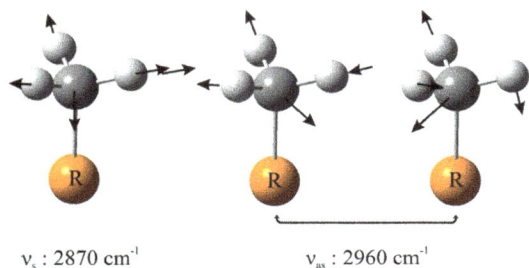

v_s : 2870 cm^{-1} v_{as} : 2960 cm^{-1}

Figure 5.3: Symmetric and antisymmetric stretching vibrations of the methyl group.

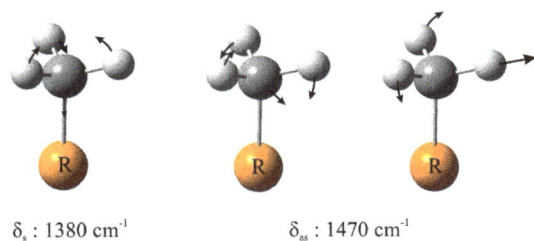

δ_s : 1380 cm^{-1} δ_{as} : 1470 cm^{-1}

Figure 5.4: Deformation vibrations of the methyl group.

The *t*-butyl group shows a very characteristic Raman band at 1,250 cm^{-1} (IR m-w).

The C–H stretching mode of methine can be found as a medium intensity Raman band at 2,890 cm^{-1} (IR w), but it is often obscured by the much stronger CH_2 and CH_3 bands. The C–C–H deformational band is found at 1,350–1,330 cm^{-1}. This Raman band is strong and very valuable for identifying the isopropyl group, for which it occurs at 1,345 cm^{-1}.

The enthalpy differences between conformations of normal alkanes were calculated in a Raman spectroscopy study of *n*-pentane (Figure 5.7) and *n*-butane by Balabin (2009). For *n*-pentane, the *trans–trans* conformer is dominating.

Skeletal carbon stretching and CCC deformation frequencies of *n*-alkanes are shown in Table 5.1.

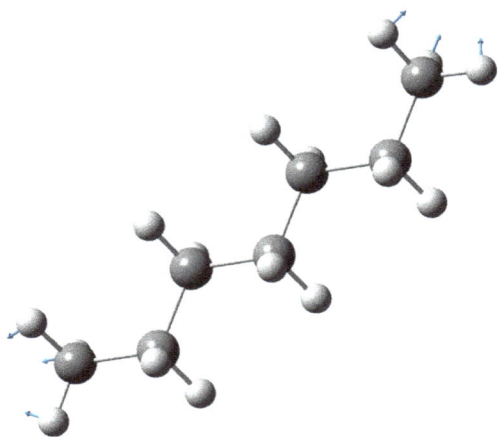

Figure 5.5: Symmetric deformation vibrations of the methyl groups of *n*-heptane (experimental 1,377 cm^{-1}, calculated 1,414 cm^{-1}: DFT, B3LYP/6-311^{++}G(d, 2p), f = 0.9648 (Hoffmann 2017–2018)).

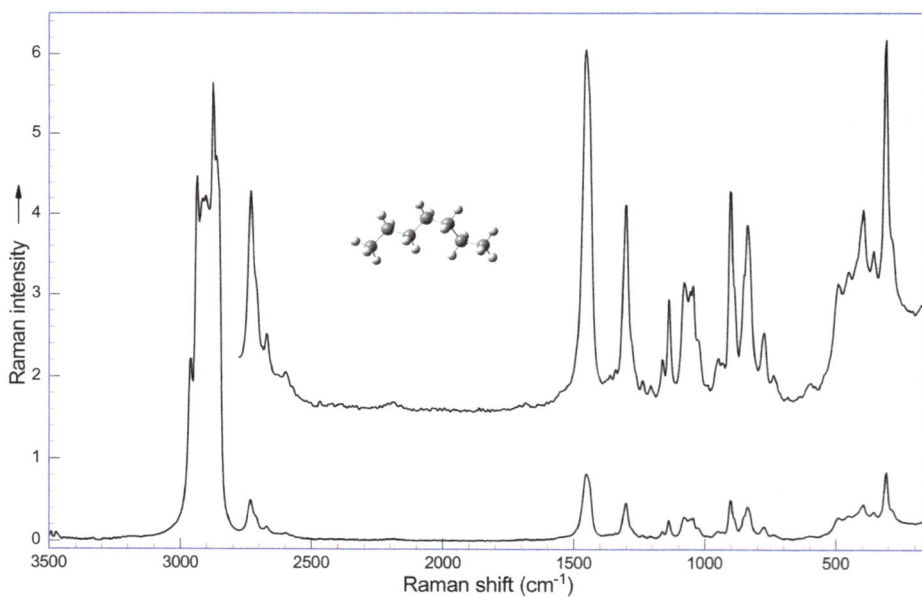

Figure 5.6: Raman spectrum of *n*-heptane (λ_{exc} = 1,064 nm, 270 mW; data courtesy of Bruker Optik GmbH, Ettlingen).

Table 5.1: Vibrational frequencies of *n*-alkanes in the liquid state (data taken from Dollish, Fateley, and Bentley (1973) and references cited therein).

n-Alkanes	Skeletal carbon stretching (cm^{-1})	CCC deformation frequency (cm^{-1})
n-Butane	837	425
n-Pentane	869	406
n-Hexane	898	373
n-Heptane	905	311
n-Octane	899	283
n-Nonane	888	249
n-Decane	886	231
n-Dodecane	892	194
n-Hexadecane	888	150
Raman intensity	s	s

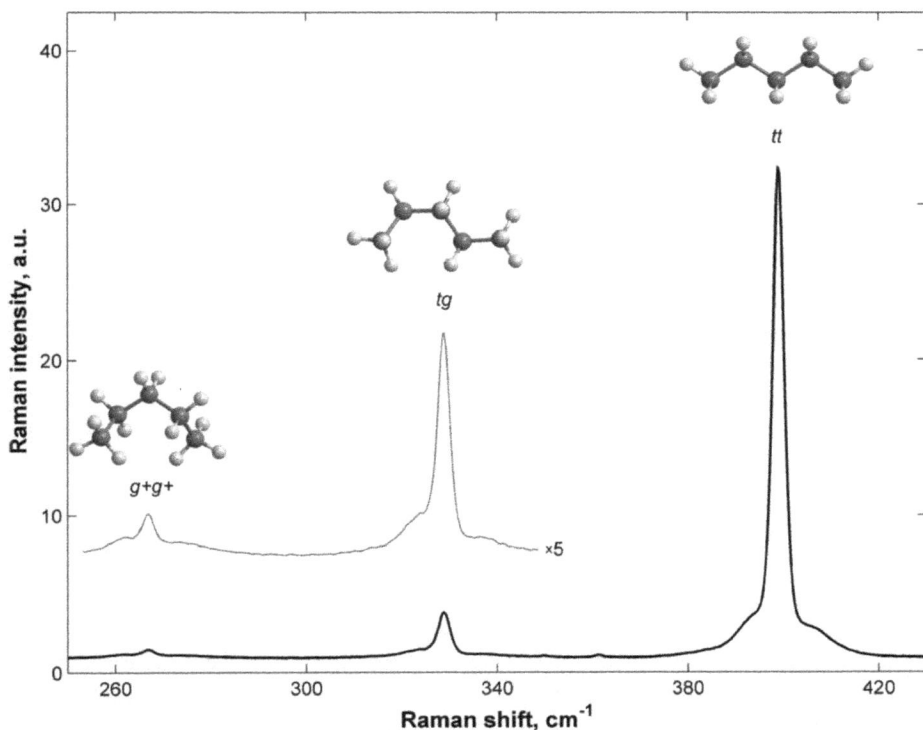

Figure 5.7: Experimental Raman spectra of *n*-pentane in the region of 250–430 cm^{-1} in the gas phase at −130.3 °C. The spectrum in the region of 253–348 cm^{-1} with a magnified intensity (×5) is also presented. The band assignment: (*tt*) *trans-trans* conformer: 399.0 cm^{-1}, (*tg*) *trans–gauche*-conformer: 328.9 cm^{-1}, (*g+g+*) *gauche*(+)*-gauche*(+)-conformer: 267.1 cm^{-1}. The ratio of *g+g+/tt* and *tg/tt* peaks is the lowest in the temperature range studied (reprinted with permission from Balabin (2009). Copyright ©2009 American Chemical Society).

5.1.1.2 Cycloalkanes

The spectra of the smaller cyclic alkanes, for example, cyclohexane (Figure 5.8)), show sharper bands as the linear hydrocarbons, as the number of conformers is limited. But ring systems with 12 carbon atoms or more show the same Raman frequencies as non-cyclic alkanes: they can adopt the same conformations as acyclic alkanes.

Takahashi *et al.* (1964) study the IR spectra of cyclohexane in the region of 4,000–150 cm^{-1} in the liquid and vapour states. The IR band due to the E_u skeletal deformation vibration was found at 248 cm^{-1}. Normal coordinate treatment using a modified Urey-Bradley force field assisted the assignment of the IR and Raman bands.

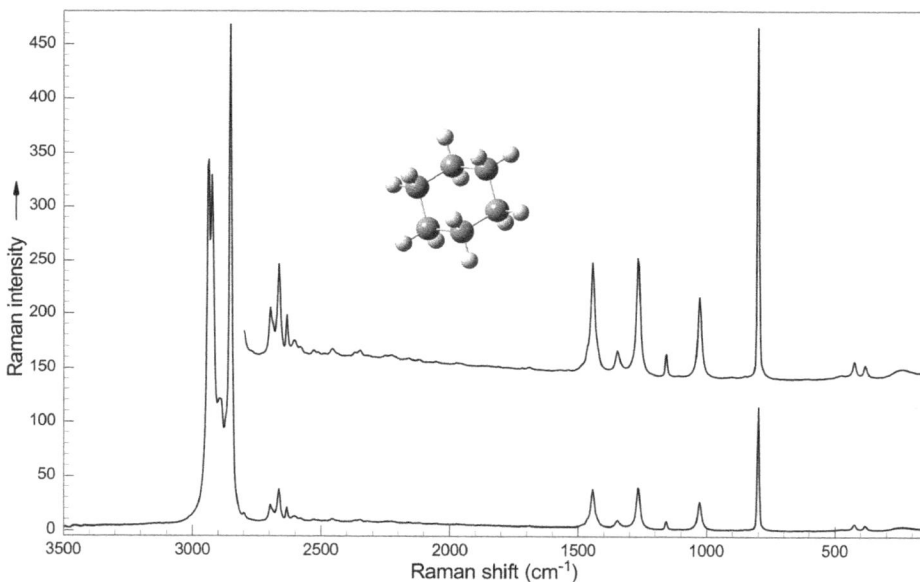

Figure 5.8: Raman spectrum of cyclohexane (λ_{exc} = 1,064 nm, 270 mW; data courtesy of Bruker Optik GmbH, Ettlingen).

Cyclopentane conformations are studied by Sakhaee, Jalili, and Darvish (2016). The authors find a spherical conformational landscape showing the fluxional nature of cyclopentane and its derivatives. This fluxional nature is confirmed by their Raman spectra. Ring twisting and bending vibrations of cyclopentane are visualized in Figure 5.9.

Other vibrational frequencies of cycloalkanes in the liquid state are summarized in Table 5.2.

In a series of papers Kirkpatrick *et al.* (Kirkpatrick *et al.* 2008, 2009; Maki, *et al.* 2010; Kirkpatrick *et al.* 2012) study the IR spectrum of [1.1.1]propellane at very high resolution of 0.002–0.0015 cm^{-1}. The research group studies more than 16,000 roto-vibrational transitions. Best described are the vibrations ν_9, ν_{10}, ν_{11}, ν_{12}, ν_{14}, ν_{15}, and ν_{18}.

Ananthakrishnan (1936) reports the Raman spectrum of the simplest of the cyclic hydrocarbons: cyclopropane

IR spectra of bicyclo[1.1.1]pentane (C_5H_8) are recorded at a resolution of 0.0015 cm^{-1} by Martin *et al.* (2010). This high resolution was sufficient to resolve individual rovibrational lines for the first time.

Perry *et al.* (2012) report a Coriolis analysis of several high-resolution IR bands of bicyclo[111]pentane-d_0 and -d_1. They found Coriolis coupling in two (a_2'') parallel bands, ν_{17} and ν_{18}, of bicyclopentane-d_0. For both isotopologues, quantum calculations (B3LYP/cc-pVTZ) done at the anharmonic level were very helpful in their interpretation of the spectra.

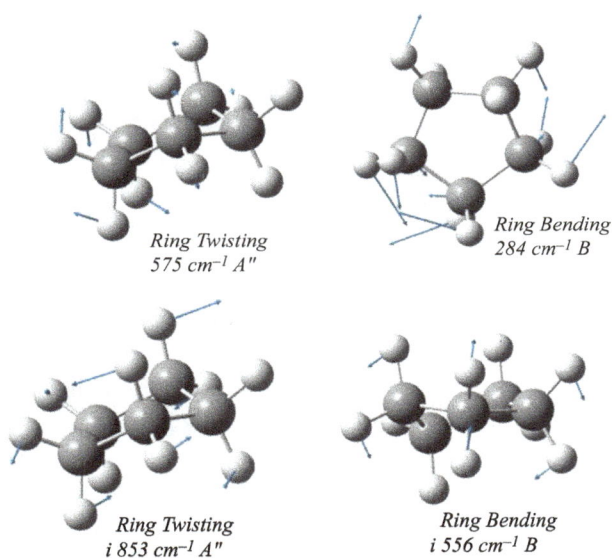

Ring Twisting
575 cm^{-1} A"

Ring Bending
284 cm^{-1} B

Ring Twisting
i 853 cm^{-1} A"

Ring Bending
i 556 cm^{-1} B

Figure 5.9: Ring twisting and bending vibrations of cyclopentane (Sakhaee, Jalili *et al.* 2016). Copyright ©2016 Elsevier B.V., with permission.

The tricyclic hydrocarbon adamantane (tricyclo(3.3.1.1.^{3,7})decane) may be considered as a hydrogenated section of the diamond lattice. Its IR spectrum (Figure 5.10) features as major bands vibrations at 2,927 cm^{-1} (methylene antisymmetric stretch), 2,901 cm^{-1} (methylene symmetric stretch), 2,847 cm^{-1}, (C–H stretching mode of methine), 2,666, 2,635, 1,450 cm^{-1} (methylene scissoring), 1,353, 1,101 cm^{-1} carbon–carbon stretching, and 798 cm^{-1} (ring "breathing") (Pachler *et al.* (1988)). A vibrational assignment of adamantane and some of its isotopomers is communicated by Bistričić *et al.* (1995). They compare the results from an empirical force field to the results from a scaled semiempirical one. Jenkins and Lewis (1980) study the Raman spectra of adamantane, diamantane, and triamantane at different temperatures. May *et al.* (1998) publish interactive

Table 5.2: Vibrational frequencies of cycloalkanes in the liquid state (data taken from Dollish, Fateley, and Bentley (1973) and references cited therein).

Compound	Vibrational frequency (cm^{-1})			
	Methylene antisymmetric stretch	Methylene symmetric stretch	Methylene scissors	Ring "breathing"
Intensities: IR, Raman				
Cyclopropane	3,101–3,090 vs, m	3,038–3,019 vs, s	1,443 m,m	1,188 w,s
Cyclobutane	2,987–2,975 vs, s	2,895–2,887 s, m	1,443 m,m	1,001 –, vs
Cyclopentane	2,959–2,952 s, s	2,866–2,853 s, s	1455 vs, m	886 –, vs
Cyclohexane (Chair)	2,933–2,915 s, s	2,897–2,852 s, s	1,452 s, m	802 –, vs
Cycloheptane	2,935–2,917	2,862–2851	1,450	733
Cyclooctane	2,925	2,855	1,467	703
n-Alkanes	2,929–2,912	2,861–2,849	1,468	–

Raman spectra of adamantane, diamantane, and diamond. They discuss relevance to diamond film deposition.

5.1.2 Haloalkanes

The Raman bands of haloalkane C–X stretching vibrations are very strong due to their large change in polarizability.

The simplest haloalkanes are, of course, the halomethanes. Three of the haloalkanes (methylene chloride, chloroform, and carbon tetrachloride) are in heavy use as solvents.

For frequencies of monohalogenated methanes, symmetry group C_{3v}, refer Table 5.3.

Wagner (1938), Welsh (1952), and Fenlon, Cleveland, and Meister (1951) report vibrational spectra, force constants, and calculated thermodynamic properties for CH_3I and CD_3I.

Raman peak frequencies of fluoromethane molecules in clathrate hydrate crystals are measured by Uchida, Ohmura, and Hori (2010) and calculated by density functional theory at the B3LYP/6-31G(d) level (Figure 5.11).

CH_3F is studied by Wu et al. (1995) and Duncan et al. (1972), CH_2F_2 by Wu et al. (1994), CHF_3 by Wu et al. (1993), and CF_4 by Clark and Rippon (1972) and Sugahara et al. (2004).

The general harmonic force field of methyl fluoride is calculated by Duncan, Kean, and Speirs (1972).

The Raman spectrum of $CHCl_3$ in argon at 4.2 K is communicated by Shirk and Claassen (1971). They report bands at 259, 364, 518, 670, 766, 1,220, and 3,050 cm^{-1}, which are close to the gas phase bands.

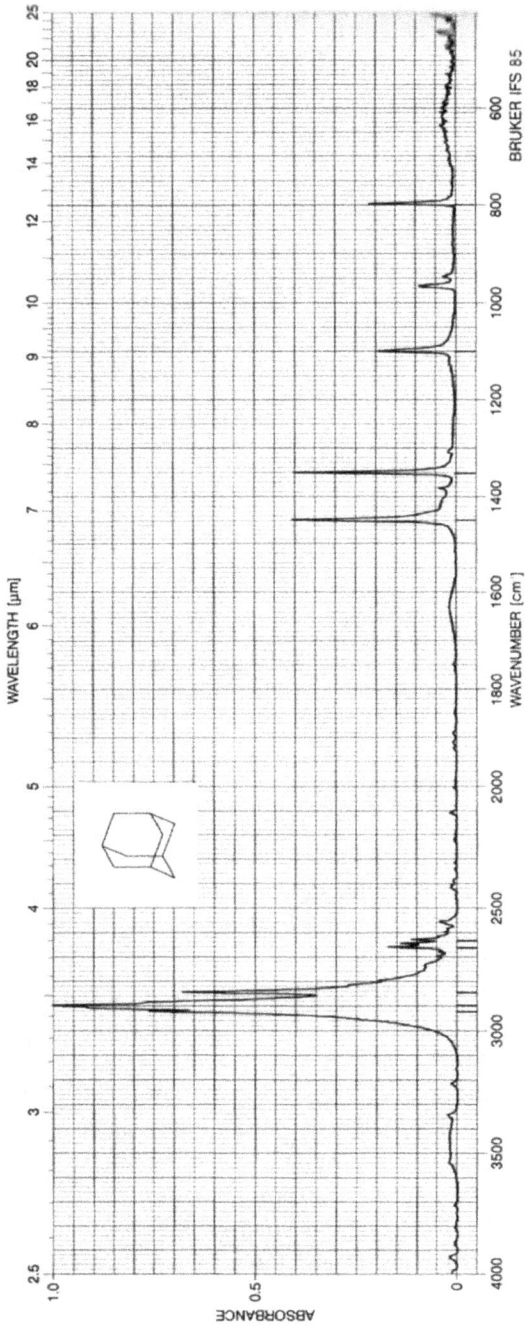

Figure 5.10: FT-IR spectrum of adamantane (Pachler, Matlok *et al.* 1988). Copyright ©1988 VCH Verlagsgesellschaft mbH, with permission.

Table 5.3: Frequencies and intensities of monohalogenated methanes, symmetry group C_{3v}.

	Vibration	$X = F$	$X = Cl$	$X = Br$	$X = I$
		Vibrational frequency (cm^{-1})			
CH_3X	C–X stretch	1,048	709	609	523
	Raman intensity	m	vs	s	vs
	Infrared intensity	s	s	s	s

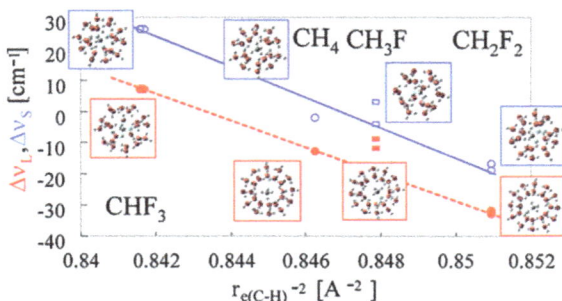

Figure 5.11: Variation in Raman peak position shift with intramolecular C–H distance. The open and solid circles indicate Δv_S and Δv_L, respectively. $r_{e(C-H)}{}^{-2}$ was calculated from the previous studies (Uchida, Ohmura *et al.* 2010). Copyright ©2010 American Chemical Society, with permission.

Wu, Onomichi, Sasaki, and Shimizu (1994) study CH_2F_2 and CHF_2Cl and CHF_3 (1993) under high pressure. Liquid and crystalline CH_2F_2 were measured up to 29 GPa at 300 K in a gasketed diamond-anvil cell. Four solid phases are found.

Clark, Ellestad, and Escribano (1976) examine the vapour phase Raman spectra, Raman band contour analyses, Coriolis coupling constants, and force constants for the molecules $F^{12}CH_3$, $F^{13}CH_3$, $H^{12}CCl_3$, and $H^{13}CCl_3$.

Clark and Rippon (1972) examine the vapour-phase Raman spectra of the spherical-top molecules CX_4 with $X =$ F, Cl, Br, and I (symmetry group T_d). The authors also calculate the force constants of the compounds.

Compton and Hammer (2014) measure the high resolution Raman spectrum of carbon tetrachloride under liquid nitrogen. They clearly separate the v_1 bands of the isotopomers around 459 cm^{-1} (Figure 5.12).

Sugahara *et al.* (2004) conduct a laser Raman spectroscopic analysis of the CF_4 hydrate in the temperature range 276–317 K and at pressures up to 454 MPa.

Shimanouchi and Suzuki (1962) report IR spectra and force constants of CH_2Cl_2, $CHDCl_2$, and CD_2Cl_2.

The frequencies of dihalogenated methanes, symmetry group C_{2v}, can be found in Table 5.4, the Raman spectrum of dichlormethane in Figure 5.13.

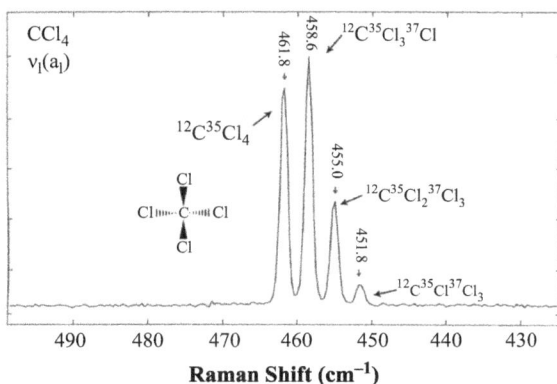

Figure 5.12: High-resolution Raman spectrum of the v_1 band of carbon tetrachloride under liquid nitrogen (Compton and Hammer 2014). Copyright ©2014 by the authors, licensed by Creative Commons (CC 3.0). Label of smallest peak corrected.

Table 5.4: Frequencies of dihalogenated methanes, symmetry group C_{2v}.

	Vibration	X = F	X = Cl	X = Br	X = I
		Frequency (cm^{-1}), intensity			
CH$_2$X$_2$	CX$_2$ antisymmetric stretch	1,111 Ra s	742 Ra w, IR vs	639 Ra wk	566 Ra vs
	CX$_2$ symmetric stretch	1,090 Ra w	703 Ra vs	577 Ra s	483 Ra vs
	CX$_2$ scissoring	529 Ra m	285 Ra s	174 Ra s	121 Ra vs

The frequencies of trihalogenated methanes and symmetry group C_{3v} are summarized in Table 5.5, and the frequencies of tetrahalogenated methanes, symmetry group T_d are listed in Table 5.6.

Palma *et al.* (1964) report Raman and IR spectral data and calculated thermodynamic properties for CH$_2$Cl$_2$, CHDCl$_2$, and CD$_2$Cl$_2$.

Wagner (1938) studies the Raman spectra of methyl derivatives.

As an example of mixed-halogen CX$_4$ compounds, the work of Zietlow, Cleveland, and Meister (1950) may be mentioned, which reports the Raman spectrum and calculates the force constants for fluorotrichloromethane (symmetry group C_{3v}).

In part 2 of their study of the infrared and Raman spectra of fluorinated ethanes, Nielsen *et al.* (1950) report their work on 1,1,1-trifluoroethane.

Klaboe (1970) studies the vibrational spectra of the halopropanes.

Marshall, Srinivas, and Schwartz (2005) calculate geometries and vibrational frequencies of the methane halogenides with quadratic configuration interaction methods at the QCISD/6-311G(*d,p*) level of theory and energies via QCISD(T)/6-311+G(3df,2*p*). The molecules studied are CH$_2$Br$_2$, CHBr$_3$, CBr$_4$, CH$_2$I$_2$, CHI$_3$, CI$_4$, CH$_2$BrI, CHBr$_2$I, and CHBrI$_2$.

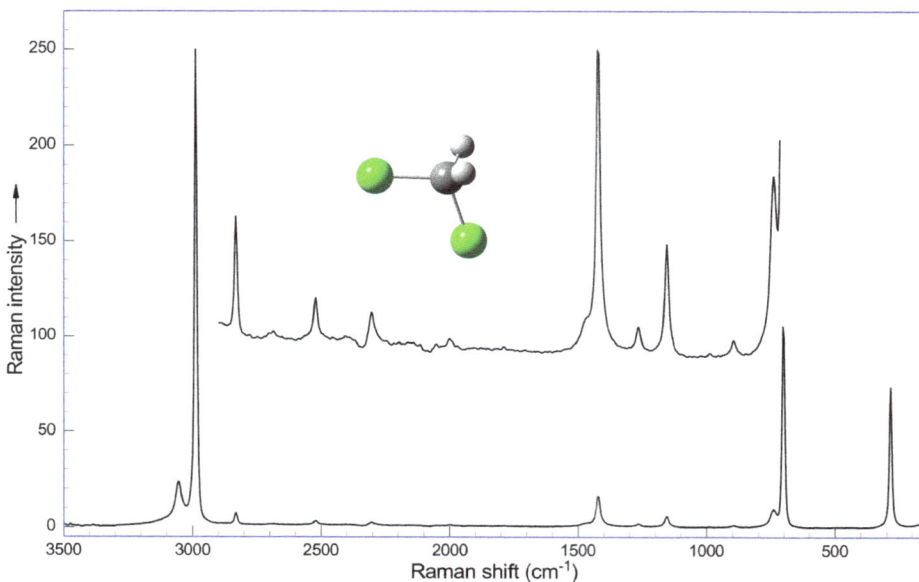

Figure 5.13: Raman spectrum of dichlormethane (λ_{exc} = 1,064 nm, 270 mW; data courtesy of Bruker Optik GmbH, Ettlingen).

Table 5.5: Frequencies of trihalogenated methanes with symmetry group C_{3v} (data from Wood (1935), Redlich (1936), Stammreich and Forneris (1956) and Schrader (1996)).

	Vibration	X = F Gas	X = Cl	X = Br	X = I
		Frequency (cm^{-1}), Raman intensity			
CX$_3$	CX$_3$ symmetric stretch	1,117	668, s	539, s	437, s
	CX$_3$ symmetric deformation	700	366, s	222, vs	153, vs
	CX$_3$ degenerate stretch	1,152	761, s	656, s	578, s
	CX$_3$ degenerate deformation	507	262, vs	154, s	105, vs

For haloethanes, the C–F stretching can be found at 1,365–1,120 cm^{-1}, the C–Cl stretching at 658 cm^{-1}, the C–Br stretching at 559 cm^{-1}, and the C–I stretching at 499 cm^{-1}.

Longer haloalkanes can occur in two conformations: *trans* and *gauche*. The *gauche*-form consists of two equivalent conformers. Figure 5.15 shows models of the conformers viewed along the C–C axis (which means that carbon 2 is covered) and their Newman projections (Figure 5.16).

The v(C–Cl) stretching vibrational frequencies and intensities of some chloroalkanes and their conformers are shown in Table 5.7.

As an example of a bromoalkane, the Raman spectrum of *t*-butyl bromide is shown in Figure 5.14.

Table 5.6: Frequencies of tetrahalogenated methanes, symmetry group T_d (data from Clark and Rippon (1972), Zietlow, Cleveland, and Meister (1950), Meister, Rosser, and Cleveland (1950), and Stammreich, Tavares, and Bassi (1961)).

Vibration	$X = F$ Gas	$X = Cl$	$X = Br$	$X = I$
CX_4		Frequency (cm^{-1}), Raman intensity		
CX_4 symmetric stretch	908	459, vs	267, s s	178, vs
CX_4 degenerate deformation	435	217, s	122, vs	90, m
CX_4 degenerate stretch	1,283 IR vs	790, 762, m	671, wk	560, m
CX_4 degenerate deformation	631 IR s	314, vs	182, m	123, m

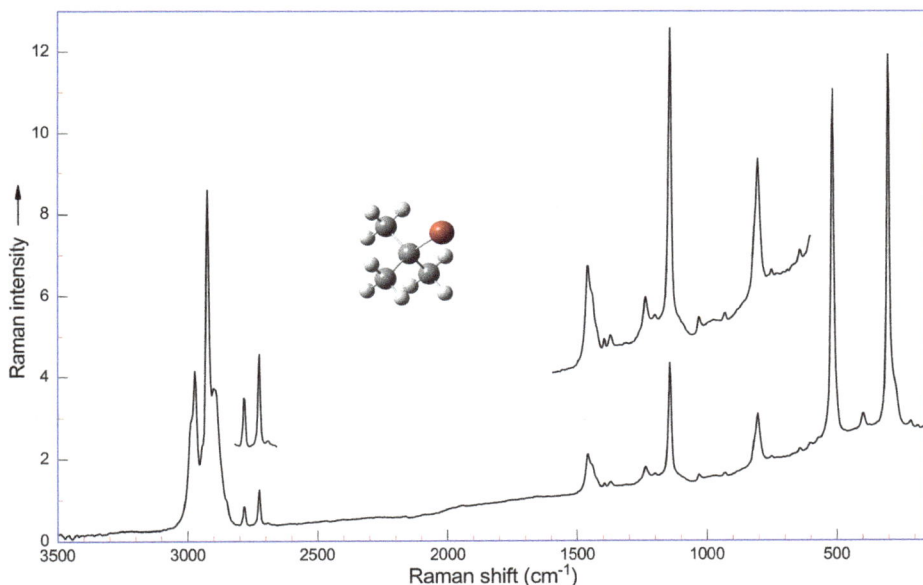

Figure 5.14: Raman spectrum of *t*-butyl bromide (λ_{exc} = 1,064 nm, 270 mW; data courtesy of Bruker Optik GmbH, Ettlingen).

Vibrational frequencies of C–X stretching vibrations in halogenoalkanes and in arylhalogenides (mainly ring-X bending) are shown in Table 5.8.

The molecular vibrations and thermodynamic functions of 2,2-dichloropropane are communicated by Green and Harrison (1971).

In their series of papers on spectra and structures of small ring compounds, fluorocyclobutane is studied by Durig, Willis, and Green (1971).

Compton and Hammer (2014) measure the Raman under nitrogen spectrum of $C_{60}Br_{24}$.

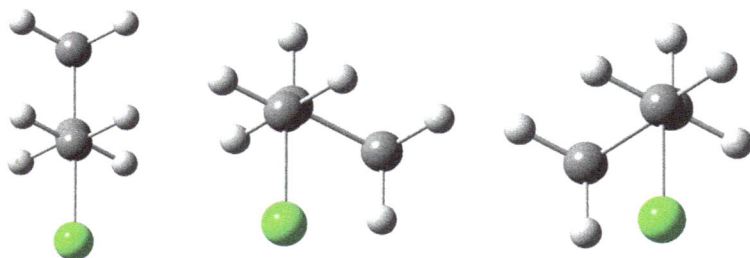

Figure 5.15: Models of the 1-chloropropane conformers (left: *trans*, right: two equivalent *gauche*-conformers) viewed along the C–C axis (one carbon covered).

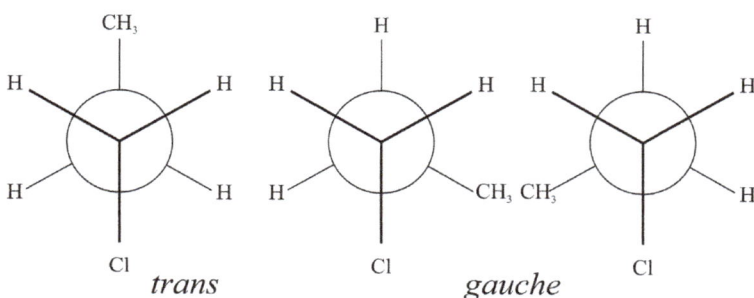

Figure 5.16: Newman projections of 1-chloropropane conformers.

Table 5.7: Raman frequencies and intensities of the v(C–Cl) of chloroalkanes (data taken from Dollish, Fateley, and Bentley (1973) and references cited therein).

Conformation	v(C–Cl) frequency (cm^{-1})		
	Trans	*Gauche* 1	*Gauche* 2
Chloroethane		658	
1-Chloropropane	725	651	
1-Chlorobutane	722	650	
1-Chloropentane	722	653	
1-Chlorohexane	724	651	
1-Chloroheptane	720	655	
1-Chlorooctane	721	651	
1-Chlorononane	729	651	
1-Chlorodecane	725	651	
1-Chloro-3-methylbutane	722	656	
1-Chloro-2-methylpropane	730		688
1-Chloro-2-methylbutane	726		680
1-Chloro-2,2-dimethylpropane	722		
Raman intensity	m-s	vs	s
IR intensity	s	s	s

Table 5.8: Vibrational frequencies of C–X stretching vibrations in halogenoalkanes and -benzenes (data taken from Dollish, Fateley, and Bentley (1973) and references cited therein).

Group	Alkyl-halogenide	Aryl-halogenide (mainly ring-X bending)
C–F	$1{,}365$–$1{,}120$ cm^{-1}	$1{,}270$–$1{,}100$ cm^{-1}
C–Cl	830–560 cm^{-1}	$1{,}096$–$1{,}034$ cm^{-1}
C–Br	680–515 cm^{-1}	$1{,}073$–$1{,}028$ cm^{-1}
C–I	610–485 cm^{-1}	$1{,}061$–$1{,}057$ cm^{-1} (*p*-substituted)
Intensities	(IR s, Ra s-vs)	(IR s, Ra s)

5.1.3 Alkenes

The parent compound, ethylene, as a liquid shows a very strong, polarized ν(C=C) band in its Raman spectrum at $1{,}619$ cm^{-1}, which is weak in the IR and belongs to the C=C stretch. Other bands are the symmetric and the antisymmetric CH_2 stretch as a strong Raman band at $3{,}074$ cm^{-1} (missing in IR) and a very strong band at $3{,}004$ cm^{-1} (IR $3{,}016$ cm^{-1}, vw), respectively. The very strong CH_2 scissoring band may be found at $1{,}339$ cm^{-1} (IR m). The CH_2 rocking at $1{,}238$ cm^{-1} is very weak both in the IR and Raman, and the medium CH_2 wagging band at 942 cm^{-1} is missing in the IR (Arnett and Crawford (1950), Brecher and Halford (1961), Feldman, Romanko, and Welsh (1956), Blumenfeld, Reddy, and Welsh (1970)).

C=C-containing compounds can be classified into monosubstituted, disubstituted, trisubstituted, and tetrasubstituted alkenes.

5.1.3.1 Monosubstituted alkenes

For these compounds, the antisymmetric carbon–hydrogen stretch of the =CH_2 group is observed as a medium-to-strong Raman band at $3{,}086$–$3{,}079$ cm^{-1} (IR m). The C–H stretch shows as a strong, polarized Raman band at $3{,}015$–$2{,}993$ cm^{-1}, which is weak band in the IR. A strong ν(C=C) stretching band can be found at $1{,}648$–$1{,}638$ cm^{-1}. A medium, polarized Raman band, which is medium weak in the IR, shows up at $1{,}419$–$1{,}415$ cm^{-1}, originating from the symmetric deformation of the =CH_2 group. The band at $1{,}309$–$1{,}288$ cm^{-1} (IR w, Ra m, p) arises from the in-plane C–H deformation, a band at 928–909 cm^{-1} (IR vs, Ra w) from the out-of-plane =CH_2 deformation. The very weak band at 634–621 cm^{-1} is identified as the out-of-plane C–H deformation and corresponds to an equally weak band in the IR. Figure 5.17 visualizes the in-plane deformational and C–H stretching vibrations and Figure 5.18 shows the out-of-plane deformational vibrations of monosubstituated alkenes.

Allyl derivatives show the same frequencies as other monoalkyl ethylenes, with a strong Raman band in the region of $1{,}649$–$1{,}625$ cm^{-1}.

Molecular polarizabilities and derivatives with respect to C–C and C–H stretching modes are obtained from *ab initio* molecular orbital calculations at the HF/6–311G

(3df,3pd) level for ethene, cyclopropene, cyclobutadiene, cyclopentadiene, *cis-* and *trans-*butadiene, and *trans-*hexatriene by Lupinetti and Gough (2002).

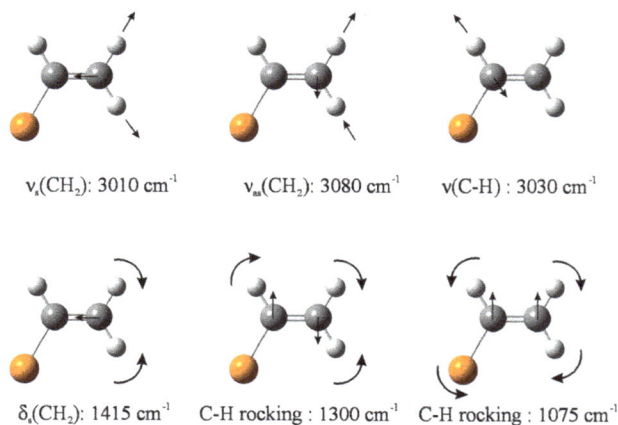

$v_s(CH_2)$: 3010 cm^{-1} $v_{as}(CH_2)$: 3080 cm^{-1} v(C-H) : 3030 cm^{-1}

$\delta_s(CH_2)$: 1415 cm^{-1} C-H rocking : 1300 cm^{-1} C-H rocking : 1075 cm^{-1}

Figure 5.17: In-plane deformational and C–H stretching vibrations of monosubstituted alkenes.

=CH$_2$ wagging   *trans* C-H wagging   *cis* C-H wagging
910 cm^{-1} 990 cm^{-1} 630 cm^{-1}

Figure 5.18: Out-of-plane deformational vibrations of monosubstituated alkenes.

5.1.3.2 Disubstituted alkenes

Disubstituted alkenes can be identified by their wagging frequencies (Figure 5.19): for 1,1-disubstituted compounds they show up at 890 cm^{-1} (IR vs, Ra w), for 1,2-*cis*-substituted alkenes at 680 cm^{-1} (IR ms, Ra vw), and for 1,2-*trans*-disubstituted alkenes at 970 cm^{-1} (IR vs, Ra vw).

5.1.3.3 1,1-Dialkyl alkenes

The antisymmetric carbon–hydrogen stretch of the =CH$_2$ group occurs as a weak Raman band at 3,092–3,073 cm^{-1} with medium intensity in the IR spectrum, the symmetric carbon–hydrogen stretch of the =CH$_2$ group as a strong, polarized Raman band at 2,990–2,983 cm^{-1}, again with medium intensity in the IR. The v(C =C) band appears at 1,658–1,644 cm^{-1} with medium intensity in the IR, but very strong and polarized in the Raman spectrum. A medium-sized, polarized Raman band shows up at 1,413–1,399 cm^{-1}, originating from the symmetric deformation of the =CH$_2$ group. At 1,309–1,288 cm^{-1} the in-plane C–H deformation can be found at 909–985 cm^{-1} (IR vs, RA w) the out-of-plane =CH$_2$ deformation and at 711–684 cm^{-1} the weak band Raman band of the out-of-plane =CH$_2$ deformation. The skeletal deformation of the C=C–C group only shows up as a very weak Raman band at 450–400 cm^{-1}.

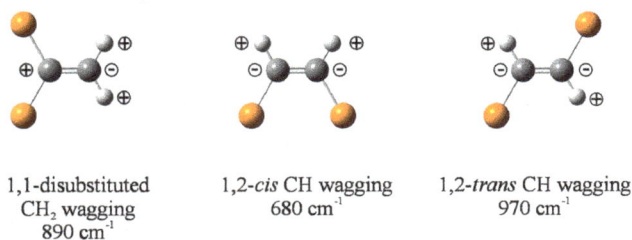

1,1-disubstituted CH$_2$ wagging 890 cm^{-1}

1,2-*cis* CH wagging 680 cm^{-1}

1,2-*trans* CH wagging 970 cm^{-1}

Figure 5.19: Methylene wagging of disubstituted alkenes.

5.1.3.4 1,2-*Cis*-di-alkyl-substituted alkenes

The carbon–hydrogen stretch can be found as IR and Raman bands of medium intensity at 3,016–3,001 cm^{-1}. A ν(C=C) band is observed at 1,660–1,654 cm^{-1}, which is very weak in the IR, but strong and polarized in the Raman spectrum. The *cis* C–H antisymmetric rocking vibration can only be detected in the IR spectrum at 1,429–1,397 cm^{-1} with medium intensity. Another strong Raman band appears at 1,270–1,251 cm^{-1} (IR w), originating from the in-plane C–H deformation, whereas we find at 970–952 cm^{-1} the C–C stretch. At 720–700 cm^{-1} the very weak out-of-plane C–H deformation Raman band (IR m) can be detected. The skeletal deformation of the C=C–C group only shows as a very weak Raman band at 592–545 cm^{-1}.

Cyclohexene is an example of a cyclic 1,2-*cis*-disubstituted alkene, its Raman spectrum is shown in Figure 5.20.

Flett *et al.* (1947) publish the IR and Raman spectrum of *cyclo*-octatetraene. Lippincott *et al.* (1948) report the IR and Raman spectra of heavy *cyclo*-octatetraene. Later, Lippincott *et al.* (1950) discuss the structure of *cyclo*-octatetraene using calculations of its frequencies. They believe that the structure of D_4 symmetry is the correct one.

5.1.3.5 1,2-*Trans*-di-alkyl substituted alkenes

The carbon–hydrogen stretch can be found as medium strong IR and Raman bands at 3,007–2,995 cm^{-1}. The ν(C=C) stretching band appears at 1,676–1,665 cm^{-1} (IR vw, Ra vs,p). The strong Raman band at 1,314–1,290 cm^{-1} (IR –) belongs to the in-plane C–H deformation. Another in-plane C–H deformation band can be found at 776–745 cm^{-1}. In the IR spectrum the very strong CH wagging bands at at 980–965 cm^{-1} are characteristic. The skeletal deformation vibrations of the C=C–C group only shows up as a very weak Raman band at 492–455 cm^{-1}.

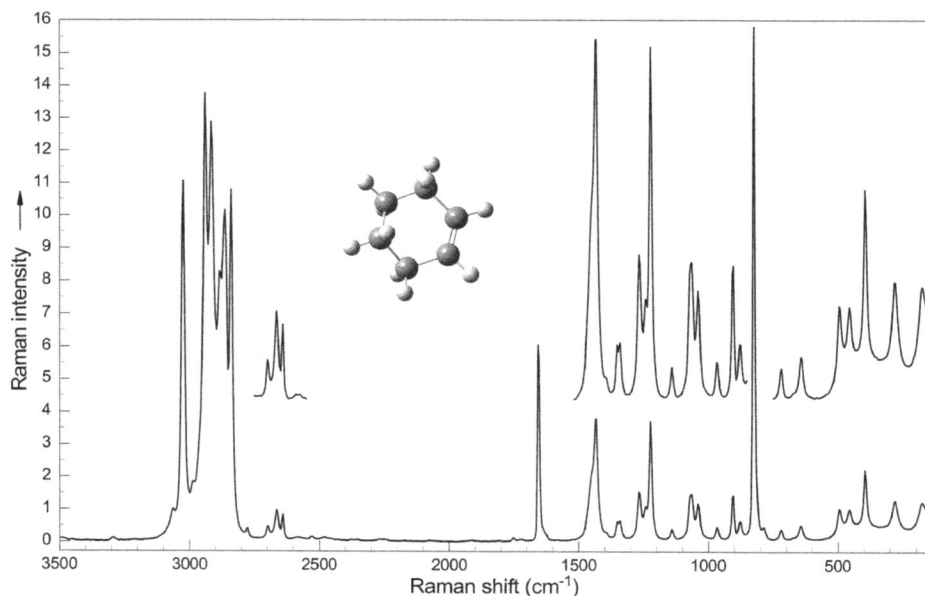

Figure 5.20: Raman spectrum of cyclohexene (λ_{exc} = 1,064 nm, 270 mW; data courtesy of Bruker Optik GmbH, Ettlingen).

Dowling *et al.* (1957) publish vibrational spectra, potential constants, and calculated thermodynamic properties of *cis*-BrHC=CHBr and *trans*-BrHC=CHBr, and *cis*-BrDC=CDBr and *trans*-BrDC=CDBr.

5.1.3.6 Tri-alkyl-substituted alkenes

The remaining C–H stretch of trisubstituted alkenes gives rise to a weak Raman and IR band at 3,040–3,020 cm^{-1}, whereas the C=C stretch is located as a medium to strong Raman band (IR w) at 1,690–1,665 cm^{-1}. At 1,360–1,322 cm^{-1}, the in-plane C–H deformation can be found as a weak Raman band, missing in the IR. The trisubstituted compounds only perform a single out-of-plane C–H deformation vibration at 830–800 cm^{-1} (IR m, Ra w), and the skeletal deformation of the C=C–C group shows up as a weak Raman band at 522–488 cm^{-1}.

5.1.3.7 Tetrasubstituted alkenes

In these compounds, the C=C stretch is located as a strong, polarized Raman band at 1,690–1,670 cm^{-1}, which is forbidden in the IR. We find a strong-to-medium Raman band at 690–675 cm^{-1}, originating from the symmetric C–C stretch. A band of medium intensity at 510–485 cm^{-1} and a weak band at 424–388 cm^{-1} can be attributed to skeletal vibrations.

Halogen substitution changes the vibrational frequencies of alkenes. As an example of a tetrahalogen substituted alkene, tetrachloroethylene (Figure 5.23) shows its C=C stretch at 1,571 cm^{-1} (Ra s), its symmetrical CCl$_2$ stretch at 447 cm^{-1} (Ra vs), and its CCl$_2$ deformation band at 237 cm^{-1} (Ra s). IR and Raman spectra of fluorinated ethylenes are published in a series of papers by Nielsen *et al.* (Nielsen *et al.* 1950; Smith *et al.* 1950).

The Raman and IR spectra, *ab initio* calculations and spectral assignments of 1,1-dicyclopropyl-2,2-dimethylethene (*cyclo*-C$_3$H$_5$)$_2$C=C(CH$_3$)$_2$, DCPDME) were reported by Andersen *et al.* (2019). Quantum chemical calculations by DFT (B3LYP/cc-pVTZ) were carried out (Figure 5.21) to identify possible stable conformers and their vibrational wavenumbers, IR and Raman intensities (Figure 5.22), and depolarization properties. A potential energy surface scan, obtained by rotating the geminal cyclopropyl-groups independently, disclosed occurrence of two dominant conformers of symmetries C_2 and C_1 with an enthalpy difference of around 4 kJ/mol and separated by a low barrier of ca. 6 kJ/mol.

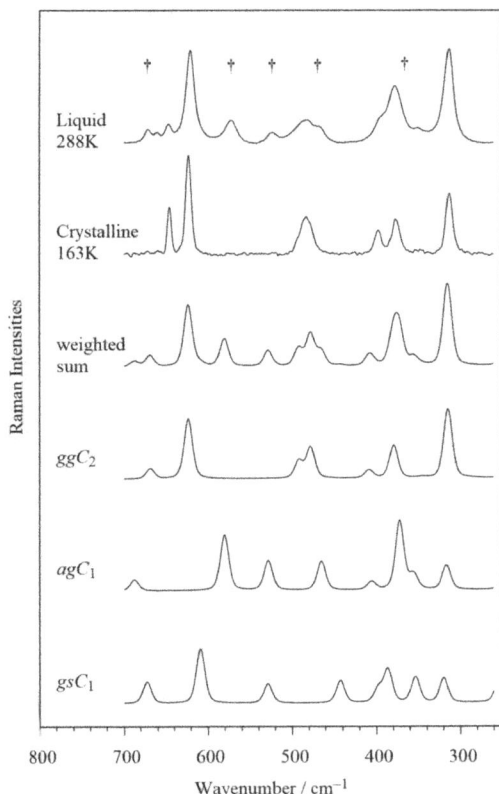

Figure 5.21: Comparison of measured and theoretical Raman spectra of DCPDME in the range 700–250 cm^{-1}. The B3LYP/cc-pVTZ calculated Raman activities have been converted to relative intensities, see text. Observed Raman bands assigned to the high-energy AgC$_1$ conformer are indicated by † (Andersen *et al.* 2019). Copyright ©2019 Elsevier B.V., with permission.

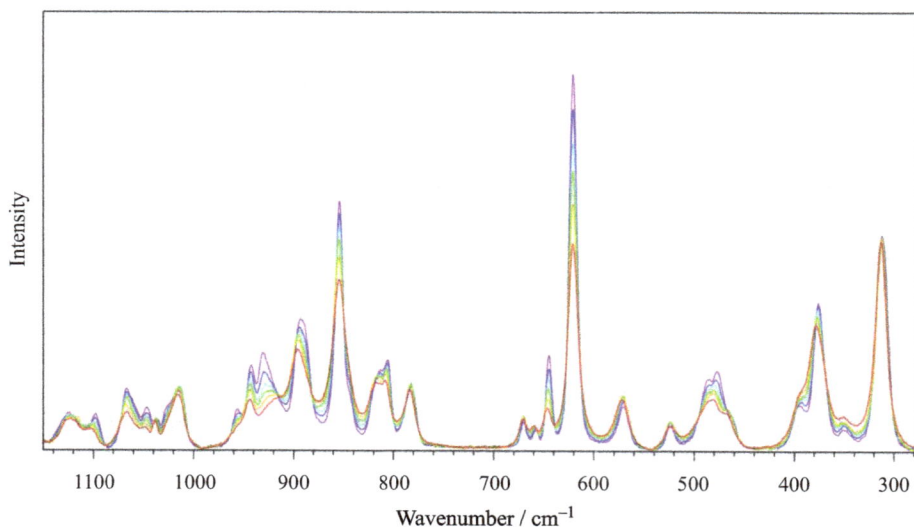

Figure 5.22: Raman spectra of liquid/amorphous DCPDME in the temperature range from 288 K (red) in steps of 30 K via the colours (gold, green, light blue, blue) to 138 K (purple), respectively, in the 1,150–270 cm^{-1} range (Andersen *et al*. 2019). Copyright ©2019 Elsevier B.V., with permission.

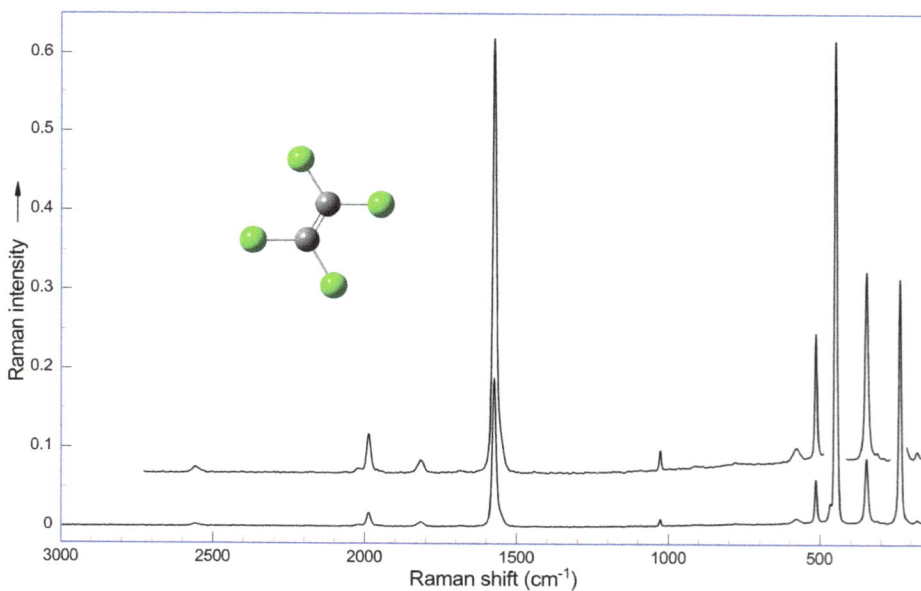

Figure 5.23: Raman spectrum of tetrachloroethylene (λ_{exc} = 1,064 nm, 270 mW (Hoffmann 2017–2018)).

5.1.3.8 Allenes

Substituted propadienes ($R_2C=C=CR_2$), the allenes, are the simplest of the cumulenes. Propadiene is studied by Lord and Ocampo (1951); later, Narayanan, Annamalai, and Keiderling (1988) publish a normal mode calculation and vibrational spectra of the compound and its deuterated isotopomer. Recording the vibrational spectra of the parent compound, the authors find that the IR activ antisymmetric $C=C=C$ stretch at 1,957 cm^{-1} is symmetry-forbidden in the Raman spectrum, but occurs as a very weak band at 1,964–1,958 cm^{-1} in the substituted derivatives. At 3,067 cm^{-1} the antisymmetric CH_2 stretch can be found at 2,996 cm^{-1} the symmetric CH_2 stretch, at 1,440 cm^{-1} the CH_2 bend, at 1,076 cm^{-1} the $C=C=C$ symmetric stretch (IR –, Ra vs, p), at 848 cm^{-1} the CH_2 wagging, and finally the $C=C=C$ bend at 356 cm^{-1}. In the IR spectra of allenes, the overtone of $=CH_2$ wagging may be observed.

5.1.4 Alkines

A triple bond between carbon atoms is stronger than a single or a double bond, so the stretching vibration of the triple bond $v(-C\equiv C-)$ can be identified as the strongest Raman band at 2,260–2,190 cm^{-1}, which is of variable intensity in the IR. We find a stretching vibration of the terminal hydrogen bond $v(\equiv C-H)$ at 3,340–3,267 cm^{-1} (Ra w, p, IR m-s) and the deformation band $\delta(C\equiv C-H)$ at 700–610 cm^{-1} (Ra w, p, IR s). Feldman, Shepherd, and Welsh (1956) study the Raman spectrum of the parent compound.

A vibrational force field of 1-butyne is published by Crowder (1987). It is calculated together with the force field of propionitrile.

Gredy (1935) applies Raman spectrography to study the acetylenic bond. Alkyl-substituted acetylenes show a strong Raman band at 340 cm^{-1} (Ra m-w, p, IR w), originating from the deformational vibration $\delta(C-C\equiv)$.

Experimental and calculated Raman spectra of 3-methyl-1-butyne are reported by Guirgis *et al.* (2005). They also show the Raman spectrum of the alkine in the gaseous state (Figure 5.24) for the first time.

Murray and Cleveland (1938) publish the Raman spectra of acetylenes. Their first paper concentrates on derivatives of phenylacetylene, $C_6H_5C\equiv CR$. In particular they study 3-chloro-1-phenylpropyne-1, 4-chloro-1-phenylbutyne-1, 5-chloro-1-phenylpentyne-1, 3-bromo-1-phenylpropyne-1, 3-phenyl-2-propynol-1, and 4-phenyl-3-butynol-1.

Nyquist *et al.* (1971) publish the IR and Raman spectra of 3-chloro- and 3-bromopropyne-1-*d*.

Brozek-Pluska *et al.* (2005) study low-temperature Raman spectra of stable and metastable structures of phenylacetylene in benzene and derive vibrational dynamics in undercooled liquid solutions, crystals, and glassy crystals from them.

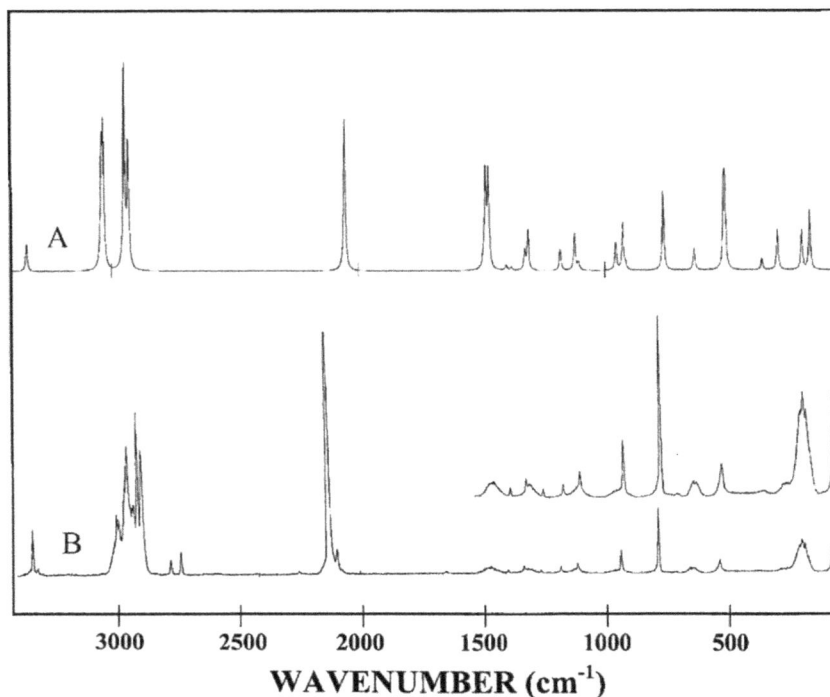

Figure 5.24: Raman spectra of 3-methyl-1-butyne with the predicted (A) and experimental (B) Raman spectrum of the gas (reprinted from Guirgis *et al.* (2005)). Copyright ©(2005) Elsevier B.V., with permission.

5.2 Oxygen-containing aliphatic compounds

5.2.1 Alcohols

The spectra of aliphatic alcohols are quite similar to the spectra of the corresponding alkanes, but of course the vibrations involving oxygen differ. These vibrations are the antisymmetric CCO stretch near 1,060 cm^{-1}, the symmetric CCO stretch near 970 cm^{-1}, the CH$_2$ twist containing δ(COH) near 1,270 cm^{-1}, a band from a vibration composed of a C–O stretch with a CH$_3$ rock and the δ(COH) at 1,100 cm^{-1}, and the δ(CCO) at 460 cm^{-1}. Secondary alcohols show a symmetric stretching band from the C$_3$O moiety, this strong band is observed around 820 cm^{-1}. Tertiary alcohols display a strong band around 760–730 cm^{-1} that is assigned to the symmetric stretch of the C$_4$O skeleton (Lin-Vien *et al.* 1991). In *t*-butanol, the band can be found at 752 cm^{-1} (Kipkemboi, Kiprono, and Sanga 2003). In corresponding hydrocarbons, the Raman band of the C$_4$ symmetric stretch shows up at 795 cm^{-1}. This vibration can be detected in neopentane at 733 cm^{-1}

(Long, Matterson, and Woodward 1954) and in 2,2-dimethylbutane at 714 cm^{-1}, both Raman bands are very strong.

Intensities of the Raman lines of 12 alcohols are measured by Venkateswarlu and Mariam (1962). Table 5.9 summarizes the vibrational frequencies of aliphatic alcohols in the liquid state.

The O–H stretching vibration of aliphatic alcohols can be found at 3,680 cm^{-1}, if we are studying very dilute solutions in apolar solvents (e.g. CCl$_4$). As the OH group can easily form hydrogen bonds, a broad band at 3,400–3,300 cm^{-1} shows up in more concentrated solutions and in polar solvents. Studying the polarized Raman spectra of water/ethanol solutions at different temperatures (Figure 5.25), Dolenko *et al.* (2015) estimate the hydrogen bonding energy and detect clathrate-like structures in solutions.

Raman spectra of methanol and ethanol at pressures up to 100 kbar are studied by Mammone, Sharma, and Nicol (1980). Populations of ethanol conformers in liquid CCl$_4$ and CS$_2$ are studied by Raman spectroscopy in the O–H stretching region by Hu *et al.* (2015). Yu *et al.* (2007) find new C–H stretching vibrational spectral features in the Raman spectra of gaseous and liquid ethanol.

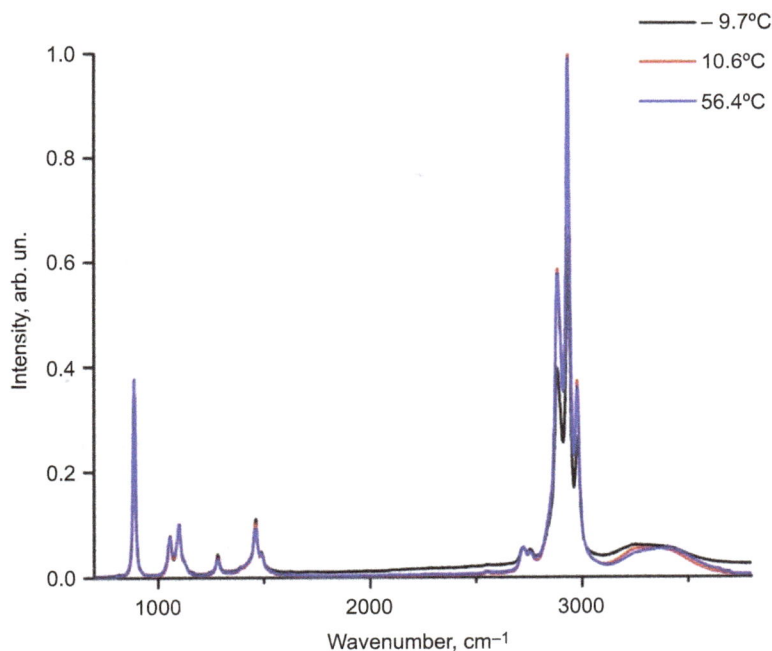

Figure 5.25: Polarized Raman spectra of ethanol at different temperatures (Dolenko, Burikov *et al.* 2015). Copyright ©2015 American Chemical Society, with permission.

Raman spectroscopic analysis of nine saturated monohydroxy alcohols is published by Liu *et al.* (2017). The vibrations are identified with the help of DFT calculations. These alcohols can be discriminated fast by principal component analysis.

Table 5.9: Vibrational frequencies of aliphatic alcohols in the liquid state (data taken from Dollish, Fateley, and Bentley (1973) and references cited therein).

	Vibrational frequency (cm^{-1})					
	O–H stretch	Antisymmetric CCO stretch	Symmetric CCO stretch	CH$_2$ twist + δ (COH)	CO stretch + CH$_3$ rock + δ (COH)	δ (CCO)
Primary alcohols						
Methanol	3,330	–	1,033	1,363	–	480
Ethanol	3,360	1,052	884	1,276	1,093	430
1-Propanol	3,340	1,058	971	1,275	1,100	460
1-Butanol	3,340	1,070	965	1,250	1,100	450
1-Pentanol	3,340	1,055	962	1,208	1,100	440
2-Methyl-1-propanol	3,360	1,050	965	1,250	1,127	435
Secondary alcohols						
2-Propanol	3,360	1,135	819	–	1,125	490
2-Butanol	3,360	1,120	822	1,260	1,115	500
Tertiary alcohols						
2-Methyl-2-propanol	3,380	1,210	753	–	–	350
2-Methyl-2-butanol	3,380	1,200	730	–	–	360
IR intensity	vs, br	s	m	w	w	
Raman intensity	w, br	s-m	vs-sw	w-m	w-m	w-m

Venkateswarlu and Mariam (1962) measure intensites of bands in Raman spectra. The intensities of prominent Raman lines in 12 alcohols are estimated.

The hydrogen bonding network in the chiral alcohol (1 *R*,2*S*,5 *R*)-(−)-menthol is studied by IR, Raman, and vibrational circular dichroism (VCD) spectroscopies by Moreno, Urena, and Gonzalez (2013). The measurements are supported by quantum chemical calculations. Zou *et al.* (2013) provide vibrational assignments in the Raman spectra of 31 fatty alcohols by calculating them by DFT (B3LYP/6-31G(*d*)).

Cryogenic Raman spectroscopy of glycerol is reported by Mendelovici *et al.* (2000). The authors observe a significant cryogenic shift to lower wavenumbers in the high-wavenumber region (3,700–2,500 cm^{-1}), which affects the bands at 3,407, 3240 (antisymmetric and symmetric O–H stretches), and the bands at 2,946, 2,886, and 2,763 cm^{-1} (C–H stretching vibrations). Cryogenic shifts to higher wavenumbers are observed mainly for the lines attributed to the COH deformation (1,342 cm^{-1}), the CCO vibration modes (1,054 and 1,115 cm^{-1}), the CH$_2$ rocking vibrations (976 and 925 cm^{-1}), and the bending vibration at 673 cm^{-1}. Real Raman intensities and the polarization effects are shown in Figures 5.26 and 5.27. To investigate the conformational relaxation behaviour, different admixtures of argon to helium were used in case of t6-6. Saturator (TS) and nozzle temperatures (TN) are provided. The simulation is based on B3LYP calculations and Boltzmann weighted according to TN and the relative energies. Wavenumber scaling factors for the free OH mode for the most stable conformer are also given.

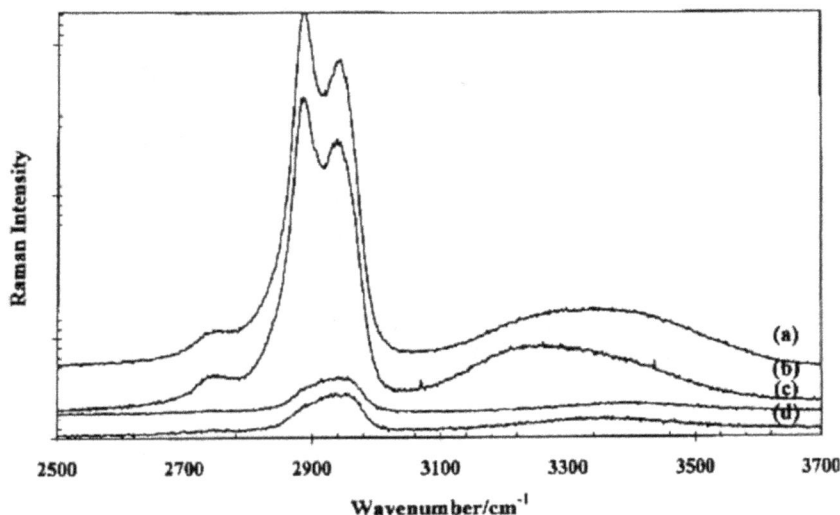

Figure 5.26: Raman spectra of the O–H and C–H region: (a) polarized at 298 K, (b) polarized at 223 K, (c) depolarized at 298 K, and (d) depolarized at 223 K (Mendelovici, Frost *et al.* 2000). Copyright ©2000 John Wiley and Sons, Ltd., with permission.

Paoloni *et al.* (2020) are aiming at fully unsupervised anharmonic vibrational computations. As examples for robust and reliable assignment and interpretation of IR and VCD spectra from mid-IR to NIR, they study 2,3-butanediol and *trans*-1,2-cyclohexanediol. The spectra (Figure 5.28) of each conformer were weighted with their respective Boltzmann population based on B2PLYP harmonic energy. The spectra were simulated assigning Gaussian distribution functions of 15 cm^{-1} half-width at half-maximum. Dashed lines represent the contribution of each conformer to the final spectra.

In the field of carbohydrate research, especially for glycans, the prediction of O–H stretching vibrations is important. To understand subtle hydrogen bonds, Hartwig and Suhm (2021) study the O–H stretching fundamentals of vicinal diols in the gas phase. They combine Raman jet spectroscopy with appropriately scaled harmonic DFT predictions and relaxation path analyses for the conformations of 16 vicinal diols. To illustrate the work, the experimental and calculated Raman spectra of ethanediol, *trans*-cyclohexanediol, and pinacol are shown in Figure 5.29.

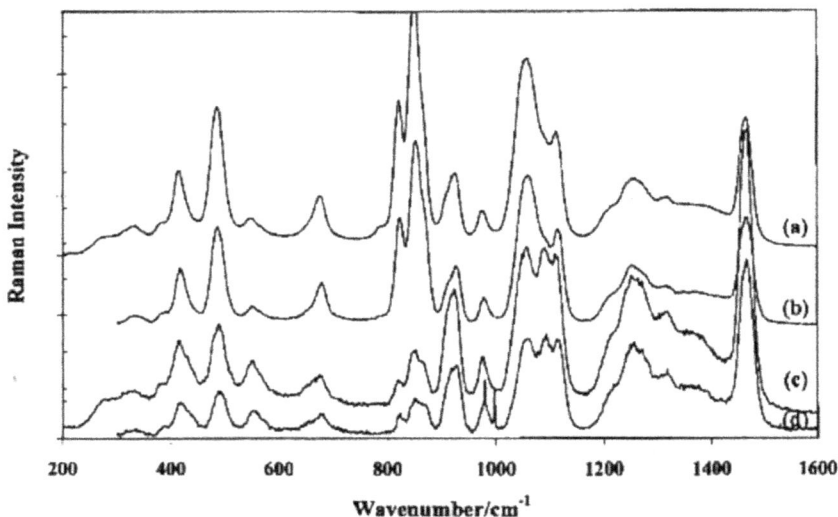

Figure 5.27: Raman spectrum of the 1,600 region: (a) polarized at 298 K, (b) polarized at 223 K, (c) depolarized at 298 K, and (d) depolarized at 223 K (Mendelovici, Frost *et al.* 2000). Copyright ©2000 John Wiley and Sons, Ltd., with permission.

5.2.2 Ethers

Upon substitution of a CH_2 group of an aliphatic chain by an oxygen bridge, a new vibrational band at approximately 1,100 cm^{-1} occurs, it is strong in the IR, medium in the Raman spectrum. This broad band belongs to the antisymmetric C–O–C stretching.

Mashiko *et al.* (1959) (1962) study methyl and ethyl ethers extensively and give some assignments.

In diethylether we can discriminate two rotamers: *trans–trans* (larger amount) and *trans–gauche*. The Raman spectrum (Figure 5.30) shows the antisymmetric C–O–C stretching giving rise to bands at 1,120 and 1,140 cm^{-1} (IR vs, Ra w), the symmetric C–O–C stretching shows up at 845 and 835 cm^{-1} (IR w, Ra vs, p), and the C–O–C symmetric deformation can be found at 440 and 499 cm^{-1} (IR –, Ra vs, p).

Wieser *et al.* (1968) study IR and Raman spectra of diethyl ether and deuterated analogues. The authors derived a valence force field for *trans–trans* (TT) and *trans–gauche* (TG) conformers. IR spectra in the vapour phase for the ethers are also shown.

5.2.3 Acetals

A totally symmetric stretching band of COCOC from 900 to 800 cm^{-1} can be found in the Raman spectrum of acetals. This band is highly polarized and strong in the Raman

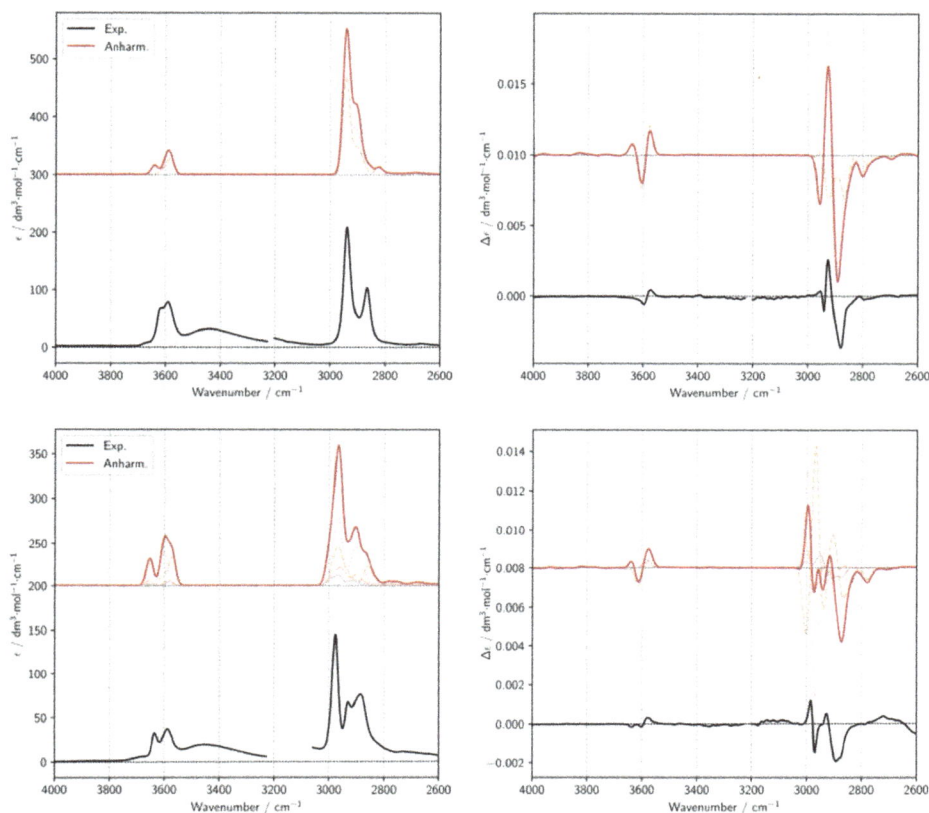

Figure 5.28: Comparison of experimental spectra of 1*R*,2*R*-cyclohexanediol (top left and top right images) and 2*R*,3*R*-butanediol (bottom left and bottom right images) with anharmonic calculations of IR and VCD spectra in the region of fundamental O–H stretching transitions (Δ*v* = 1) and C–H stretching region (Paoloni, Mazzeo *et al*. 2020). Copyright ©2020 by the authors, licensed by Creative Commons (CC 4.0).

spectrum and also strong in the IR. The Raman band of the antisymmetric stretch occurs as a medium band at 1,145–1,129 cm^{-1} (IR s). The three skeletal deformational vibrations of the COCOC group can be detected as a weak band at 654–556 cm^{-1} (IR s), a strong band at 537–370 cm^{-1} (IR s), and a medium band at 396–295 cm^{-1} (IR s) (Mashiko *et al*. (1959)).

5.2.4 Peroxides

The simplest peroxide, hydrogen peroxide, H–O–O–H, shows the v(O–O) at 877 cm^{-1} (IR w, Ra s, p). Our DFT (B3LYP/6-311^{++}G(d,p)) calculation using Gaussian16 (Frisch *et al*. 2016) yields 934 cm^{-1} (scaled 901 cm^{-1}) as vibrational frequency (Figure 5.31).

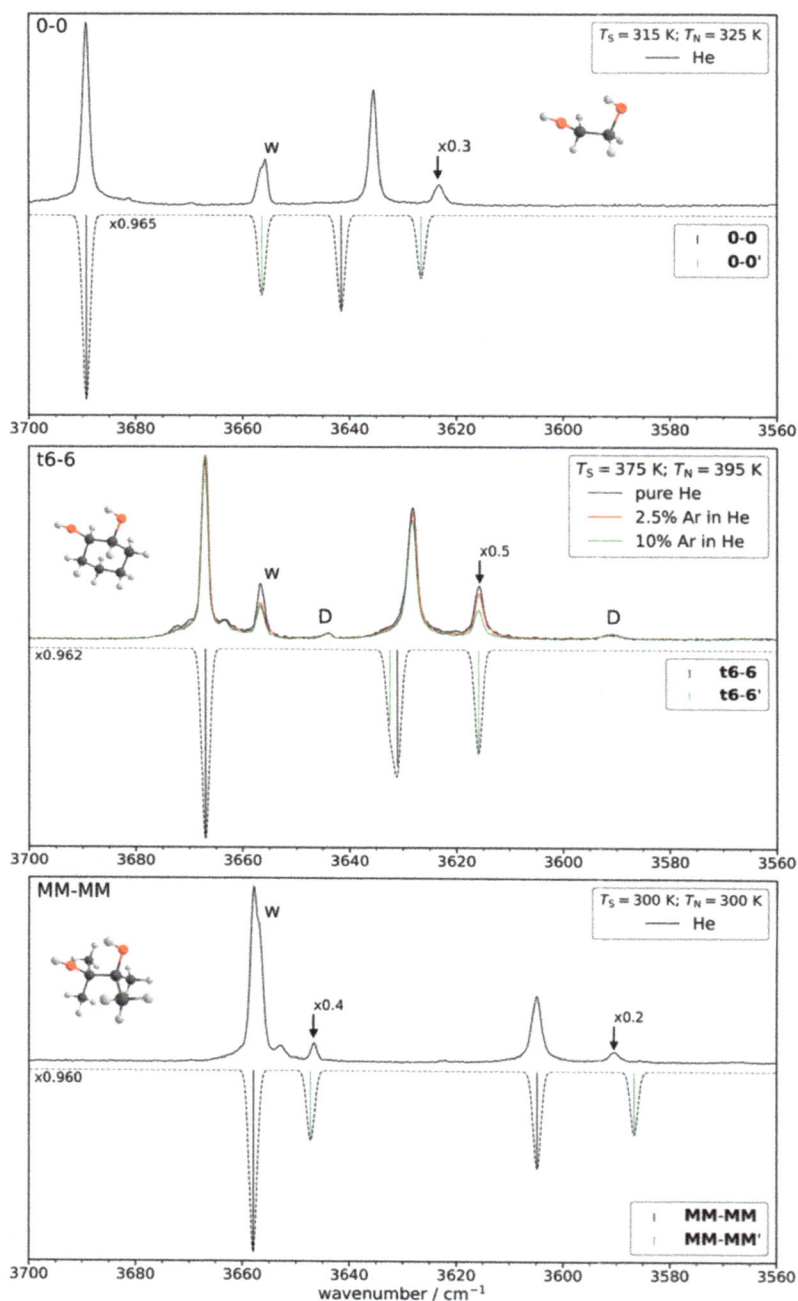

Figure 5.29: Experimental spectra (plotted upwards) and simulated spectra (plotted downwards) of ethanediol (0–0), *trans*-cyclohexanediol (t6-6), and pinacol (MM-MM) (Hartwig and Suhm 2021). Copyright ©2021 the PCCP Owner Societies, with permission.

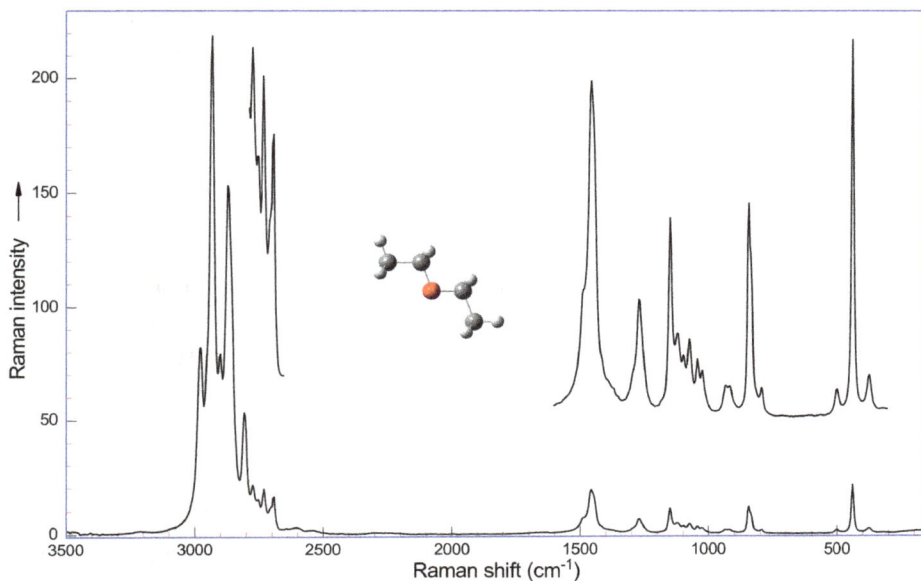

Figure 5.30: Raman spectrum of diethyl ether (λ_{exc} = 1,064 nm, 275 mW; data courtesy of Bruker Optik GmbH, Ettlingen).

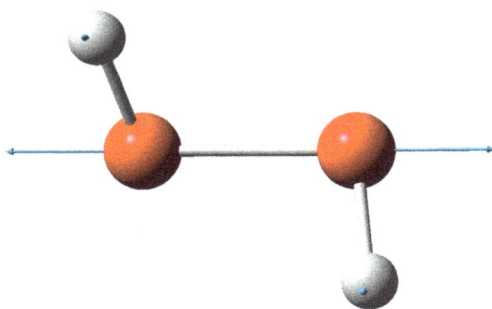

Figure 5.31: Structure and stretching vibration ν(O–O) of hydrogen peroxide (experimental 880 cm^{-1}, calculated 901 cm^{-1}: DFT, B3LYP/6-311^{++}G(d,p), f = 0.9648 (Hoffmann 2022).

Baraban, Changala, and Stanton (2018) calculate (at the CCSD(T)/cc-pCVTZ-5Z level) the equilibrium structure of hydrogen peroxide. Their results are: $r_{O–O}$ = 1.4524 Å, $r_{O–H}$ = 0.9617 Å, $\angle$OOH = 99.76°, and $\angle$HOOH = 113.6°.

The stretching frequencies of O–O for other peroxides range from 943 cm^{-1} for CF_3OOCl to 771 cm^{-1} for $(CH_3)_3COOC(CH_3)_3$, as reported by Leadbeater (1950). Dimethyl peroxides are studied by Christe (1971). Depending upon the electron withdrawing strength of adjacent groups, the frequency of the band is increased. Usually the ν(O–O) is polarized and the strongest band in the Raman spectrum.

In some fluoroperoxides (e.g. $(CF_3)COOC(CF_3)$) Melveger *et al.* (1972) identified the O–O stretching frequency by Raman spectroscopy at 781 cm^{-1}. Far outside the normal range of the O–O stretching frequencies, Gardiner, Lawrence, and Turner (1971) found the $v(O-O)$ for FOOF at 1,306 cm^{-1}.

5.2.5 Aldehydes and ketones

The $v(C=O)$ of simple aldehydes is found with medium Raman intensity at 1,725–1,705 cm^{-1} (see Figure 5.32 for propionic aldehyde), very strong in the IR. Dong *et al.* (2011) conduct a study of several aldehyde molecules by Raman spectroscopy.

Vibrational frequencies of aliphatic aldehydes in the liquid state can be found in Table 5.10.

The table also shows the bands with Fermi resonance: if an overtone or combination band has a frequency near that of an intense fundamental band, its intensity under special circumstances (same symmetry species, mechanical coupling of vibrating groups) may be enhanced. Enrico Fermi explained the phenomenon and gave it its name. The expected frequencies of the two bands are also slightly changed to new values.

Table 5.10: Vibrational frequencies of aliphatic aldehydes in the liquid state (data taken from Dollish, Fateley, and Bentley (1973) and references cited therein).

Compound	Fermi resonance		C=O	C–H	C–C	C–C=O
	dublet		cm^{-1}			
Non-α-carbon-branched						
Acetaldehyde	2,838	2,732	1,715	1,392	1,111	512
Propionaldehyde	2,823	2,724	1,722	1,392	1,089	512
n-Butyraldehyde	2,817	2,732	1,718	1,389	1,112	511
n-Valeraldehyde	2,820	2,722	1,721	1,390	1,116	522
Isovaleraldehyde	2,825	2,726	1,721	1,390	1,125	515
n-Caproaldehyde	2,821	2,710	1,720	1,388	1,112	516
n-Heptylaldehyde	2,820	2,712	1,721	1,388	1,120	520
n-Octylaldehyde	2,825	2,708	1,723	1,388	1,120	525
n-Nonylaldehyde	2,810	2,722	1,721	1,397	1,111	523
n-Decylaldehyde	2,810	2,710	1,722	1,388	1,118	522
α-Carbon-branched						
Isobutyraldehyde	2,811	2,716	1,721	1,392	796	543
sec-Valeraldehyde	2,821	2,719	1,718	1,392	771	540
tert-Valeraldehyde	2,816	2,723	1,720	1,406	762	594
tert-Capronaldehyde	–	–	1,725	1,389	749	585
IR intensity	m	m	vs	m		
Raman intensity	m (sh)	vs	vs-s	s-m	m-vs	m-w

In the case of aldehydes, the fundamental band is the C–H stretching and the first overtone of the in-plane C–H bending vibration.

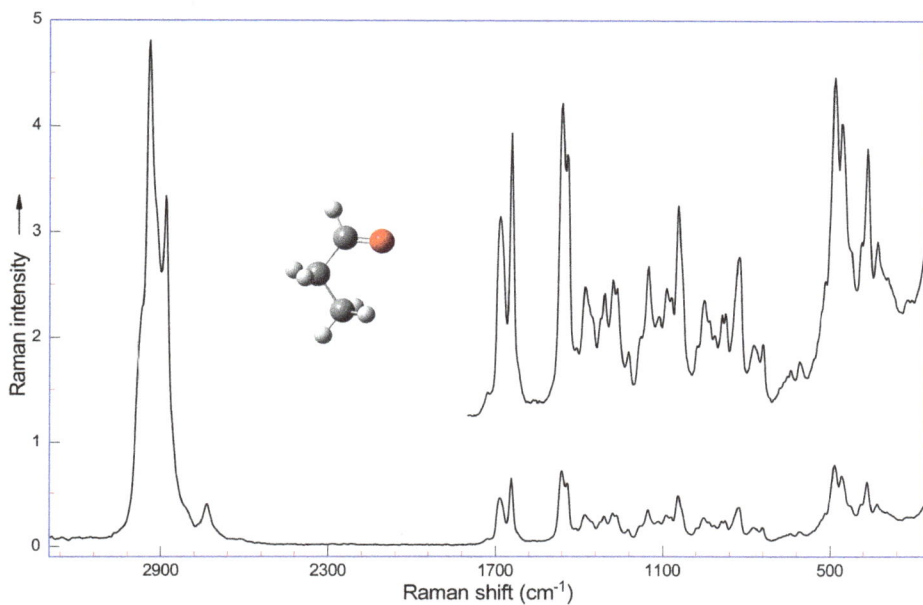

Figure 5.32: Raman spectrum of propionic aldehyde (λ_{exc} = 1,064 nm, 268 mW; data courtesy of Bruker Optik GmbH, Ettlingen).

Table 5.11: Vibrational v(C=O) frequencies of halogen-containing aliphatic aldehydes in the liquid state (data taken from Dollish, Fateley, and Bentley (1973) and references cited therein, e.g. Lucazeau and Novak 1969, 1970; Hagen 1971).

Aldehyde	v(C=O)	IR	Raman intensity
CF$_3$CHO	1,770 cm^{-1}	s	s
CH$_2$ClCHO	1,731 cm^{-1}	s	m
(CH$_3$)$_2$CClCHO	1,754 cm^{-1}	s	m
CHCl$_2$CHO	1,747 cm^{-1}	s	m
CCl$_3$CHO	1,757 cm^{-1}	s	m
CBr$_3$CHO	1,745 cm^{-1}	s	m

Vibrational v(C=O) frequencies of halogen-containing aliphatic aldehydes are shown in Table 5.11

Hagen (1971) reports the IR and Raman spectra of acetaldehyde and some deuterologues, fluoral, chloral, and bromal with band assignments.

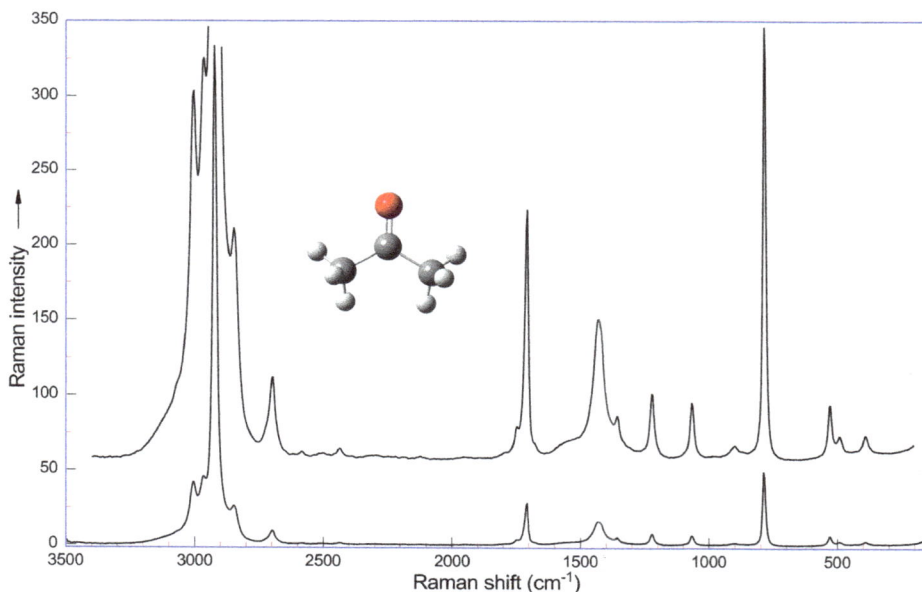

Figure 5.33: Raman spectrum of acetone (λ_{exc} = 1,064 nm, 270 mW; data courtesy of Bruker Optik GmbH, Ettlingen).

Dipolar repulsion in α-halocarbonyl compounds is detailed by Silva *et al.* (2021).

Characteristic for ketones (Table 5.12) is the ν(C=O) band around 1,725–1,700 cm^{-1} (see Figure 5.33 for acetone), a band of medium Raman intensity (IR s). At 1,370–1,350 cm^{-1} (Ra w, IR s) we find the δ_s(CH$_3$), at 1,440–1,405 cm^{-1} (Ra m-w, IR s) the δ_s(CH$_2$), at 1,230–1,100 cm^{-1} (Ra m-s, IR s) the ν_{as}(C–C). The ν_{as}(C–C) occurs at 1,170 cm^{-1} (Ra m–w, IR s) if carbonyl is neighbouring a methyl and an alkyl group. The rest of the Raman spectrum is quite complicated in the case of longer linear ketones, as a large number of conformers are possible. If the ketones are cyclic, this number is restricted and the ν(C=O) changes with the number of carbons in the ring (Table 5.13).

Datta and Kumar (2005) study methyl ethyl ketone, and Devi, Das, and Kumar (2004) methyl isobutyl ketone, as Durig *et al.* (2003) do.

The four-membered ring molecule cyclobutanone in the crystalline state is studied by Cataliotti (1975) using vibrational spectroscopy, and Durig and Green (1968) study its derivatives 2-chlorocyclobutanone and 2-chloro-2,4,4-trideuterocyclobutanone.

Kartha, Mantsch, and Jones (1973) show IR and Raman spectra of cyclopentanone, $\alpha\alpha\alpha'\alpha'$-$d_4$-cyclopentanone, $\beta\beta\beta'\beta'$-d_4-cyclopentanone, and d_8-cyclopentanone and publish a normal co-ordinate analysis.

Fuhrer *et al.* (1972) show IR and Raman spectra of cyclohexanone and deuterated species. They give a comprehensive analysis of the spectra.

Table 5.12: Characteristic frequencies of aliphatic ketones in the liquid state (data taken from Günzler and Gremlich (2003), Dollish, Fateley, and Bentley (1973), and references cited therein).

Functional Group	cm^{-1}	Assignment	IR	Raman Intensity
O ‖ -C-	1725–1705	v(C=O)	s	m
O ‖ -C-CH$_3$	1370–1350	δ_s(CH$_3$)	s	m - w
O ‖ -C-CH$_2$-	1440–1405	δ_s(CH$_2$)	s	m - w
O ‖ -CH$_2$-C-CH$_2$-	1230–1100	v_{as}(C-C)	m - w	m - s
O ‖ Alkyl-C-CH$_3$	1170–700	v_{as}(C-C)	m - w	m - w

Table 5.13: Vibrational v(C=O) frequencies of cyclic aliphatic ketones in the liquid state (data taken from Roberts and Sauer (1949), Frei and Günthard (1961), Kartha (1973), Fuhrer *et al.* (1972), Dollish, Fateley, and Bentley (1973), and references cited therein).

Ketone	v(C=O)	IR	Raman
Cyclobutanone	1,782 cm^{-1}	vs	s
Cyclopentanone	1,744 cm^{-1}	vs, sh	s, dublett
Cyclohexanone	1,709 cm^{-1}	vs	s
Cycloheptanone	1,699 cm^{-1}	vs	s
Cyclooctanone	1,701 cm^{-1}	vs	s
Cyclononanone	1,702 cm^{-1}	vs	s
Cyclodecanone	1,708 cm^{-1}	vs	s
Cycloundecanone	1,708 cm^{-1}	vs	s
Cyclododecanone	1,712 cm^{-1}	vs	s
Cyclotridecanone	1,711 cm^{-1}	vs	s
Cyclopentadecanone	1,712 cm^{-1}	vs	s

A quantum mechanical study on the structure and vibrational spectra of cyclobutanone and 1,2-cyclobutanedione is reported by Pandey *et al.* (2013). The study uses DFT on the B3LYP, LSDA, B3PW91, and MP2 level with the 6-31G(*d, p*) basis set.

Aamouche *et al.* (2000) study conformations of chiral molecules in solution. The flexible molecules 4,4a,5,6,7,8-hexahydro-4a-methyl-2(3H)-naphthalenone and 3,4,8,8a-tetrahydro-8a-methyl-1,6(2H,7H)-naphthalenedione were studied theoretically using *ab initio* density functional theory (DFT), hybrid functionals (B3PW91 and B3LYP), gauge-invariant atomic orbitals (GIAOs), and the 6-31G* basis set. The calculational results for the different conformers are then compared to the experimental spectra.

5.2.6 Carboxylic acids

The additional –O–H group makes the distinction between the ketones and the organic acids. In the acetic acid's vapour the O–H stretch is found at 3,583 cm^{-1} (IR m) and the C–O stretch at 1,182 cm^{-1}. For propionic acid the v(C–O) of the monomer is visualized in Figure 5.34.

Through their ability to form hydrogen bridges between the OH and the C=O moiety, the acids form dimers even in dilute solutions. The optimized geometry of the dimer of propionic acid is shown in Figure 5.35. As the double-bound character of the C=O bond is diminished, its frequency is shifted to longer wavelength: from 1,715 to 1,705 cm^{-1} (IR s).

The vibrational band of free O–H is found at 3,550–3,500 cm^{-1} (IR s) and the bonded O–H at 3,300–2,500 cm^{-1} (IR w). All O–H groups show their out-of-plane deformational bend at 955–890 cm^{-1} (IR v).

In a series of papers, Haurie and Novak (1965, 1965, 1965, 1966, 1967, 1967) study acetic acid as monomer, dimer, and as crystal by IR and Raman spectroscopy.

The dimers of carboxylic acids give rise to an antisymmetric and a symmetric combination of the two C=O stretches. Only the symmetric form is Raman active (IR –), while the antisymmetric form shows a band in the IR spectrum only. The Raman spectrum of propionic acid is shown in Figure 5.36.

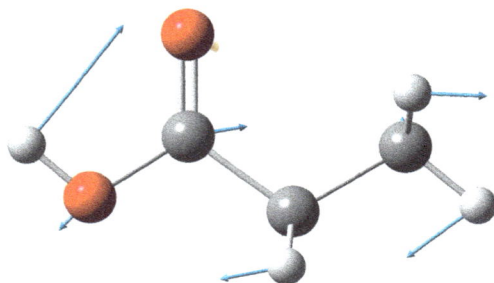

Figure 5.34: Vibration v(C–O) of propionic acid at calculated 1,160 cm^{-1} (DFT, B3LYP/6-311^{++}G(d,p), exp. 1,182 cm^{-1}) (Hoffmann 2022).

In a series of papers, Tanaka *et al.* publish a Raman spectroscopic study of hydrogen bonding in aqueous carboxylic acid solutions. Acetic, propionic, and *n*-butyric acid (Tanaka *et al.* 1990), their deuterated analogues (Tanaka *et al.* 1991), glutaric acid-d_1, and poly(acrylic acid) (Tanaka *et al.* 1991) are examined. The studies are enhanced by normal coordinate analysis.

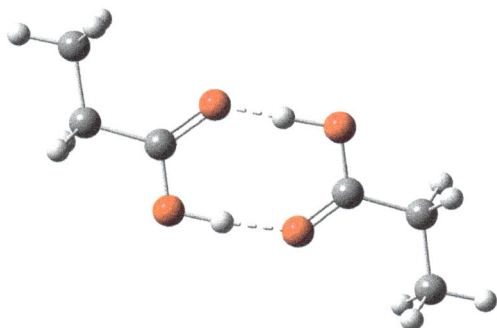

Figure 5.35: Optimized geometry of the dimer of propionic acid (DFT, B3LYP/6-311^{++}G(*d,p*)) (Hoffmann 2017–2018).

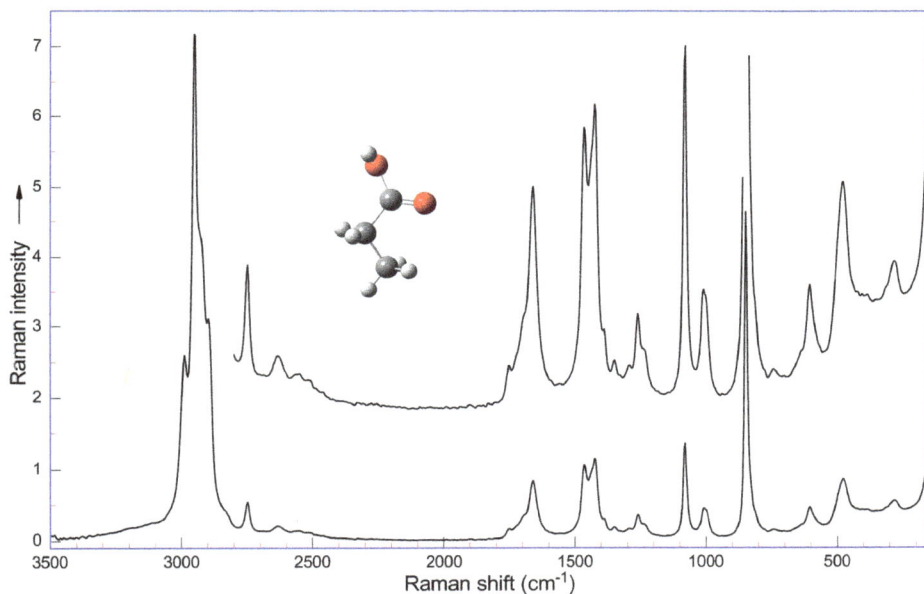

Figure 5.36: Raman spectrum of propionic acid (λ_{exc} = 1,064 nm, 275 mW; data courtesy of Bruker Optik GmbH, Ettlingen).

An *ab initio* study of the ground-state vibrations of the tribasic citric acid and the citrate trianion is reported at the HF/4-21G level by Tarakeshwar and Manogaran (1994).

Long-chained fatty acids show so-called 'progression bands' in the solid state as Meiklejohn *et al.* (1957) report. The bands can only be observed if the acid has more than 12 C atoms in its chain. Attributed to twisting and rocking of *trans*-oriented $-CH_2-$ groups, they occur at 1,350–1,180 cm^{-1} in the IR spectrum and their number is chain-length-dependent. For the straight-chain fatty acids with an even number of

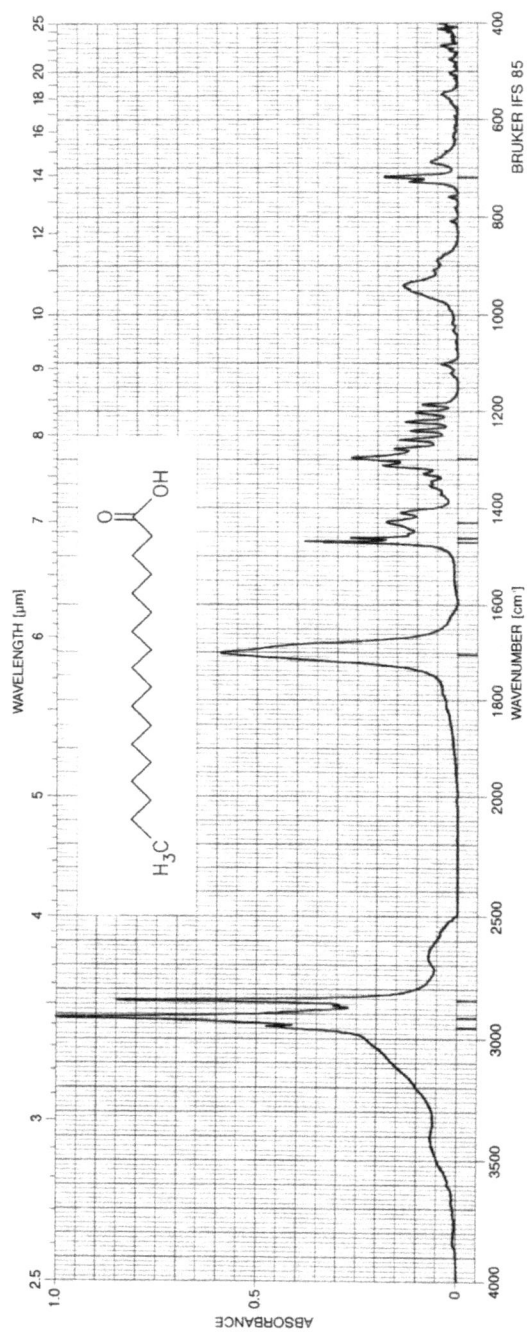

Figure 5.37: FT-IR spectrum of stearic acid. Merck atlas (Pachler *et al.* 1988). Copyright ©1988 VCH Verlagsgesellschaft mbH, with permission.

carbons, the number of carbons in chain/2 equals the number of bands, but for the acids with an odd number of carbons, the number of carbons in chain +1/2 equals the number of bands in progression series. As an example, the IR spectrum of stearic acid is shown in Figure 5.37. The strongest bands in the spectrum can be found at 2,955, 2,917, 2,850, 1,703, 1,473, 1,464, 1,431, 1,298, and 719 cm^{-1}.

Unfortunately, these bands are missing in the Raman spectrum. Raman spectra of some saturated, unsaturated, and deuterated C18 fatty acids in the H–C–H deformation and C–H stretching regions are published by Verma and Wallach (1977). The Raman spectra of longer chain fatty acids are shown in the paper of Czamara *et al.* (2015) (Figure 5.38).

The Raman spectra of long-chain fatty acids (FAs) show a typical spectral profile in three spectral regions: Czamara *et al.* find the stretching vibrations v(C–C) at 1,060–1,090 and at 1,110–1,180 cm^{-1}, the twisting vibrations δ(CH$_2$) at approx. 1,300 cm^{-1}, and the deformational vibrations δ(CH$_3$) or δ(CH$_2$) at 1,400–1,500 cm^{-1}. Most valuable for the analysis of saturated FAs are bands in the 1,000–1,200 cm^{-1} spectral range. For all investigated saturated FAs, the position of a Raman band due to vibrations of the chain in the *gauche* conformation varies from 1,091 (myristic acid, MA) to 1,111 cm^{-1} (arachidic acid, ARA) depending on the number of carbon atoms in the chain. Other features due to the skeletal C–C stretching optical modes for SFAs are observed at very similar wavenumbers, i.e. ~1,130 (high intensity) and 1,178 cm^{-1} (low intensity). Bands in this region for unsaturated fatty acids are relatively less intense, while more characteristic regions for this group of compounds are the region with a characteristic doublet and the region with the intense band due to the C=C stretching vibrations at 1,653–1,672 cm^{-1}. Fatty acids also differ in the spectral region associated with the C–H stretching vibrations, where the bands of variable intensity at 2,832–2888 (due to CH$_2$ groups) and 2,909–2,967 cm^{-1} (due to CH$_3$ groups) are seen. Related to unsaturated fatty acids are the stretching modes of the v(=C–H) observable at ~3,000 cm^{-1}.

Czamara *et al.* (2015) give a recipe for the analysis of lipid unsaturation based on the intensities of bands corresponding to the =C–H deformation (1,260 cm^{-1}) and the C=C (1,655 cm^{-1}) and =C–H stretching (3,000 cm^{-1}) vibrations. Additionally, unsaturation is manifested by the bands due to the *gauche*-C–C stretching (~1,086 cm^{-1}) and *trans*-C–C stretching (~1,068 and 1,120 cm^{-1}) vibrations. A proportionality between the intensity of the band at ~3,000 cm^{-1} and the number of the C=C double bonds in the lipid molecule can be noticed. The spectroscopic data also provide the iodine value for vegetable oils or their isolated fractions with a precision of ~1% for a single determination, as Bailey and Horvat (1972) report.

Unsaturated fatty acids selected for Czamara and co-worker's study differ in the position and number of the double bonds as calculated from their Raman spectra. The ratio of n(C=C)/n(CH$_2$), calculated using integral intensities of Raman bands at 1,655/1,444 cm^{-1}, provides information about a degree of unsaturation of fatty acids. Raman bands near 1,656 and 1,670 cm^{-1} are useful to estimate the level of Z and E unsaturation, respectively. The band at ~1,420 cm^{-1} observed for saturated and Z-unsaturated fatty

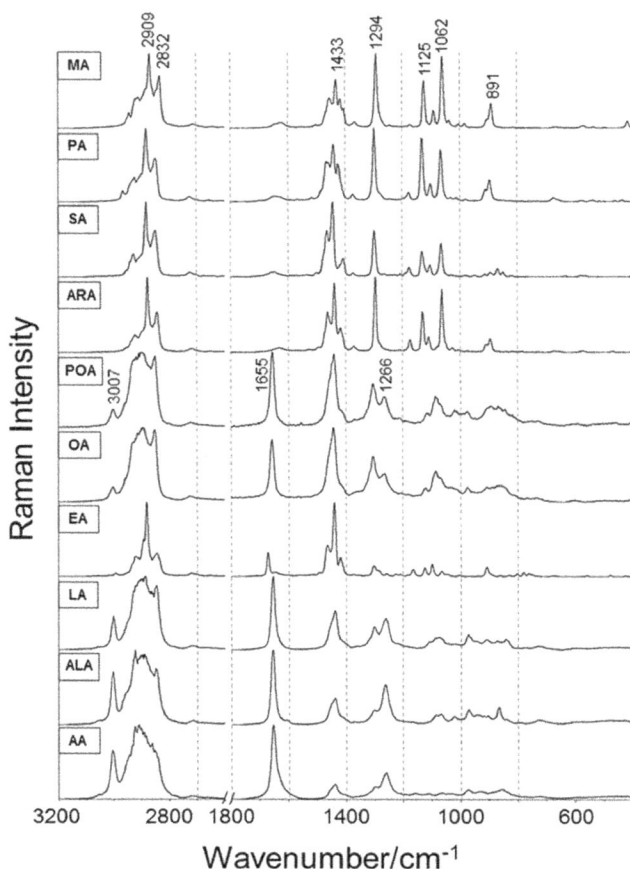

Figure 5.38: Raman spectra of saturated (MA–ARA) and unsaturated fatty acids at 532 nm excitation: myristic acid (MA, solid), plmitic acid (PA, solid), stearic acid (SA, solid), arachidic acid (ARA, solid), palmitoleic acid (POA, liquid), oleic acid (OA, liquid), elaidic acid (EA, solid), linoleic acid (LA, liquid), α-linolenic acid (ALA, liquid), and arachidonic acid (AA, liquid) (Czamara, Majzner *et al*. 2015). Copyright ©2014 John Wiley & Sons, Ltd., with permission.

acids is only seen as a shoulder at approx. 1,422 cm^{-1} in the spectra of the unsaturated *E* compounds, as Verma and Wallach (1977) report. Genin, Quiles, and Burneau (2001) publish an IR and Raman spectroscopic study of carboxylic acids in heavy water.

The polyunsaturated omega 6 fatty acid linoleic acid (*cis, cis*-9,12-octodecadienoic acid, LA, from, e.g. sunflower oil) is studied by *ab initio* calculations (DFT, B3LYP/6-311^{++}G(*d,p*)) by Gocen, Bayari, and Haluk Guven (2017) in the gas phase. Fourier-transform IR (FT-IR, Figure 5.39) and micro-Raman spectra of pure LA in liquid form are also published.

Figure 5.39: IR spectrum of linoleic acid (Gocen *et al*. 2017). Copyright ©2017 Elsevier B.V., with permission.

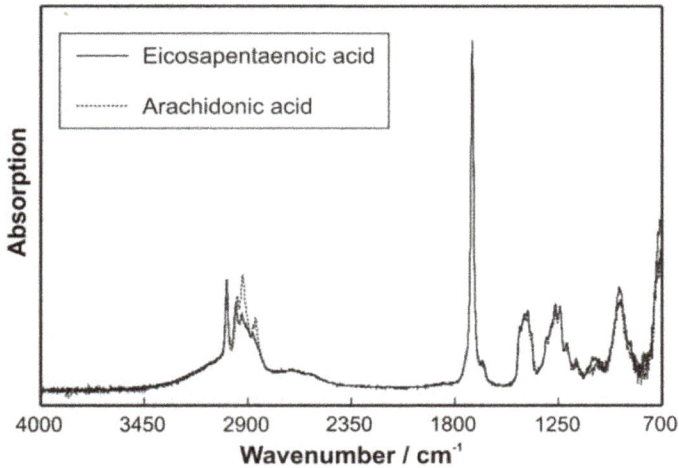

Figure 5.40: Infrared spectra of eicosapentanoic acid and arachidonic acid (Kiefer, Noack *et al*. 2010). Copyright ©2010 Elsevier B.V., with permission.

Kiefer *et al.* (2010) study the vibrational structure of the polyunsaturated fatty acids 5,8,11,14,17-eicosapentaenoic acid and 5,8,11,14-eicosatetraenoic acid (arachidonic acid) (Figure 5.41) by IR spectroscopy. The fingerprint regions of the recorded spectra are found to be almost identical (Figure 5.40), while the C–H stretching mode regions around 3,000 cm^{-1} (Figure 5.42) show such significant differences as a result of electronic and molecular structure alterations based on the different degrees of saturation that both fatty acids can be clearly distinguished from each other.

Eicosapentanoic acid	Arachidonic acid	Assignment
925 cm^{-1}	925 cm^{-1}	O–H bending
1,156 cm^{-1}	1,156 cm^{-1}	C–C stretching
1,207 cm^{-1}	1,207 cm^{-1}	C–C stretching
1,240 cm^{-1}	1,240 cm^{-1}	C–O bending
1,265 cm^{-1}	1,265 cm^{-1}	CH$_2$ twisting
1,313 cm^{-1}	1,313 cm^{-1}	C–O stretching, CH$_2$ wagging
1,413 cm^{-1}	1,413 cm^{-1}	Symmetric CH$_3$ bending
1,429 cm^{-1}	1,429 cm^{-1}	O–H bending
1,456 cm^{-1}	1,456 cm^{-1}	CH$_2$ bending
1,656 cm^{-1}	1,656 cm^{-1}	C=C stretching
1,708 cm^{-1}	1,708 cm^{-1}	C=OC stretching
	2,857 cm^{-1}	Symmetric CH$_2$ stretching
2,874 cm^{-1}	2,872 cm^{-1}	Symmetric CH$_3$ stretching
2,932 cm^{-1}	2,928 cm^{-1}	Antisymmetric CH$_2$ stretching
2,964 cm^{-1}	2,958 cm^{-1}	Antisymmetric CH$_3$ stretching
3,013 cm^{-1}	3,013 cm^{-1}	C–H stretching

IR spectral bands of eicosapentanoic acid and arachidonic acid (Kiefer, Noack *et al.* 2010).

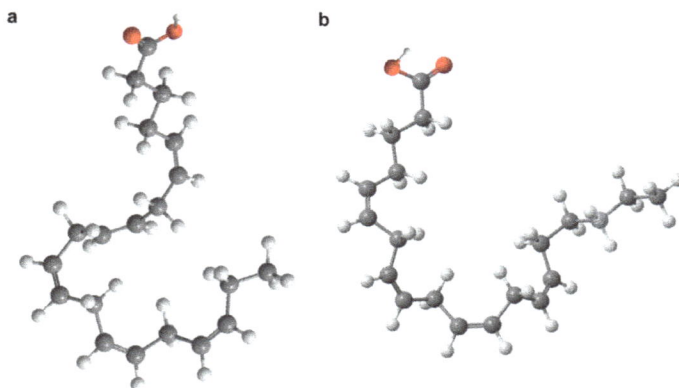

Figure 5.41: Eicosapentanoic acid and arachidonic acid (Kiefer, Noack *et al.* 2010). Copyright ©2010 Elsevier B.V., with permission.

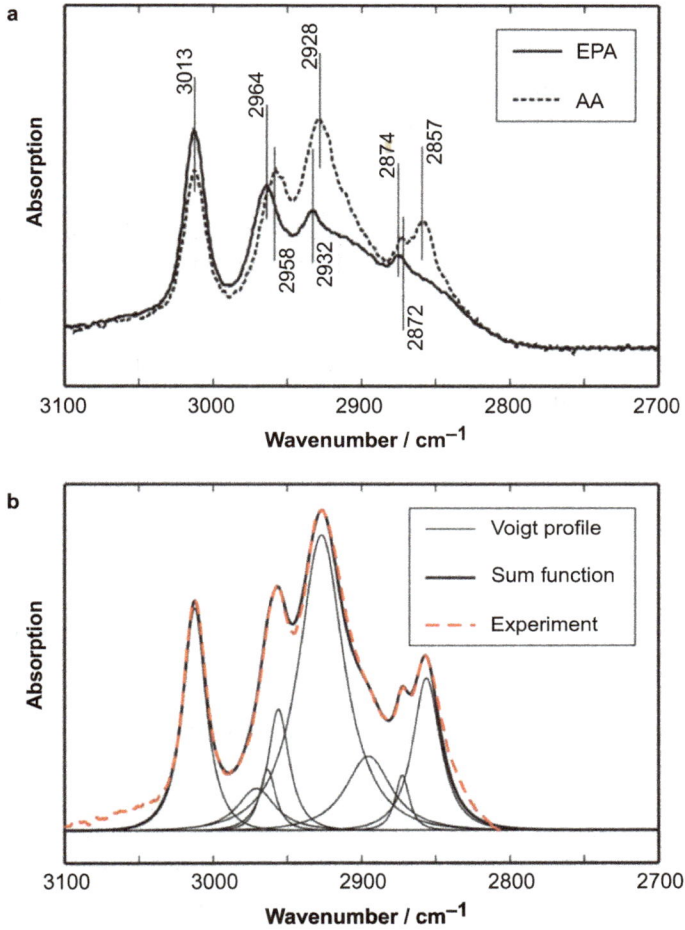

Figure 5.42: Enlargement of the CAH stretching region of the infrared spectra: (a) comparison of both fatty acids and (b) deconvoluted C–H stretching modes of arachidonic acid (Kiefer, Noack *et al.* 2010). Copyright ©2010 Elsevier B.V., with permission.

5.2.7 Acid halides

The characteristic stretching frequencies of the acid halides, R–C(=O)–X are given in the following table:

X	ν(C–X) cm^{-1}		ν(C=O) cm^{-1}	
F	832 (CH$_3$COF)		1,800	
Cl	731–565 (IR s, Ra m-w, p)		1,790	IR vvs
Br	700–535 (IR s)		1,800	Ra m-w, p
I	542 (CH$_3$COI)		1,785	

In compounds R–(C=O)–Cl, the OCX in-plane deformation shows up at 450–420 cm^{-1} as an IR band of medium intensity, very strong, and polarized in the Raman spectrum.

Vibrational assignments and *ab initio* calculations of chloroacetyl fluoride are given by Durig *et al.* (1988).

Chloroacetyl chloride and bromide show their ν(C=O) at 1,825/1,790 and 1,825/1,798 cm^{-1}, as Klaeboe *et al.* (1991) report. The authors study the equilibria of the *gauche/anti*-conformers in an argon matrix and in nitrogen using IR and Raman spectroscopy.

5.2.8 Cyclic anhydrides

Dicarboxylic acids can form cyclic anhydrides on loosing water. This means that the ν(C=O) splits into two bands.

Succinic anhydride shows its antisymmetric ν(C=O) stretching vibration at 1,782 cm^{-1}, whereas its symmetric ν(C=O) stretching occurs at 1,854 cm^{-1} (Figure 5.43).

In general, in non-cyclic anhydrides the higher wavenumber IR absorption is more intense and the bands can be found at 1,825–1,815 cm^{-1} and 1,755–1,745 cm^{-1}. If the compounds are conjugated, these regions change to 1,780–1,770 cm^{-1} and 1,725–1,715 cm^{-1}. Cyclic anhydrides, with their five-membered rings, absorb at 1,870–1,845 cm^{-1} and 1,800–1,775 cm^{-1}, and this changes to 1,860–1,850 cm^{-1} and 1,780–1,760 cm^{-1} on conjugation. In cyclic anhydrides the lower wavenumber absorption is more intense.

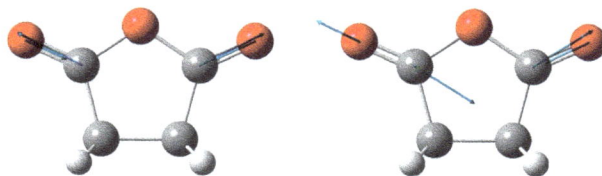

Figure 5.43: Symmetric and antisymmetric ν(C=O) stretching vibration of succinic anhydride (DFT, B3LYP/6-311^{++}G(d,p)) (Hoffmann 2017–2018).

5.2.9 Esters

Kohlrausch, Köppel, and Pongratz (1933) study the Raman spectra of 16 methyl and 14 ethyl esters of mono-basic fatty acids. The range of the fatty acids in the esters ranges from formic to decanoic acid.

In the Raman spectrum of esters, the v(C=O) is normally found as a Raman band of medium intensity (IR s) in the range of 1,750–1,720 cm^{-1} and the v(C–O) as a small Raman band (IR s) around 1,230 cm^{-1} (see Table 5.14).

The vibrations giving rise to the small Raman bands at 1,250 and 1,050 cm^{-1} are attributed to v_{as}(C–(C=O)–O–) and v_{as}(O–CH$_2$–C–) (exemplarily shown for ethylacetate in Figure 5.44).

Table 5.14: Vibrational frequencies of aliphatic esters in the liquid state (data taken from Günzler and Gremlich (2003) and references cited therein).

Ester class	v(C=O) cm^{-1}	v(C–O) cm^{-1}
Formiates	1,720–1,715 (IR s, Ra m)	1,200–1,180 (IR s)
Acetates	1,741–1,734 (IR s, Ra m)	1,260–1,230 (IR s)
Higher aliphatic esters	1,738–1,731 (IR s, Ra m)	1,210–1,160 (IR s)
Acrylates	1,722–1,740 (IR s, Ra m)	1,300–1,200 (IR s)
Benzoates	1,720 (IR s, Ra m)	1,310–1,250 (IR s) 1,150–1,100
Acetates with primary alcohols	1,741–1,734	1,260–1,230 1,060–1,035
Acetates with secondary alcohols	1,741–1,734	1,260–1,230 1,100
Acetates with phenols	1,741–1,734	1,205

5.2.10 Lactones

In cyclic esters (lactones), the frequency of the v(C=O) is higher than in the open chain esters. Therefore, we find the v(C=O) band of β-lactones around 1,841 cm^{-1}, γ-lactones at 1,774 cm^{-1} (IR s), δ-lactones at 1,750 cm^{-1} (IR s), and ε-lactones at 1,727 cm^{-1}.

In γ-lactones, the carbonyl frequencies are found around 1,770 cm^{-1}, as reported by Mecke, Mecke, and Lüttringhaus (1957). Jones *et al.* (1957, 1959) study unsaturated lactones. If a double bond is conjugated to the carbonyl group, the band is split into two: one at 1,778 cm^{-1} and the other at 1,745 cm^{-1}. The v(C=C) bands are observed at 1,618 and 1,601 cm^{-1}. A phenyl group conjugated to the double bond, as in β-phenyl $\Delta^{\alpha\beta}$-butenolide, shifts the carbonyl frequencies to 1,789 and 1,751 cm^{-1}. The v(C=C)

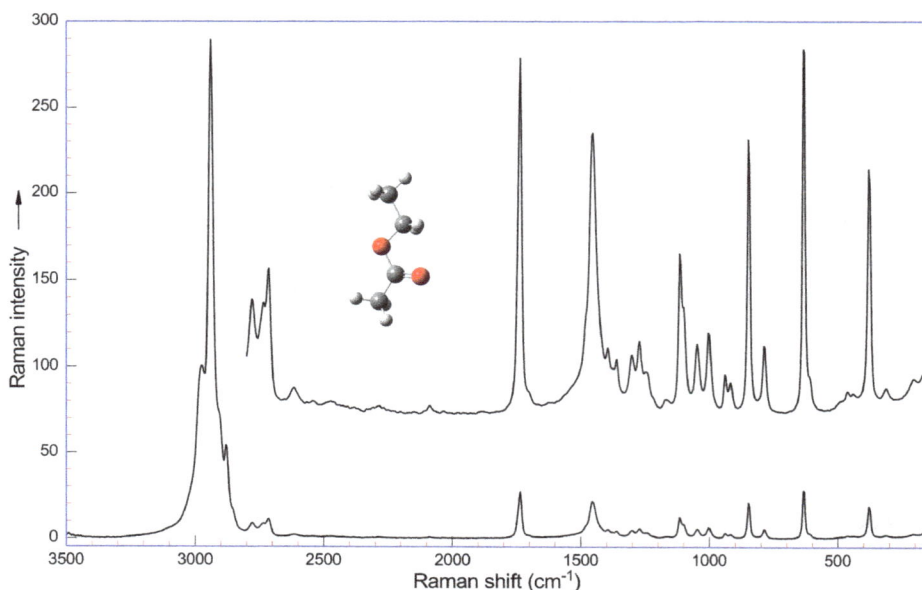

Figure 5.44: Raman spectrum of ethyl acetate (λ_{exc} = 1,064 nm, 275 mW; data courtesy of Bruker Optik GmbH, Ettlingen).

bands are shifted to 1,694 and 1,628 cm^{-1}. In the cyclic anhydride phthalide (isobenzo-furan-1(3 H)-one, $C_8H_6O_2$) the C=O band is found at 1,753 cm^{-1}.

Chain *et al.* (2014) publish an experimental study of the structural and vibrational properties of the sesquiterpene lactone cnicin using FT-IR, FT-Raman, UV-visible, and NMR spectroscopies. Cnicin was isolated from Centaurea (Asteraceae). The spectra were also calculated by DFT on the B3LYP/6-31G level. The lactone ring has half-chair configuration.

Jia *et al.* (2009) study the alkali-hydrolysis of D-glucono-δ-lactone by Raman optical activity (ROA) and VCD spectroscopies.

5.3 Nitrogen-containing aliphatic compounds

5.3.1 Amines

Six vibrations are possible for the C–NH$_2$ group: the antisymmetric N–H stretch, the symmetric N–H stretch, the NH$_2$ deformation or scissoring, a carbon–nitrogen stretch, a NH$_2$ wagging, and a NH$_2$ twist.

Table 5.15 shows four of those frequencies for a number of amines with primary, secondary, and tertiary α-carbon.

Dellepiane and Zerbi (1968, 1968, 1968, 1969) publish experimental and calculated values of some simple aliphatic amines and their deuterated analogues.

Durig, Bush, and Baglin (1968) investigate condensed phases of methylamine and its deuterium derivatives by Raman spectroscopy.

Trimethylamine and its deuterated analogue are studied by Clippard and Tailor (1969), who also assign the bands.

In amines, the $v_{as}(NH_2)$ is found as a strong Raman band near 3,370 cm^{-1}. The very strong Raman band of the $v_s(NH_2)$ is detected nearby around 3,320 cm^{-1}. The δ_s (NH_2) can be observed as a weak band around 1,620 cm^{-1}, whereas the $v(C-N)$ vibration shows up as a medium band in the region from 1,040 to 1,085 cm^{-1}. Figure 5.45 shows the Raman spectrum of dimethyl amine in water.

Table 5.15: Vibrational frequencies of aliphatic amines in the liquid state (data taken from Dollish, Fateley, and Bentley (1973) and references cited therein).

Compound	Vibrational frequency (cm^{-1})			
	$v_{as}(NH_2)$	$v_s(NH_2)$	$\delta(NH_2)$	$v(C-N)$
Primary α-carbon				
Methyl amine	3,372	3,320	1,600	1,040
Ethyl amine	3,371	3,318	1,619	1,085
n-Propyl amine	3,378	3,325	1,624	1,072
n-Butyl amine	3,379	3,320	1,610	1,083
n-Hexyl amine	3,375	3,320	1,620	1,075
Isobutyl amine	3,380	3,330	1,616	1,061
Secondary α-carbon				
Isopropyl amine	3,363	3,308	1,622	1,034, 1,135
sec-Butyl amine	3,364	3,321	1,605	1,044, 1,145
Tertiary α-carbon				
tert-Butyl amine	3,360	3,310	1,623	1,040, 1,240
IR intensity				
	m-w	m-w	m	m-w
Relative Raman intensity				
	s	vs, p	w	m, dp

The heterocyclic amine Tröger's base (Figure 5.46) is studied by Aamouche *et al.* (2000). Using spectroscopic data, especially VCD, the absolute configuration of the compound is deduced. *Ab initio* DFT, using the B3PW91 and B3LYP functionals and the 6-31G* basis set is used to calculate the theoretical VCD spectra for comparison to the experiments.

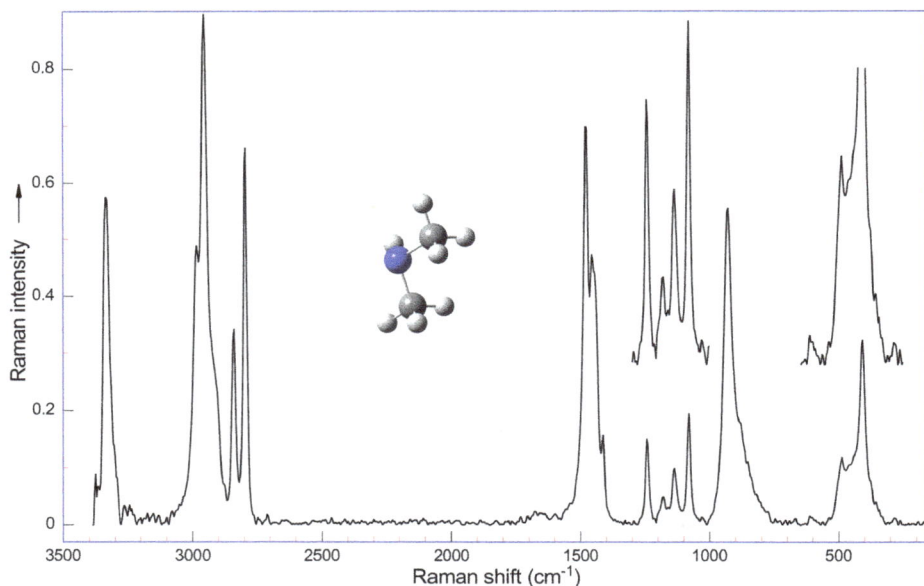

Figure 5.45: Raman spectrum of dimethyl amine (40% in water, λ_{exc} = 1,064 nm, 275 mW; data courtesy of Bruker Optik GmbH, Ettlingen).

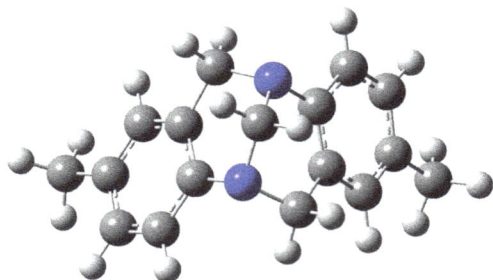

Figure 5.46: The heterocyclic amine Tröger's base (5*R*, 11*R*)-enatiomer.

5.3.2 Amides

Miyazawa, Shimanouchi, and Mizushima (1956) report characteristic IR bands of monosubstituted amides and establish the amide I to VI nomenclature for the characteristic bands. Suzuki (1960) studies formamide and deuterated analogues. Suzuki (1962) reports spectra of acetamide and deuterated analogues. Primary alkyl amides feature two N–H stretches, whose frequencies depend upon whether they are hydrogen-bonded or free. The antisymmetric N–H stretch can be found at 3,350 cm^{-1}, the symmetric stretch at 3,180 cm^{-1}, when bonded, or at 3,500 and 3,400 cm^{-1}, when free. Other amide bands have been labelled amide I to amide VI (Table 5.16). The amide I band arises from a mixture of carbonyl stretch and C–N stretch in the ratio of 2 to 1. In acetamide (see Figure 5.47), for example, we find the N–H antisymmetric stretch at 3,355 cm^{-1}, the N–H symmetric stretch at 3,165 cm^{-1}, the amide I band can be observed at 1,660 cm^{-1}, the amide II band at 1,615 cm^{-1}, the C–N stretch at 1,389 cm^{-1}, the NH$_2$ rocking band at 1,120 cm^{-1}, the O=C–N deformation at 568 cm^{-1}, and finally the CCO deformation at 446 cm^{-1}.

Table 5.16: Vibrational frequencies of aliphatic amides in the liquid state (data taken from Günzler and Gremlich (2003), Dollish, Fateley, and Bentley (1973), and references cited therein).

Designation	Range in solution (cm^{-1})	Raman intensity	IR intensity	Assignment
Amide I	≈1,690	w-m	s	ν(C=O) of primary amides
	1,700–1,670	w-m	s	ν(C=O) of secondary amides
	1,670–1,630	m	s	ν(C=O) of tertiary amides
Amide II	1,620–1,590	m		ν(C–N) + δ(O–C–N) of primary amides
	1,550–1,510	m	s	ν(C–N) + δ(O–C–N) of secondary amides
Amide III	1,330–1,200	m	m	δ(N–H) + δ(O–C–N) of secondary amides
Amide IV	≈620 630–600	m-s		Primary amides ν(C–C) + δ(O–C–N) of secondary amides
Amide V	≈720		m, br	δ(N–H … bridge) of secondary amides
Amide VI	1,440–1,400	m-s	m	Skeletal vibration of primary amides

A FT-IR and FT-Raman vibrational spectra and molecular structure investigation of nicotinamide is published as a combined experimental and theoretical study by Ramalingam *et al.* (2010).

Ureas (H$_2$N-CO-NH$_2$) are special cases of amides. We find the ν(C=O) at 1,705 to 1,635 cm^{-1}, N–H stretches occur in the region of 3,440 to 3,200 cm^{-1}. The Raman bands are of medium to weak intensity. The amide II band can be detected as a weak band at 1,605 to 1,515 cm^{-1}. N–C–N stretches can be found as medium bands from 1,360 to 1,300 cm^{-1} for the asymmetrical vibration and from 1,190 to 1,140 cm^{-1} in the symmetrical case.

According to Stewart (1957), all vibrations of urea are Raman active. The author gives frequencies and intensities of all vibrations and performed a normal coordinate analysis.

Otvos and Edsall (1939) report the Raman spectra of urea and deuterated urea in water solution. The authors also detect the polarization of the lines. The data of the study are given in Table 5.17 together with some solid phase data from Schrader (1996).

IR spectra of amorphous and crystalline urea ices are studied by Timon *et al.* (2021).

Amino acids and proteins are discussed later in this work (Chapter 6 on natural products).

Table 5.17: Raman data of urea and deuterated urea (data from Stewart (1957), Otvos and Edsall (1939), and Schrader (1996)).

$CO(NH_2)_2$ in H_2O	Polarization	$CO(NH_2)_2$ (pellet)	Intensity IR Raman		$CO(ND_2)_2$ in D_2O	Polarization
cm^{-1}		cm^{-1}			cm^{-1}	
		58		vs		
		102		s		
		135		s		
534	d	549	m	s	458	d
601	d				548	d
1,008	p	1,011	w	vs	890	p
1,167	p		m		997	p
1,478	d	1,542	s	m	1,049	
1,604	p		s		1,164	d
1,680	p	1,649	s	m	1,201	
3,235	p	3,232	s	m	1,247	p
3,385	p	3,352	s	m	1,613	p
3,496	d	3,436	s	m	2,421	p
					2,506	p
					2,603	d

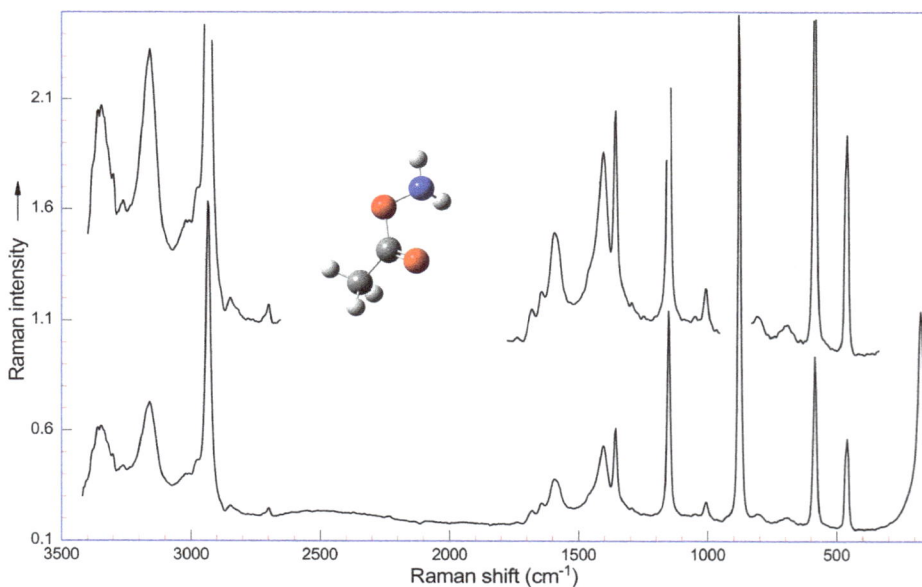

Figure 5.47: Raman spectrum of acetamide (λ_{exc} = 1,064 nm, 250 mW; data courtesy of Bruker Optik GmbH, Ettlingen).

5.3.3 Lactams

In cyclic lactams, the stretching vibration $v(C=O)$ is dependent upon the ring size, ranging from 1,669 to 1,850 cm^{-1} (see Table 5.18).

In the cyclic compounds only the *cis*-configuration is possible. That means that contrary to open-chain secondary amides, only the deformational vibration *cis*-δ(NH) can be observed at 1,490–1,440 cm^{-1}, the band of the stretching vibration $v(C-N)$ can be located at 1350–1,310 cm^{-1}, and the broad wagging vibrational band δ(NH) at 800 cm^{-1}.

Table 5.18: Vibrational frequencies of aliphatic lactams in the liquid state (data taken from Günzler and Gremlich (2003) and references cited therein).

Lactams	cm^{-1}	Raman intensity	IR intensity
O=C=NH	(2,336 in CCl$_4$)	w	s
(CH$_2$) | | NH–C=O	1,850	w	s
(CH$_2$)$_2$ | | NH–C=O	1,750	w	s
(CH$_2$)$_3$ | | NH–C=O	1,717	w	s
(CH$_2$)$_4$ | | NH–C=O	1,673	w	s
(CH$_2$)$_5$ | | NH–C=O	1,669	w	s

For the γ-lactam pyrrolidone the carbonyl frequency is detected at 1,707 cm^{-1} for the monomeric form (Figure 5.48). Additional Raman bands at 1,659 cm^{-1} and at 1,695 cm^{-1} originate from the symmetric and the antisymmetric C=O stretch of the dimer formed by hydrogen bonding. The very strong Raman band at 888 cm^{-1} belongs to the ring breathing vibration. Mdluli *et al.* (2009) report a surface-enhanced Raman spectroscopy (SERS) and DFT study of 2-pyrrolidone and *N*-methyl-2-pyrrolidinone. The DFT calculations at the B3LYP/LANL2DZ level also focus on the interaction of the compounds with gold and silver.

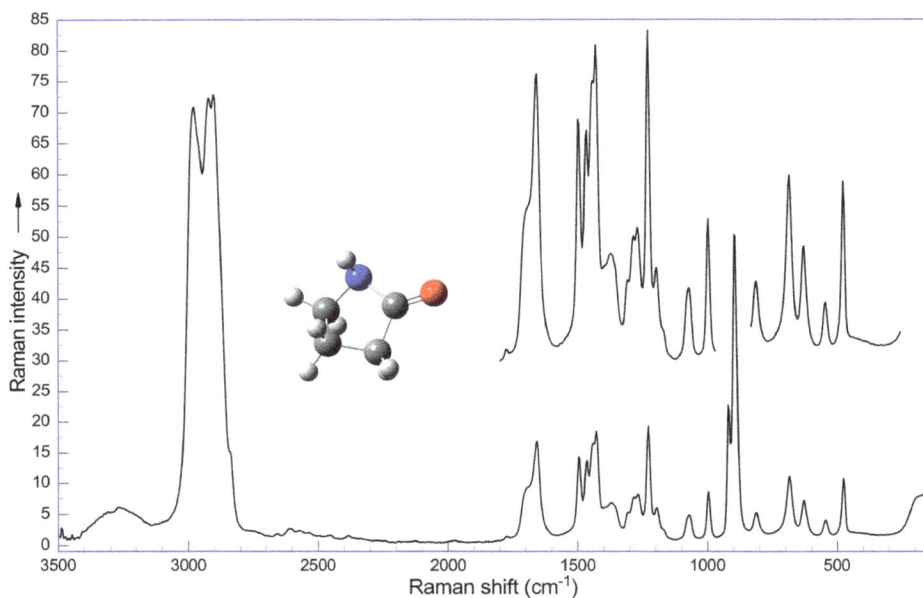

Figure 5.48: Raman spectrum of 2-pyrrolidone (λ_{exc} = 1,064 nm, 265 mW; data courtesy of Bruker Optik GmbH, Ettlingen).

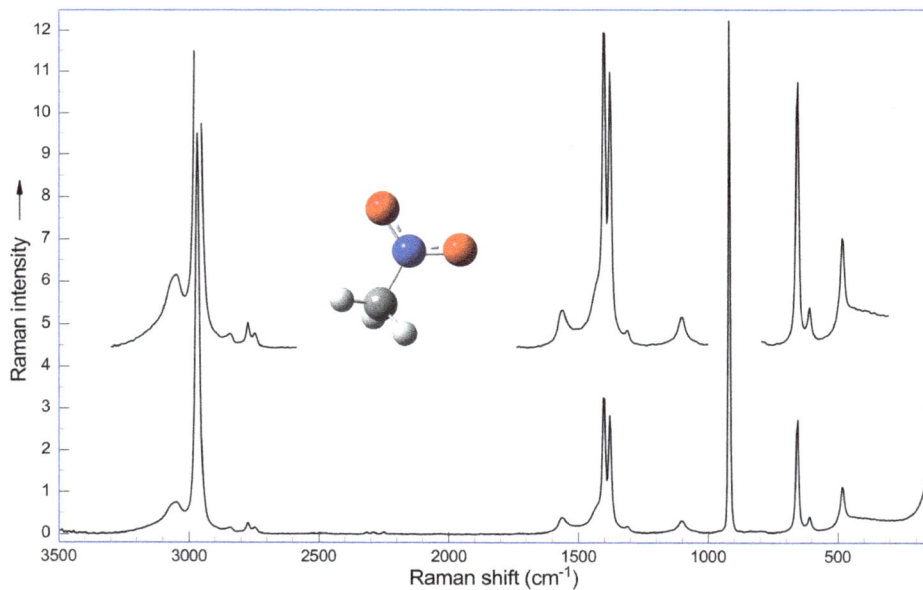

Figure 5.49: Raman spectrum of nitromethane (λ_{exc} = 1,064 nm, 250 mW; data courtesy of Bruker Optik GmbH, Ettlingen).

5.3.4 Nitroalkanes

The nitro group NO_2 has two antisymmetrically vibrating N=O bonds. Nitroalkanes, $R-NO_2$, show six characteristic bands. See, for example, the Raman spectrum of nitromethane in Figure 5.49. We observe a weak Raman band (IR vs) at 1,561 cm^{-1} from the NO_2 antisymmetric stretch, a medium band at 1,376 cm^{-1} from the NO_2 symmetric stretch, a very strong band at 920 cm^{-1} from the C–N stretch, a weak Raman band at 609 cm^{-1} from the C–NO_2 out-of-plane deformation, a medium band at 656 cm^{-1} from the NO_2 symmetric deformation, and a weak-to-medium band at 483 cm^{-1} originating from the NO_2 in-plane deformation. The vibrations are visualized in Figure 5.50.

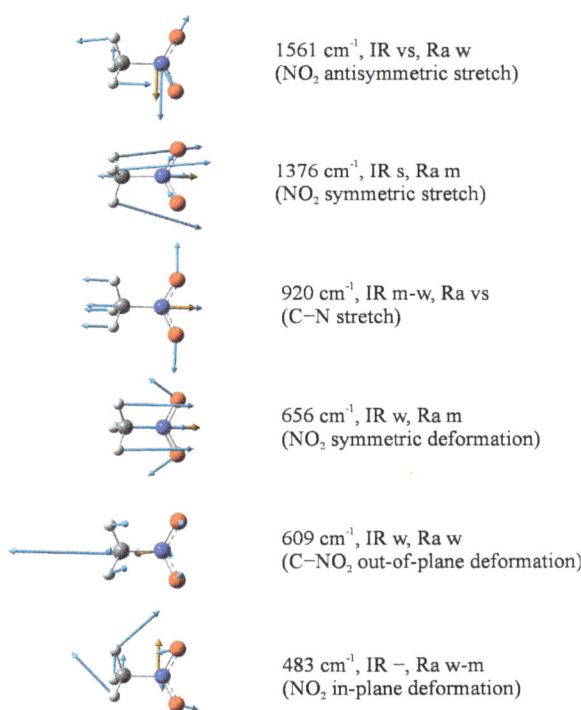

1561 cm^{-1}, IR vs, Ra w
(NO_2 antisymmetric stretch)

1376 cm^{-1}, IR s, Ra m
(NO_2 symmetric stretch)

920 cm^{-1}, IR m-w, Ra vs
(C–N stretch)

656 cm^{-1}, IR w, Ra m
(NO_2 symmetric deformation)

609 cm^{-1}, IR w, Ra w
(C–NO_2 out-of-plane deformation)

483 cm^{-1}, IR –, Ra w-m
(NO_2 in-plane deformation)

Figure 5.50: Calculated vibrations of nitromethane (DFT, B3LYP/6-311^{++}G(d,p), dipole moment vector in orange) (Hoffmann 2022).

Mathieu and Massignon (1941), Smith *et al.* (1950), and Geiseler and Kessler (1966) are the first to publish Raman spectra of nitroalkanes and halogenated nitroalkanes.

The Raman spectra of higher nitroalkanes are a little different, also rotational isomers start to occur. So in the Raman spectrum of 1-nitropropane we find a medium band at 1,553 cm^{-1} from the NO_2 antisymmetric stretch, a strong band at 1,382 cm^{-1} (NO_2 symmetric stretch), a medium-to-strong band at 900 cm^{-1} (C–N stretch of the *gauche-*

rotamer), a medium-to-strong band at 876 cm^{-1} (C–N stretch of the *trans*-rotamer), a weak band at 633 cm^{-1} (NO$_2$ out-of-plane deformation), a weak band at 618 cm^{-1} (NO$_2$ symmetric deformation), and a weak band at 478 cm^{-1} (NO$_2$ in-plane deformation).

Using the wavelength of the NO$_2$ in-plane deformation, it is possible to distinguish between primary and secondary nitroalkanes: a primary nitro group gives rise to a broad band at 490–470 cm^{-1}, for a secondary nitro group a strong and sharp band can be found at 560–520 cm^{-1}.

5.3.5 Nitrates

Dadieu (1931), Lecomte and Mathieu (1941), and Wittek (1942) study the Raman spectra of organic nitrates. In organic nitrates, RONO$_2$, we find the asymmetric stretch of the nitro group, ν_{as}(NO$_2$), at 1,660–1,625 cm^{-1}, while the symmetric ν_s(NO$_2$) stretch can be detected at 1,315–1,260 cm^{-1}. The NO$_2$ in plane deformation gives rise to a strong polarized Raman band at 610–562 cm^{-1}. In the IR, both stretching bands are strong.

5.3.6 Nitrites

Examining five alkyl nitrites (RO–N=O), Dadieu *et al.* (1931) found a strong Raman band of the N=O stretching vibration in the region 1,648–1,640 cm^{-1}. In the IR we detect two very strong bands of the N=O stretching at 1,681–1,653 cm^{-1} and at 1,625–1,613 cm^{-1}, belonging to the *cis*-resp. *trans*-form of the nitrite structure.

5.3.7 Imines

Aldimines show a stretching frequency ν(C=N) at approximately 1,670 cm^{-1} (Fabrian, Legrand, and Poirier (1956)), whereas ketimines as diphenylketimine show a C=N stretching band near 1,598 cm^{-1} (Perrier-Datin and Lebas (1969)).

5.3.8 Amidines

As reported by Shigorin and Syrkin (1949), in *N*-monosubstituted amidines two ν(C=N) bands (Figure 5.51) can be found in the range of 1,658–1,632 cm^{-1}. This can be attributed to the presence of a *syn* and an *anti* form of the compounds. *N*-disubstituted amidines only exist as the *syn* form as is concluded by one ν(C=N) band only. Waver (1990) studies the hindered rotation in amidines and reported the Raman frequencies of ν(C=N) of formamidines (R=H) and acetamidines (R=CH$_3$).

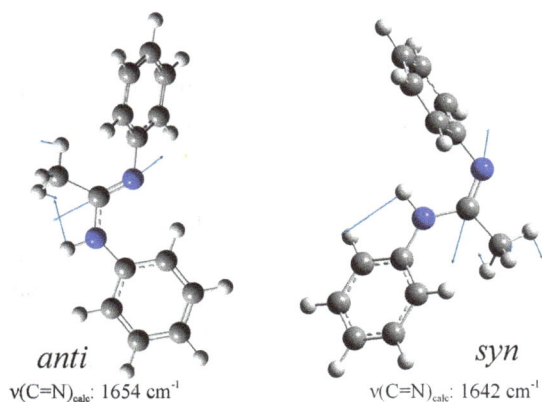

Figure 5.51: Stretching vibrations v(C=N) of the *syn* and *anti* form of the *N*-phenyl-monosubstituted amidine at 1,654 and 1,642 cm^{-1} (calculated by DFT, B3LYP/6-31^{++}G(d,2p), f = 0.9648 (Hoffmann 2017–2018).

5.3.9 Oximes

Aliphatic aldoximes show the strong band of the v(C=N) stretching vibration at 1,660–1,649 cm^{-1}, and a strong Raman band at 1,335–1,330 cm^{-1}, contributed to the OH in plane deformational vibration. Aliphatic ketoximes show the strong band of the v(C=N) stretching vibration at 1,666–1,652 cm^{-1}. Detailed vibrational assignments in the vibrational spectra of acetone oxime, $(CH_3)_2$C=N-OH, are published by Harris and Bush (1972). The Raman spectra of the compounds are studied, for example, by Nyquist, Lo, and Evans (1964), and by Nyquist *et al.* (1971).

5.3.10 Hydrazones

Kitaev, Buzykin, and Troepol'skayaand (1970) studied the structure of hydrazones (R_2C=N–NR_2). The v(C=N) of the compounds can be found in the range of 1,680–1,570 cm^{-1} as a strong Raman band and of medium intensity in the IR spectrum.

5.3.11 Azines

Methanal azine CH$_3$CH=N–N=CHCH$_3$ gives rise to a band of the symmetric stretching vibration v(C=N–N=C) at 1,615 cm^{-1}. The asymmetric stretching vibration is inactive in Raman (1,637 cm^{-1} in the IR). For higher compounds we find the asymmetric stretching vibration in the range of 1,670–1,635 cm^{-1} (weak, only in azines without centre of symmetry, else missing); in the range of 1,625–1,600 cm^{-1} we observe the symmetric stretching vibration as a strong Raman band.

5.3.12 Azo compounds

Azo compounds are studied by Kübler, Lüttke, and Weckherlin (1960). The ν(N=N) vibration gives rise to a band at 1,576 cm^{-1}. Azomethane, $CH_3N=NCH_3$, has C_{2h} symmetry, which means that the ν(N=N) is only active in the Raman. Azodicarboxylic acid esters show a =N–C band at 1,555 cm^{-1}.

Brandmüller, Hacker, and Schrötter (1966) as well as Kellerer, Brandmüller, and Hacker (1971) find the N=N stretching vibration of azobenzene and derivatives as a band at 1,440–1,380 cm^{-1} (Ra s, p, IR vw to –). Fliegl *et al.* (2003) publish an *ab initio* calculation of the vibrational and electronic spectra of *trans-* and *cis-*azobenzene.

Raman frequencies of C≡N triple bond containing compounds are summarized in Table 5.19.

5.3.13 Nitriles

Nitriles show a ν(–C≡N) band at 2,260–2,240 cm^{-1}. Fumaronitrile is studied by Jensen (2003) and the Raman spectra are reported by Reitz *et al.* (Reitz and Sabathy 1937a, 1937b; Reitz and Skrabal 1937). Jesson (1958) studies vibrational band intensities of the C=N group in aliphatic nitriles. A study of the vibrations of the C≡N bond in nitriles is published by Hidalgo (1962). Klaboe (Klaboe 1970) studies the vibrational spectrum of cyano propane.

The Raman spectrum of acetonitrile is shown in Figure 5.52.

5.3.14 Cyanides

From the ν(C≡N)$^-$ vibration of cyanides a band is observed in the range of 2,200 to 2,070 cm^{-1}. Its IR band varies from vs to vw due to neighbouring inductive effects. In the Raman spectrum the band is always strong.

5.3.15 Rhodanides

In alkyl thiocyanates (rhodanides) a strong Raman band of ν(–S–C≡N) can be detected in the range of 2,170–2,135 cm^{-1}, and it is also strong in the IR. Dadieux, Kohlrausch, and Pongratz (1931) study ethyl rhodanide, Hibben (1932), Kahovec and Kohlrausch (1937) examine the spectra of methyl rhodanide. Hirschmann, Kniseley, and Fassel (1964) as well as Lieber, Rao, and Ramachandran (1959) study aromatic thiocyanates. *n*-Alkyl rhodanides show two different C–S stretching bands: one of ν(**–S–C**≡N) at 680–650 cm^{-1} and one of ν(**C–S**–C≡N) at 700–621 cm^{-1}.

Caldow and Thompson (1958) report the vibrational bands of isothiocyanates, thiocyanates and isocyanates.

5.3.16 Azides

For azides Sheinker and Syrkin (1959) report the region of 2,169–2,080 cm^{-1} for the ν_{as} (N$_3$) vibration and 1,343–1,177 cm^{-1} for the ν_s(N$_3$) vibration. The Raman band of the antisymmetric N=N=N stretch is medium to strong and polarized (IR vs), and the Raman band of the symmetric stretch is strong and polarized (IR w).

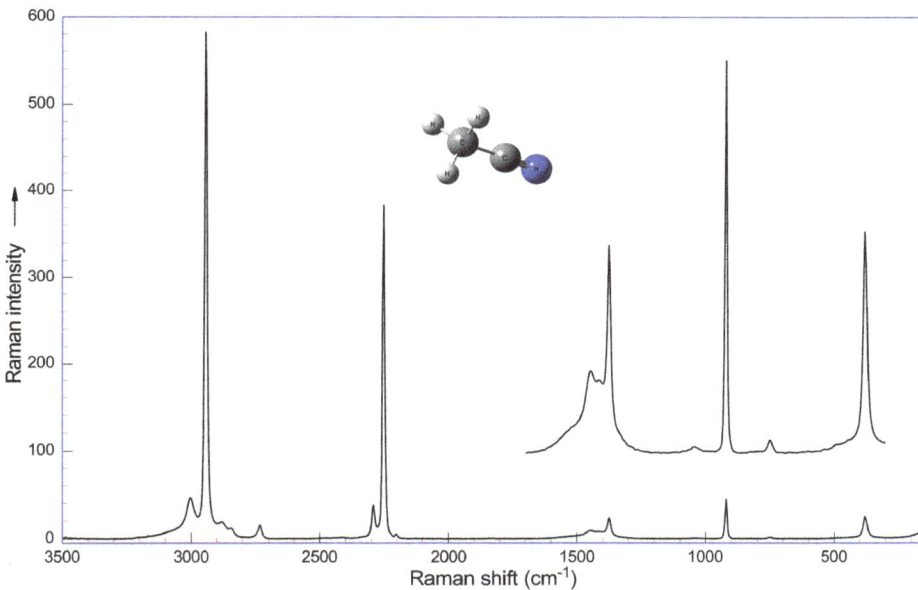

Figure 5.52: Raman spectrum of acetonitrile (λ_{exc} = 1,064 nm, 262 mW; data courtesy of Bruker Optik GmbH, Ettlingen).

5.4 Sulfur-containing aliphatic compounds

The stretching band ν(R$_2$C=S) is normally found in the range of 1,075–1,030 cm^{-1}.

5.4.1 Thiols

In organic thiols the ν(S–H) vibration has a band at 2,590–2,560 cm^{-1}, it is strong and polarized in the Raman spectrum, weak in the IR.

Table 5.19: Raman frequencies of C≡N triple bond containing compounds (data taken from Dollish, Fateley, and Bentley (1973) and references cited therein).

C≡N triple bond containing compound		cm^{-1}	IR intensity	Raman intensity
$-CH_2-C≡N$	Nitriles	2,260–2,240	s	vs, p
$-C=C-C≡N$	Acrylnitriles	2,235–2,215	s	vs
Aryl-C≡N	Benzonitriles	2,240–2,220	m-s	s, p
C≡N→O	Nitrileoxides	2,304–2,288		
$-S-C≡N$	Rhodanides	2,170–2,135	s	s, p
N-C≡N	Aminonitriles	2,225–2,175	s	s
$-N^+≡C^-$	Isonitriles	2,165–2,110	s	
C≡N$^-$	Cyanides	2,200–2,070	s	

The band of the C–S stretching is found at 735–590 cm^{-1}. It is very strong to strong and polarized in the Raman spectrum, but weak in the IR.

Propanethiol is studied by Pennington *et al.* (1956). The force field of six alkanethiols is derived by Scott and el-Sabban (1969): methanethiol, ethanethiol, 1-propanethiol, 2-propanethiol, 2-methyl-2-propanethiol, and cyclohexanethiol. Hsu (1974) reports the vibrational spectrum of allyl mercaptan (2-propen-1-thiol).

5.4.2 Sulfides

Sufides or thiaalkanes show the band of their C–S stretching at 740–585 cm^{-1}, it is strong and polarized in the Raman spectrum (see, for example, the Raman spectrum of dimethylsulphide, Figure 5.53), but weak in the IR.

The in-plane deformation band of the –C–C–S–R moiety is found at 380–320 cm^{-1}, and the polarized Raman band is of medium intensity.

The force field of five thiaalkanes is calculated by Scott and el-Sabban (1969): 2-thiapropane, 2-thiabutane, 3-thiapentane, 3-methyl-2-thiabutane, and 3,3-dimethyl-2-thiabutane.

Vibrational spectra, gas phase structure, and conformational properties of perfluorodimethyl trithiocarbonate, $(CF_3S)_2C=S$, are studied by Hermann *et al.* (2000).

Szasz *et al.* (1998) investigate the molecular structure and molecular vibrations of 1,3,5,7,-tetramethyl-2,4,6,8,9,10-hexathiaadamantane.

5.4.3 Disulfides

Herzog *et al.* (1969) calculate the force field of four molecules: 2,3-dithiabutane, 2,3-dithiapentane, 3,4-dithiahexane, and 2,2,5,5-tetramethyl-3,4-dithiahexane. Green *et al.*

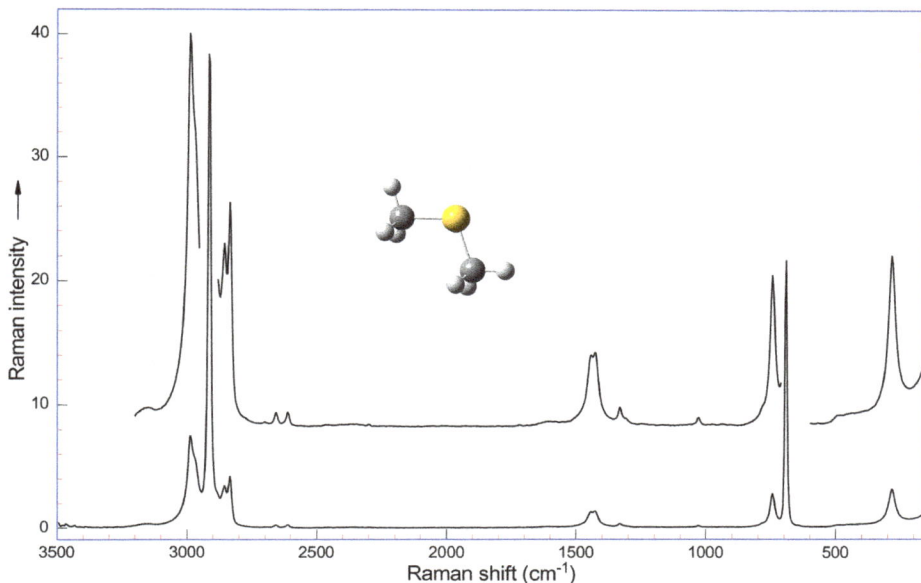

Figure 5.53: Raman spectrum of dimethylsulphide (λ_{exc} = 1,064 nm, 262 mW; data courtesy of Bruker Optik GmbH, Ettlingen).

(1969) report the vibrational spectrum of 2,2,5,5-tetramethyl-3,4-dithiahexane. A valence force field for dithiaalkanes is calculated by Scott and Elsabban (1969). Feher and Kruse (1958) study di-, tri-, and tetrasulfanes.

Krafft *et al.* (2007) publish Raman spectra of dialkyldisulfides (Figure 5.55). The Raman spectrum of dimethyldisulfide contains bands due to antisymmetric and symmetric CH_3 valence vibrations at 2,984 and 2,913 cm^{-1}, respectively, bands due to CH_3 deformation vibrations at 1,423 cm^{-1}, due to symmetric C–S valence vibrations at 693 cm^{-1}, and due to symmetric S–S valence vibrations at 509 cm^{-1} (Figure 5.54). The number of bands increases with increasing number of atoms in diethyldisulfide and dipropyldisulfide. For example, valence and deformation vibrations of CH_2 groups can be assigned near 2,927/2,930, 2,871, 1,446, 1,256/1,289 cm^{-1}, and skeletal vibrations of C–C groups near 1,050 and 1,029 cm^{-1}.

The C–S stretching is generally found at 715–620 cm^{-1} as a strong, polarized Raman band which is weak in the IR spectrum.

The S–S stretching produces two bands at 525–510 cm^{-1}, both bands are strong and polarized in the Raman spectrum, medium to weak in the IR spectrum.

Diaryl disulfides feature the S–S stretching vibration at 540–520 cm^{-1}, the Raman band is strong and polarized.

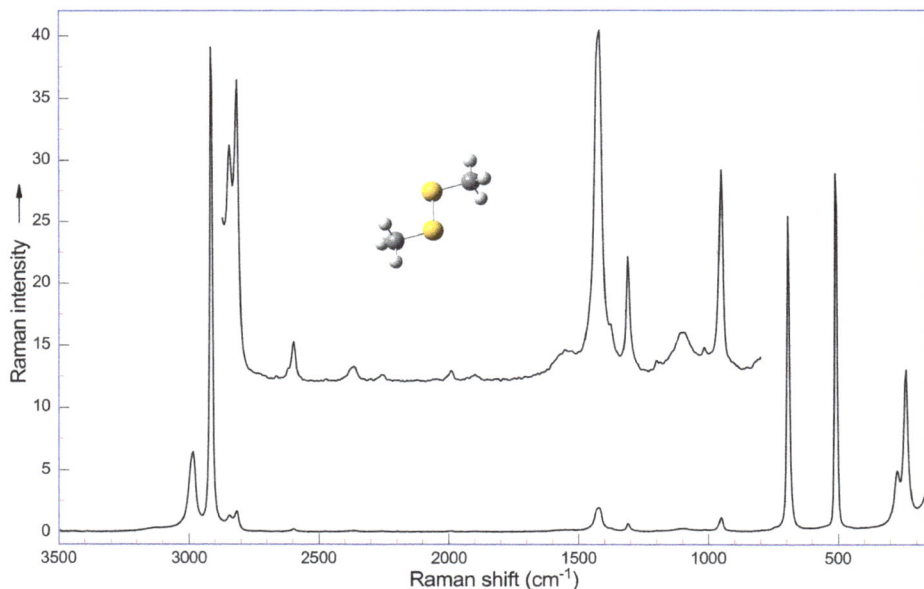

Figure 5.54: Raman spectrum of dimethyldisulphide (λ_{exc} = 1,064 nm, 250 mW; data courtesy of Bruker Optik GmbH, Ettlingen).

5.4.4 Polysulfides

Krafft *et al.* (2007) determine configurational isomers in cyclic polysulfides by Raman spectroscopy analysing 2,5-dimethyl-1,3,4-trithiolane, 2,5-diethyl-1,3,4-trithiolane, 3,6-dimethyl-1,2,4,5-tetrathiane, 3,6-diethyl-1,2,4,5-tetrathiane, and 4,6-dimethyl-1,2,3,5-tetrathiane (Figure 5.56).

In trisulfides a strong S–S band can be detected in the range of 510–450 cm^{-1} (Feher and Kruse (1958)).

5.4.5 Thiocarboxylic acids and derivatives

Thiolacetic acid (Sheppard (1949)) shows the S–H stretch at 2,568 cm^{-1} (Ra strong), the C=O stretch at 1,694 cm^{-1} (Ra medium, IR strong), and the C–S stretch at 626 cm^{-1} (Ra strong). In thiobenzoic acid the Raman frequencies are 2,591 cm^{-1}, 1,676 cm^{-1} (IR strong), and 608 cm^{-1}, respectively.

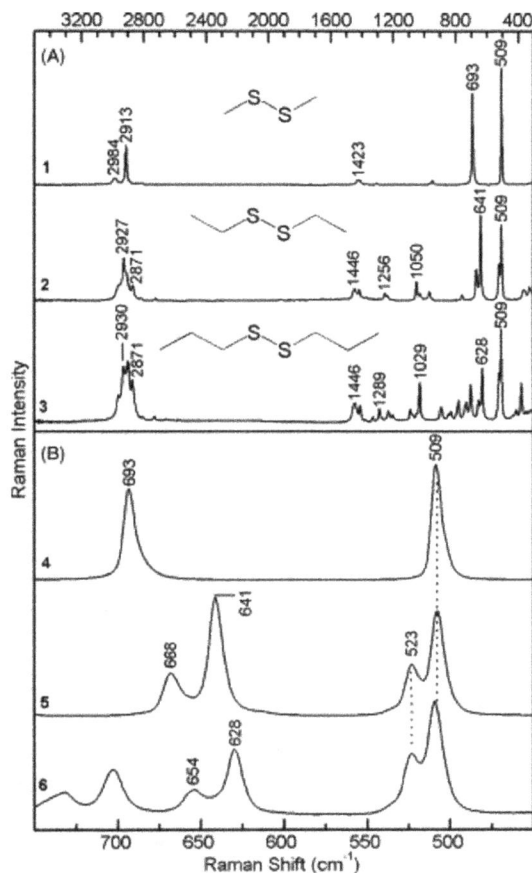

Figure 5.55: Raman spectra from 3,500 to 300 cm^{-1} (A) and from 750 to 450 cm^{-1} (B) of dimethyldisulfide (traces 1 and 4), diethyldisulphide (traces 2 and 5), and dipropyldisulfide (traces 3 and 6) (Krafft, Pigorsch *et al.* 2007). Copyright ©2007 Elsevier B.V., with permission.

5.4.6 Thionyl compounds

Kriegsmann (1955) conducts a Raman spectroscopic investigation on isomeric and tautomeric forms of the organic derivatives of sulphureous acid.

Steudel (1970) reports the S=O stretching frequency in thionyl compounds. The S=O stretching frequency is dependent of intra and intermolecular influences in thionyl compounds X – SO – Y. A report is given by literature data and own results.

In the Raman spectra of thionyl compounds R$_2$S=O the S=O stretching band occurs at 1,070–1,040 cm^{-1} (IR vs, Ra m-w).

For the most important thionyl halide SOCl$_2$ the v(S=O) is found at 1,233 cm^{-1} (IR vs), the stretching band v_s(Cl–S–Cl) at 490 cm^{-1} (IR vs), and the v_{as}(Cl–S–Cl) at 442 cm^{-1}

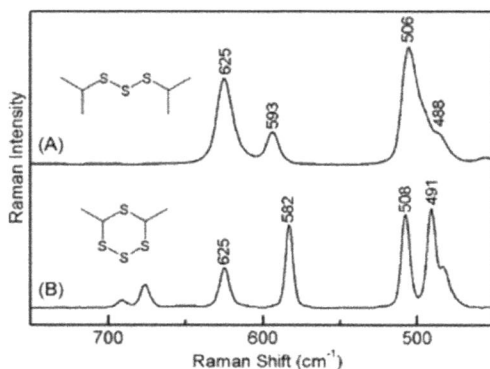

Figure 5.56: Raman spectra (750–450 cm⁻¹) of diisopropyltrisulfide (trace A) and 4,6-dimethyl-tetrathiane (trace B) (Krafft, Pigorsch *et al.* 2007). Copyright ©2007 Elsevier B.V., with permission.

(IR vs). For thionyl fluoride the S=O stretching occurs at 1,308 cm^{-1} (IR vs, Ra vs) and for thionyl bromide at 1,121 cm^{-1} (Ra w).

Dimethyl sulfoxide shows the stretching band v(S=O) at 1,042 cm^{-1} (Ra m,p), and the band at 384 cm^{-1} (Ra m,p) belongs to the deformational vibration δ(C–S=O).

5.4.7 Sulfonyl compounds

Sulfonyl compounds show a symmetrical and an antisymmetrical O=S=O stretching vibration: $v_{as}(SO_2)$ at 1360–1,290 cm^{-1} (Ra v, IR vs) and $v_s(SO_2)$ at 1,170–1,120 cm^{-1}. The symmetrical stretching is very strong and polarized in the Raman spectrum, but of variable intensity in the IR.

Gayathri and Arivazhagan (2012) publish an experimental (FT-IR and FT-Raman) and theoretical (HF and DFT) investigation on 2,4,5-trichlorobenzene sulfonyl chloride.

In sulfonamides the symmetrical S=O vibration $v_s(SO_2)$ ranges from 1,163 to 1,138 cm^{-1} (IR s), and the antisymmetrical S=O vibration $v_{as}(SO_2)$ can be found from 1,370 to 1,300 cm^{-1} (IR s). For example, the symmetrical S=O vibration shows up as a very strong Raman band at 1,138 cm^{-1} in $CH_3–SO_2–NH_2$. Another band of medium intensity in the spectrum of the compound is the antisymmetrical S=O vibration $v_{as}(SO_2)$ at 1,322 cm^{-1} (Ra m) and $\delta(SO_2)$ at 524 cm^{-1} (Ra s).

5.5 Benzene and its derivatives

Although aromatic compounds have in common the sp^2 hybridization of their ring carbon atoms with the carbon atoms of alkenes, they are chemically so different that they have to be treated in separate chapter.

In principle in benzene derivatives we find the stretching vibrations v(C–H) and v(C=C) and the out-of-plane γ(C–H) and in plane δ(C–H) deformational bands (Figure 5.57). Overtone and combination bands also occur, but in the Raman spectrum (Figure 5.58) they are of minor intensity and importance.

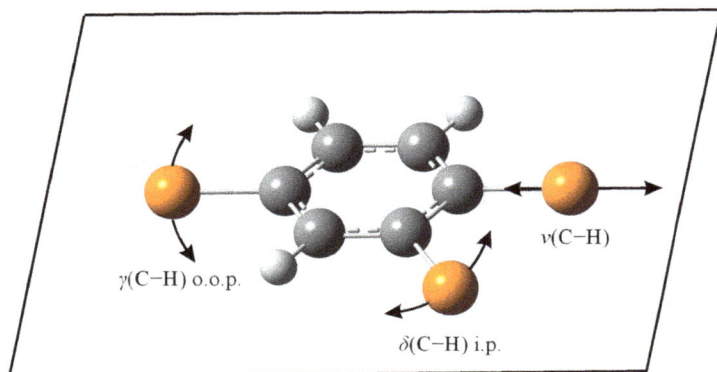

Figure 5.57: Different vibrational modes of the hydrogens in benzene and its derivatives.

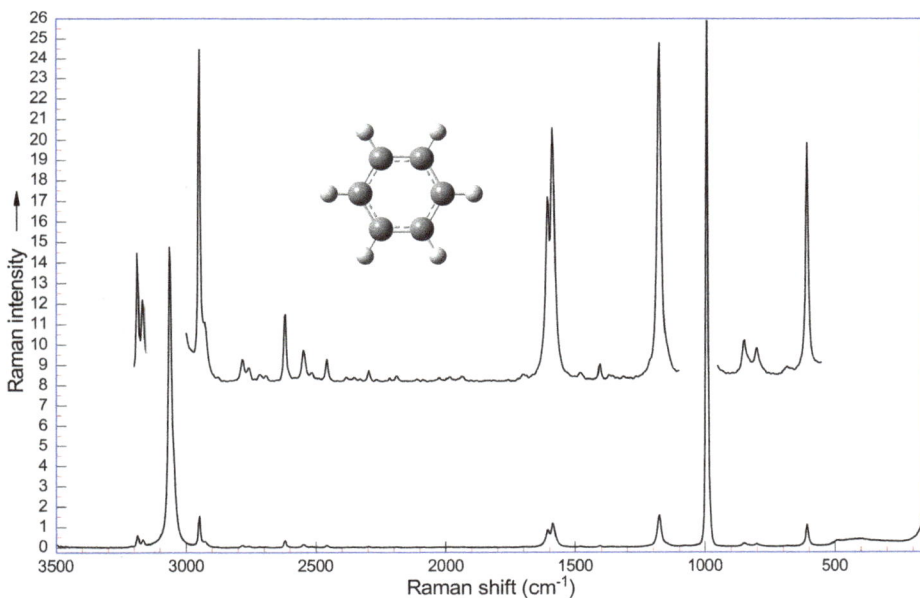

Figure 5.58: Raman spectrum of benzene (λ_{exc} = 1,064 nm, 275 mW; data courtesy of Bruker Optik GmbH, Ettlingen).

(a)

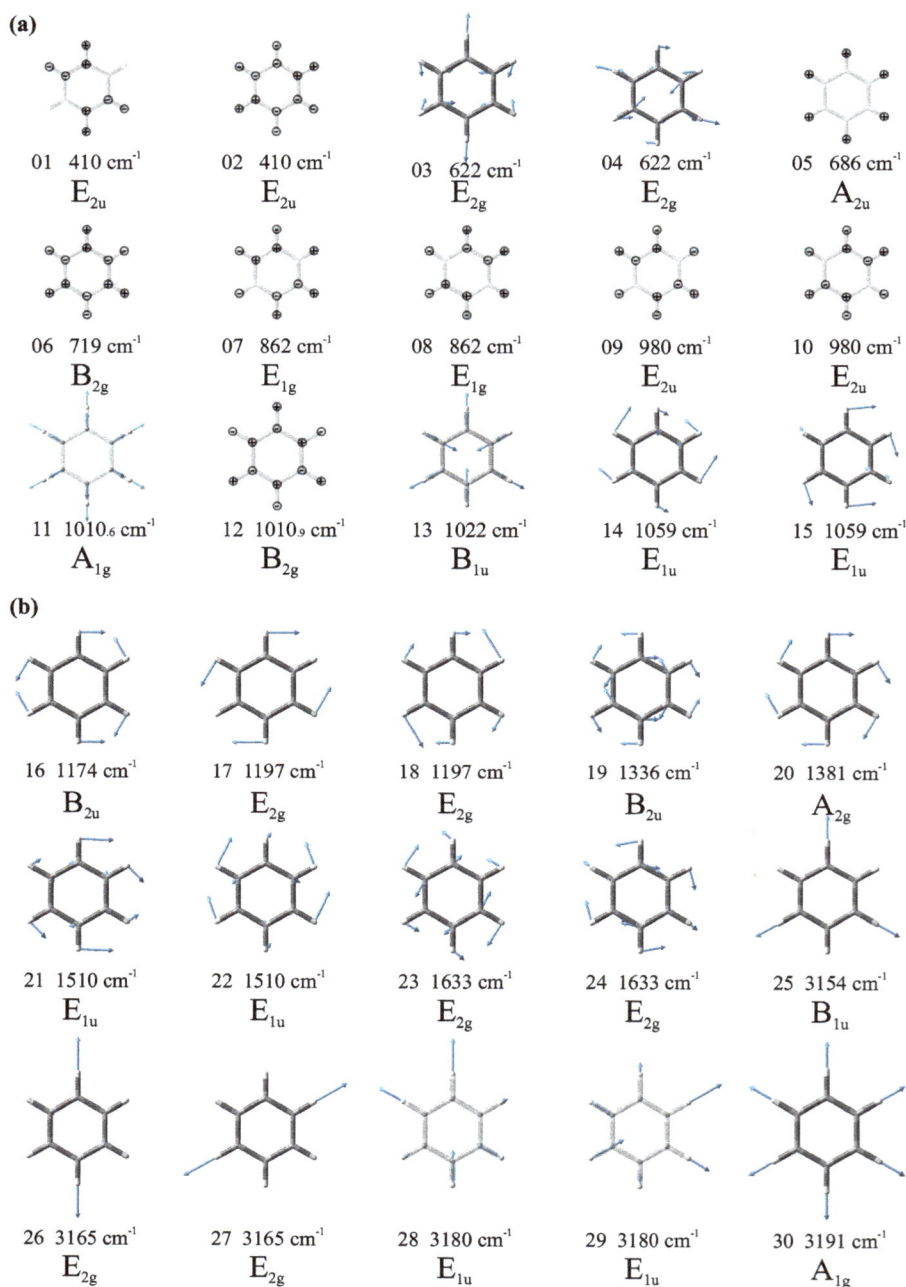

01 410 cm⁻¹
E_{2u}

02 410 cm⁻¹
E_{2u}

03 622 cm⁻¹
E_{2g}

04 622 cm⁻¹
E_{2g}

05 686 cm⁻¹
A_{2u}

06 719 cm⁻¹
B_{2g}

07 862 cm⁻¹
E_{1g}

08 862 cm⁻¹
E_{1g}

09 980 cm⁻¹
E_{2u}

10 980 cm⁻¹
E_{2u}

11 1010.6 cm⁻¹
A_{1g}

12 1010.9 cm⁻¹
B_{2g}

13 1022 cm⁻¹
B_{1u}

14 1059 cm⁻¹
E_{1u}

15 1059 cm⁻¹
E_{1u}

(b)

16 1174 cm⁻¹
B_{2u}

17 1197 cm⁻¹
E_{2g}

18 1197 cm⁻¹
E_{2g}

19 1336 cm⁻¹
B_{2u}

20 1381 cm⁻¹
A_{2g}

21 1510 cm⁻¹
E_{1u}

22 1510 cm⁻¹
E_{1u}

23 1633 cm⁻¹
E_{2g}

24 1633 cm⁻¹
E_{2g}

25 3154 cm⁻¹
B_{1u}

26 3165 cm⁻¹
E_{2g}

27 3165 cm⁻¹
E_{2g}

28 3180 cm⁻¹
E_{1u}

29 3180 cm⁻¹
E_{1u}

30 3191 cm⁻¹
A_{1g}

Figure 5.59: (a) Calculated vibrational modes 1–15 of benzene [B3LYP/6-311⁺⁺G(*d,p*), (Hoffmann 2022)]. (b) Calculated vibrational modes 16–30 of benzene [B3LYP/6-311⁺⁺G(d,p), (Hoffmann 2022)].

To illustrate the vibrational modes in benzene (Figure 5.59 a and b), they have been calculated using the Gaussian program with DFT. The functional B3LYP has been used together with the 6-311^{++}G(d,p) basis set. Shown with its image (arrows indicate the vectors of movement of the atoms, $\oplus$ and $\ominus$ indicate movement above or below the paper plane, resp.) are the number of the calculated modes (6*12−6 = 30), calculated vibrational frequencies, and the symmetry species to which the mode belongs.

Bond flexing, twisting, anharmonicity, and responsivity for the IR-active modes of benzene is studied by Yang *et al.* (2020).

Benzene derivatives have to be classified according to their symmetry (Figure 5.60). The symmetry group D_{6h} is reduced upon substitution, with the exception of hexasubstituted derivatives with equal substituents.

Herzfeld, Ingold, and Poole (1946) study the Raman spectra of benzene and its deuterated isotopomer.

Schrader and Meier (1972) give a scheme to determine the substitution pattern of benzene derivatives using Raman spectroscopy.

Aromatic compounds show strong Raman bands of the =C–H stretching in the region of 3,105–3,000 cm^{-1} (IR w-m). The number of peaks in this region decreases with decreasing number of hydrogens due to substitution.

In the region of C=C stretching vibrations a depolarized Raman band of medium or strong intensity can be found at 1,625–1,590 cm^{-1} (IR v, usually this band is located around 1,600 cm^{-1}), and a medium intensity band at 1,590–1,575 cm^{-1} (IR v), which is the strongest Raman and IR band if the substituent is conjugated to the ring. This band is frequently located around 1,580 cm^{-1}. A weak band shows up in the region from 1,525 to 1,470 cm^{-1} (IR v). It occurs usually at 1,470 cm^{-1} (if the substituent is an electron acceptor) and at 1,510 cm^{-1} (if the substituent is an electron donor). Another weak Raman band can be observed at 1,470–1,430 cm^{-1} (IR v).

Molecular polarizabilities and derivatives with respect to C–C and C–H stretching modes are obtained from *ab initio* molecular orbital calculations at the HF/6–311G (3df,3pd) level for benzene and naphthalene by Lupinetti and Gough (2002).

Yurtseven and Iseri (2013) study the Raman spectrum of benzene near the melting point and Yurtseven and Özdemir (2017) calculate Raman frequencies as functions of temperature and pressure using volume data for solid phase I of benzene.

Bender, Coasne, and Fourkas (2015) apply polarizability models for the simulation of low-frequency Raman spectra of benzene.

Pinan *et al.* (1998) publish a high-resolution Raman study of phonon and vibron bandwidths in isotopically pure and natural benzene crystals. They report the frequency and the bandwidth of the 12 Raman active lattice modes and of more than 45 internal vibrons of natural and ^{12}C isotopically pure benzene. The Raman spectrum of the ν_6 vibration in natural benzene at T = 57 K is shown.

In a large series of papers Bogomolov report the characteristic vibrations of substituted benzenes (Kovner and Bogomolov 1958, Kovner and Bogomolov 1959, Bogomolov 1960, Bogomolov 1961, Bogomolov 1962, Bogomolov 1962, Bogomolov 1962).

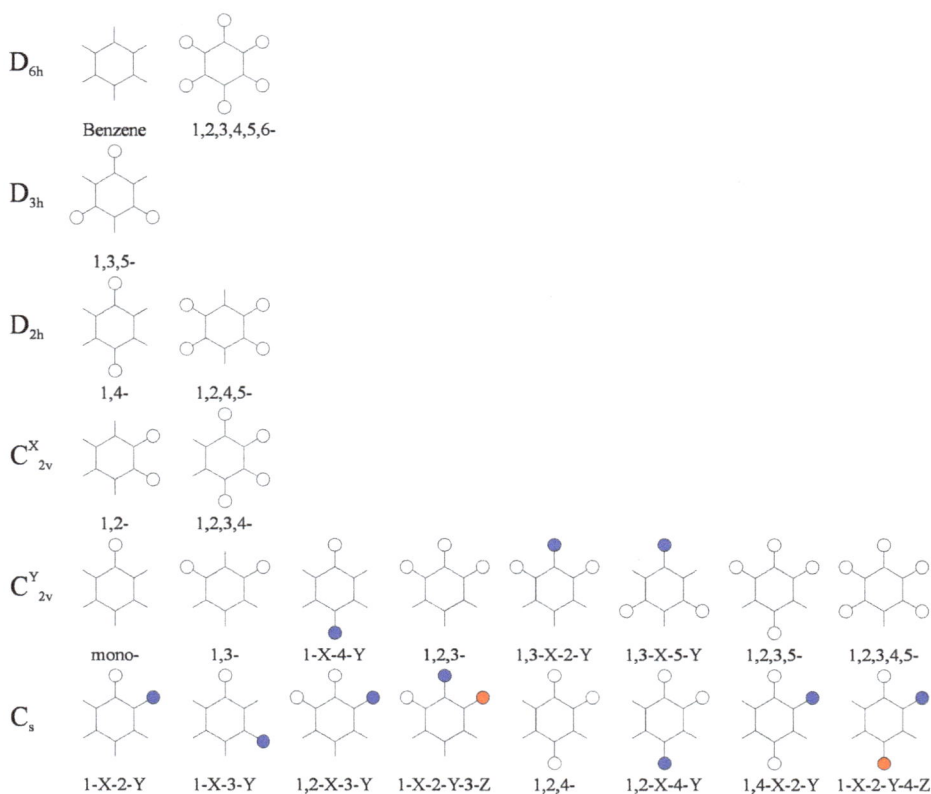

Figure 5.60: Symmetry classes of substituted benzenes (Redrawn from Dollish, Fateley, and Bentley (1973)).

An overview of Raman bands of benzene and substituted benzenes in the range of 1,900–100 cm^{-1} is shown in Figure 5.61.

5.5.1 Aromatic halogen compounds

Bands of variable Raman intensity can be found for vibrations mainly attributed to ring-halogen bending. The ring-F vibrations can be observed at 420–375 cm^{-1}, the ring-Cl vibrations at 390–270 cm^{-1}, the ring-Br vibrations at 320–255 cm^{-1}, and the ring-I vibrations at around 200 cm^{-1}.

In the IR spectrum, weak overtones and combination bands of the C–H out-of-plane deformation vibrations in the range 2,000–1,650 cm^{-1} may be observed in concentrated solutions. These patterns are consistent and characteristic of the different substitutions of the benzene ring.

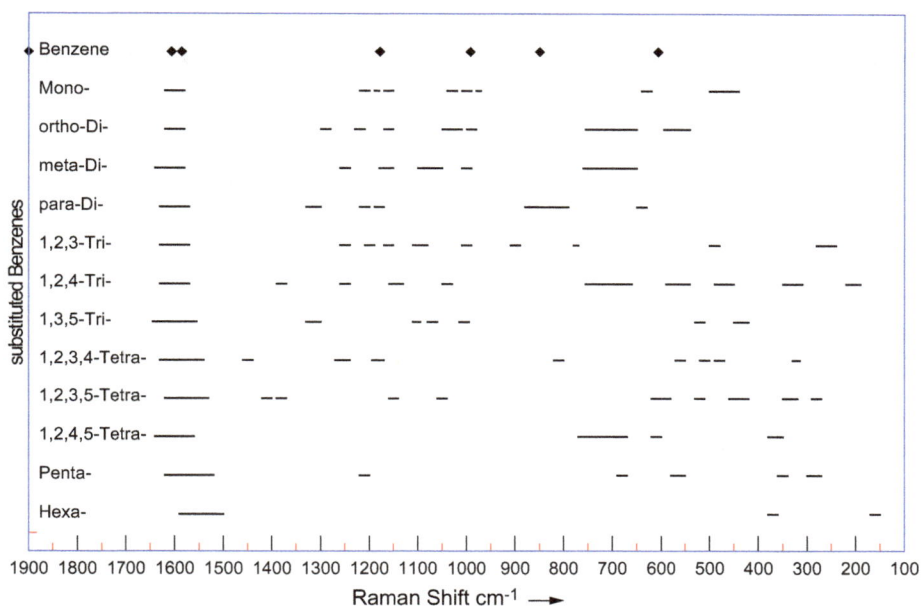

Figure 5.61: Raman bands of substituted benzenes in the range 1,900–100 cm^{-1} (data taken from Dollish, Fateley, and Bentley (1973) and references cited therein).

5.5.2 Monosubstituted benzene derivatives

The monosubstituted compounds have C_{2v}^Y symmetry.

Vibrational assignments of some monosubstituted benzenes have been reported by Bogomolov (1960).

The strong Raman band of the aromatic C–H stretch occurs at 3,070–3,030 cm^{-1}, it is medium to weak in the IR spectrum.

The aromatic in-plane ring deformation vibrations of the compounds occur at 630–605 cm^{-1} (IR w, Ra m), whereas the out-of-plane vibrations give rise to bands at 560–415 cm^{-1} (Ra w).

The aromatic =C–H in-plane deformation vibrations can be found as bands at 1,195–1,165 cm^{-1} (IR w, Ra w).

Weak Raman bands of the aromatic =C–H out-of-plane deformation vibrations of five hydrogens can be found at 900–860 cm^{-1}, at 820–720 cm^{-1}, and a ring out-of-plane deformational vibration at 710–670 cm^{-1} (IR s).

For monosubstituted benzene derivatives see the Raman spectrum of toluene (Figure 5.62) as an example.

Yamakita, Isogai, and Ohno (2006) find large Raman-scattering activities for the low-frequency modes of substituted benzenes: induced polarizability and stereo-specific ring-substituent interactions.

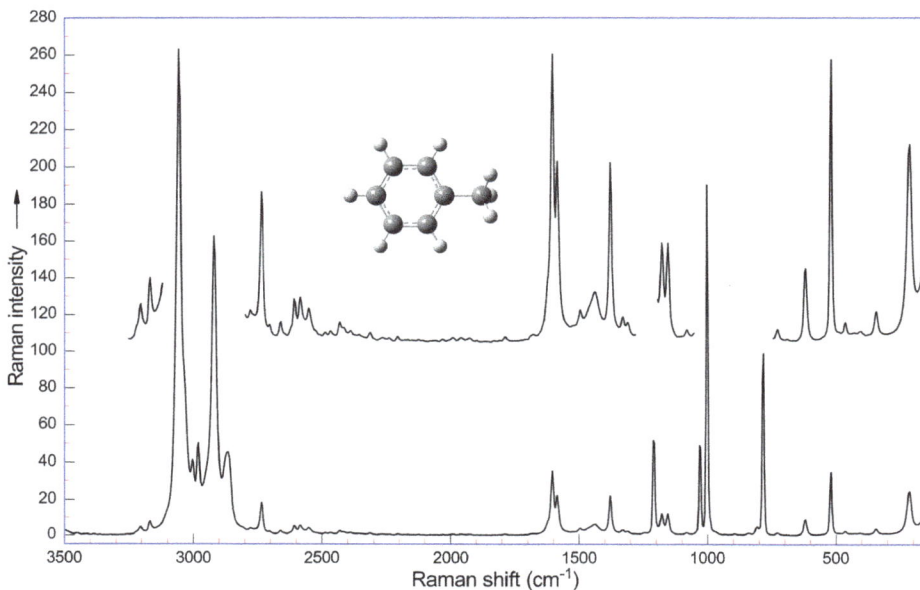

Figure 5.62: Raman spectrum of toluene (λ_{exc} = 1,064 nm, 270 mW; data courtesy of Bruker Optik GmbH, Ettlingen).

Durocher (1969) studies the IR and Raman spectra of toluene in the 150–1,600 cm^{-1} spectral range.

Brozek-Pluska *et al.* (2005) investigate low-temperature Raman spectra of stable and metastable structures of phenylacetylene in benzene. They derived the vibrational dynamics in undercooled liquid solutions, crystals, and glassy crystals.

A spectroscopic (FT-IR and FT-Raman) and molecular modelling (MM) study of the benzene sulfonamide molecule is conducted by Vinod, Periandy, and Govindarajan (2016) using quantum chemical calculations.

Fleming *et al.* (2008) study the surface-enhanced Raman spectra of benzene and benzonitrile and DFT. Fleming *et al.* (2008) report density functional theoretical SERS modelling of *o*-, *m*-, and *p*-methoxybenzonitrile using Becke's three-parameter hybrid method using and the Lee-Yang-Parr correlation functional (B3LYP) together with the LANL2DZ basis set.

Palafox *et al.* (2002) give an example for the accurate scaling of vibrational spectra. Using semi-empirical, SCF, post-SCF, and DFT calculations of aniline and several derivatives, the best values are determined by three different methods.

5.5.3 Disubstituted benzene derivatives

Scherer and Evans (1963) measure the vibrational spectra of the dichlorobenzenes o-$C_6Cl_2H_4$, o-$C_6Cl_2D_4$, m-$C_6Cl_2H_4$, p-$C_6Cl_2H_4$, p-$C_6Cl_2D_4$, and assign the bands using a modified Urey-Bradley force field to predict the planar vibrational frequencies and a valence force field to predict the non-planar frequencies.

To compare the substitution bands of disubstituted benzene derivatives, the reader is referred to the Raman spectra of the xylenes (Figures 5.64, 5.66, and 5.67).

5.5.4 *ortho*-Disubstituted benzene derivatives

Bogomolov (1961) reports the characteristic vibrations of *ortho*-disubstituted benzenes, as Green does (1970).

The aromatic in-plane ring deformation vibrations of the compounds occur at 555–495 cm^{-1} (Ra w), whereas the out-of-plane vibrations give rise to bands at 470–415 cm^{-1} (Ra w).

Aromatic =C–H in-plane deformation vibrations are found at 1,290–1,250 cm^{-1} (Ra w), 1,230–1,215 cm^{-1} (Ra m, alkyl benzenes), 1,170–1,150 cm^{-1} (IR w, Ra w), 1,150–1,110 cm^{-1} (IR w, Ra v), and 1,055–1,020 cm^{-1} (IR w, Ra s).

Weak Raman bands of the aromatic =C–H out-of-plane deformational vibrations of four hydrogens can be found at 960–900 cm^{-1}, at 850–810 cm^{-1}, and at 790–720 cm^{-1} (IR s).

ortho-Dialkylbenzenes show additional strong bands in the range of 740–715 cm^{-1} and at 600–560 cm^{-1}. See, for example, the calculated vibrations of o-xylene in Figures 5.63 and its Raman spectrum in Figure 5.64. Calculated vibrations of m- and p-xylene are shown in Figure 5.65.

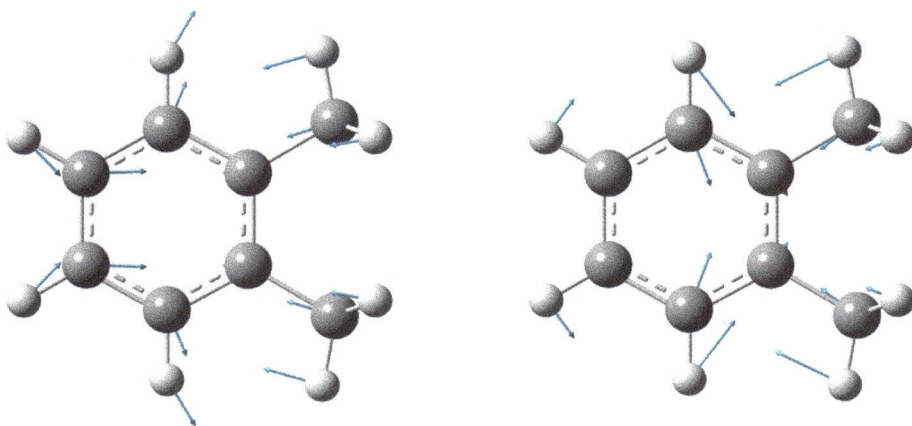

Figure 5.63: Calculated vibrations of o-xylene at 569 and 719 cm^{-1} (DFT, B3LYP/6-311^{++}G(d, 2p), f = 0.9648 (Hoffmann 2017–2018).

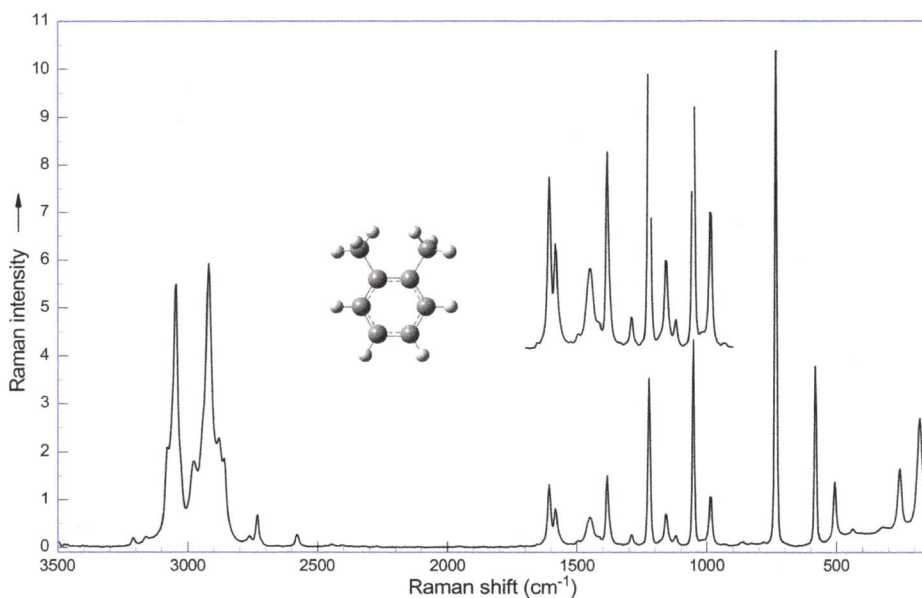

Figure 5.64: Raman spectrum of *o*-xylene (λ_{exc} = 1,064 nm, 274 mW; data courtesy of Bruker Optik GmbH, Ettlingen).

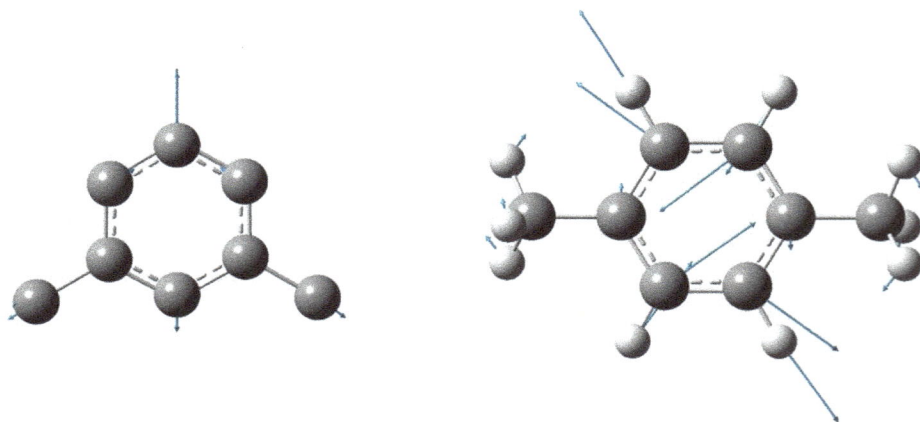

Figure 5.65: Calculated vibrations of *m*-xylene at 710 cm^{-1} (hydrogens omitted) and *p*-xylene at 719 cm^{-1} (DFT, B3LYP/6-311^{++}G(*d*, 2*p*), *f* = 0.9648 (Hoffmann 2017–2018)).

Prabhu *et al.* (2011) publish an FT-IR and FT-Raman investigation, a computed vibrational intensity analysis, and computed vibrational frequency analysis on *m*-xylene using *ab initio* HF and DFT calculations.

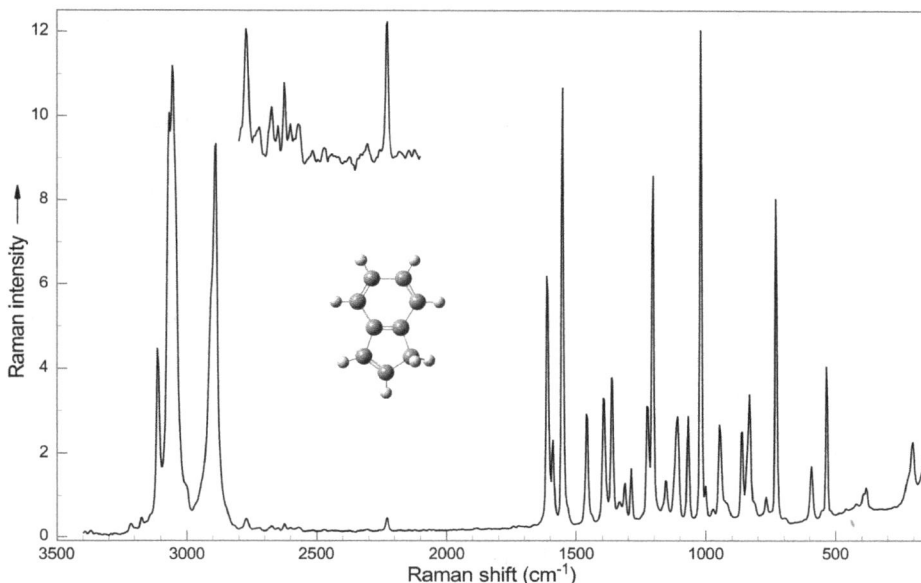

Figure 5.66: Raman spectrum of indene (λ_{exc} = 1,064 nm, 262 mW; data courtesy of Bruker Optik GmbH, Ettlingen).

The lines of the Raman spectrum of indene (bicyclo[4.3.0]nona-1,3,5,7-tetraene, Figure 5.66) are assigned by Klots (1995). The vibrational lines are nearly equally spaced and can be used advantageously for the calibration of IR as well as for the calibration of Raman spectra.

Soliman *et al.* (2007) report the conformational stability, vibrational assignments, barriers to internal rotations, and *ab initio* calculations of 2-aminophenol (d_0 and d_3). Using MP2/6-31G(*d*) calculations and experimental vibrational spectra, the authors find that *cis*- and *trans*-2-aminophenol coexist in solution, but the *cis*-isomer should be present in the crystalline state.

Fleming *et al.* (2008) publish a density functional study of the surface-enhanced Raman spectra of *o*-, *m*-, and *p*-methoxybenzonitrile.

Arjunan *et al.* (2015) conduct comprehensive quantum chemical and spectroscopic (FT-IR, FT-Raman, ^{1}H, ^{13}C NMR) investigations of (1,2-epoxyethyl)benzene and (1,2-epoxy-2-phenyl)propane.

5.5.5 *meta*-Disubstituted benzene derivatives

Bogomolov (1962) reports the characteristic vibrations of *meta*-disubstituted benzenes that have also been studied by Green (1970).

The intense polarized Raman band of the compounds at 1,000 cm^{-1} originates from the trigonal ring breathing vibration, it is weak, but very sharp, in the IR spectrum. The band is accompanied by weak Raman bands at 1,260–1,210 cm^{-1} and 1,180–1,160 cm^{-1}.

The aromatic in-plane ring deformation vibrations of the compounds occur at 560–505 cm^{-1} (Ra m), whereas the out-of-plane vibration gives rise to bands at 490–415 cm^{-1} (Ra w). An additional Raman band of medium intensity shows up at 765–645 cm.

Aromatic =C–H in-plane deformation vibrations can be found at 1,300–1,240 cm^{-1} (Ra w), 1,170–1,150 cm^{-1} (Ra w), 1,105–1,085 cm^{-1} (IR w, Ra w), 1,085–1,065 cm^{-1} (Ra w), and 1,010–990 cm^{-1} (Ra s, p, sh).

Weak Raman bands of the aromatic =C–H out-of-plane deformational vibrations can be found at 960–900 cm^{-1} (1 H), at 880–830 cm^{-1} (3 H), at 820–765 cm^{-1} (IR s, 3 H), and at 710–680 cm^{-1} (IR m, ring). An additional band of medium intensity occurs at 650–630 cm^{-1}.

meta-Dialkylbenzenes show additional strong Raman bands in the range of 740–700 cm^{-1} and 540–520 cm^{-1}.

5.5.6 *para*-Disubstituted benzene derivatives

Bogomolov (1962) examines the characteristic vibrations of *para*-disubstituted benzenes, as Green (1970) does.

The aromatic in-plane ring deformation vibrations of the compounds occur at 650–615 cm^{-1} (Ra m, *p*), whereas the out-of-plane vibrations give rise to bands at 520–445 cm^{-1} (Ra w).

Aromatic =C–H in-plane deformation vibrations are located at 1,270–1,250 cm^{-1} (Ra w), 1,230–1,215 cm^{-1} (Ra s-m, alkyl benzenes), 1,185–1,165 cm^{-1} (Ra m), 1,130–1,110 cm^{-1} (Ra v), and 1,025–1,005 cm^{-1} (Ra w).

A weak Raman band of the aromatic =C–H out-of-plane deformational vibration of two hydrogens can be found at 860–780 cm^{-1} (IR s) and another weak band at 710–680 cm^{-1}.

para-Dialkylbenzenes show a very strong additional band in the range of 830–780 cm^{-1} and a very strong to strong band at 1,230–1,200 cm^{-1} (*p*-C$_6$H$_4$C$_2$ moiety).

Mahadevan, Periandy, and Ramalingam (2011) analyse the vibrational spectra of 1,2-dinitro benzene and 1-fluoro-3-nitro benzene experimentally (FT-IR and FT-Raman) as well as theoretically using DFT and B3LYP/B3PW91.

Mahadevan et al. (2011) investigate FT-IR, FT-Raman, and UV spectra of 1-bromo-3-fluorobenzene and compare them to DFT (B3LYP, B3PW91, and MPW91PW91) calculations.

FT-IR and FT-Raman spectra, assignments, *ab initio* HF and DFT analysis of 4-nitrotoluene are published by Ramalingam *et al.* (2010).

All disubstituted benzenes show a Raman doublet (sometimes ill resolved) near 1,600 cm^{-1}. This feature can be clearly seen in the spectra of *ortho-*, *meta-*, and *para*-xylene (Figures 5.64, 5.67, 5.68). The C–H stretch can be observed as a strong Raman band in the range of 3,100–3,050 cm^{-1} (IR v-m).

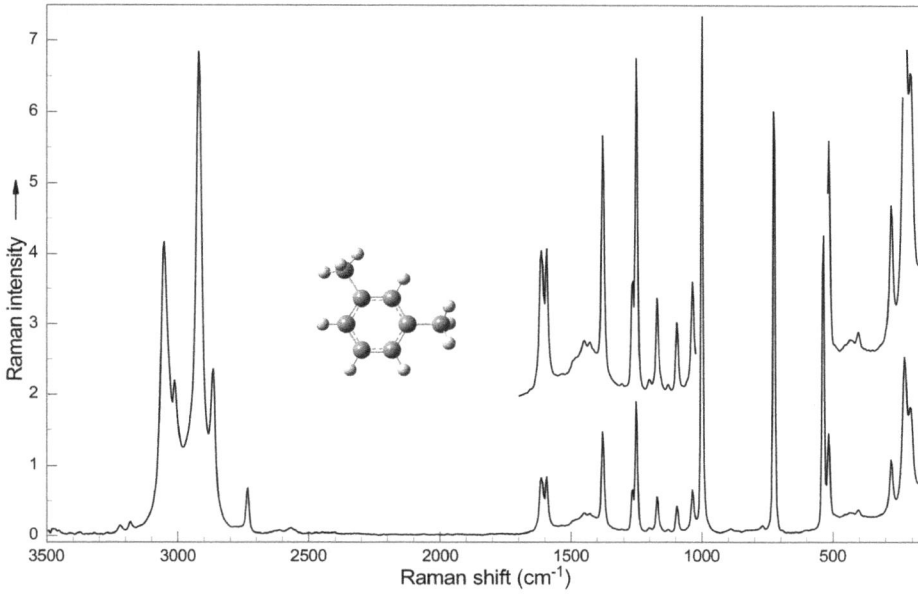

Figure 5.67: Raman spectrum of *m*-xylene (λ_{exc} = 1,064 nm, 274 mW; data courtesy of Bruker Optik GmbH, Ettlingen).

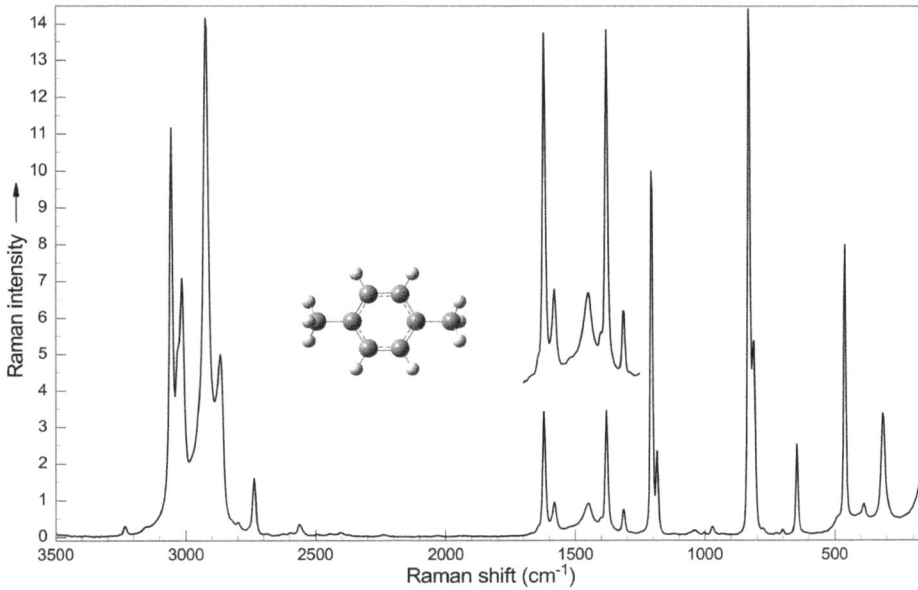

Figure 5.68: Raman spectrum of *p*-xylene (λ_{exc} = 1,064 nm, 274 mW; data courtesy of Bruker Optik GmbH, Ettlingen).

Apart from the doublet at 1,600 cm^{-1} (Ra m, IR v), *para*-disubstituted benzene derivatives show Raman bands of medium intensity at 1,180–1,150 cm^{-1} and at 650–630 cm^{-1}. A very strong band is observed in the range of 830–720 cm^{-1}; if only alkyl substituted benzene derivatives are taken into account, this range is reduced to 830–780 cm^{-1}. The *p*-C$_6$H$_4$C$_2$ moiety has a very strong to strong Raman band at 1,230–1,200 cm^{-1} that is specific for this structural arrangement of carbon atoms.

5.5.7 Trisubstituted benzene derivatives

5.5.7.1 1,2,3-Trisubstituted benzene derivatives

Bogomolov (1962) examines the characteristic vibrations of 1,2,3-trisubstituted benzenes.

For 1,2,3-trisubstituted benzene derivatives aromatic =C–H in-plane deformation vibrations can be found at 1,170–1,150 cm^{-1} (IR w, Ra w), at 1,085–1,065 cm^{-1} (IR w, Ra w), and at 1,030–1,010 cm^{-1} (IR w, Ra s-m).

Weak Raman bands of the aromatic =C–H out-of-plane deformational vibrations of three hydrogens occur at 965–950 cm^{-1}, at 900–885 cm^{-1}, at 830–760 cm^{-1}, and at 740–685 cm^{-1}.

Scherer and Evans (1963) measure the vibrational spectra of the trichlorobenzenes 1,2,3-C$_6$Cl$_3$H$_3$, and 1,2,3-C$_6$Cl$_3$D$_3$ and assign the bands. They use a modified Urey-Bradley force field to predict the planar vibrational frequencies and a valence force-field to predict the non-planar frequencies.

A very strong band can be found in the Raman spectra of 1,2,3-trisubstituted benzenes in the range of 670–500 cm^{-1}. If the substituents are alkyl groups, this band occurs around 650 cm^{-1}. The polarized band is dependent on the position and mass of the substituents. The dependency of the frequency of the band on light substituents (e.g. methyl) and heavy substituents (e.g. chlorine) is summarized in Table 5.20. The band is also strong in the infrared spectrum.

Table 5.20: Vibrational frequencies of 1,2,3-trisubstituted benzenes (data taken from Bogomolov (1962) and Dollish *et al.* (1973)).

Position	1	2	3	
	CH$_3$	CH$_3$	CH$_3$	658 cm^{-1}
	CH$_3$	Cl	CH$_3$	589 cm^{-1}
	CH$_3$	Cl	Cl	568 cm^{-1}
	Cl	CH$_3$	Cl	589 cm^{-1}
	Cl	Cl	Cl	516 cm^{-1}

5.5.7.2 1,2,4-Trisubstituted benzene derivatives

The compounds show a strong polarized Raman band in the range of 750–650 cm^{-1}, with alkyl substituents it can be found around 740 cm^{-1}, with chlorine substituents around 677 cm^{-1}. A band of medium intensity occurs in the range of 1,280–1,200 cm^{-1}. Additional bands can be found with variable intensity at 580–540 cm^{-1} and at 500–450 cm^{-1}.

We find aromatic =C–H in-plane deformation vibrations at 1,220–1,200 cm^{-1} (IR w, Ra w), at 1,160–1,140 cm^{-1} (IR w, Ra w), and at 1,040–1,020 cm^{-1} (IR w, Ra w).

Weak Raman bands of the aromatic =C–H out-of-plane deformational vibrations of one or two hydrogens can be found at 940–885 cm^{-1} (1H, IR m), at 860–840 cm^{-1} (2H, IR s), at 780–760 cm^{-1}, and at 740–690 cm^{-1}.

A vibrational analysis and a valence force field of 1,2,4-trimethylbenzene is reported by Ojha, Venkatram Reddy, and Ramana (2012).

The Raman spectrum of 1,2,4-trimethylbenzene is shown in Figure 5.69.

Scherer and Evans (1963) measure the vibrational spectra of the trichlorobenzene 1,2,4-$C_6Cl_3H_3$ and assign the bands. They use a modified Urey-Bradley force field to predict the planar vibrational frequencies and a valence force-field to predict the non-planar frequencies.

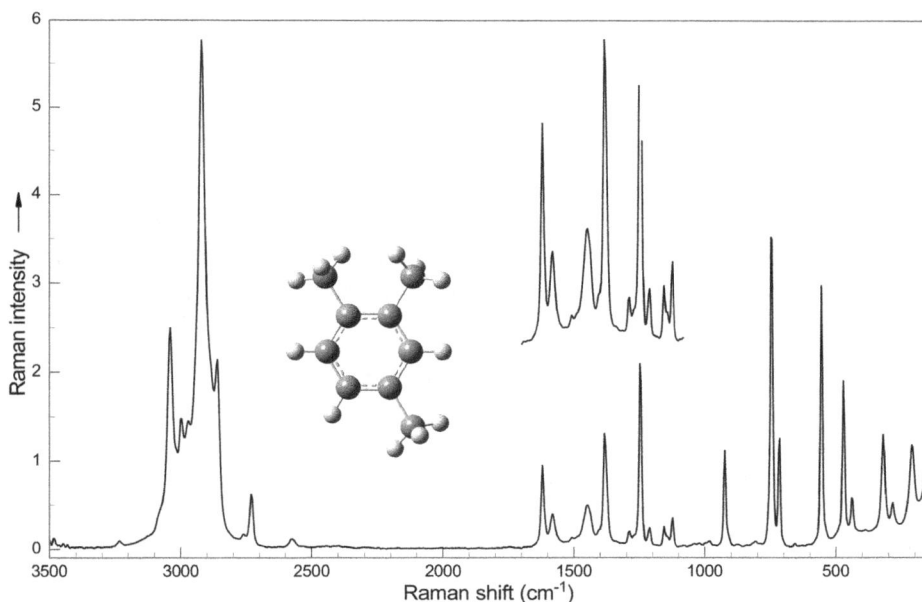

Figure 5.69: Raman spectrum of 1,2,4-trimethylbenzene (λ_{exc} = 1,064 nm, 275 mW; data courtesy of Bruker Optik GmbH, Ettlingen).

5.5.7.3 1,3,5-Trisubstituted benzene derivatives

1,3,5-Trisubstituted benzene derivatives show their trigonal ring breathing vibration as a very strong polarized Raman band around 1,000 cm^{-1} (IR w).

Aromatic =C–H in-plane deformation vibrations have bands at 1,275–1,255 cm^{-1} (w), 1,180–1,160 cm^{-1} (Ra w, IR w), and at 1,040–995 cm^{-1} (Ra vs, IR w).

Weak Raman bands of the aromatic =C–H out-of-plane deformational vibrations of one hydrogen can be found at 890–830 cm^{-1} (IR m), at 865–810 cm^{-1} (IR s), and at 730–660 cm^{-1} (IR s).

Scherer, Evans, and Muelder (1962) report vibrational spectra and Urey-Bradley force constants of *sym*-trifluoro and tribromo benzene (1,3,5-C$_6$X$_3$H$_3$, D_{3h}).

Trialkyl-substituted benzene derivatives of D_{3h} symmetry show a medium intensity Raman band at 1,600 cm^{-1} (degenerate vibration) and bands at 580–510 cm^{-1} (Ra s), at 520–480 cm^{-1} (Ra m), and at 280–250 cm^{-1} (Ra m).

A vibrational analysis and a valence force field of 1,3,5-trimethylbenzene is reported by Ojha, Venkatram Reddy, and Ramana (2012).

Braden and Hudson (2000) study structure and vibrations of C$_6$F$_6$ and *sym*-C$_6$F$_3$H$_3$ by *ab initio* and DFT methods.

Nomoto *et al.* (2007) study excited-state structure and dynamics of 1,3,5-tris(phenylethynyl)benzene by Raman and time-resolved fluorescence spectroscopy.

5.5.8 Tetrasubstituted benzene derivatives

Scherer and Evans (1963) measure the vibrational spectra of the tetrachlorobenzenes 1,2,3,4-C$_6$Cl$_4$H$_2$, 1,2,3,5-C$_6$Cl$_4$H$_2$, 1,2,4,5-C$_6$Cl$_4$H$_2$, 1,2,4,6-C$_6$Cl$_4$D$_2$, and 1,2,4,5-C$_6$Cl$_4$HD and assigned the bands. They use a modified Urey-Bradley force field to predict the planar vibrational frequencies and a valence force-field to predict the non-planar frequencies.

5.5.8.1 1,2,3,4-Tetrasubstituted benzene derivatives

Weak Raman bands of the aromatic =C–H out-of-plane deformational vibrations of two hydrogens can be found at 860–780 cm^{-1}.

The most prominent Raman bands in the spectrum of 1,2,3,4-tetramethylbenzene are a very strong band at 648 cm^{-1}, a strong band at 1,260 cm^{-1}, and two medium bands at 1,390 cm^{-1} and at 470 cm^{-1}.

Steele (1962) reports the vibrational spectra of tetrafluorobenzenes.

In the Raman spectrum of 1,2,3,4-tetrafluorobenzene a strong band occurs at 1,328 cm^{-1}, a very strong band at 684 cm^{-1}, a medium band at 489 cm^{-1}, and a medium-strong band at 374 cm^{-1} (associated mainly with ring-F bending).

The similar chloro compound 1,2,3,4-tetrachlorobenzene shows three very strong bands at 515, 1,175, and at 332 cm^{-1} (mainly ring-Cl bending), in its Raman spectrum. An additional strong band is detected at 1,130 cm^{-1}.

5.5.8.2 1,2,3,5-Tetrasubstituted benzene derivatives

A weak Raman band of the aromatic =C–H out-of-plane deformational vibration of a single hydrogen can be found at 850–840 cm^{-1} (IR m).

Varsanyi *et al.* (1977) show spectra of 1,2,3,5-tetrasubstituted benzene derivatives.

Durocher (1969) study IR and Raman spectra of isodurene (1,2,3,5-tetramethylbenzene) in the 150–1,600 cm^{-1} spectral range.

In the Raman spectrum of 1,2,3,5-tetramethylbenzene a very strong band at 648 cm^{-1}, a strong band at 1,260 cm^{-1}, and medium bands at 1,390 and 470 cm^{-1} can be detected.

1,2,3,5-Tetrafluorobenzene shows a very strong band at 684 cm^{-1}, a strong band at 1,328 cm^{-1}, a medium-strong band at 374 cm^{-1} (mainly ring-F bending), and a medium band at 489 cm^{-1} in its Raman spectrum.

Three very strong bands are detected in the Raman spectrum of 1,2,3,5-tetrachlorobenzene at 515, 1,175, and 332 cm^{-1} (mainly ring-Cl bending), and another strong band can be found at 1,130 cm^{-1}.

5.5.8.3 1,2,4,5-Tetrasubstituted benzene derivatives

A weak Raman band of the aromatic =C–H out-of-plane deformational vibration of one hydrogen can be found at 870–860 cm^{-1} (IR m) and another weak band at 820–790 cm^{-1}.

Durocher (1969) studies IR and Raman spectra of durene (1,2,4,5-tetramethylbenzene) in the 150–1,600 cm^{-1} spectral range. 1,2,4,5-Tetramethylbenzene shows a very strong band at 648 cm^{-1}, a strong band at 1,260 cm^{-1}, and two medium bands at 1,390 and 470 cm^{-1}.

Aromatic =C–H in-plane deformation vibrations can be found at 1,280–1,260 cm^{-1} (w) and 1,205–1,185 cm^{-1} (w).

The Raman spectrum of 1,2,4,5-tetrafluorobenzene features a very strong band at 684 cm^{-1}, a strong band at 1,328 cm^{-1}, a medium-strong band at 374 cm^{-1}, and a band of medium intensity at 489 cm^{-1}.

In the Raman spectrum of 1,2,4,5-tetrachlorobenzene three very strong bands at 515 cm^{-1}, 1,175 cm^{-1}, and at 332 cm^{-1} can be found alongside a strong band at 1,130 cm^{-1}.

1,2,4,5-Tetraisopropylbenzene shows a very strong Raman band at 352 cm^{-1}, strong bands at 1,264 cm^{-1}, and at 884 cm^{-1}, and a band of medium intensity at 1,056 cm^{-1}.

5.5.9 Pentasubstituted benzene derivatives

A weak Raman band of the aromatic C–H out-of-plane deformational vibration of the single hydrogen can be found at 900–860 cm^{-1}, which is medium to strong in the IR.

The IR and Raman spectra of pentamethylbenzene are studied by Durocher (1969) in the 150–1,600 cm^{-1} spectral range.

A vibrational analysis and a valence force field of pentamethylbenzene is reported by Ojha, Venkatram Reddy, and Ramana (2012).

Pentamethylbenzene (Figure 5.70) shows Raman bands at 567 (very strong), 1,289 (very strong), 1,379 (very strong), and a strong band at 481 cm^{-1}. Steele and Whiffen (1960) report the vibrational frequencies of pentafluorobenzene.

The most prominent Raman bands in the spectrum of pentafluorobenzene are two strong bands at 578 and 719 cm^{-1} and two medium-strong bands at 1,410 and 1,648 cm^{-1}.

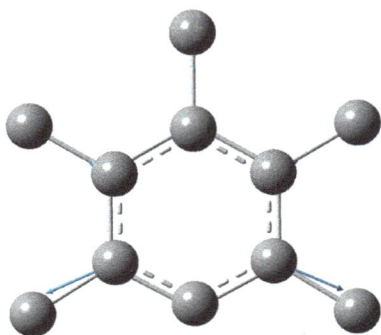

Figure 5.70: Vibration at 1,265 cm^{-1} of the calculated Raman spectrum of pentamethylbenzene (hydrogens omitted, DFT, B3LYP/6-311^{++}G(d,2p), f = 0.9648 (Hoffmann 2017–2018)).

Scherer and Evans (1963) measure the vibrational spectra of the pentachlorobenzenes C_6Cl_5H and C_6Cl_5D and assign the bands. They use a modified Urey-Bradley force field to predict the planar vibrational frequencies and a valence force-field to predict the non-planar frequencies.

Pentachlorobenzene shows very strong bands at 387, 1,208, 344 (mainly ring–Cl bending), and 200 cm^{-1}. The C–H stretching occurs at 3,065 cm^{-1}.

5.5.10 Hexasubstituted benzene derivatives

IR and Raman spectra of hexamethylbenzene (Figure 5.71) are studied by Durocher (1969) in the 150–1,600 cm^{-1} spectral range. The compound shows very strong Raman bands at 515, 1,175, and 332 cm^{-1}, also a strong band at 1,130 cm^{-1}.

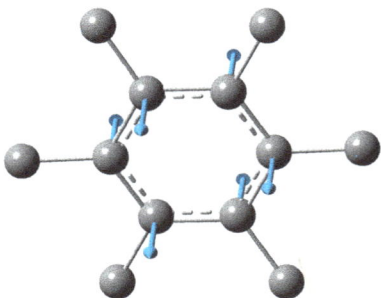

Figure 5.71: Vibration calculated at 712 cm^{-1} in the Raman spectrum of hexamethylbenzene (hydrogens omitted, DFT, B3LYP/6-311^{++}G(d,p), f = 0.9648 (Hoffmann 2017–2018)).

Scherer and Evans (1963) measure the vibrational spectra of hexachlorobenzene C_6Cl_6 and assign the bands. They use a modified Urey-Bradley force field to predict the planar vibrational frequencies and a valence force-field to predict the non-planar frequencies.

Steele and Whiffen (1959) measure the vibrational frequencies of hexafluorobenzene (C_6F_6) and find very strong bands at 515, 1,175, and 332 cm^{-1}, plus a strong band at 1,130 cm^{-1}. Steele and Whiffen (1960) cite an IR band at 315 cm^{-1} (e_{1u}).

Nielsen and Brandt (1965) measure the Raman spectrum of 1,3,5-trifluoro-2,4,6-trichlorobenzene. The four strongest bands of $C_6F_3Cl_3$ are at 382, 579, and 184 cm^{-1}, plus a band of medium intensity at 1,600 cm^{-1}.

In hexachlorobenzene, the 372 cm^{-1} Raman band is very strong, the 323 cm^{-1} band is strong, as is the 219 cm^{-1} band, whereas the 1,222 cm^{-1} band is of medium intensity.

Hyams, Lippincott, and Bailey (1966) report the Raman and low frequency IR spectra of C_6F_5Cl, C_6F_5Br, and C_6F_5I. The four strongest Raman bands of the compounds are as follows:

For pentafluorochlorobenzene the authors find very strong bands at 516, 1,643, and 1,595 cm^{-1}, plus a strong band at 116 cm^{-1}.

For pentafluorobromobenzene Hyams, Lippincott, and Bailey find very strong bands at 496 and 114 cm^{-1} and strong bands at 1,639 and at 583 cm^{-1}.

For pentafluoroiodobenzene the authors find a very strong Raman band at 204 cm^{-1} (mainly ring-I bending) and strong bands at 489, 1,633, and 581 cm^{-1}.

Krishnakumar and Mathammal (2009) conduct a joint FTIR, FT-Raman, and scaled quantum mechanical study of 1,3-dibromo-2,4,5,6-tetrafluoro benzene and 1,2,3,4,5-pentafluoro benzene.

5.5.11 Naphthalenes

The naphthalene molecule belongs to the point group D_{2h}. The strongest Raman bands in powder form can be found at 3,056, 1,576, 1,464, 1,383, 1,147, 1,021, 764, and 513 cm^{-1}. These bands, attributed to the planar a_g modes, are polarized in the single crystal and in the liquid state. The Raman spectrum of a naphthalene single crystal is reported by Suzuki, Yokoyama, and Ito (1968).

Molecular polarizabilities and derivatives with respect to C–C, and C–H stretching modes were obtained from *ab initio* molecular orbital calculations at the HF/6–311G (3df,3pd) level for benzene and naphthalene by Lupinetti and Gough (2002).

Compton and Hammer (2014) record a RUN (Raman under liquid nitrogen) spectrum of naphthalene.

Prabhu, Periandy, and Mohanc (2011) publish a combined experimental and theoretical study on molecular and vibrational structure of 2,3-dimethylnaphthalene. The FTIR and FT Raman spectra of 2,3-dimethylnaphthalene are recorded in the region 4,000–100 cm^{-1}. The optimized geometries were calculated by HF and DFT (B3LYP) methods with 6-31^{++}G(d,p), 6-311G(d,p), and 6-311^{++}G(d,p) basis sets, the harmonic vibrational frequencies, IR intensities, and Raman activities of the 2,3-dimethylnaphthalene are evaluated with these methods as well.

5.5.12 Anthracenes and phenanthrenes

The Raman bands of anthracene (Figure 5.72) can be found around 1,630 cm^{-1} (m) and around 1,550 cm^{-1} (m, disappears upon 9,10 substitution), while phenanthrenes show two bands around 1,600 cm^{-1} (m-s) and a band around 1,500 cm^{-1} (m).

For example, Suzuki, Yokoyama, and Ito (1968) publish the polarized Raman spectrum of an anthracene single crystal with bands at 1,403, 748, and 396 cm^{-1}. Anthracene measured as a polycrystalline solid (Figure 5.72) also shows bands at 1,403, 748, and 396 cm^{-1} as the strongest bands.

5.5.13 Higher polycyclic aromatic hydrocarbons

Raman frequencies of other polycyclic aromatic compounds and derivatives can be found in Dollish, Fateley, and Bentley (1973) and the original literature cited therein.

In a series of paper on Raman spectroscopy of mineral oil products, Schrader (1991) publishes NIR-FT Raman spectra of polycyclic aromatic hydrocarbons (PAHs, Figure 5.73).

Zhang *et al.* (1996) measure far-IR emission spectra of selected gas-phase PAHs. Spectroscopic fingerprints of naphthalene, pyrene, and chrysene are shown.

Cloutis *et al.* (2016) use Raman spectroscopy for the identification and discrimination of PAHs.

Onchoke (2020) publishes a review covering the 1960–2019 papers on vibrational spectroscopic studies of nitrated PAHs (NPAHs).

Lehmann *et al.* (1979) report vibrational assignments for the Raman and the phosphorescence spectra of 9,10-anthraquinone and 9,10-anthraquinone-d_8. Later, Simeral and Hafner (2022) discuss the Raman-active vibrational modes of anthraquinones supported by TDDFT calculations.

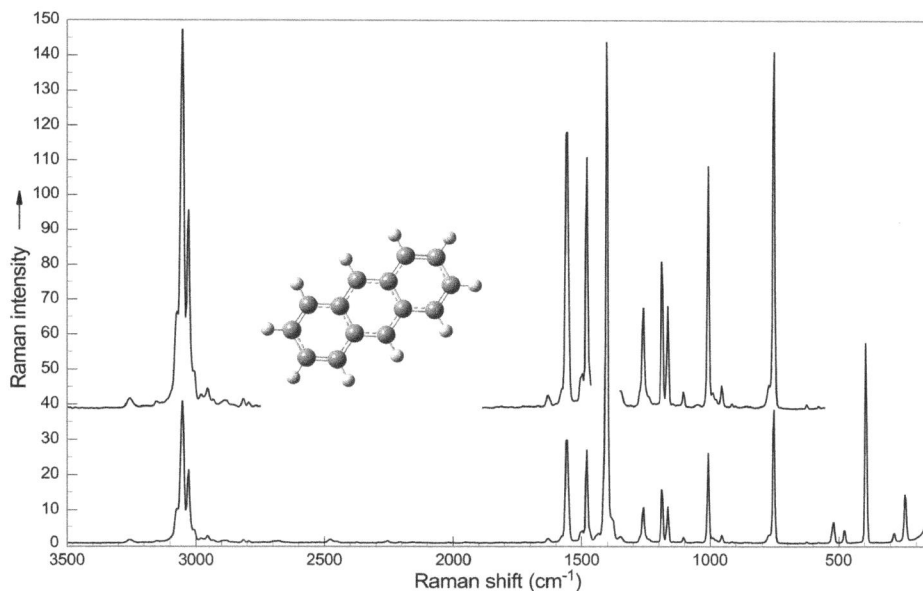

Figure 5.72: Raman spectrum of anthracene (λ_{exc} = 1,064 nm, 274 mW; data courtesy of Bruker Optik GmbH, Ettlingen).

5.6 Heterocycles

5.6.1 Three-membered rings

Three-membered rings show five Raman bands: the antisymmetric and the symmetric C–H stretch of the methylene group and three ring vibrations. The parent isocyclic compound, cyclopropane, shows the methylene C–H stretches at 3,090 and 3,039 cm^{-1}, whereas the ring vibrations (in this case, only two) can be detected at 1,188 and 740 cm^{-1}. The degenerate ring vibration at 740 cm^{-1} splits into a symmetric and an antisymmetric mode when a CH$_2$ group is replaced by a heteroatom or a heteroatom containing group. In these cases, the symmetric mode is more intense than the antisymmetric mode.

Ethylene oxide (oxirane) shows strong antisymmetric and symmetric ring methylene stretches at 3,063 and 3,005 cm^{-1}, respectively. The very strong band of the C–O stretch can be found at 1,266 cm^{-1}. The symmetric and the antisymmetric ring vibrations are medium to weak Raman bands at 877 and at 892 cm^{-1}.

Nyquist (1986) studies the group frequencies of 1,2-epoxyalkanes. In his table we find monosubstituted oxiranes from the parent compound (R = H) over R = CH$_3$ to R = C$_{16}$H$_{33}$. The ring breathing frequencies of these compounds are 1,268, 1,265, and 1,255 cm^{-1}, respectively (absorbance 0.862, 0.403–0.023), the symmetric in plane

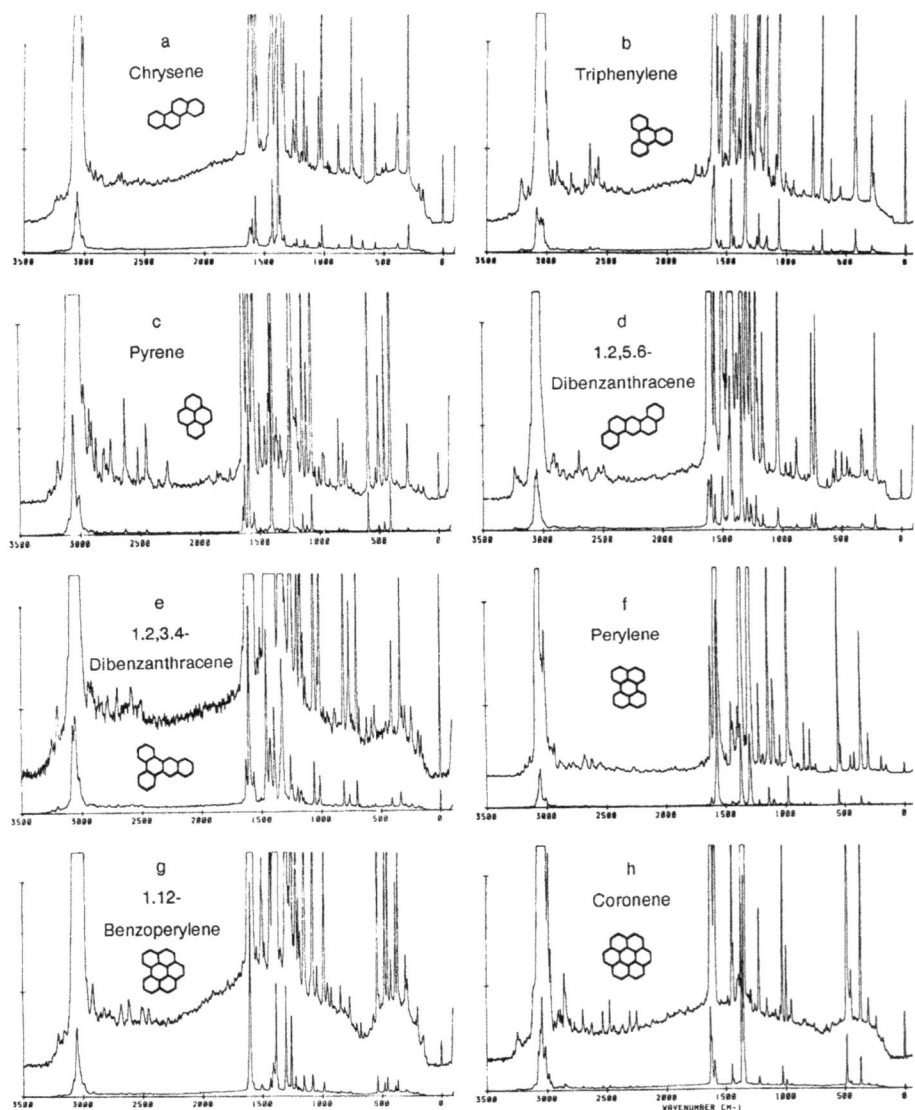

Figure 5.73: Raman spectra of PAHs (Schrader 1991). Copyright ©1991 John Wiley and Sons, with permission.

deformation frequencies are 877, 829, and 830 cm^{-1} (absorbance 1.677, 1.755–0.065), and the antisymmetric in plane deformation frequencies are 892, 893, and 909 cm^{-1} (absorbance 0.156 to 0.047). For a linear chain substituent R, the values are in line, only branched substituents (*iso*-propyl, *t*-butyl) show deviations.

For ethylene imine (aziridine) we find antisymmetric and symmetric ring methylene stretches at 3,059 and 2,999 cm^{-1}. Apart from those two stretching bands, the most intense band is the band of the C–O stretch at 1,212 cm^{-1}. The symmetric ring vibration is a Raman band of medium to weak intensity at 857 cm^{-1}, whereas the antisymmetric ring vibration at 904 cm^{-1} can only be detected in the IR.

The antisymmetric and symmetric ring methylene stretches of ethylene sulfide (thiirane) occur at 3,085 and 3,010 cm^{-1}. The sulfide's ring breathing mode shows up at 1,112 cm^{-1}, and the symmetric and antisymmetric ring vibration can be found at 625 and 660 cm^{-1}.

Spiekermann, Bougeard, and Schrader (1982) perform coupled calculations of vibrational frequencies and intensities in the IR and Raman spectra of ethylene oxide and ethylene sulfide.

5.6.2 Four-membered rings

Kohlrausch and Reitz (1940) examine the Raman bands of oxetane. The antisymmetric CH$_2$ stretch is found as a strong band at 3,001 cm^{-1}. The band at 1,140 cm^{-1} of medium intensity is the symmetric ring mode, the band at 1,029 cm^{-1} is the ring "breathing" vibration, it is very strong. The antisymmetric ring vibration can be found with medium intensity at 931 cm^{-1}.

Lippert and Prigge (1963) examine the influence of the ring size on the vibrational spectra of ethers, thioethers, and imines. In going from unstrained rings to those in which strain is present, the CH-valence frequencies increase in all three classes of compounds and the NH-valence frequencies of the free imine molecules decrease. Whereas the s-contribution to the CH-bond grows with increasing ring-strain, the s-contribution to the NH-bond appears to decrease. In the case of planar or quasi-planar skeletal vibrations, the frequencies of strongly polarized Raman lines of the totally symmetrical pulsation vibration and the intense IR bands of an antisymmetric ring deformation vibration are greatly influenced by the size of the ring.

The normal vibrations of trimethylene sulfide and trimethylene imine are empirically assigned by comparison with data in the newer literature.

The Raman spectrum of acetidine (trimethylene imine) shows a very strong band at 1,026 cm^{-1}, the ring "breathing" vibration, the band at 989 cm^{-1} (medium) is the antisymmetric ring vibration.

Thietane has Raman bands at 991 cm^{-1} (the medium out-of-phase C–C stretch), at 932 cm^{-1} (the strong in-phase C–C stretch), at 699 cm^{-1} (the strong in-phase C–S stretch), at 670 cm^{-1} (the very strong out-of-phase C–S stretch), and finally a band at 528 cm^{-1} (the strong in-phase ring deformation).

The vibrational spectra and a normal-coordinate analysis of silacyclobutane are published by Laane (1970). The author detects the Raman bands of the compound at 932 cm^{-1} (the out-of-phase C–C stretch, strong), at 903 cm^{-1} (the in-phase C–C stretch, strong), at 876 cm^{-1} (the in-phase C–Si stretch, very strong), at 652 cm^{-1} (the out-of-phase C–Si stretch, very strong), and at 539 cm^{-1} the very strong ring deformational band. Three of silacyclobutane's derivatives, $(CH_2)_3SiX_3$ where X = D, F, and Cl, are studied as well.

Trimethylene selenide is studied by Harvey, Durig, and Morrissey (1969). The authors publish a vibrational analysis and the ring puckering vibration of trimethylene selenide and trimethylene selenide-d_4. Raman bands of the undeuterated compound are found at 937 cm^{-1} (in-phase ring deformation, medium), at 650 cm^{-1} (in-phase C–C stretch, strong), 563 cm^{-1} (out-of-phase C–Se stretch, strong), and at 416 cm^{-1} the very strong in-phase C–Se stretching band appears.

5.6.3 Five-membered rings

The saturated five-membered rings show pseudorotation that means that the atoms of the ring are successively out of the ring's plane. While in cyclopentane the barrier is non-existent, tetrahydrofurane exists as a number of well-defined conformers.

Tetrahydrofurane is studied by Kohlrausch and Reitz (1940), Luther *et al.* (1950), and Barrow and Searles (1953).

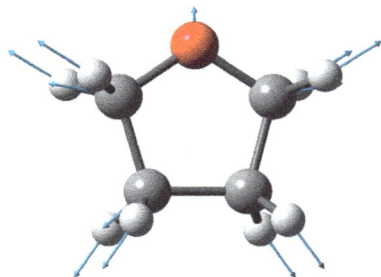

Figure 5.74: Ring breathing vibration of tetrahydrofurane at the calculated frequency of 891 cm^{-1} (experimental 914 cm^{-1}, DFT, B3LYP/6-311^{++}G(d,2p), f = 0.9648, Hoffmann (2017–2018)).

In the Raman spectrum of tetrahydrofurane (Figure 5.75) the three-ring stretching bands are located at 1,174 cm^{-1} with medium intensity, at 1,028 cm^{-1} as a strong band, and at 1,071 cm^{-1} with medium intensity. The ring breathing can be found at 914 cm^{-1} as a very strong band (Figure 5.74). The in-plane ring deformation at 596 cm^{-1} is weak, as is the out-of-plane ring deformation at 276 cm^{-1}. Another out-of-plane ring deformation shows up very weakly at 215 cm^{-1}.

Using Raman spectroscopy and *ab initio* calculations, Wu *et al.* (2011) investigate intermolecular hydrogen bonds in a binary mixture (tetrahydrofurane plus water).

5.6.3.1 Pyrrolidine

The heterocyclic compound shows ring stretches with medium Raman intensity at 980 and 925 cm^{-1}. Another band at 872 cm^{-1} is barely detectable. A ring breathing band with very high intensity occurs at 899 cm^{-1}. The in-plane ring deformations at 593 and 822 cm^{-1} are weak, and the out-of-plane ring deformations at 300 and 219 cm^{-1} are even weaker.

Matysik *et al.* (1997) perform a Raman spectroscopic analysis of isomers of biliverdin dimethyl ester. The compound has four pyrrol rings.

The constitutional isomers of biliverdin dimethyl ester, IX alpha and XIII alpha, are studied by resonance Raman spectroscopy. Matysik *et al.* report: "The far-reaching spectral similarities suggest that despite the different substitution patterns, the compositions of the normal modes are closely related. This conclusion does not hold only for the parent state (ZZZ, sss configuration) but also for the configurational isomers which were obtained upon double-bond photoisomerization. Based on a comparison of the resonance Raman spectra, a EZZ configuration is proposed for one of the two photoisomers of biliverdin dimethyl ester IX alpha, while a ZZE, ssa configuration has been assigned previously to the second isomer." Copyright 1997 ©Elsevier B.V., with permission.

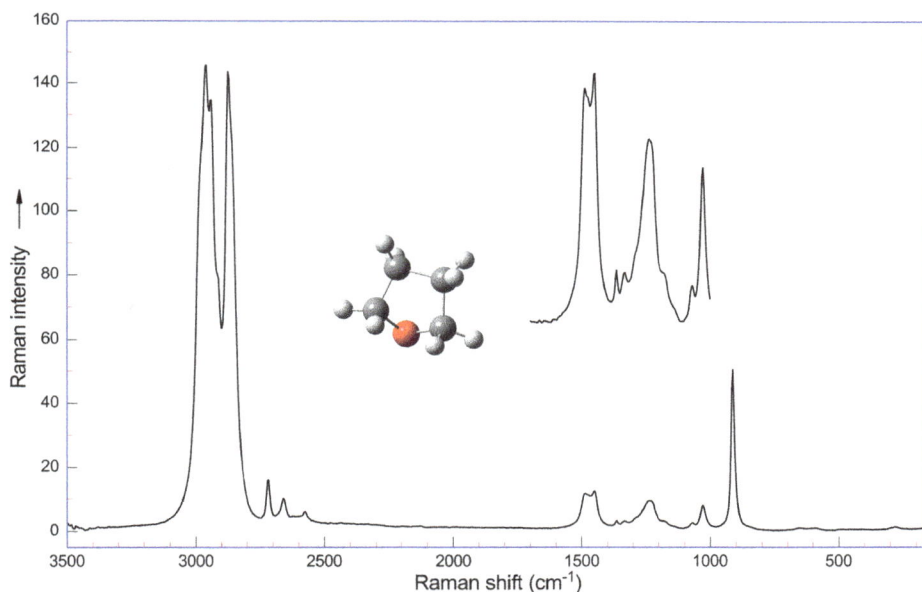

Figure 5.75: Raman spectrum of tetrahydrofurane (λ_{exc} = 1,064 nm, 270 mW; data courtesy of Bruker Optik GmbH, Ettlingen.).

Hildebrandt *et al.* (1994) study halocyanin from *Natronobacterium pharaonis*. The blue copper protein is examined by Raman spectroscopic techniques. The authors report: "Near-IR FT- Raman spectra display the Raman bands of the protein and, in the oxidized state, also the preresonance-enhanced bands of the copper centre. The frequency of the amide I band at 1,676 cm^{-1} indicates a predominant β-sheet protein structure, which is typical for small blue copper proteins. Resonance Raman spectra of the oxidized protein obtained upon excitation close to the 600 nm absorption band are measured in the pH range between 7.7 and 4.5. The vibrational band pattern in the Cu-cysteine stretching region is closely related to that of azurin, indicating far-reaching similarities of the coordination geometry of the copper centre in both proteins. Significantly lower frequencies, however, are noted for the Cu-histidine stretches, which appear as a closely spaced doublet at ca. 260 cm^{-1}. Lowering the pH to 4.5 leads to an increase of this band splitting with one component shifting down to 247 cm^{-1} This downshift is attributed to the rupture of a hydrogen bond between one of the histidine ligands and a nearby carboxyl group, which becomes protonated at such a low pH. On the other hand, no major changes in the Cu-cysteine stretching region are noted at pH 4.5, implying that the coordination geometry remains largely unchanged." Copyright © 1994 American Chemical Society, with permission.

IR and Raman spectra and assignments for tetrahydrothiophene, tetrahydroselenophene, and tetrahydrotellurophene are reported by Giorgini *et al.* (1977).

5.6.3.2 Tetrahydrothiophene
Four ring stretches of the compound can be observed: one Raman band with medium intensity at 1,036 cm^{-1} and three strong bands at 960 cm^{-1}, at 822 cm^{-1}, and at 884 cm^{-1}. The ring breathing at 688 cm^{-1} is very strong (Figure 5.76).

The two in-plane ring deformational bands show up with medium intensity at 520 cm^{-1} and as a strong band at 471 cm^{-1}, whereas the out-of-plane ring deformation gives rise to a weak band at 297 cm^{-1}.

5.6.3.3 Silacyclopentane
Durig and Willis (1970) study silacyclopentane. The IR and Raman spectra of the compound are taken in the gas phase as well as in the liquid phase.

5.6.3.4 Germylcyclopentane
Durig and Willis (1969) also study germylcyclopentane. They detect four-ring stretching vibrations at 848 cm^{-1} (Ra strong), at 761 cm^{-1} (Ra medium), at 946 cm^{-1} (Ra medium), and at 588 cm^{-1} (Ra medium). The ring breathing at 635 cm^{-1} is very strong, whereas the in-plane ring deformations at 481 cm^{-1} and at 345 cm^{-1} are only weak. Another weak band is the out-of-plane ring deformation at 273 cm^{-1}.

FT-IR and FT-Raman spectra of 2-(adamantan-1-yl)-5-(4-nitrophenyl)-1,3,4-oxadiazole are recorded and analysed. The vibrational frequencies were computed using quantum chemical calculations (DFT, B3LYP/6-311^{++}G(5D,7F)). The data obtained from wavenumber calculations are used to assign the vibrational bands obtained experimentally. The energy barriers of the internal rotations about the C–C bonds connecting the oxadiazole to the adamantane and benzene rings are reported by Haress *et al.* (2015).

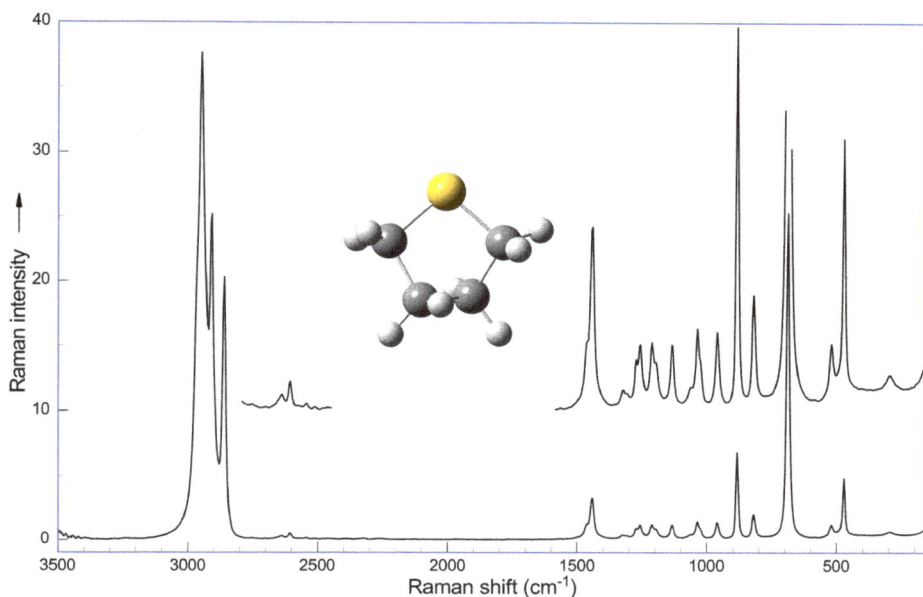

Figure 5.76: Raman spectrum of tetrahydrothiophene (λ_{exc} = 1,064 nm, 275 mW; data courtesy of Bruker Optik GmbH, Ettlingen).

5.6.3.5 Selenacyclopentane

Green, Harvey, and Greenhouse (1971) determine the pseudorotation barrier in selenacyclopentane (tetrahydroselenophene) by vibrational spectroscopy. The strongest vibration of the compound is visualized in Figure 5.77.

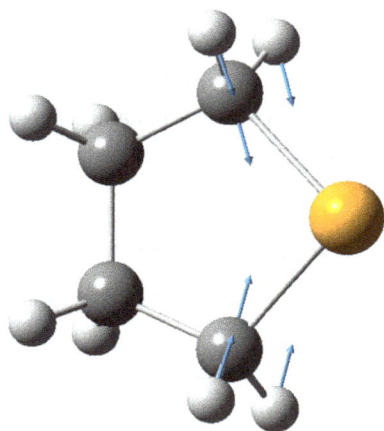

Figure 5.77: Vibration at calculated 641 cm^{-1} of tetrahydroselenophene (DFT, B3LYP/6-311^{++}G(d,2p), f = 0.9648, Hoffmann (Hoffmann 2017–2018)).

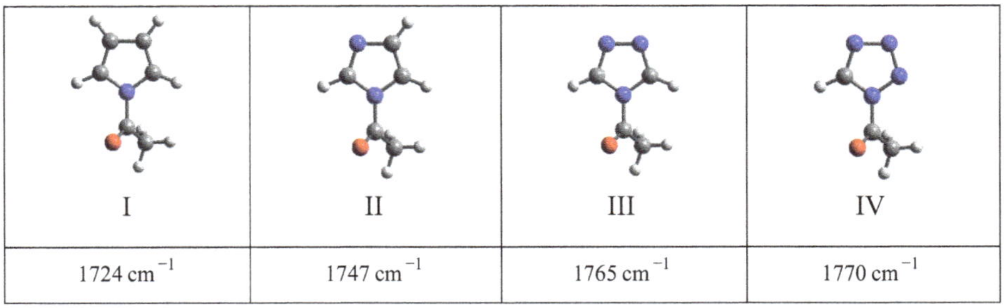

I	II	III	IV
1724 cm^{-1}	1747 cm^{-1}	1765 cm^{-1}	1770 cm^{-1}

Figure 5.78: C–N frequencies of *N*-acetyl-pyrrol, *N*-acetyl-imidazol, *N*-acetyl-triazol, and *N*-acetyl-tetrazol.

N-Acetyl-pyrrol (I)	1,325 cm^{-1}		1,305 cm^{-1}
N-Acetyl-imidazol(II)	1,294 cm^{-1}		1,268 cm^{-1}
N-Acetyl-triazol (III)	1,282 cm^{-1}	1,245 cm^{-1}	1,212 cm^{-1}
N-Acetyl-tetrazol (IV)	1,205 cm^{-1}		1,198 cm^{-1} (sh)
IR intensity	s	m	s

Otting (1956) describes inductive effects in the *N*-acyl derivatives of nitogen-containing heterocycles.

The IR spectra of *N*-acetyl-pyrrol, *N*-acetyl-imidazol, *N*-acetyl-triazol, and *N*-acetyl-tetrazol (Figure 5.78) in the given sequence show an increase of the C=O-frequency from 1,732 cm^{-1} to 1,779 cm^{-1} and an increase of the C–N-frequency of approximately 100 cm^{-1}.

5.6.4 Six-membered rings

5.6.4.1 Pyridine

Pyridine is studied by Urena *et al.* (2003). The heterocycle belongs to the point group C_{2v}. Though this is a greatly reduced symmetry compared to benzene, the Raman spectrum is very similar to benzene as N is isoelectronic to C–H.

The Raman spectrum of pyridine (Figure 5.79) features C–H stretching bands in the range of 3,100–3,000 cm^{-1} (Ra s, IR s), aromatic ring C–C stretches as a doublet at 1,640–1,560 cm^{-1} (Ra w, IR m), trigonal ring breathing at 1,030 cm^{-1}, and whole-ring breathing at 992 cm^{-1}. These last two bands are the most intense in the Raman spectrum and strong in the IR.

Raman and IR spectra of pyridine under high pressure are investigated by Zhuravlev *et al.*(2010).

The vibrational spectra of methylpyridines are published by Arenas *et al.* (1999). The authors report the vibrational assignments of the monosubstituted derivatives 2-methyl-, 3-methyl-, and 4-methyl-pyridine, and the disubstituted derivatives 2,6-dimethyl- and 3,5-

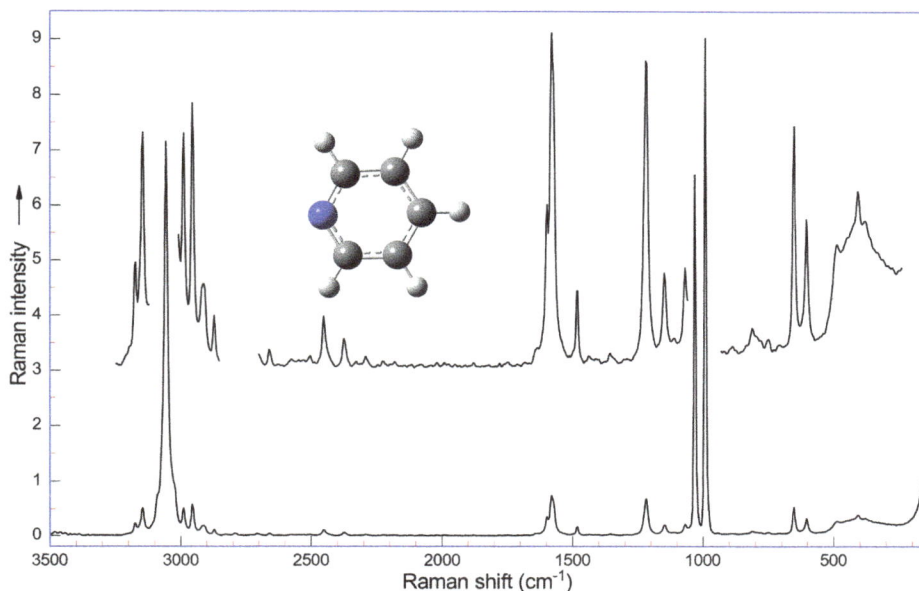

Figure 5.79: Raman spectrum of pyridine (λ_{exc} = 1,064 nm, 275 mW; data courtesy of Bruker Optik GmbH, Ettlingen).

dimethyl-pyridine as well as the trisubstituted derivative 2,4,6-trimethylpyridine using Pulay's scaled quantum mechanical force field (SQMFF) methodology at the RHF/3-21 G level.

Experimental (FT-IR and FT-Raman) and theoretical investigation of some pyridine-dicarboxylic acids are reported by Naik, Reddy, and Prabavathi (2015).

Sundaraganesan *et al.* (2008) report FT-IR, FT-Raman spectra, and *ab initio* HF and DFT calculations of 4*N*,*N*'-dimethylamino pyridine.

Karabacak and Kurt (2008) compare the experimental and density functional computational (B3LYP with different basis sets) molecular structure, IR and Raman spectra, and vibrational assignments of 6-chloronicotinic acid.

Karabacak *et al.* (2008) report the FT Raman and FT-IR spectra of 2-chloronicotinic acid (2-CNA) in the solid phase. The authors also study the molecular structure and vibrational spectra of 2-chloronicotinic acid by DFT and *ab initio* Hartree-Fock calculations and provided vibrational assignments.

Spectroscopic investigations (FT-IR and FT-Raman) and molecular docking analysis of 6-[(1-methyl-4-nitro-1H-imidazol-5-yl) sulfanyl]-7H-purine (azothioprine) are done by Prasath, Govindammal, and Sathya (2017).

Prasad and Dube (1987) report the IR and Raman spectra of the three isomeric pyridine carboxylic acids: picolinic, nicotinic, and iso-nicotinic acid.

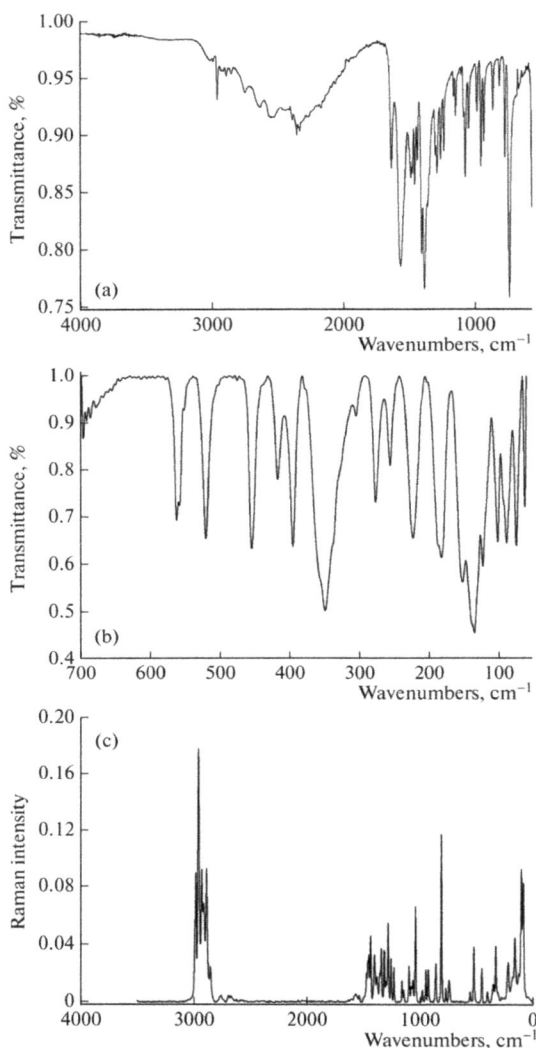

Figure 5.80: Experimental (a) mid-IR, (b) far-IR, and (c) Raman spectra of piperidine-3-carboxylic acid (Yurdakul, Yasayan *et al*. 2014). Copyright ©2014 Springer Nature, with permission.

γ-Aminobutyric acid is a major inhibitory neurotransmitter in mammals, whose uptake in glial cells is inhibited by piperidine-3-carboxylic acid. FT-IR and FT-Raman spectra of piperidine-3-carboxylic acid (Figure 5.80), its tautomers, and isomers are studied by Yurdakul *et al*. (2014). The authors complement their experimental study with MP2/6-311^{++}G(d,p) and B3LYP/6-311^{++}G(d,p) calculations, giving full vibrational assignments.

5.6.4.2 Triazine

s-Triazine is studied by Lancaster, Stamm, and Colthup (1961) The IR and Raman spectra of *s*-triazine and *s*-triazine-d_3 are measured and an assignment of the fundamental vibrational modes is given. Overtones are identified as well. The values of some force constants are calculated from a simple valence force field normal co-ordinate analysis.

1,3,5-*s*-Triazine (symmetry group D_{3h}) shows its C–H stretch as a medium intensity Raman band at 3,042 cm^{-1} (polarized), its ring stretches at 1,555 cm^{-1} (Ra m-w, "quadrant stretching", IR s), and at 1,410 cm^{-1} (Ra w, depolarized, "semicircle stretching", IR vs), the ring breathing vibrations at 1,132 (Ra m, p, IR vw) and 992 cm^{-1} (Ra m, p), the in-plane C–H deformation at 1,176 cm^{-1} (Ra mw, IR s), the in-plane ring deformation at 676 cm^{-1} (Ra m, IR s), and finally the out-of-plane ring deformation at 340 cm^{-1} (Ra m, IR –).

Stagi *et al.* (2017) characterize the vibrational and optical properties of three *s*-triazine derivatives (Figure 5.81). The compounds were obtained starting from three different amines, allowing to achieve a structure with a triazine core, surrounded by three ammine group and terminated with cyano or methyl or oxy-methyl functional group. Experimental vibrational and optical data were interpreted in view of the electronic insights gathered by means of DFT simulations using the Becke three-parameter Lee-Yang

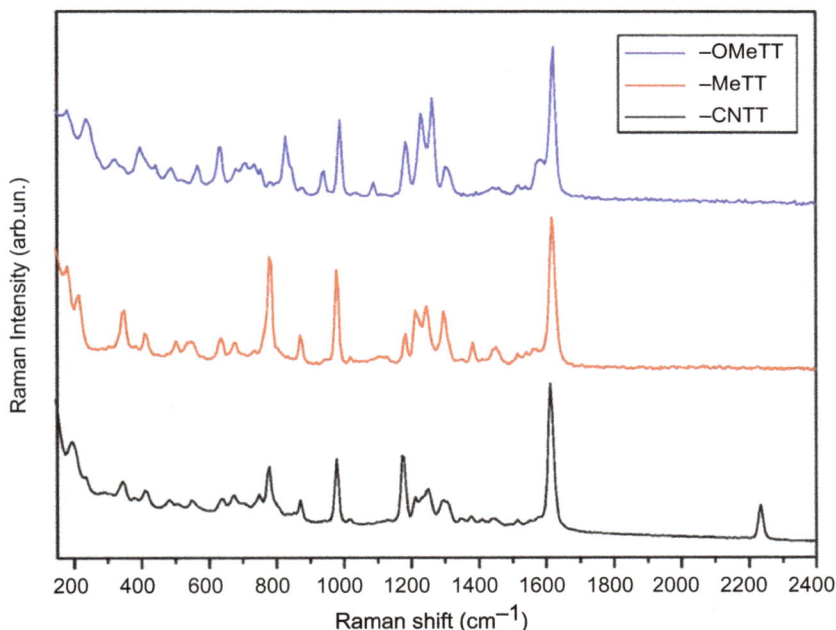

Figure 5.81: Raman spectra of the *s*-triazine derivatives N_2,N_4,N_6-tris(4-methoxyphenyl)-1,3,5-triazine-2,4,6-triamine (OMeTT), N_2,N_4,N_6-tri-*p*-tolyl-1,3,5-triazine-2,4,6-triamine (MeTT), and 4,4′,4″-((1,3,5-triazine-2,4,6-triyl)tris(azanediyl))tribenzonitrile (CNTT) (Stagi, Chiriu *et al.* 2017). Copyright ©2017 Elsevier B.V., with permission.

-Parr (B3LYP) hybrid exchange-correlation functional in combination with the 6-31$^+$G* basis set, a valence double-ζ set augmented with d polarization functions and s and p diffuse functions on base compounds. The cyano compound shows the most promising optical features for photonics and lighting applications.

The triamine of s-triazine, melamine (1,3,5-triazine-2,4,6-triamine), has a sharp band of its ring breathing vibration at 982 cm^{-1} (Figures 5.83 and 5.84). Mircescu *et al.* (2012) study melamine by FTIR, FT-Raman, SERS, and DFT calculations. An *et al.* (2016) publish an experimental and DFT study on surface enhanced Raman scattering of melamine on silver substrate (Figure 5.82). Melamines in general show their NH$_2$ stretches as medium to weak Raman bands at 3,500–3,100 cm^{-1}, the weak band of NH$_2$ deformations occurs at 1,680–1,640 cm^{-1}, the medium-to-strong stretches can be found at 1,600–1,500 cm^{-1}, and a medium band can be found at 1,450–1,350 cm^{-1}. A sharp band of medium intensity in the IR at 825–800 cm^{-1} is not detectable in the Raman spectrum. As melamines can also exist in tautomeric forms, in their *iso*-form with a double bond external to the ring the band occurs at 795–750 cm^{-1}.

5.6.4.3 Indole

Dieng and Schelvis (2010) measure, analyse, and calculate Raman spectra of indole (Figure 5.85), 3-methylindole, and tryptophan (Figure 5.86). In addition to observed and predicted isotope shifts of indole of natural isotopic composition, the shifts of indole 2-^{13}C in H$_2$O and indole 2-^{13}C in D$_2$O are taken into account.

Computational
approach

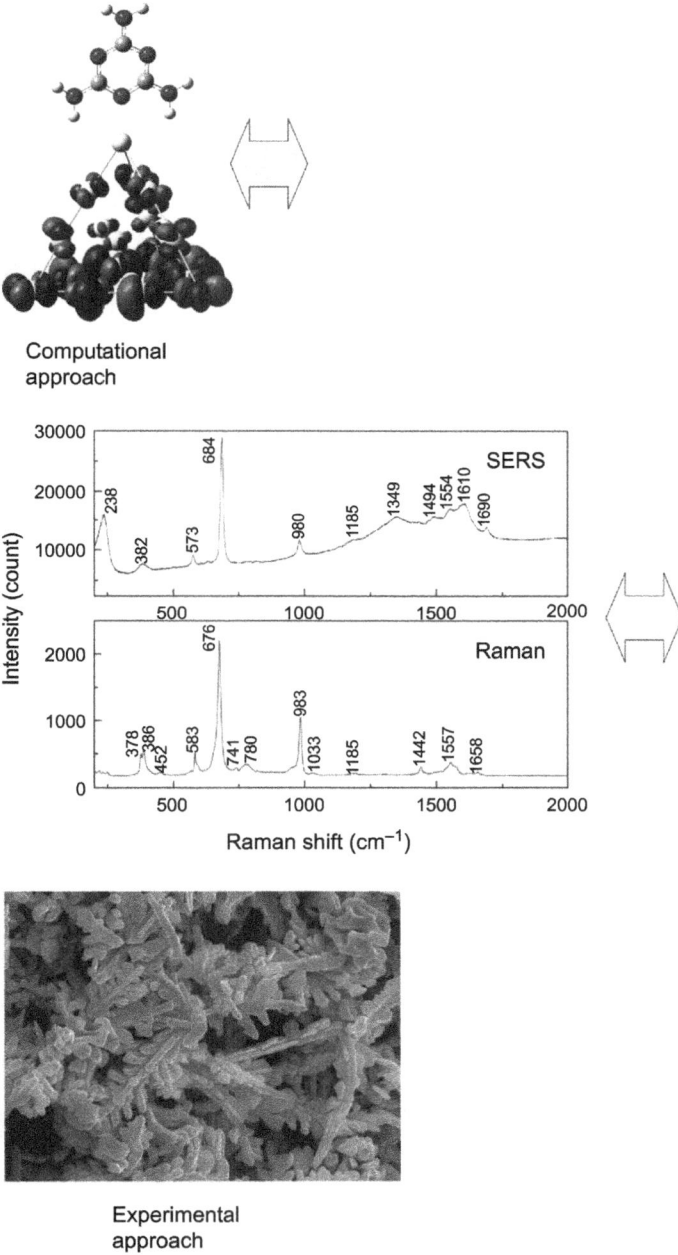

Figure 5.82: Melamine: computed structure of the molecule on silver, its Raman and SERS spectra, and the silver substrate (An, Dao *et al*. 2016). Copyright ©2016 Elsevier B.V., with permission.

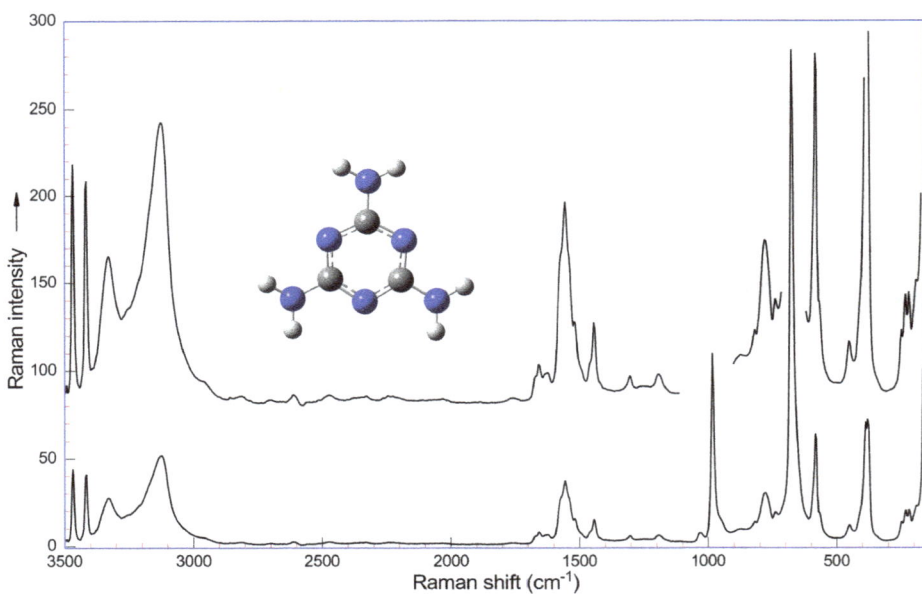

Figure 5.83: Raman spectrum of melamine (λ_{exc} = 1,064 nm, 275 mW; data courtesy of Bruker Optik GmbH, Ettlingen).

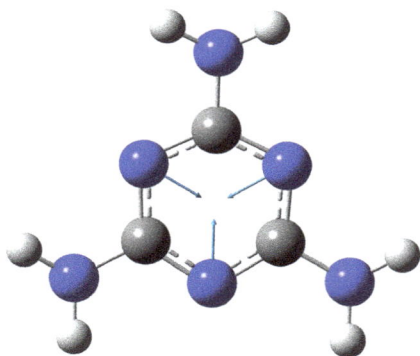

Figure 5.84: Ring breathing vibration of melamine (1,3,5-triazine-2,4,6-triamine) (experimental 982 cm^{-1}, calculated 989 cm^{-1}: DFT, B3LYP/6-311^{++}G(d,p), Hoffmann, 2022).

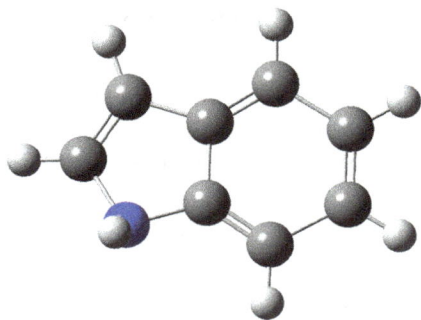

Figure 5.85: Structure of indole.

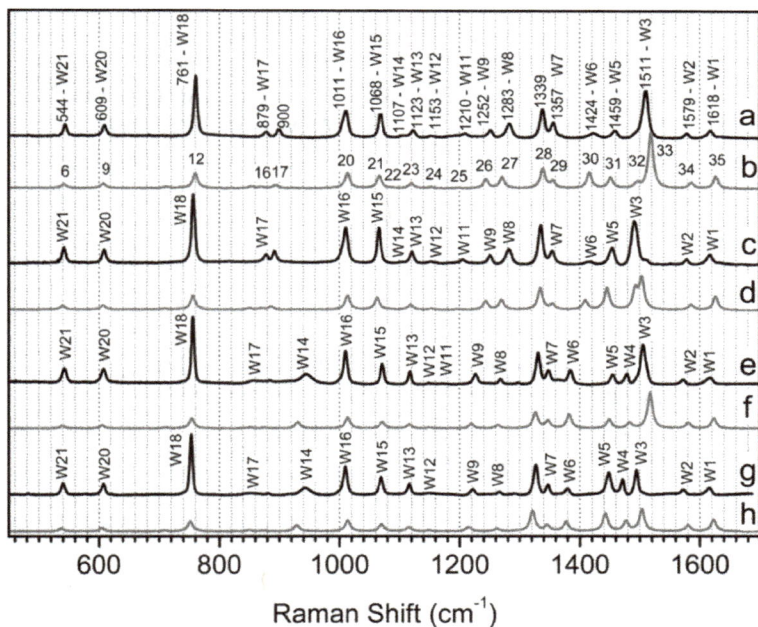

Figure 5.86: Raman spectra of indole (a) and indole 2-^{13}C (c) in H_2O; indole (e), and indole 2-^{13}C (g) in D_2O; and the calculated Raman spectra of indole (b), indole 2-^{13}C (d), indole-ND (f), and indole-ND 2-^{13}C (h) after scaling with a factor of 0.9824 (Dieng and Schelvis 2010). Copyright ©2010 American Chemical Society, with permission.

6 Vibrational spectroscopy of natural products

The complicated structure of most natural products makes the analysis of vibrational spectra complicated as well. Especially IR spectra are difficult to interpret, as they are more complicated than Raman spectra because of the additional lines from overtone and combination bands. A complete analysis of a vibrational spectrum is only possible by using density functional theory (DFT) and potential energy distribution (PED) analysis.

Koenig (1972) published an early review on Raman spectroscopy of biological molecules. The groups of compounds treated are amino acids, polypeptides, proteins, carbohydrates, nucleosides, nucleotides, polynucleotides, DNA, and RNA.

In the very informative and entertaining books of Berger and Sicker (2009) and Sicker *et al.* (2019) about history, isolation, and structure elucidation of natural products, we find IR spectra of alkaloids, dyes, carbohydrates, glycosides, terpenoids including steroids, and aromatic compounds together with UV, CD, NMR, and mass spectra. The chapters may also be read in German as papers in the journal *Chemie in unserer Zeit*.

Many more interesting papers about vibrational spectroscopy of natural products may be found in Chapter 18.

6.1 Amino acids

Amino acids possess both the functional groups of amines and organic acids. The spectra of these groups of compounds have been discussed in Chapter 5.

Amino acids normally exist as the zwitterion. We find characteristic bands at 1,665–1,585 cm^{-1} (asymmetric NH_3^+ deformation, IR w) and at 1,605–1,555 cm^{-1} (asymmetric CO_2^- stretch, IR s, Ra w)

Suzuki *et al.* (1963) give band assignments for the vibrational spectra of zwitterionic glycine as follows:

502 cm^{-1}	CO_2^- rock, NH_3^+ torsion (Ra w, IR s)
605 cm^{-1}	(Ra s, IR m)
700 cm^{-1}	(Ra w, IR s)
895 cm^{-1}	C–C stretch (Ra vs, IR s)
1,035 cm^{-1}	C–N stretch (Ra w, IR m)
1,113 cm^{-1}	NH_2 twist, NH_3^+ rock (Ra w, IR m)
1,327 cm^{-1}.	CH_2 wagging (Ra vs, IR vs)
1,413 cm^{-1}	CO_2^- symmetric stretch (Ra m, IR vs)
1,515 cm^{-1}	(Ra w, IR vs)
1,595 cm^{-1}	CO_2^- asymmetric stretch (Ra w, IR vvs)
2,977 cm^{-1}	Symmetric CH_2 stretch (Ra s, IR s)
3,011 cm^{-1}	Antisymmetric CH_2 stretch (Ra m, IR s)
3,146 cm^{-1}	v(N–H) (Ra w, IR s)

https://doi.org/10.1515/9783110717556-006

The authors also study deuterated glycine.

The Raman spectrum of glycine is shown in Figure 6.1.

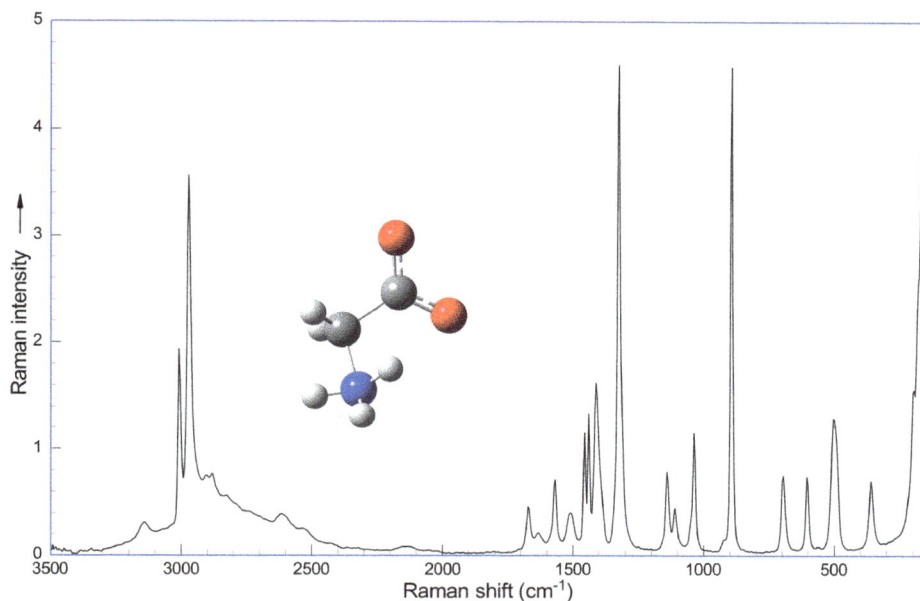

Figure 6.1: Raman spectrum of glycine (λ_{exc} = 1,064 nm, 270 mW; data courtesy of Bruker Optik GmbH, Ettlingen).

Cao and Fischer (1999) report infrared (IR) spectral, structural, and conformational studies of zwitterionic L-tryptophan (Figure 6.2). A complete assignment of the spectral bands is given on the basis of SCRF (solvent reaction field) *ab initio* molecular orbital calculations at the HF/6-31G(d,p) level, using KBr as the continuum.

dos Santos *et al.* (2018) study high-pressure Raman spectra and perform DFT calculations of L-tyrosine hydrochloride in crystalline form.

Tasumi *et al.* (1982) study the Raman spectra of L-histidine, and the related compounds imidazole, 4-methylimidazole, and *N*-acetyl-L-histidine methylamide as solutions in H_2O and D_2O. The solutions are studied at various pH and pD values, and the ^{13}C-labeled L-histidine was included in the study. A strong Raman band, assigned to a mode characteristic of the *N*-deuterated imidazolium ring (Im+) in which the N_1–C_2–N_3 symmetric stretching and N–D bending vibrations are mixed, was found at around 1,410 cm^{-1} in acidic D_2O solutions. The authors propose to use this band as a probe for studying the ionization states of histidine residues in proteins.

Raman spectra of amino acid crystals are published by Freire *et al.* (2017). The spectra of glycine, L-alanine, L-valine, L-leucine, L-isoleucine, L-phenylalanine, L-methionine, L-proline,

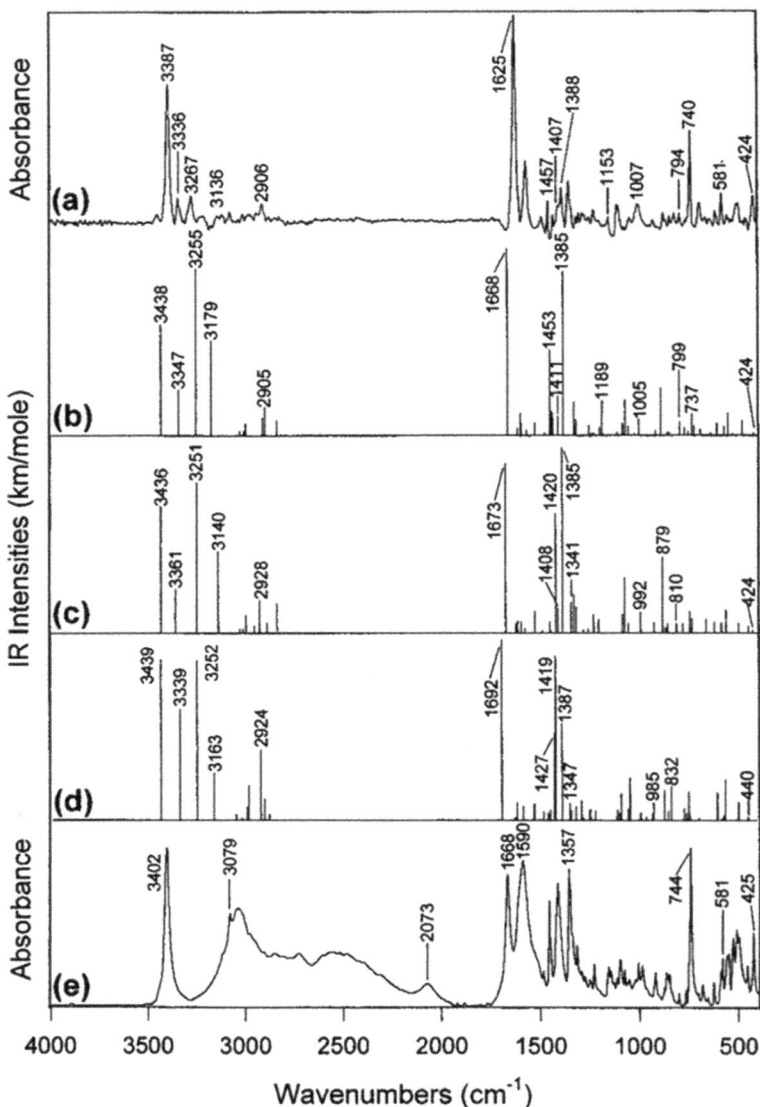

Figure 6.2: Mid-IR spectra of zwitterionic tryptophan obtained by different methods: (a) DSD (dissolution–spray–deposition); SCRF/HF predictions for conformers G (b), G* (c), and G+ (d), respectively; (e) KBr pellet (Cao and Fischer 1999). Copyright ©1999 American Chemical Society, with permission.

L-tryptophan, L-cysteine, L-serine, L-glutamine, L-asparagine, L-tyrosine, L-threonine, L-aspartic acid, L-glutamic acid, and L-histidine (Figures 6.3–6.5) are shown (see also the spectra in Figures 6.6–6.9 for comparison). The Raman spectrum of L-histidine hydrochloride monohydrate is measured in three scattering geometries.

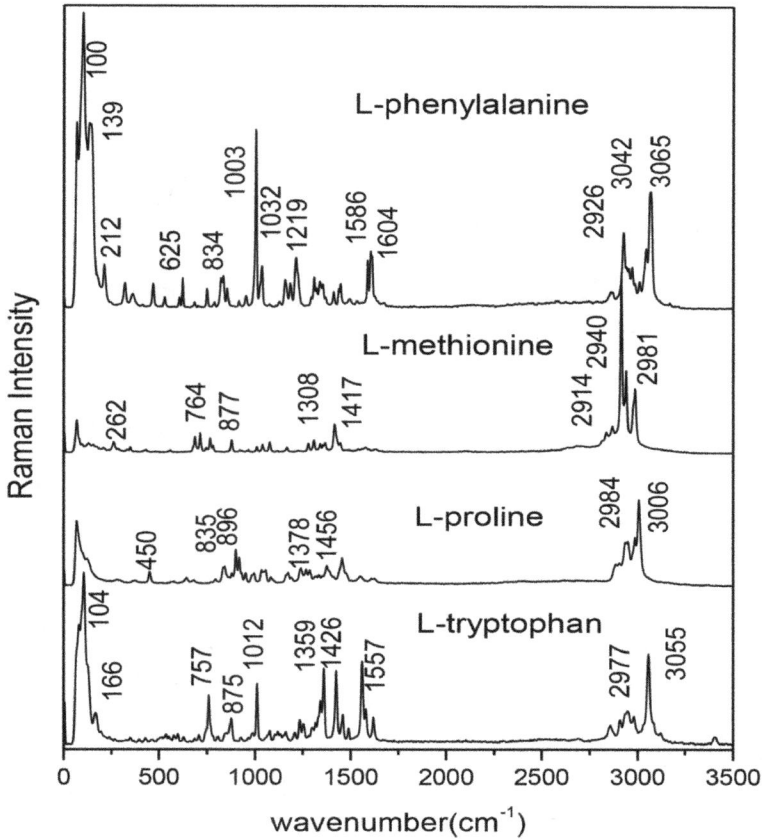

Figure 6.3: FT-Raman spectra of ʟ-phenylalanine, ʟ-methionine, ʟ-proline, and ʟ-tryptophan (Freire *et al.* 2017). Copyright ©2017 by the authors. Licensed by Creative Commons (CC 3.0).

6.2 Peptides

The Raman spectrum of glycylglycine is first studied by Garfinkel and Edsall (1958) together with the spectra of glycyl-ᴅʟ-serine, glycyl-ᴅʟ-valine, ʟ-alanyl-ʟ-alanine, and ᴅ-alanyl-ʟ-alanine. Glycylglycine is studied as the dipolar ion, anion and cation; the other compounds as the dipolar ion and cation only, except that asparagine is observed only as the cation in acid solution.

Schweitzer-Stenner (2001) reviews various visible and UV-resonance Raman experiments on model peptides such as *N*-methylacetamide and di- and triglycine, exploring the structure and dynamics of these molecules. Whereas visible Raman spectroscopy probes the electronic ground state, the UV Raman technique may be utilized additionally to obtain structural information about excited electronic states.

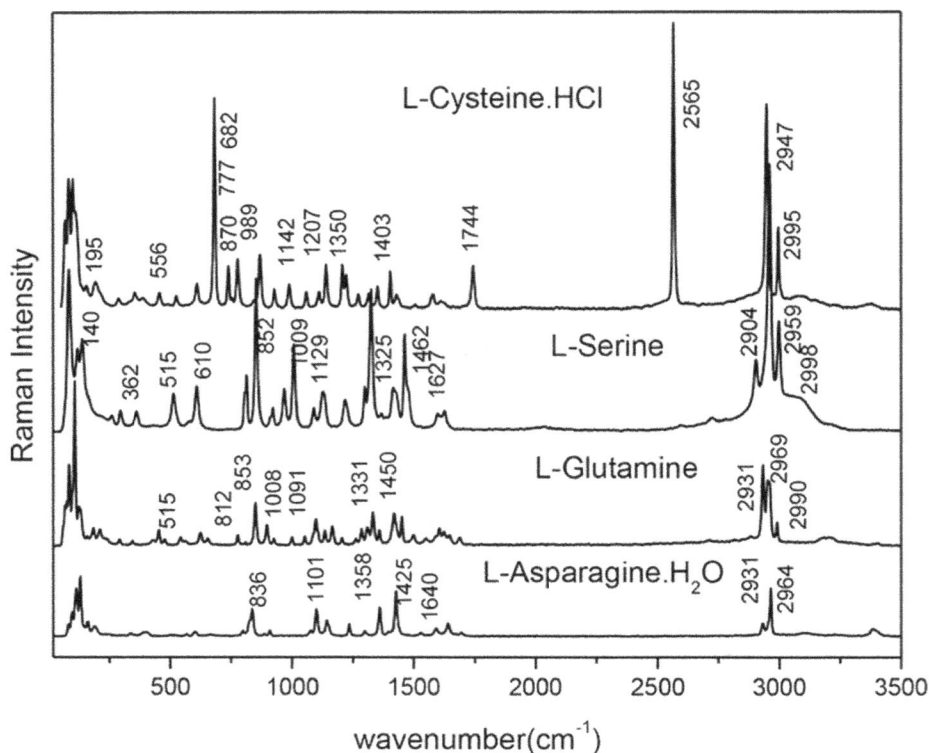

Figure 6.4: FT-Raman spectra of polar (neutral) amino acids L-serine and L-glutamine and compounds L-cysteine·HCl and L-asparagine·H₂O. In the spectrum of L-cysteine·HCl stands out an intense band associated with the S–H stretching, v(S–H), similar to the monoclinic phase of L-cysteine, but different from the orthorhombic phase, where broad split bands are observed (Freire *et al.* 2017). Copyright ©2017 by the authors. Licensed by Creative Commons (CC 3.0).

Schweitzer-Stenner *et al.* (2002) report the structure analysis of dipeptides in water by exploring and utilizing the structural sensitivity of the amide III band. The methods they use are polarized visible Raman and Fourier-transform (FT)-IR spectroscopy, enhanced by DFT-based normal coordinate analysis.

Structure, spectra, and the effects of twisting of β-sheet peptides are studied by DFT by Bour and Keiderling (2004). The dimer (Ac–Ala₃–NHMe)₂ was chosen as a test case. The parallel and both antiparallel β-sheet structures were found to be stable in vacuum at the BPW91/6-31G** level.

The structures of unfolded peptides are studied by Schweitzer-Stenner *et al.* (2012) by Raman spectroscopy. The authors describe how high-quality polarized Raman spectra of peptides can be recorded with a Raman microspectrometer and how the structure-sensitive amide I band profiles of isotropic and anisotropic Raman scattering can be analysed in conjunction with the respective IR and vibrational

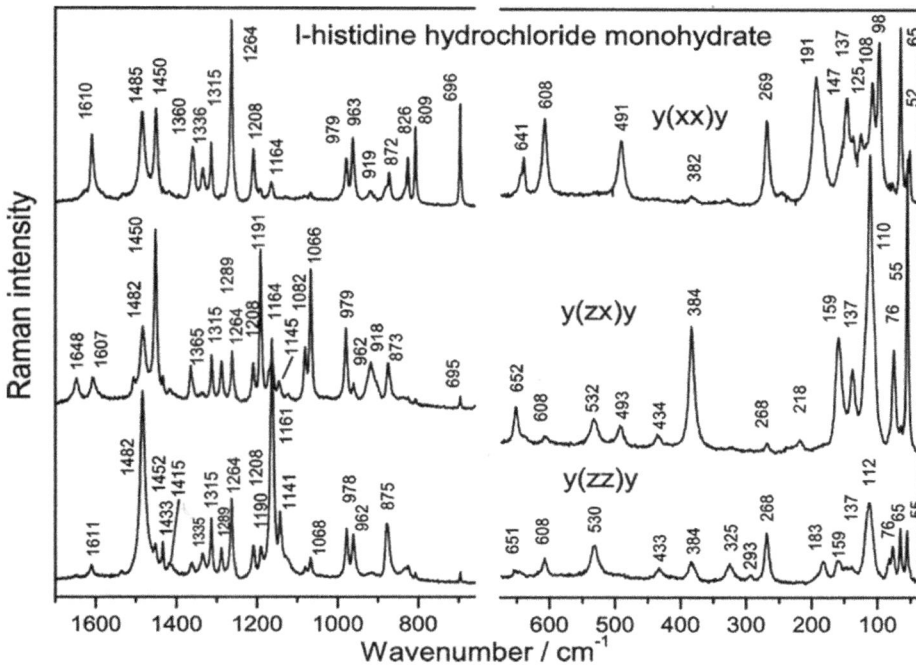

Figure 6.5: Raman spectra of ʟ-histidine hydrochloride monohydrate in three scattering geometries (Freire *et al.* 2017). Copyright ©2017 by the authors. Licensed by Creative Commons (CC 3.0).

circular dichroism (VCD) profiles to obtain conformational distributions of short unfolded peptides.

Ramanauskaite and Snitka (2015) report surface-enhanced Raman spectroscopy (SERS) of ʟ-alanyl-ʟ-tryptophan dipeptide adsorbed on a silicon substrate decorated with triangular silver nanoplates. The dipeptide's orientation on the nanostructured surface is studied.

6.3 Proteins

The review of *Raman* spectroscopy of proteins by Tuma (2005) discusses spectral assignments, protein folding and assembly. Raman spectroscopy is used to characterize intrinsically unstructured proteins. Even enzyme–substrate interactions may be characterized by Raman difference spectroscopy as the authors report. Some applications of Raman spectroscopy to protein analysis in biotechnology and food industry are discussed as well.

Another review by Barth (2007) highlights *infrared* spectroscopy of proteins. After discussing instruments and theory, the IR absorptions of amino acid side chains and

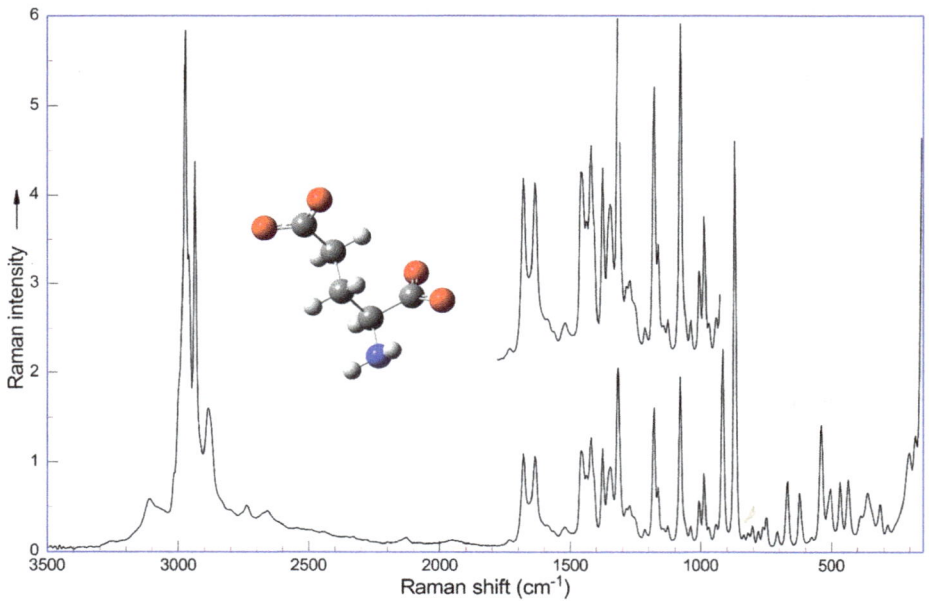

Figure 6.6: Raman spectrum of L-glutamic acid (λ_{exc} = 1,064 nm, 270 mW; data courtesy of Bruker Optik GmbH, Ettlingen).

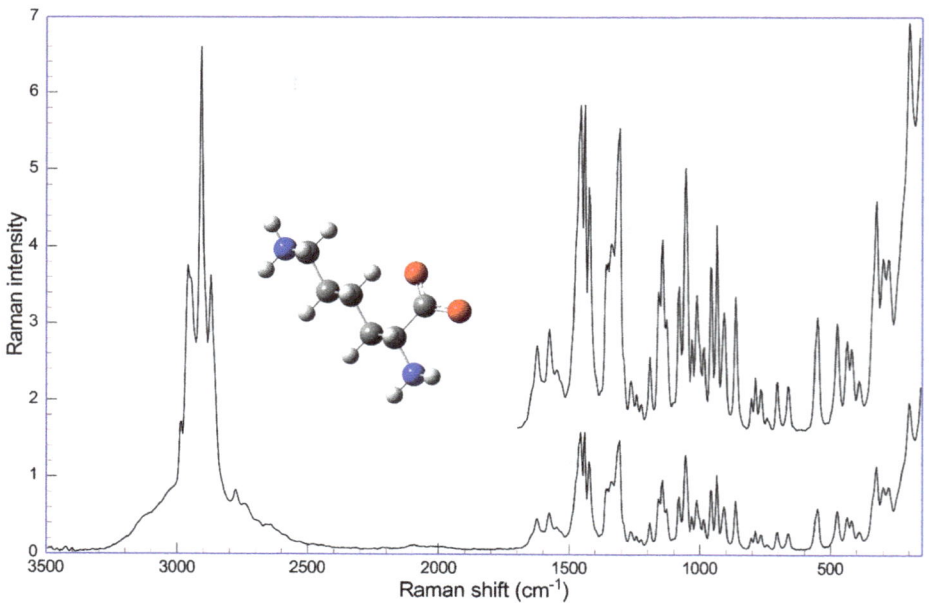

Figure 6.7: Raman spectrum of L-lysine as hydrochloride (λ_{exc} = 1,064 nm, 270 mW; data courtesy of Bruker Optik GmbH, Ettlingen).

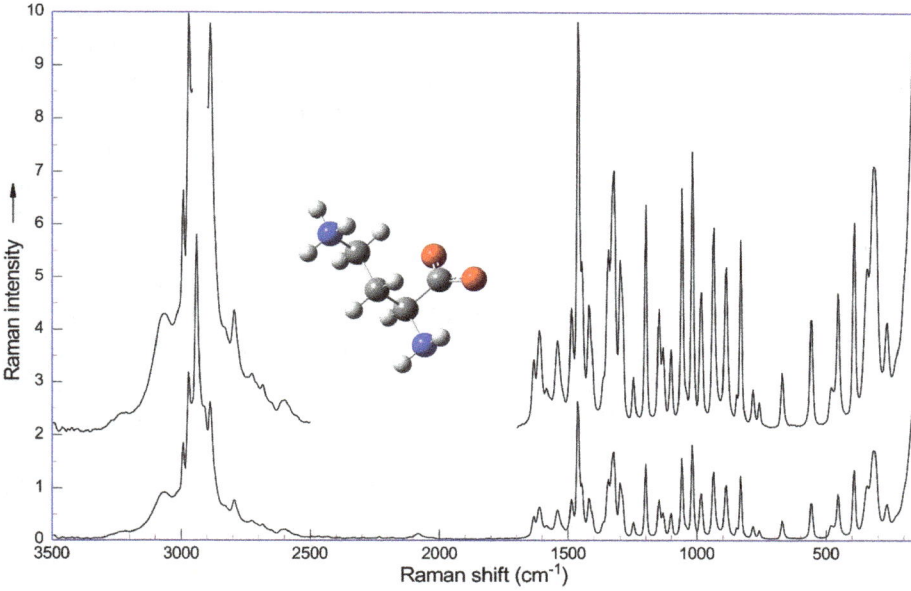

Figure 6.8: Raman spectrum of L-ornithine as hydrochloride (λ_{exc} = 1,064 nm, 270 mW; data courtesy of Bruker Optik GmbH, Ettlingen).

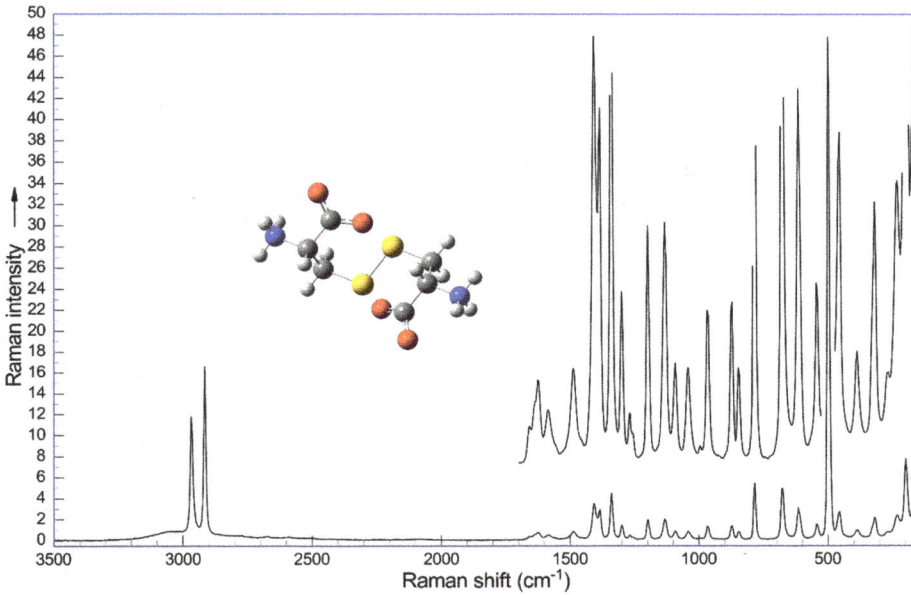

Figure 6.9: Raman spectrum of L-cysteine (λ_{exc} = 1,064 nm, 270 mW; data courtesy of Bruker Optik GmbH, Ettlingen).

the backbone are detailed in comprehensive tables, and the characteristic bands of secondary structures are discussed. Barth also puts a focus on the mid-IR spectral region and the study of protein reactions by reaction-induced IR difference spectroscopy.

In the review of Rygula *et al.* (2013), 26 proteins of different structures, functions, and properties are investigated by Raman spectroscopy with 488, 532, and 1,064 nm excitation. The effect of choosing different excitation lines is nicely shown. For the α-helix, the authors identify mostly amide I (1,650–1,680 cm^{-1}) and amide III (1,230–1,280 cm^{-1}) as diagnostic regions. The most typical difference of the α-helix compared to both the β-sheet and random coil is the absence of spectral intensity at 1,235–1,240 cm^{-1}, characteristic for the α-helix. One such example is the band at 1,345 cm^{-1} assigned to C–C$_\alpha$–H bending and C$_a$–C stretching as well as the band at ca. 740 cm^{-1}, which is attributed to amide IV (the carbonyl in-plane bending motion). Complete band assignments are given. Some Raman spectra (only those taken at 532 nm excitation) copied from the paper of Rygula *et al.* are shown in Figures 6.10–6.12.

Raman spectra of the α-helix proteins are shown in Figure 6.10: albumin (ALB), collagen (COLL), ferritin (FTN), and glutathione transferase (GST). Rygula *et al.* describe the spectra of the following β-sheet proteins (Figure 6.11): carbonic anhydrase (CAH), α-chymotrypsinogen A (type II) (CTG), elastase (EST), lectin (LCL), pepsin (PEP), pepsinogen (PGN), superoxide dismutase (SOD), thaumatin (TMT), trypsin (TPO), trypsinogen (TGN), and xylanase (XYN). The next part of proteins contains molecules whose folds are not essentially formed by α- or β-structure (Figure 6.12). Among these, the α- and β-class (α/β) is composed of proteins with α-helices and β-strands largely interspersed. To this class belong glucose oxidase (UOX), proteinase (subtilisin Carlsberg) (PRT), and triosephosphate isomerase (TPI). The next group, α plus β class (α + β), classifies those proteins, in which α-helices and β-strands are largely segregated, that is, α-lactalbumin (ALA), papain (PAP), and ubiquitin (UBQ). As an example of a small protein, trypsin inhibitor (TPO) is shown. The Raman spectra of the iron porphyrin containing heme proteins haemoglobin (Hb), myoglobin (Mb), cytochrome c (cyt c), and horseradish peroxidase (HRP) are shown in Figure 6.13.

The strengths of Raman spectroscopy for studying proteins and nucleoproteins are illustrated by Nemecek *et al.* (2013), using examples from their own work. The scientists show work on subunit folding and recognition in the assembly of icosahedral bacteriophages, orientations of subunit main chains and side chains in native filamentous viruses, roles of cysteine hydrogen bonding in the folding, assembly, and function of virus structural proteins, and structural determinants of protein/DNA recognition in gene regulatory complexes.

Three major classes of plasma lipoproteins are characterized by Ricciardi *et al.* (2020) by Raman spectroscopy: the triacylglycerol-rich very-low-density lipoproteins; the more cholesterol-rich low-density lipoproteins; and the high-density lipoproteins. The lipoproteins were obtained from human plasma of six fasting healthy volunteers. The extracted lipoproteins were dried on CaF$_2$ slides and analysed using a 633 nm laser line not in resonance with the carotenoids present in the sample.

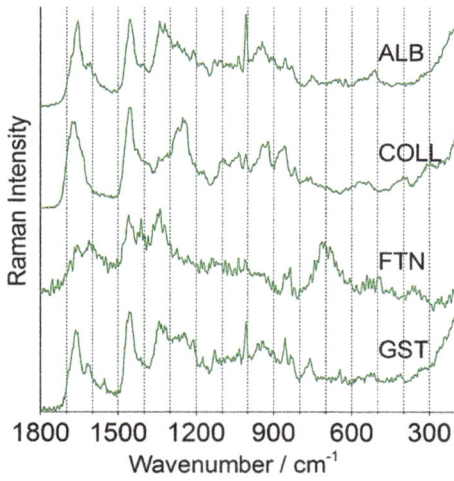

Figure 6.10: Raman spectra of the α-helix proteins albumin, collagen, ferritin, and glutathione transferase at 532 nm in the 1,800–200 cm^{-1} range (Rygula *et al.* 2013). Copyright ©2013 John Wiley and Sons, with permission.

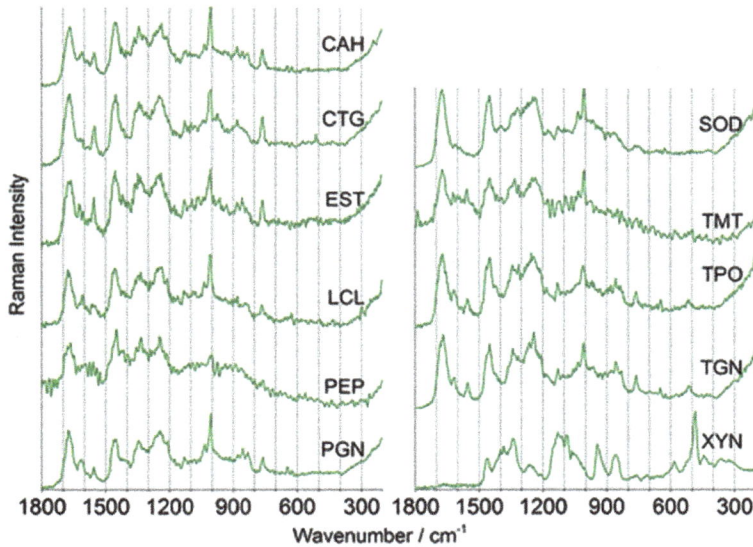

Figure 6.11: Raman spectra of the β-sheet proteins at 532 nm in the 1,800–200 cm^{-1} range (Rygula *et al.* 2013). Copyright ©2013 John Wiley and Sons, with permission.

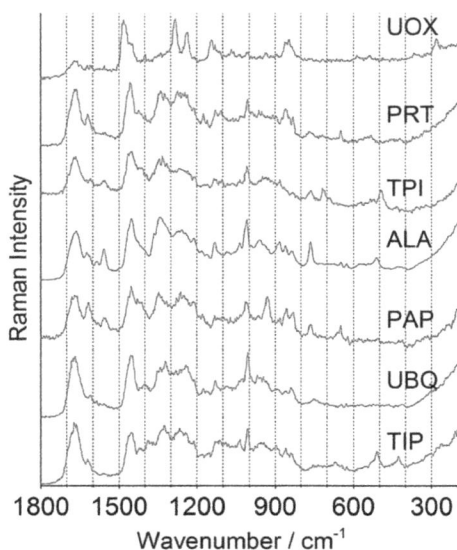

Figure 6.12: Raman spectra of the α/β, α + β, and the small proteins at 532 nm in the 1,800–200 cm^{-1} range (Rygula *et al.* 2013). Copyright ©2013 John Wiley and Sons, with permission.

6.3.1 Enzymes

The Raman spectrum of lysozyme is reported by Garfinkel and Edsall (1958). The position of Raman bands and intensities is given together with the IR spectrum from earlier works and some assignments.

Tobin (1968) studies Raman spectra of crystalline lysozyme, pepsin, and α-chymotrypsin. Minor differences allow the identification of compounds. Bands assignable to specific groupings are noted.

Koenig *et al.* (1972) report the Raman spectra of three globular proteins: beef pancreas chymotrypsinogen A, beef pancreas ribonuclease, and hen egg white ovalbumin in the solid state and aqueous solution. Spectral changes on solution and thermal denaturation are discussed.

Brown *et al.* (1972) study the conformational dependency of low-frequency motions of proteins by laser Raman spectroscopy.

6.4 Aromatic compounds

Sinensetin can be isolated from cold-pressed orange oil (*Citrus sinensis* L., Rutaceae). Its IR spectrum shows bands at 3,050 and 2,850 cm^{-1}, and characteristic bands of the unsaturated carbonyl system at 1,650 cm^{-1} (C=O) and 1,600 cm^{-1} (C=C), as reported by Steinke *et al.* (2013, 2019).

Rosmarinic acid is extracted from the dried leaves of lemon balm (*Melissa officina-lis* L., Lamiaceae). The herb was used against the plague and is used as a Mediterranean seasoning as well. Gao *et al.* (2015) and Sicker *et al.* (2019) show its IR spectrum. As expected, the broad band of the different OH groups dominates the spectrum. Other prominent bands are the C=O band at 1,700 cm^{-1} and a C=C band at 1,600 cm^{-1}.

The psychoactive tetrahydrocannabinol from marijuana (dried flowers and leaves near the flowers of the female plants of *Cannabis sativa* L. var. *sativa*, Cannabaceae) is studied by Berger and Sicker (2009). Its IR spectrum shows phenolic O–H bands at 3,300 cm^{-1}, broad, strong C–H bands from 3,050 to 2,800 cm^{-1}, and two absorptions near 1,620 cm^{-1} and at 1,580 cm^{-1} from aromatic and olefinic C=C vibrations.

Eugenol is isolated as a pale yellow liquid from cloves (*Syzygium aromaticum* (L.) MERR. and L.M. PERRY, Myrtaceae), which is used for the production of vanillin. Its IR spectrum is published by Berger and Sicker (2009). It shows IR bands from O–H stretching at 3,250 cm^{-1} and C–H bands from sp^2-hybridized carbon as well as from sp^3-hybridized carbon atoms and two bands of olefinic and aromatic C=C vibrations near 1,620 cm^{-1}.

6.5 Alkaloids

Strychnine is described by Narayanan *et al.* (1999) as a hard, white, crystalline alkaloid, isolated from the seeds of *Strychnos nux vomica* L. and similar plants. It forms very bitter orthorhombic prisms from a solution in alcohol.

Strychnine's vibrational spectrum is analysed by the authors as follows:

Skeletal molecular deformation bands of rings	300–540 cm^{-1}
Several ring breathing bands, such as the piperidine ring pulsation band	770–820 cm^{-1}
The pyrrolidine ring breathing band	890–920 cm^{-1}
The benzene ring breathing band	980 cm^{-1}
The pyrrole ring breathing band	1,150–1,160 cm^{-1}
An in-plane C–H deformation band	1,000–1,250 cm^{-1}
The out-of-plane C–H deformation band	640–880 cm^{-1}
A carbon–carbon stretching vibration band	1,479–1,670 cm^{-1}
The C–N band of the indole ring	1,650 cm^{-1}
The C=O vibration band of the amide group	1,665 cm^{-1}
The C–H stretching modes	2,800–3,074 cm^{-1}

Roman *et al.* (2015) study the vibrational spectra of cinchona alkaloids in the solid state and aqueous solutions. The spectra of the antimalarial compounds quinine, quinidine, cinchonine, and cinchonidine (Figure 6.14) are interpreted on the basis of theoretical calculations (DFT, B3LYP, and aug-cc-pVDZ basis sets).

Raman spectra of the cinchona alkaloids in solution exhibit very similar patterns. Therefore, Roman *et al.* use Raman optical activity (ROA) for the differentiation between the derivatives (e.g. quinine and cinchonidine) and corresponding pseudoenantiomers (e.g. quinine and quinidine; see also Section 18.2).

Strong IR bands of cinchonine (Figure 6.13) in KBr occur at 3,068 cm^{-1}, 2,942 cm^{-1}, 2,918 cm^{-1}, 2,876 cm^{-1}, 1,592 cm^{-1} (C=C stretch), 1,113 cm^{-1}, 906 cm^{-1}, 763 cm^{-1}, and 635 cm^{-1}.

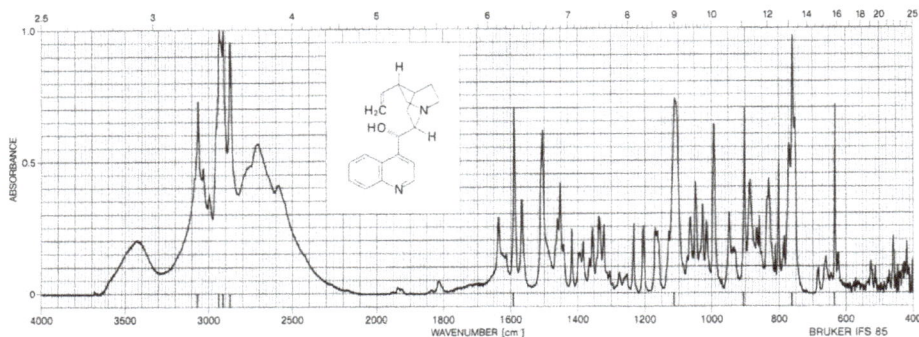

Figure 6.13: FT-IR spectrum of cinchonine in KBr. Merck atlas, spectrum 2,938 (Pachler *et al.* 1988). Copyright ©1988 VCH Verlagsgesellschaft mbH, with permission.

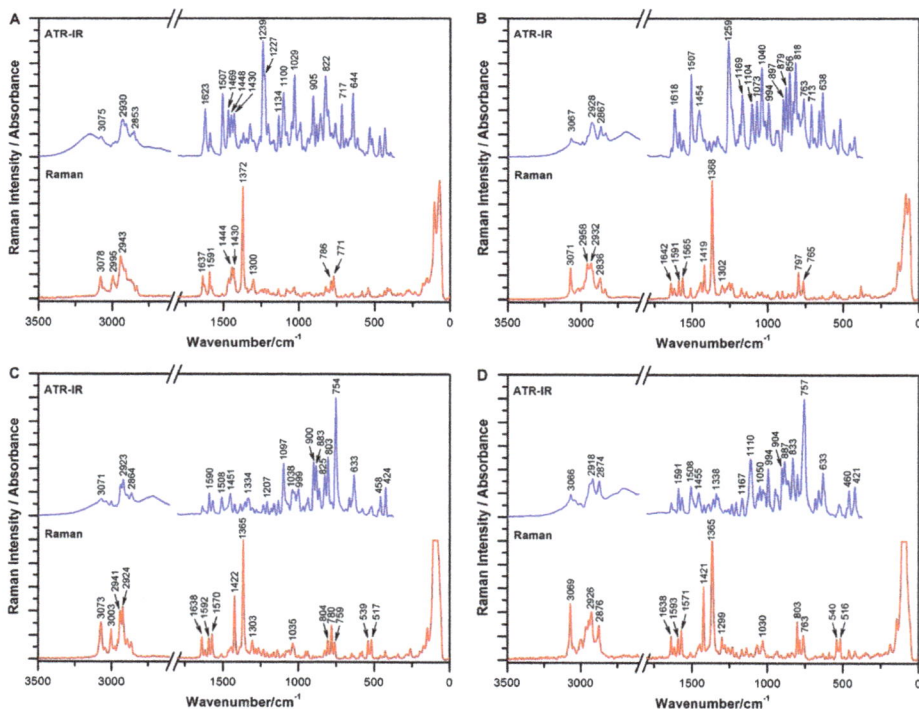

Figure 6.14: FT-IR and FT-Raman spectra of quinine (A), quinidine (B), cinchonine (C), and cinchonidine (D) (Roman *et al.* 2015). Copyright ©2015 John Wiley & Sons, Ltd., with permission.

Paclitaxel (Figure 6.15; see also Section 18.2) is a substance from the taxane group isolated from the bark of the Pacific yew (*Taxus brevifolia* NUTT., Taxaceae). It is a mitotic spindle poison used in cancer treatment. Renuga Devi and Gayathri (2010) analyse paclitaxel drugs using FT-IR and Raman spectroscopies.

Ancistrocladus heyneanus is a tropical liana known to produce pharmacologically interesting naphthylisoquinoline alkaloids as secondary metabolites. Urlaub *et al.* (1998) use the micro-FT-Raman technique to localize and identify some of these alkaloids in different parts of the plant.

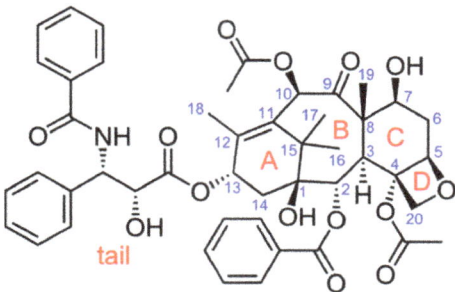

Figure 6.15: Structure of paclitaxel, with labelled rings and the numbering scheme of the [9.3.1.0] pentadecene system (Profant *et al.* 2017). Copyright ©2017 American Chemical Society, with permission.

Volochansky *et al.* (2019) use SERS for the detection and identification of medically significant alkaloids. SERS spectra of atropine and pergolide were measured in the concentration range of 10^{-3}–10^{-9} mol/L using gold or silver SERS-active layers (Figure 6.16). DFT calculations and Raman spectra of pure compounds aided the identification of the compounds. Concentrations were calculated with chemometrical methods.

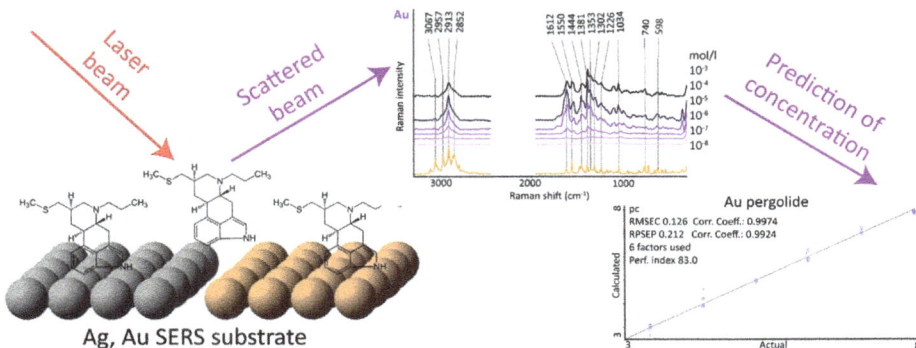

Figure 6.16: Raman spectra of pergolide and analysis of concentrations (Volochanskyi *et al.* 2019). Copyright ©2019 Elsevier B.V., with permission.

Gudi *et al.* (2015) use IR and Raman spectroscopic methods for the characterization of *Taxus baccata* L. The authors could improve taxane (Figure 6.15) isolation by accelerating quality control and process surveillance.

Baranska and Schulz (2009) dedicate a chapter in volume 67 of *The Alkaloids* to the analysis of alkaloids by IR and Raman spectroscopies.

Nitrogen bridgehead compounds in the IR at 2,700–2,800 cm^{-1} show the Bohlmann bands, which may be used to distinguish between the *cis*- and the *trans*-configurations of the quinolizidine moiety: the bands occur only in the *trans*-compound, and they are missing in the *cis*-compound.

A few examples of characteristic vibrational bands of alkaloids (as given by Baranska and Schulz) are shown as follows.

6.5.1 Purine alkaloids

Xanthine, with its two carbonyl moieties, shows an IR band at 1,658 cm^{-1} (s) from the out-of-phase and at 1,567 cm^{-1} (s) from the in-phase carbonyl stretches in the IR, at 1,600/1,560 cm^{-1} (m/s) in the Raman spectrum.

The Raman spectra of theobromine show a characteristic peak at 620 cm^{-1}, for caffeine a strong band at 556 cm^{-1} and a medium doublet at 643 cm^{-1} and 741 cm^{-1}, and for theophylline a single band at 668 cm^{-1}. FT-IR and FT-Raman spectral band positions and intensities together with assignments are shown in a table.

6.5.2 Isoquinoline alkaloids

Morphine, codeine, and thebaine show similar bands because of their similar structure, but can be differentiated by bands in the fingerprint region. The strongest bands are at 1,050 cm^{-1} from –C–O–C–, at 1,600–1,650 cm^{-1} from the ring stretching, and at 630–650 cm^{-1} from the ring deformation vibration. Full FT-Raman spectra are shown in the book chapter.

Noscapine shows strong bands at 1,765 cm^{-1} in the IR and at 1,784 cm^{-1} in the Raman spectrum from the C=O stretching vibration of the lactone ring.

6.5.3 Tropane alkaloids

The attenuated total reflection (ATR) IR spectrum of scopolamine shows an intense band at 1,725 cm^{-1} of the C=O stretching with minor bands at 1,251 cm^{-1}, 1,165 cm^{-1}, and 853 cm^{-1}. Anisodamine (6β-hydroxy hyoscyamine) from *Datura* and other plant species is studied by IR and VCD measurement and DFT calculations by Muñoz *et al.* (2006). See also Section 18.1.

Band positions for *pyrrolidine alkaloids, piperidine alkaloids, quinolizidine alkaloids, indole Alkaloids*, and other alkaloids are also given.

For the treatment of liver ailment, the tree boldo (*Peumus boldus* MOL.) is used in Chilean folk medicine. Its main alkaloid and active principle is boldine, an antioxidant. Srivastava *et al.* (2011) study the vibrational spectra of the molecule. DFT calculations of the optimized geometry and the vibrational spectra of boldine using the B3LYP functional and the 6-311G(d,p) basis set result in a complete vibrational assignment.

Cantharimide and its derivatives, obtained from the blister beetle *Mylabris phalerata* PALLA (Figure 6.17), are studied by Zeng *et al.* (2016). The planar structures and absolute configurations of compounds 1–14 were fully elucidated on the basis of spectroscopic analysis (Figure 6.18). Some of the compounds were found to be potent inhibitors of the HBV virus.

Figure 6.17: Blister beetle *Mylabris phalerata* and cantharimide derivatives (Zeng *et al.* 2016). Copyright ©2016 American Chemical Society, with permission.

Baranska and Kaczor (2012) calculate the geometry of morphine conformers and their vibrational spectra using DFT with the B3LYP functional and the 6-311^{++}G(d,p) basis set. The authors demonstrate that a change in morphine conformation does not significantly affect its Raman spectrum (Figure 6.19). Based on the calculations, the vibrational spectra of the morphine metabolite morphine 3-*O*-glucoronide can be interpreted as well.

Králík *et al.* (2020) reveal the structure of heroin in a D_2O solution by chiroptical spectroscopy. Four stable conformers of heroin (Figures 6.20–6.22) are found by calculations at the ωB97X-D/def2TZVP level using the conductor-like polarizable continuum model with defined dihedral angles α_1–α_4.

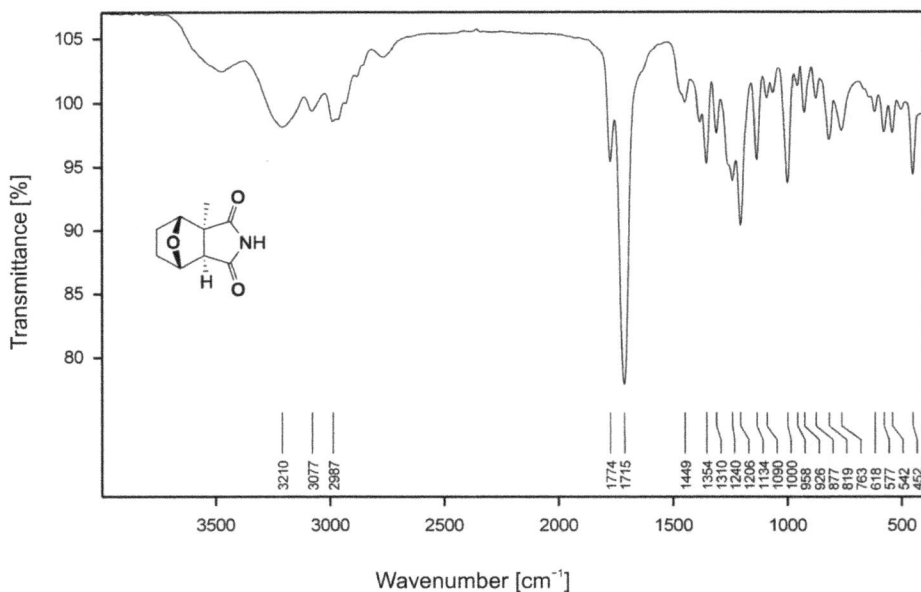

Figure 6.18: IR spectrum of the cantharimide methyl derivative 13 (Zeng *et al.* 2016). Copyright ©2016 American Chemical Society, with permission.

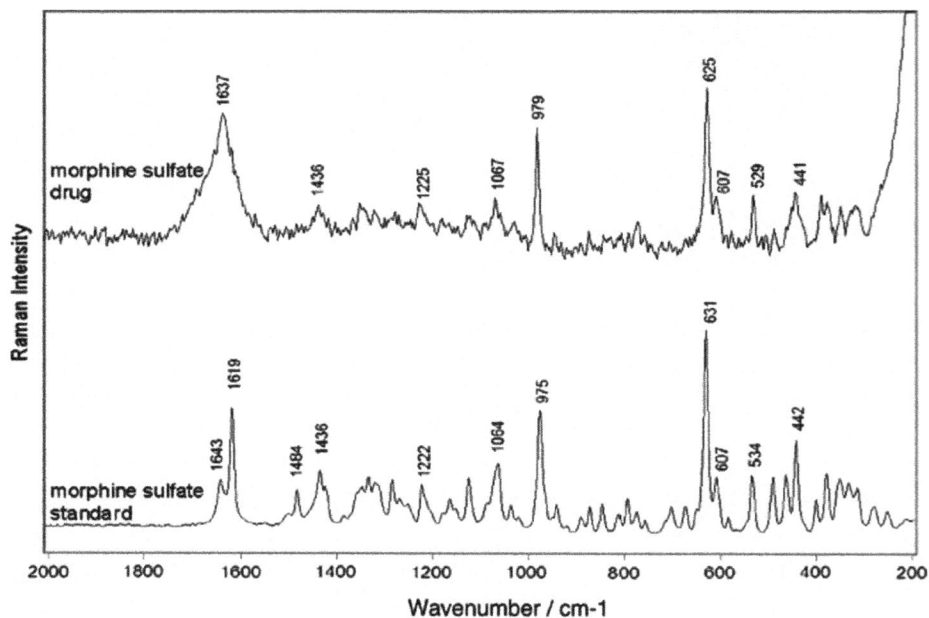

Figure 6.19: FT-Raman spectra of a pharmaceutical product containing morphine and the corresponding morphine standard (Baranska and Kaczor 2012). Copyright ©2012 John Wiley and Sons, with permission.

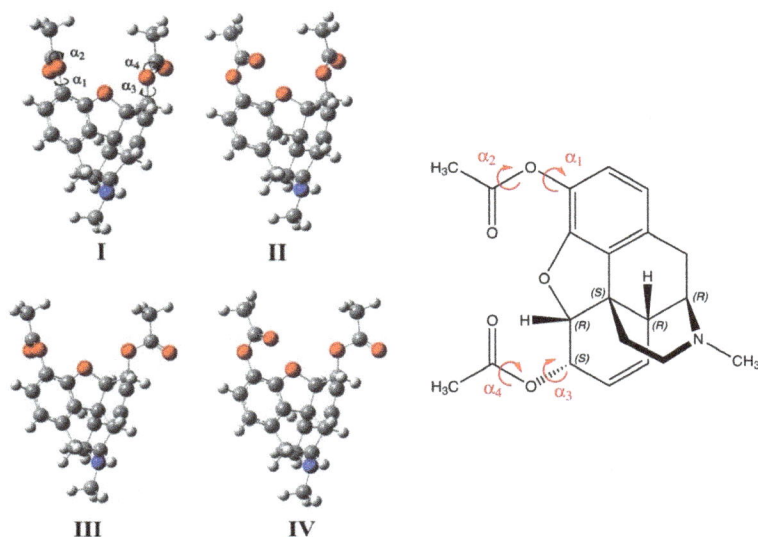

Figure 6.20: Four stable conformers of heroin with defined dihedral angles α_1–α_4 (Králík *et al.* 2020). Copyright ©2020 John Wiley and Sons, with permission.

Cocaine, the tropane alkaloid found in the leaves of the coca plant (*Erythroxylum coca* LAM., Erythroxylaceae), is studied by Steinke *et al.* (2013). Among other physical data, they publish its IR spectrum (Figure 6.23).

The very strong poison nicotine may be extracted from tobacco (*Nicotiana tobacum* L., Solanaceae), and its IR spectrum shows pyridine vibrations: slightly split C=C bands at 1,600 cm^{-1}, bands at 700 and 800 cm^{-1}, and the aliphatic part at 3,000–2,800 cm^{-1} shows very strong C–H valence vibrations (Berger and Sicker 2009).

Tewari *et al.* (2011) review molecular spectroscopic studies on theobromine and related alkaloids. Vibrational assignments are given.

Caffeine may be obtained either from green tea leaves (*Camellia sinensis* L., Theaceae) or from green coffee beans (*Coffea arabica* L., Rubiaceae). The IR spectrum (Figure 6.24) shows very clearly separated C=O valence vibrations and very strong, very clearly separated sp^2 and sp^3 C–H valence vibrations, as Berger *et al.* (2009) communicate.

The mild and lasting stimulant theobromine is isolated from cocoa beans of the cacoa tree (*Theobroma cacoa* L., Sterculiaceae). It is a diuretic, a myocardial stimulant, and a vasodilator, and is used for treating asthma. In its IR spectrum, Berger *et al.* (2009) find the N–CH$_3$ asymmetric stretch at 1,000 cm^{-1}. At 1,600 cm^{-1}, the C=C band is very strong.

Berger and Sicker (2009) study piperine from black pepper (*Piper nigrum* L., Piperaceae). In its IR spectrum, they find at 1,250 cm^{-1} a strong C–O–C vibration of the dioxolane ring. The C=O valence vibrations are very clearly separated from different C=C bands, both are very strong, whereas the sp^2 and sp^3 C–H valence vibrations are of medium strength.

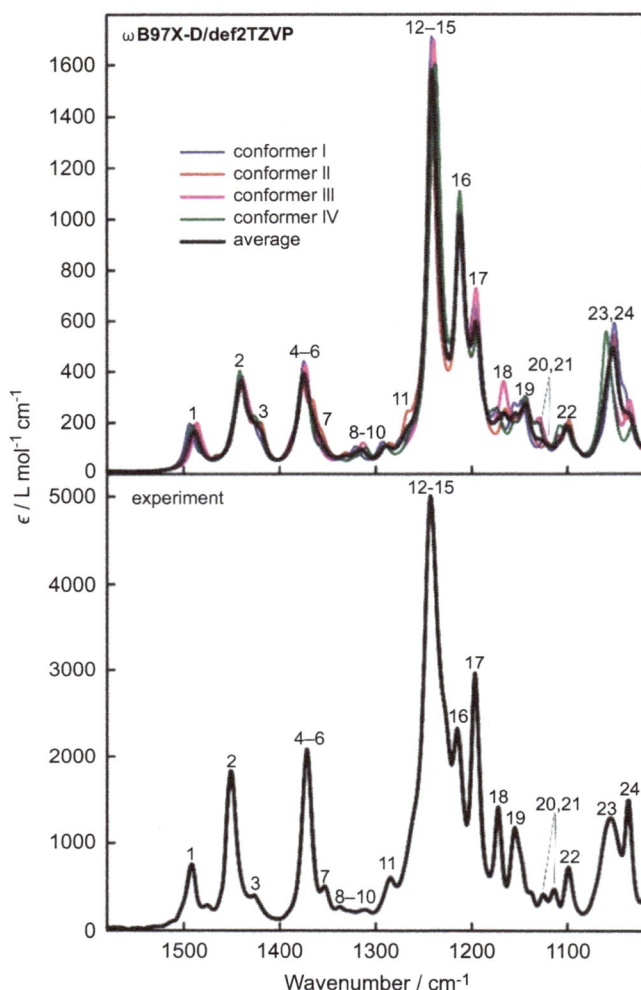

Figure 6.21: Experimental (bottom) and calculated [top, ρB97X-/def2TZVP/conductor-like polarizable continuum model (CPCM)] infrared (IR) spectra of heroin (Králík *et al.* 2020). Copyright ©2020 John Wiley and Sons, with permission.

Cytisine, the strong poison from the seeds of the golden chain tree (*Laburnum anagyroides* FABR., Fabaceae), shows sharp and split N–H valence vibrations at 3,300 cm^{-1}, amide C=O band at 1,650 cm^{-1} and 1,570 cm^{-1}, and very weak sp^2 C–H valence vibrations at 3,100–3,000 cm^{-1} followed by sp^3 aliphatic C–H stretching at 2,950–2,750 cm^{-1} (Berger and Sicker 2009).

Berger and Sicker (2009) isolate galanthamine from the bulbs of daffodils (*Narcissus pseudonarcissus* L., Amaryllidaceae); it is used to treat poliomyelitis in children. The O–H valence band of galanthamine is found at 3,300–3,400 cm^{-1} in the IR spectrum, the C–H

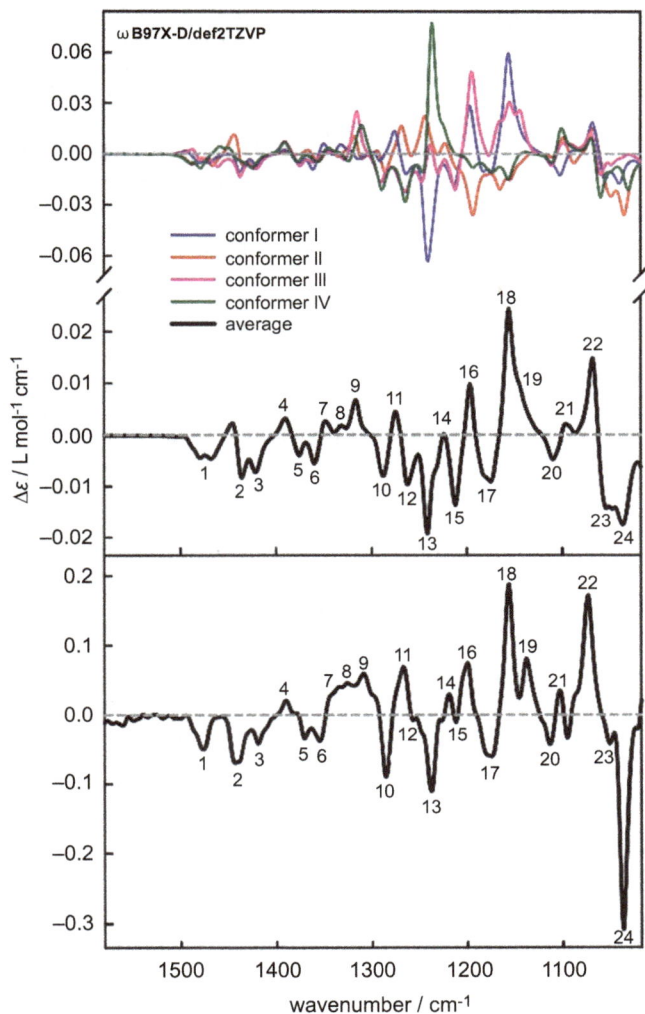

Figure 6.22: Experimental (bottom) and calculated [top, ωB97X-D/def2TZVP/conductor-like polarizable continuum model (CPCM)] vibrational circular dichroism (VCD) spectra of heroin (Králík *et al.* 2020). Copyright ©2020 John Wiley and Sons, with permission.

stretching vibration gives rise to a broad, strong absorption at 3,050–2,800 cm^{-1}, and two absorptions near 1,600 cm^{-1} are attributed to the stretching vibration of the C=C moiety. The strongest band of the spectrum at 1,430 cm^{-1} is attributed to C–H deformations.

New medicinal compounds are looked for to treat Alzheimer's disease. As it is speculated that this disease is connected to inflammatory processes, new natural products are tested for *anti*-acetylcholinesterase and *anti*-inflammatory activities. As a folk medicine in mainland China, the whole plant of *Zephyranthes candida* (LINDL.)

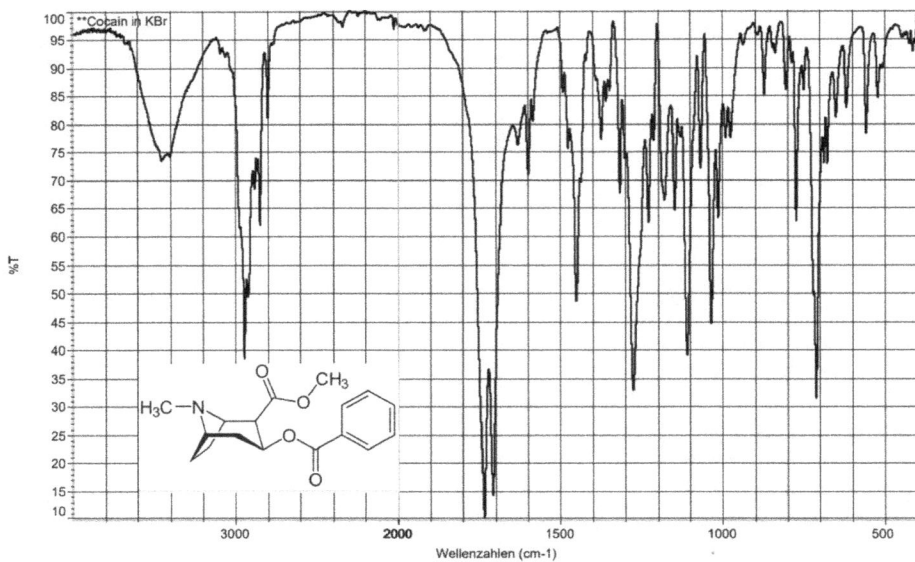

Figure 6.23: IR spectrum of cocaine (Steinke *et al.* 2013). Copyright ©2013 WILEY-VCH Verlag GmbH & Co. KgaA, with permission.

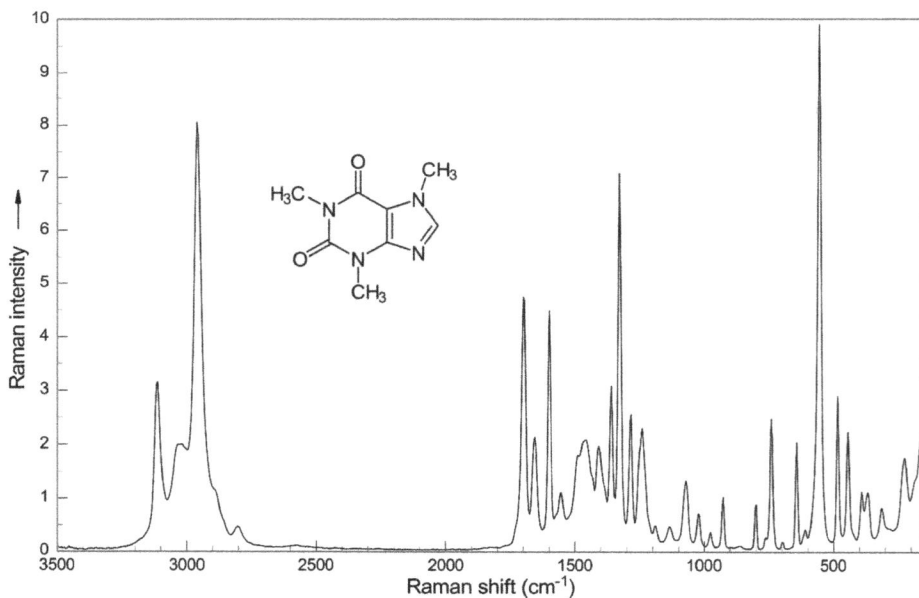

Figure 6.24: Raman spectrum of caffeine (λ_{exc} = 1,064 nm, 270 mW; data courtesy of Bruker Optik GmbH, Ettlingen).

has been used to treat infantile convulsions and epilepsy, suggesting the presence of alkaloids with potent effects on the central nervous system. Galanthamine, plicamine, and secoplicamine alkaloids (Figure 6.25) are extracted from the plant by Zhan *et al.* (2016). Spectral data ($[a]_D^{25}$, UV, ECD, IR (KBr), ^{1}H NMR (400 MHz), ^{13}C NMR, and HRESIMS) are given as numbers in the main text and as spectra in the supporting information, which also contains single-crystal X-ray diffraction data. As an example, the IR spectrum of 9-de-*O*-methyl-11β-hydroxygalanthamine (3) (colourless oil) is given in Figure 6.26.

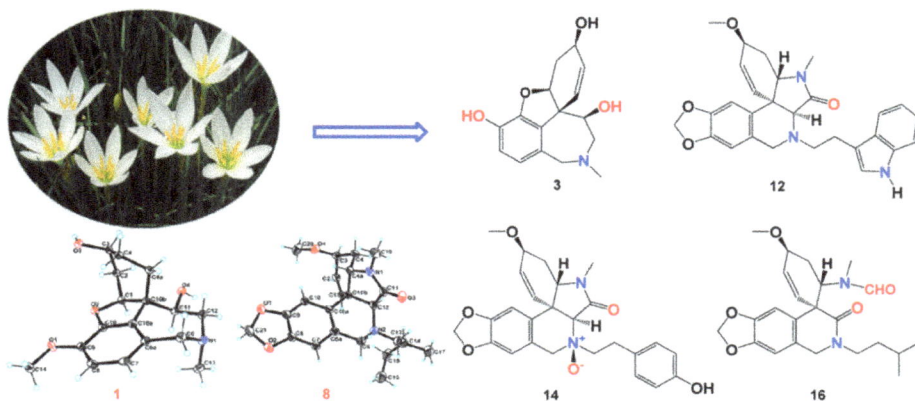

Figure 6.25: Sixteen new alkaloids belonging to the galanthamine (1–6), plicamine (7–14), and secoplicamine (15 and 16) classes, together with eight known analogues (17–24), were isolated from *Zephyranthes candida* (Zhan *et al.* 2016). Copyright ©2016 American Chemical Society, with permission.

In a review about industrial application of Raman spectroscopy, Pinzaru *et al.* (2004) discuss the alkaloids nicotine, cocaine, and papaverine hydrochloride.

Heß *et al.* (2016) and Sicker *et al.* (2019) report the IR spectrum of pseudopelletierine. The compound may be isolated from the root bark of the pomegranate tree, *Cortex punica granatum* L. (Lythraceae).

From the seeds of the autumn crocus (*Colchicum autumnale* L., Colchicaceae), colchicine, a mitotic poison, may be isolated. Appun *et al.* (2014) publish an IR spectrum (Figure 6.27) of the compound. Siddiqui *et al.* (2009) give a complete analysis of the IR and Raman spectra using *ab initio* and DFT. A PED analysis is also given.

Drosky *et al.* (2015) and Sicker *et al.* (2019) isolate capsaicin (the "hot" compound from peppers and chili) from *Capsicum frutescens* L. (Solanaceae) and publish the IR spectrum of the compound. El-Kaaby *et al.* (2016) detect capsaicin from callus and seedling of *Capsicum annuum* L. *in vitro* and provides assignments of the IR bands: N–H stretch at 3,315 cm^{-1}, aliphatic C–H stretch at 2,926 cm^{-1} and 2,864 cm^{-1}, C=O stretch (amide I) at 1,633 cm^{-1}, aromatic C=C stretch at 1,556 cm^{-1}, aromatic out-of-plane CH bending at 804 cm^{-1}, N–H bending and C–N stretch (amide II) at 1,514 cm^{-1}, asymmetric C–O–C stretch at 1,278 cm^{-1}, and C–O stretch at 1,203 cm^{-1}.

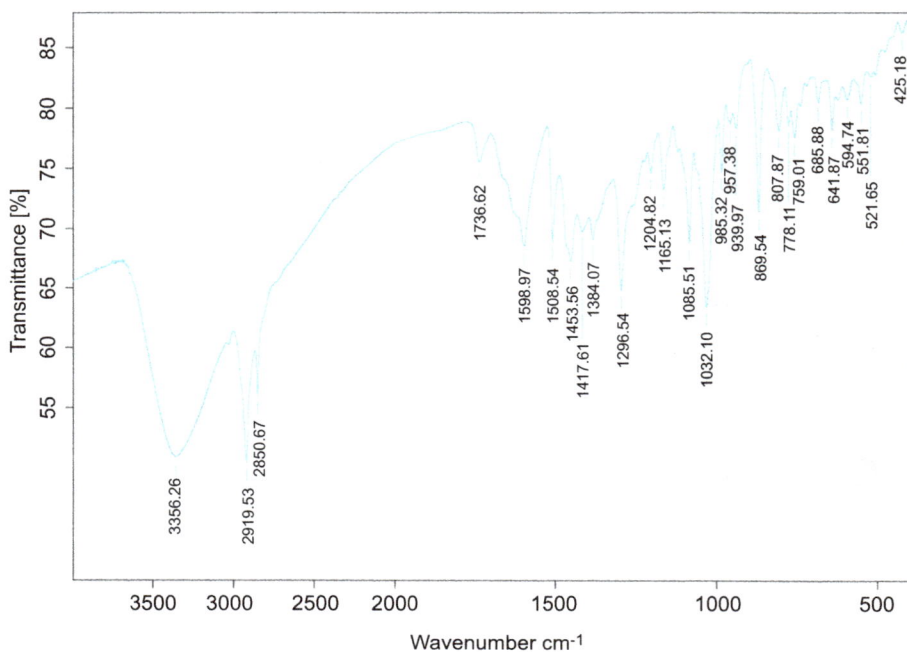

Figure 6.26: IR spectrum of 9-de-*O*-methyl-11β-hydroxygalanthamine (3) in KBr (Zhan *et al.* 2016). Copyright ©2016 American Chemical Society, with permission.

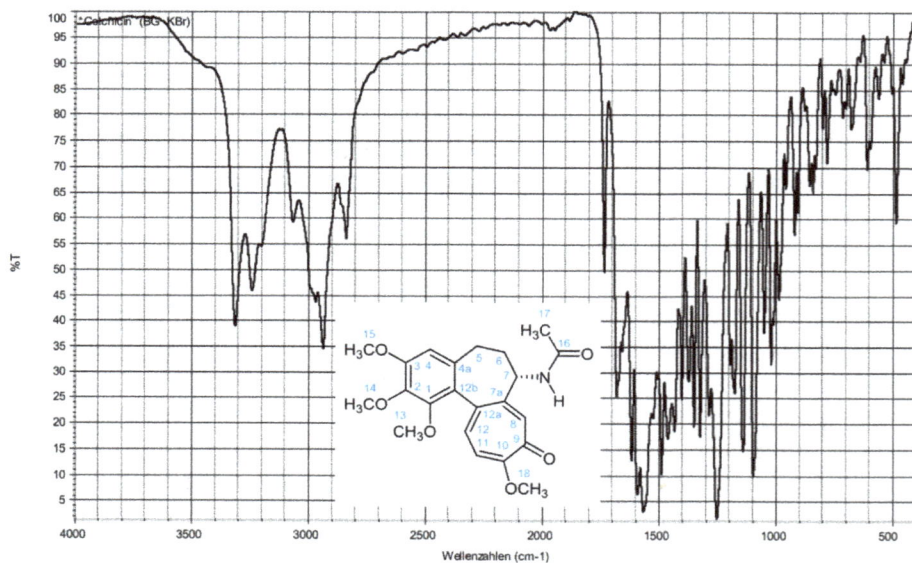

Figure 6.27: IR spectrum of colchicin (Appun *et al.* 2014). Copyright ©2014 WILEY-VCH Verlag GmbH & Co. KgaA, with permission. For a detailed band analysis see Siddiqui *et al.* (2009).

6.6 Lipids

Near-IR Raman spectra of human brain lipids are studied by Krafft *et al.* (2005). The Raman spectra of the major brain lipids (phosphatidylcholine, phosphatidylethanol-amine, phosphatidylserine, phosphatidylinositol, phosphatidic acid, sphingomyelin, galactocerebrosides, gangliosides, sulfatides, and cholesterol) and the minor lipids (cholesterol ester and triacylglycerides) were excited by a 785 nm laser.

Czamara *et al.* (2015) publish a review on lipids. Saturated and unsaturated fatty acids have been discussed in Section 5.2.6. The authors also review triacylglycerols, cholesterol, cholesteryl esters, and phospholipids.

6.7 Steroids, terpenes, and terpenoids

In a series of papers, Jones and co-workers, starting as early as 1948, apply IR spectrometry to the analysis of steroid structure.

Jones *et al.* (1948) characterize the carbonyl (Figure 6.28) and other functional groups in steroids by IR spectrometry.

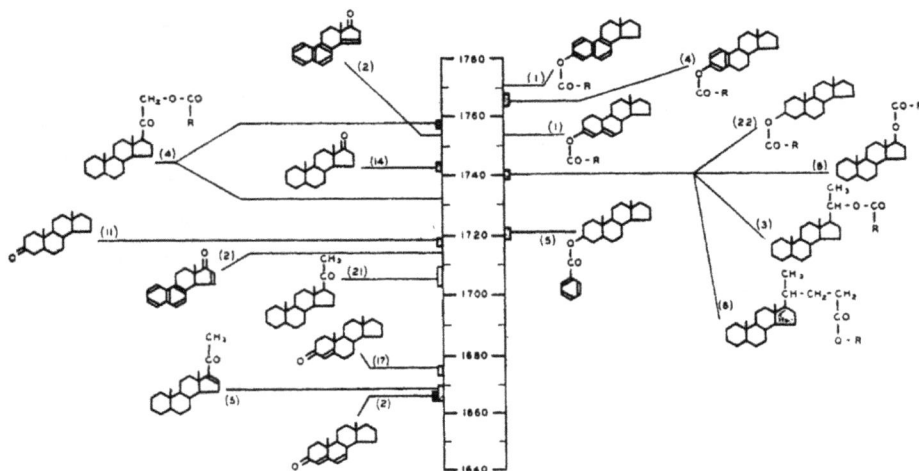

Figure 6.28: Carbonyl stretching bands in the infrared spectra of steroids (Jones *et al.* 1948). Copyright ©1948 American Chemical Society, with permission.

Jones *et al.* (1949) apply IR spectrometry to steroid structure and metabolism, using deuteration of the compounds and microtechniques.

Jones *et al.* (1957, 1959) study the carbonyl stretching bands in the IR spectra of unsaturated lactones.

The Raman spectra of steroids are studied by Schrader and Steigner (1970). As an example of the complexity of steroid spectra, the Raman spectra of estrone by Hoffmann (Figures 6.29 6.31 and Figure 6.32 for comparison) are given. Vedad *et al.* (2018) in the supplement to their paper give a complete DFT-based assignment of the vibrational bands of estrone, estradiol, estriol, and ethynylestradiol.

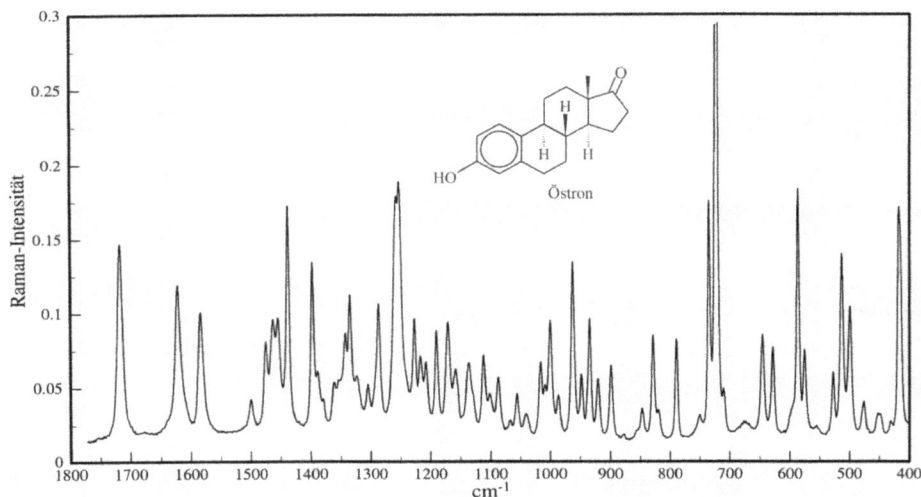

Figure 6.29: Raman spectrum of estrone in the crystalline state (Hoffmann, unpublished).

Minaeva *et al.* (2008) report the vibrational spectra of the steroid hormones estradiol and estriol, calculated by DFT using the B3LYP/6-31G** approximation. The roles of low-frequency vibrations in hormone–receptor interaction and in energy transfer processes are discussed.

Brandán (2019) is reporting a structural and vibrational spectroscopic study of the steroids equilenin, equilin, and estrone. Performing DFT calculations of the molecules in the gas phase at the B3LYP/6-31G* level, she gives a complete assignment of their vibrational spectra. The experimental Raman and IR spectra of the solids are shown as well.

Other examples of steroid Raman spectra are given for two hormones, in Figure 6.33 for cortisone and in Figure 6.34 for progesterone.

Berger and Sicker (2009) isolate cnicin from blessed thistle leaves (*Cnicus benedictus* L., Asteraceae). In the IR spectrum, two carbonyl bands at 1,760 and 1,700 cm^{-1} and two C=C bond vibrations at 1,660 and 1,620 cm^{-1} are found.

Abietic acid is isolated from colophony of pine trees (*Pinus silvestris*, Pinaceae) by Berger and Sicker (2009). The carboxylic group of the acid shows a broad IR band of O–H absorptions, overlapping with the aliphatic C–H vibration band at 2,950 cm^{-1}. At 2,100 cm^{-1}, a small sharp band of an overtone vibration may be observed. The C=O

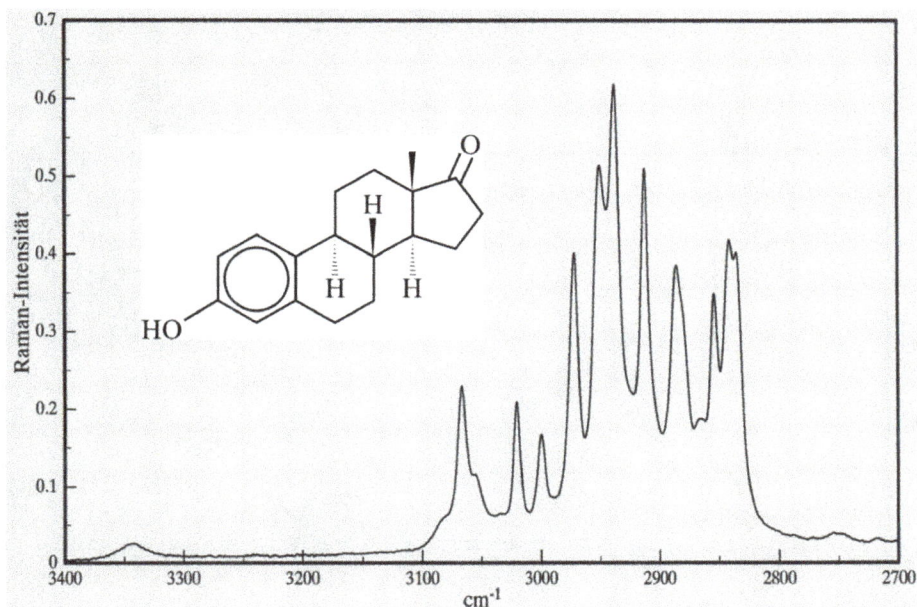

Figure 6.30: Raman spectrum of estrone in the crystalline state (Hoffmann, unpublished).

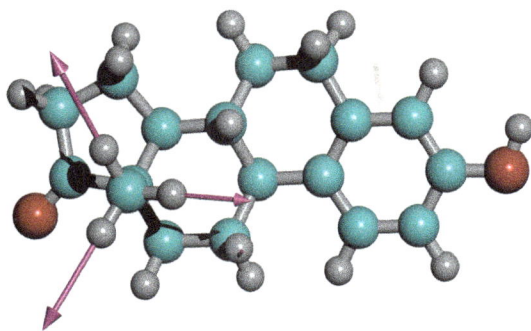

Figure 6.31: Methyl symmetric stretching vibration of estrone calc. (DFT, BPW91, 6-311G*) 3067.7 cm^{-1} exp. 2,880 cm^{-1} (Hoffmann, unpublished).

band at 1,700 cm^{-1} is broad and overlaps with the C=C vibration, which is only barely seen as a shoulder at 1,600 cm^{-1}.

The triterpenoid betulinic acid (BA) is effective as an HIV inhibitor. The Raman technique is used by Pinzaru *et al.* (2002) for evidencing betulin in birch bark. Raman and IR spectra of BA (3β-hydroxy-20(19)-lupaen-28-oic acid) and of natural birch bark (Figure 6.35) are recorded and discussed.

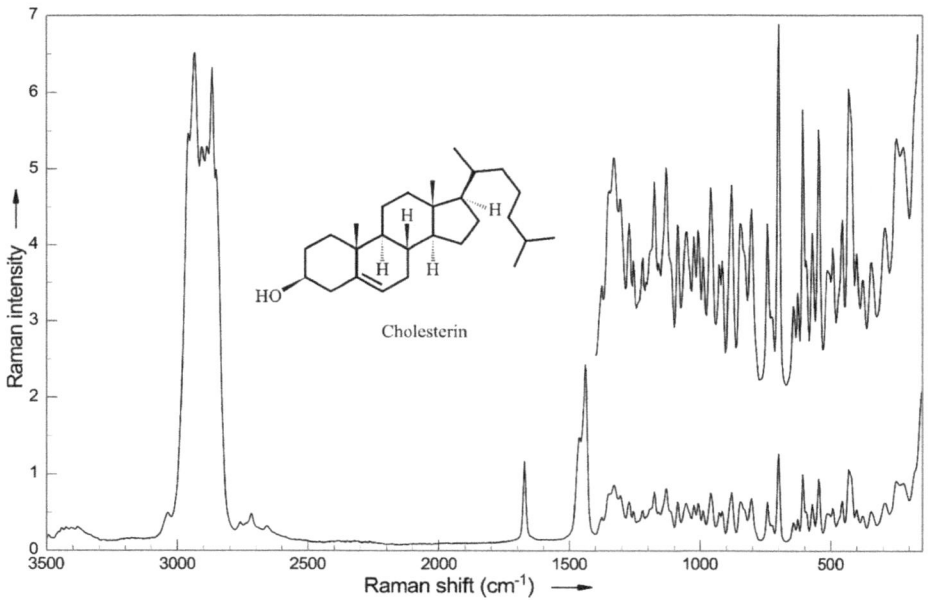

Figure 6.32: Raman spectrum of cholesterin (λ_{exc} = 1,064 nm, 270 mW; data courtesy of Bruker Optik GmbH, Ettlingen).

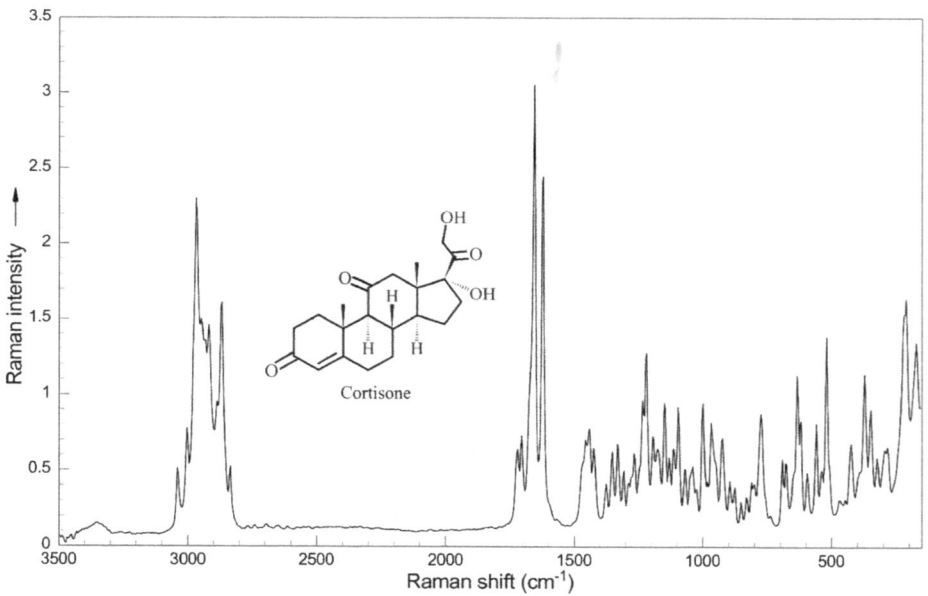

Figure 6.33: Raman spectrum of cortisone (λ_{exc} = 1,064 nm, 270 mW; data courtesy of Bruker Optik GmbH, Ettlingen).

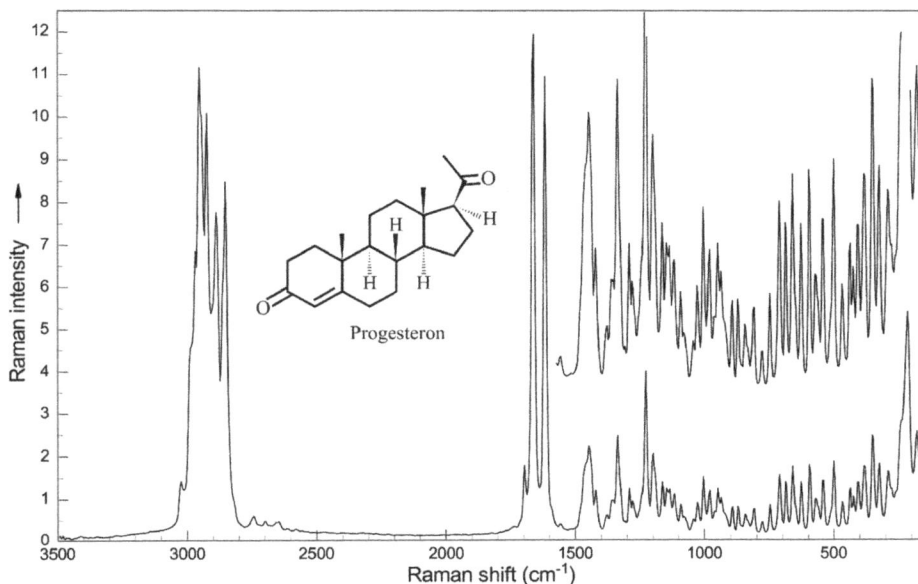

Figure 6.34: Raman spectrum of progesterone (λ_{exc} = 1,064 nm, 270 mW; data courtesy of Bruker Optik GmbH, Ettlingen).

Betulinic acid from plane tree bark (*Platanus hybrida* L., Platanaceae) is studied by Berger and Sicker (2009). IR bands of O–H occur around 3,450 cm^{-1}, aliphatic and olefinic C–H stretching, C=O at 1,700 cm^{-1}, and C=C stretching at 1,650 cm^{-1}. A sharp band of medium intensity from end-located olefinic C–H valence vibrations can be found at 3,100 cm^{-1}.

The terpenoid thujone can be found in nature as both enantiomers: (+)-β-thujone from wormwood (*Artemisia absinthium* L.) and (–)-α-thujone from common sage (*Salvia officinalis*). Both enantiomers show the same IR spectrum: strongest band of C=O at 1,740 cm^{-1} and a strong valence band of aliphatic C–H stretching at 2,900–3,000 cm^{-1}. From the three-membered ring, two small but significant C–H stretching bands above 3,000 cm^{-1} may be observed, as reported by Berger and Sicker (2009).

Raman spectroscopic discrimination of the four estrogens estrone (E1), estradiol (E2), estriol (E3), and ethynylestradiol is achieved by Vedad *et al.* (2018). The components of oestrogen mixtures could be quantified using the intensities of identifying Raman bands. Bands at 581 cm^{-1} for estrone, 546 cm^{-1} for estradiol, 762 cm^{-1} for estriol, and 597 cm^{-1} for ethynylestradiol are compared and normalized against the intensity of a common peak at 783 cm^{-1}, resulting in an error of 20%.

Chen *et al.* (2018) analyse the content of diethylstilbestrol in chicken parts. Oestrogens in human food can have negative health effect, so their detection is very important. Using SERS coupled with multivariate analysis, they were able to detect the synthetic estrogen.

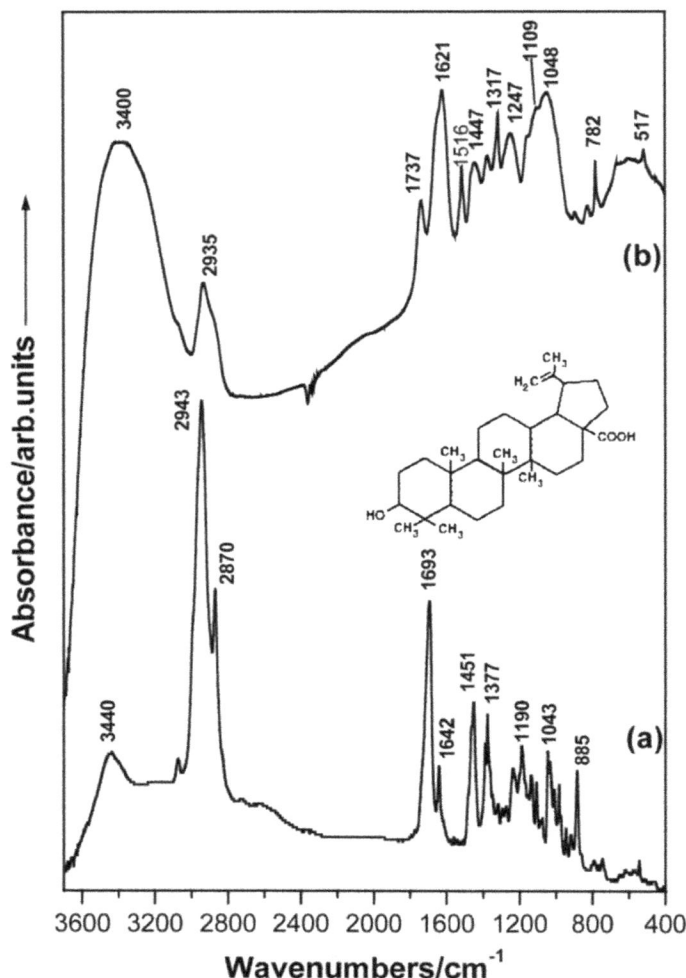

Figure 6.35: FT-IR spectra of betulinic acid (a) and of the birch bark (b) (Pinzaru *et al.* 2002). Copyright ©2002 Elsevier B.V., with permission.

From the dried leaves of the annual mugwort *Artemisia annua* L., artemisinin, a sesquiterpenoid used for the treatment of malaria, can be extracted. Its IR spectrum by Eckhardt *et al.* (2016) and Sicker *et al.* (2019) shows the band of C–H stretching of sp^3-hybridized carbon and the lactone band at 1,730 cm^{-1}.

Diosgenin from the root of the Mexican yams (*Dioscorea mexicana* SCHEIDW., Dioscoreaceae) is known as the precursor for the synthesis of hormonal contraceptives. Rudo *et al.* (2015) and Sicker *et al.* (2019) publish its IR spectrum in KBr (Figure 6.36).

From the camphor tree (*Cinnamomum camphora* (L.) SIEB., Lauraceae) oil, camphor can be obtained by careful distillation. The IR spectrum is dominated by

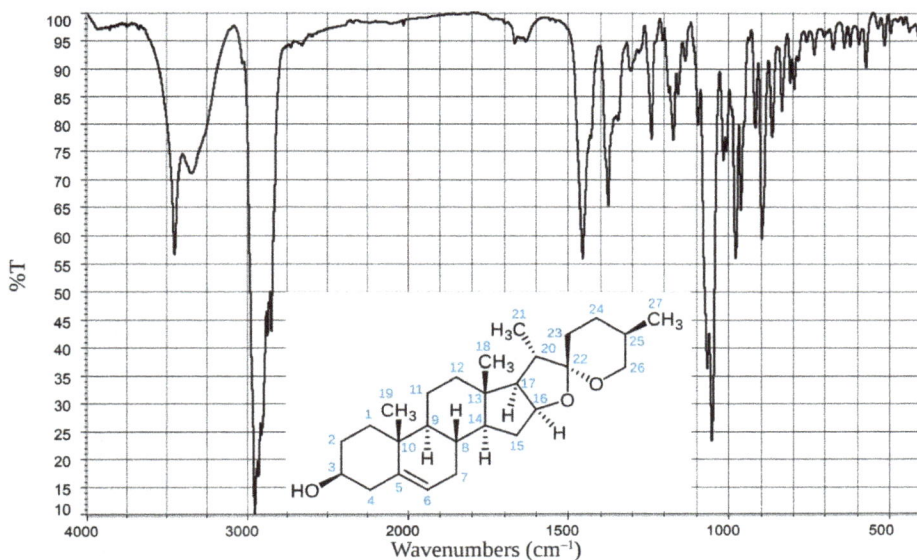

Figure 6.36: IR spectrum of diosgenin in KBr (Rudo *et al.* 2015), Copyright ©2015 WILEY-VCH Verlag GmbH & Co. KGaA, with permission.

the C=O vibrational band at 1,737 cm^{-1}, as found out by Steinke *et al.* (2013) and Sicker *et al.* (2019).

Cantharidin, the poison of the "Spanish fly" (the beetle *Lytta vesicatoria*, L. Meloidae), is extracted from the beetle by Rudo *et al.* (2013) and Sicker *et al.* (2019). In the IR spectrum, the most prominent bands are the double C=O bands at 1,780 cm^{-1} and 1,850 cm^{-1}.

The terpenoid friedelin can be extracted from the bark of the cork-oak (*Quercus suber* L., Fagaceae). Its IR spectrum, published by Seupel *et al.* (2015) and Sicker *et al.* (2019) (Figure 6.37), lacks a characteristic profile. Naturally, a strong C–H stretching band for sp^3-hybridized C-atoms is to be found. The strong C=O stretching band at 1,720 cm^{-1} is very sharp. The broad band at 1,460 cm^{-1} can be attributed to a C–H deformation.

From frankincense, the resin of Arabian olibanum tree (*Boswellia sacra* FLUECK, Burseraceae), the terpenoid 3-*O*-acetyl-11-keto-β-boswellic acid is obtained in pure form. Its IR spectrum shows the O–H band at 3,400 cm^{-1}, a split C=O band at 1,740/1,700 cm^{-1}, and a C=C band at 1,600 cm^{-1}, as is communicated by Leutzsch *et al.* (2013) and Sicker *et al.* (2019).

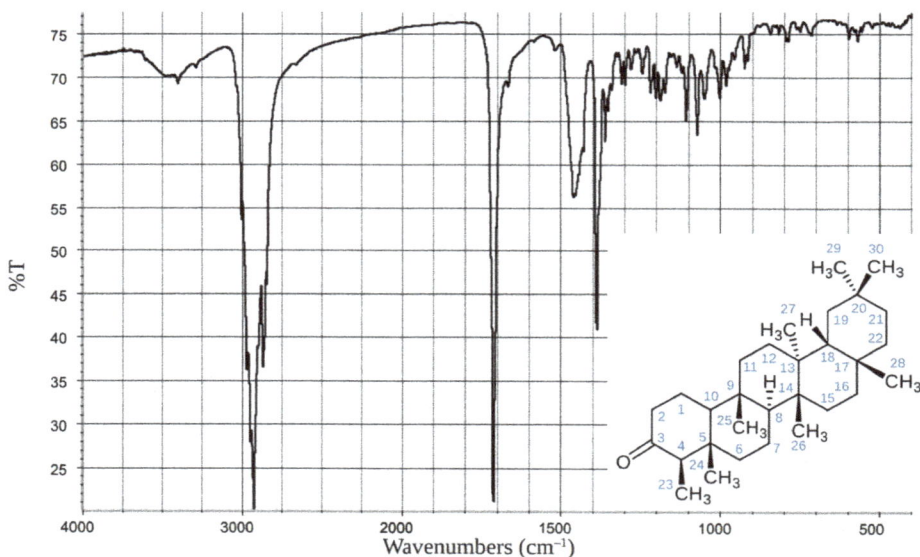

Figure 6.37: IR spectrum of friedelin (Seupel *et al.* 2015), Copyright ©2015 WILEY-VCH Verlag GmbH & Co. KgaA, with permission.

6.8 Dyes

The neoflavanoid dye brazileine is extracted from Pernambuco wood (*Caesalpina ecinata* LAM., Caesalpinia) by Berger and Sicker (2009). In the IR spectrum, it shows a split O–H valence band at 3,300–3,400 cm^{-1} from its three hydroxyl groups, IR methylene below 3,000 cm^{-1}, a weak C=O band at 1,750 cm^{-1} split from its *ortho*-OH group, and an intense C=C band at 1,600 cm^{-1}.

Berger and Sicker (2009) isolate the blue dye stuff indigo from dyer's woad (*Isatis tinctoria* L., Brassicaceae). The main bands shown in the IR spectrum are: a sharp N–H valence band at 3,200 cm^{-1}, very small C–H bands at 3,100 cm^{-1}, and a strongly split amide band at 1,620 cm^{-1}.

The yellow food dyestuff and spice curcumin is isolated by Berger and Sicker (2009) from turmeric (*Curcuma longa* L., Zingiberaceae). In its IR spectrum, we see among other bands the broad band of phenolic OH, and two absorptions near 1,630 cm^{-1} and at 1,600 cm^{-1} from aromatic and olefinic C=C vibrations.

The yellow dyestuff capsanthin acts as a protective filter for chlorophyll. Berger and Sicker (2009) isolate it from sweet pepper powder (*Capsicum annuum* L., Solanaceae). The O–H stretching band occurs at 3,400 cm^{-1}; further, a minor C–H band from sp^2-hybridized carbon and strong C–H stretching bands from sp^3-hybridized carbon atoms are observed. The carbonyl band at 1,700 cm^{-1} and the C=C band at 1,600 cm^{-1} are split into three sharp bands each, caused by the long all-*trans* conjugated chain.

Onocerin, a diuretic compound from the leaves of the spiny restharrow (*Ononis spinosa* L., Fabaceae), is studied by Berger and Sicker (2009). A C=C stretching vibration shows up at 1,650 cm^{-1} in its IR spectrum. A sharp band of medium intensity from terminal olefinic C–H valence vibrations can be found at 3,100 cm^{-1}.

Aleuritic acid from shellac, the excretions of lac insects (*Kerria lacca*, Coccoidae), is obtained by Berger and Sicker (2009) by saponification of its esters. Prominent bands in the IR spectrum are at 3,250 cm^{-1} and 1,700 cm^{-1} for the O–H and C=O vibrations of the carboxylic acid group, and a band of the C–H valence vibration of sp^3-hybridized carbon atoms splits and ranges down to 2,900 cm^{-1}.

Thymoquinone, a yellow compound from the oil of the black caraway (*Nigella sativa* L., Ranunculaceae), is studied by Spiller *et al.* (2014) and Sicker *et al.* (2019). The IR spectrum shows C–H stretching at sp^2 and sp^3 carbons at 3,100–2,900 cm^{-1}. Additionally, we see splitting of the C=O band at 1,660 cm^{-1} by Fermi resonance, as it is often observed in *p*-benzoquinone derivatives. Neighbouring C=C groups are the cause of the observed frequency shift.

The yellow dyestuff berberine chloride may be obtained from the bark of the common barberry (*Berberis vulgaris* L., Berberidaceae). Its IR spectrum by Eckhardt *et al.* (2017) and Sicker *et al.* (2019) shows at 3,400 cm^{-1} the O–H stretching band of four molecules of crystal water, C–H stretching bands from sp^3- and sp^2-hybridized C-atoms, and a sharp band at 1,600 cm^{-1} from C=C stretching of aromatic rings.

From dried cochineal (*Dactylopius coccus* Costa, Dactylopiidae), the anthraquinone derivative carminic acid can be obtained, which is used as E120 in lipsticks and food stuffs. The IR spectrum by Schulze *et al.* (2013) and Sicker *et al.* (2019) is characterized by a huge, broad O–H band (Figure 6.38); on its shoulder the C–H bands are riding at around 2,940 cm^{-1}. The C=O bands of the anthraquinone unit and the side chain which should be at 1,665 cm^{-1} (Lehmann *et al.* 1979) cannot be assigned without doubt, as there is a broad band with peaks at 1,520–1,700 cm^{-1}. DFT calculations of the IR and Raman spectra of anthraquinone dyes and lakes together with experimental values may be found in the paper of Pagliai *et al.* (2018).

The natural dye safflomin A can be isolated from flowers of the safflower (*Carthamus tinctorius* L., Asteraceae). A strong, broad band at 3,400 cm^{-1} can be attributed to the vibrations of the 12 O–H groups (Sicker *et al.* 2019).

Klaproth *et al.* (2016) and Sicker *et al.* (2019) purify chlorophyll a from frozen spinach leaves (*Spinacia oleracea* L., Amaranthaceae). The IR spectrum of the compound is shown in Figure 6.39.

As the sample contains residual humidity, a broad band between 3,500 and 3,200 cm^{-1} may be observed. A complex signal from C–H vibrations with a small band of sp^2-hybridized C-atoms is detected at 3,050 cm^{-1}. The main part may be found in the aliphatic region from 3,000 to 2,900 cm^{-1}. Two clearly different C=O bands at 1,750 and 1,700 cm^{-1} may be attributed to C=O group C-20 and the ester groups C-21^1 and C-3^3, whereas the fairly broad band at 1,620 cm^{-1} should belong to the many C=C double bonds. Assignments in the fingerprint region were published by Lutz and Mäntele (1991).

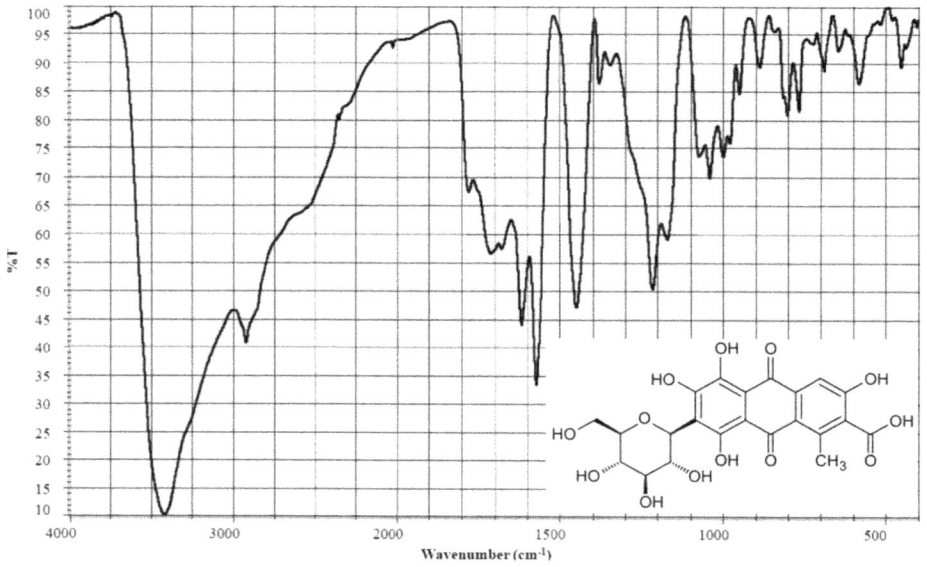

Figure 6.38: IR spectrum of carminic acid in KBr (Schulze *et al.* 2013; Sicker *et al.* 2019). Copyright ©2013 WILEY-VCH Verlag GmbH & Co. KGaA, with permission.

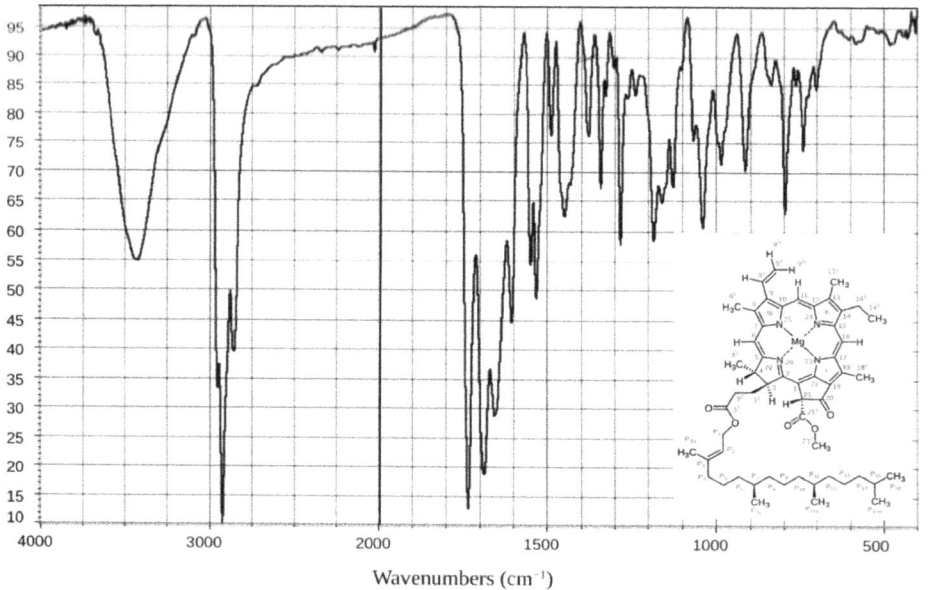

Figure 6.39: IR spectrum of chlorophyll a in KBr (Klaproth *et al.* 2016; Sicker *et al.* 2019). Copyright ©2019 WILEY-VCH Verlag GmbH & Co. KGaA, with permission.

Chlorophyll d is the major pigment of the cyanobacterium *Acaryochloris marina* MIYASHITA and CHIHARA (Acaryochloridaceae). Studies of the pigment using both Raman spectroscopy and DFT are published by Chen *et al.* (2004). They report observed Raman shifts (in cm^{-1}) at 77 K of chlorophyll together with the DFT (B3LYP/6-31*) calculated frequencies and assignments. The peaks of chlorophyll d (Figure 6.40) are compared with observed peaks of chlorophyll a, chlorophyll b, and bacteriochlorophyll a.

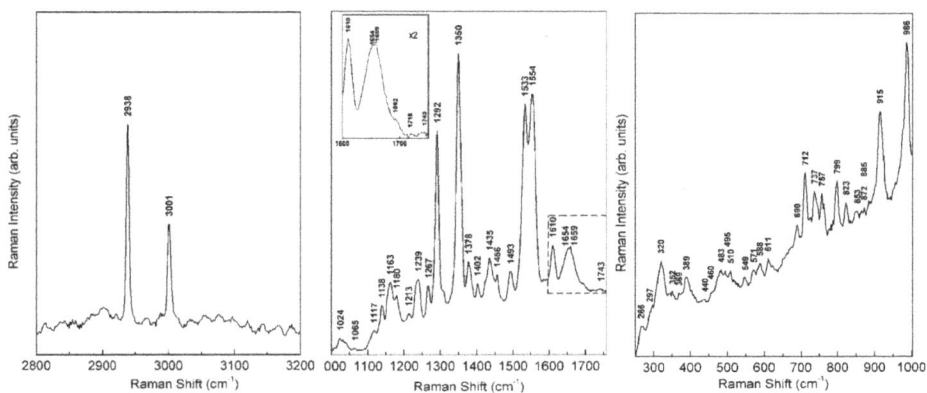

Figure 6.40: Raman shifts of chlorophyll d observed at 77 K (Chen *et al.* 2004). Copyright ©2004 Elsevier B.V., with permission.

Lutz and Mäntele (1991) publish a review of vibrational analysis of chlorophylls as a book chapter.

A density functional normal mode calculation of methyl bacteriochlorophyll a is reported by Ceccarelli *et al.* (2000). Antisymmetric $\nu C_a C_m$ ring modes are shown in Figure 6.41.

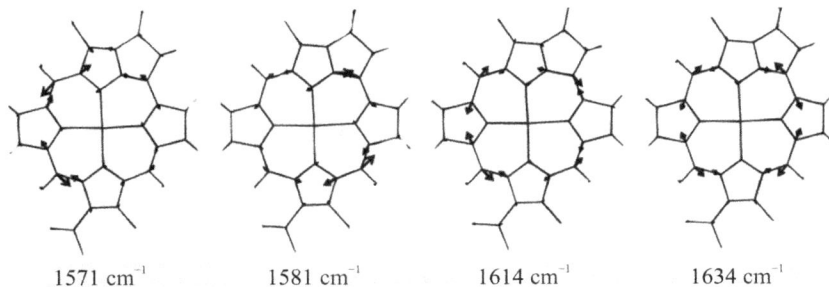

| 1571 cm^{-1} | 1581 cm^{-1} | 1614 cm^{-1} | 1634 cm^{-1} |

Figure 6.41: Antisymmetric $\nu C_a C_m$ ring modes of methyl bacteriochlorophyll a as computed by density functional theory (DFT) (Ceccarelli *et al.* 2000). Copyright ©2000 American Chemical Society, with permission.

Ramos *et al.* (2019) analyse extracts and epicuticular waxes from cashew (*Anacardium occidentale* L.) leaves from Amazon by FT-Raman and GC/MS. The leaves are used in traditional medicine for the treatment of gastrointestinal and skin diseases. The FT-Raman spectra show stretching bands of CH_2 as characteristic bands in epicuticular plant waxes.

A FT-Raman spectroscopic study of the surface of Norway spruce (*Picea abies* (L.) KARST.) needles is communicated by Křížová *et al.* (1999). The spectra from different sites and seasons are treated by cluster analysis.

Patchouli alcohol is the main component in the essential oil of the patchouli plant (*Pogostemon cablin* BENTH., Lamiaceae), used in perfumes and fragrances. As the O–H group in the tertiary alcohol is isolated, we find a strong, sharp peak at 3,500 cm^{-1} in its IR spectrum, as reported by Berger and Sicker (2009). The aliphatic C–H stretching bands occur below 3,000 cm^{-1} and the C–H bending bands below 1,500 cm^{-1}. The VCD of the compound is discussed in Section 18.1.

The diterpenes ginkgolides, isolated from the maidenhair tree *Ginkgo biloba* L., exhibit strong *anti*-inflammatory and neuroprotective properties. In a paper about ginkgolide synthesis, Hébert *et al.* (2022) report their IR spectra in the supporting information.

6.9 Carbohydrates

Wiercigroch *et al.* (2017) publish a review on vibrational spectra of carbohydrates, measuring and assigning FT-Raman and ATR FT-IR spectra of 14 carbohydrates in the solid state. The carbohydrates studied are the monosaccharides (D-(–)-ribose, 2-deoxy-D-ribose, L-(–)-arabinose, D-(+)-xylose, D-(+)-glucose, D-(+)-galactose, and D-(–)-fructose), the disaccharides ((D)-(+)-sucrose, D-(+)maltose, and D-(+)-lactose), the trisaccharide (D-(+)-raffinose), and the polysaccharides (amylopectin, amylose, and glycogen). The authors discuss the stretching vibrations of the O–H, C–H/CH_2, and C–O/C–C groups, and the deformational modes of the C–H/CH_2 and CCO groups.

The Raman spectra of the most important saccharides, the monosaccharide glucose and the disaccharide saccharose, are shown in Figures 6.42 and 6.43.

The distribution of the cyanogenic glycoside amygdalin in apricot (*Prunus armeniaca*, L. Rosaceae) seeds is studied by Raman microscopic imaging by Krafft *et al.* (2012). The subcellular distribution of amygdalin in thin apricot seed sections is probed by high-resolution Raman imaging with a step size of 2.5 μm, using the well-resolved band of the nitrile group near 2,245 cm^{-1}. Spectral contributions of amygdalin, lipids, and cellulose were identified.

D-Glucosamine hydrochloride may be isolated from the chitin of the shells of common shrimps (*Crangon crangon* L., Crangonidae). Berger and Sicker (2009) report its IR spectrum. It is dominated by the very large and broad band at 3,500 cm^{-1} of the four O–H groups, with many small C–H peaks riding on its flank around 3,000 cm^{-1}. The second largest band is found at 1,030 cm^{-1}, a sharp peak of C–O stretching.

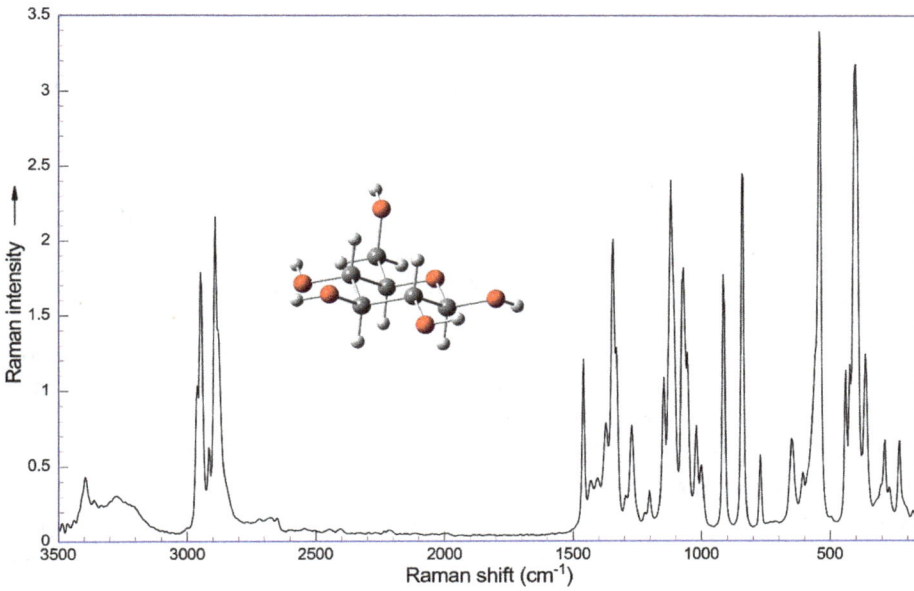

Figure 6.42: Raman spectrum of glucose (λ_{exc} = 1,064 nm, 270 mW; data courtesy of Bruker Optik GmbH, Ettlingen).

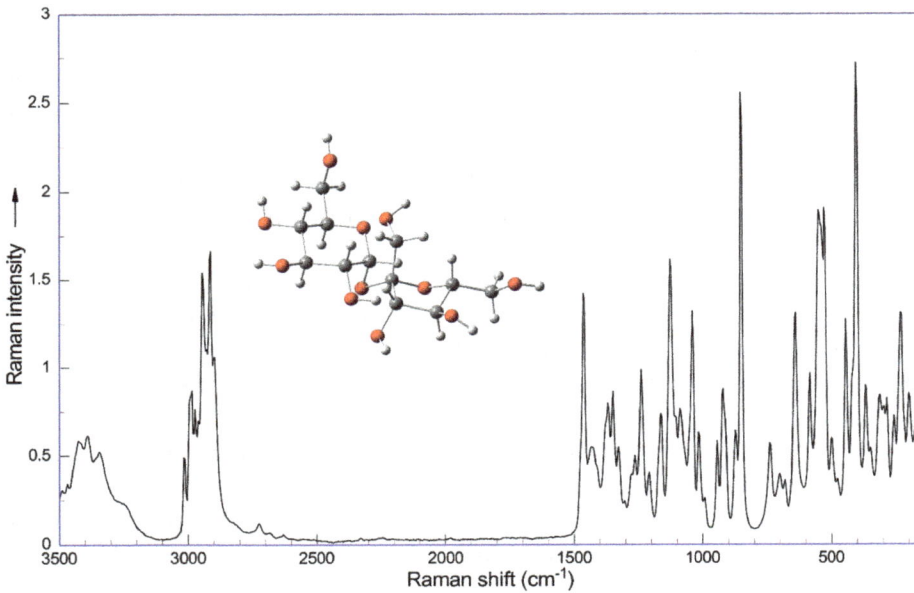

Figure 6.43: Raman spectrum of saccharose (λ_{exc} = 1,064 nm, 270 mW; data courtesy of Bruker Optik GmbH, Ettlingen).

The disaccharide D-lactose is obtained from cow's milk (*Bos taurus* L., Bovidae). The IR spectrum, published by Berger and Sicker (2009), shows the huge broad bands at 3,500 cm^{-1} of eight O–H groups. Small sharp signals from C–H are found at 2,900 cm^{-1}, and the C–O single bond vibrations contribute many bands in the region from 1,200 to 1,000 cm^{-1}.

The flavanone glycoside hesperidin from the peel of mandarin oranges (*Citrus sinensis* (L.) OSBECK, Rutaceae) is helpful for the integrity of blood vessels. Its IR spectrum is shown by Berger and Sicker (2009). IR bands from O–H and C=O are dominant, strong are aromatic C=C vibrations at 1,600 cm^{-1}, and weak aromatic overtones occur around 2,000 cm^{-1}.

The poisonous amygdalin from bitter almonds (*Prunus dulcis* WEBB var. *amara*, Roseaceae) shows in its IR spectrum a characteristic small sharp band at 2,200 cm^{-1} for the cyano group, as reported by Berger and Sicker (2009).

The red protein dye stuff lawsone, in use as hair colourant, is extracted from the dried leaves of the henna plant (*Lawsonia inermis* L., Lythraceae). Its IR spectrum with the characteristically split carbonyl stretching band at 1,680–1,640 cm^{-1} and the C=C stretching band at 1,580–1,600 cm^{-1} is shown by Berger and Sicker (2009). DFT calculations and IR studies on 2-hydroxy-1,4-naphthoquinone (lawsone) and its 3-substituted derivatives are published by Satoh *et al.* (2007). The trisaccharide raffinose can be found in the seeds of blue lupins (*Lupinus angustifolius* L., Fabaceae) and other leguminoses. Winkler *et al.* (2014) and Sicker *et al.* (2019) show the IR spectrum of the compound. As expected from the large number of OH groups in the molecule, the IR spectrum (Figure 6.44) of raffinose is dominated by O–H valence bands around 3,400 cm^{-1}. The C–H valence bands and the skeleton bands are of minor intensity. A few strong bands can be found in the O–C region around 1,100 cm^{-1}.

The glycoside fraxin from the bark of the ash tree (*Fraxinus excelsior* L.) is isolated by Drosky *et al.* (2014) and Sicker *et al.* (2019). The IR spectrum of fraxin in KBr is shown.

The natural sweetener stevioside is extracted from the dried leaves of *Stevia rebaudiana* BERTONI (Asteraceae) by Reuß *et al.* (2016) and Sicker *et al.* (2019). The authors report the O–H valence vibrations of the glucose moieties between 3,100 and 3,600 cm^{-1}, followed by the valence vibrations of the sp^3-hybridized C–H bonds in the region of 2,800–3,000 cm^{-1}. The C–H stretching vibration of the terminal C=CH$_2$ group is detected as a small shoulder at 3,000 cm^{-1}. The broad vibrational bands at 2,500 cm^{-1} are overtone and/or combination bands. The C–O stretching band of the esterified carboxylic acid group shows up at 1,750 cm^{-1}. The band at 1,640 cm^{-1} is identified as the C=C stretching band of the C=CH$_2$ group.

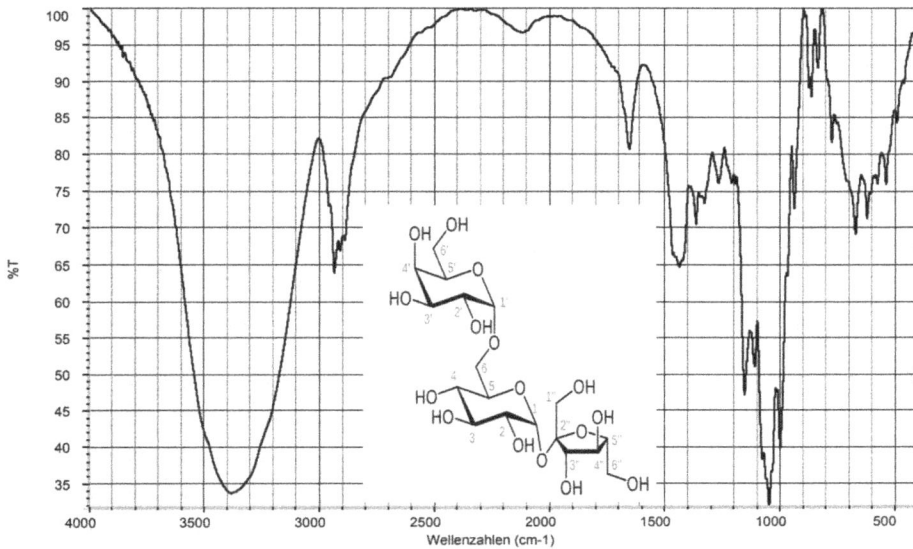

Figure 6.44: IR spectrum of raffinose in KBr (Winkler *et al.* 2014). Copyright ©2014 WILEY-VCH Verlag GmbH & Co. KGaA, with permission.

6.10 Nucleic acids and components

Nucleic acids consist of nucleic bases (adenine, cytosine, guanine, thymine, and uracil), carbohydrates (deoxyribose and ribose), and phosphate. The bases are attached to a monosaccharide. This combination is called a nucleoside. The nucleosides are then connected to phosphate, forming the nucleotides, and the resulting molecules are polymerized to the two strands of the double helix.

The Raman spectra of adenine, thymine, guanine, and uracil are shown in Figures 6.45–6.48.

The basis set dependence using DFT/B3LYP calculations to model the Raman spectrum of thymine is studied by Bielecki and Lipiec (2016) using 6-31$^+$(d,p), 6-31^{++}(d,p), and 6-311^{++}G(d,p) basis sets. Adding diffuse functions or polarized functions for small basis set or use of a medium or large basis set without diffuse or polarized functions is not sufficient to reproduce Raman intensities correctly.

The Raman spectrum of a thymidine crystal under low temperature (Figure 6.49) is studied by Freire *et al.* (2017). The analysis of the Raman spectra of the compound showed that the wavenumber positions of several bands shift at about 160 K. Freire *et al.* suggest the occurrence of a conformational modification due to the change of hydrogen bonds.

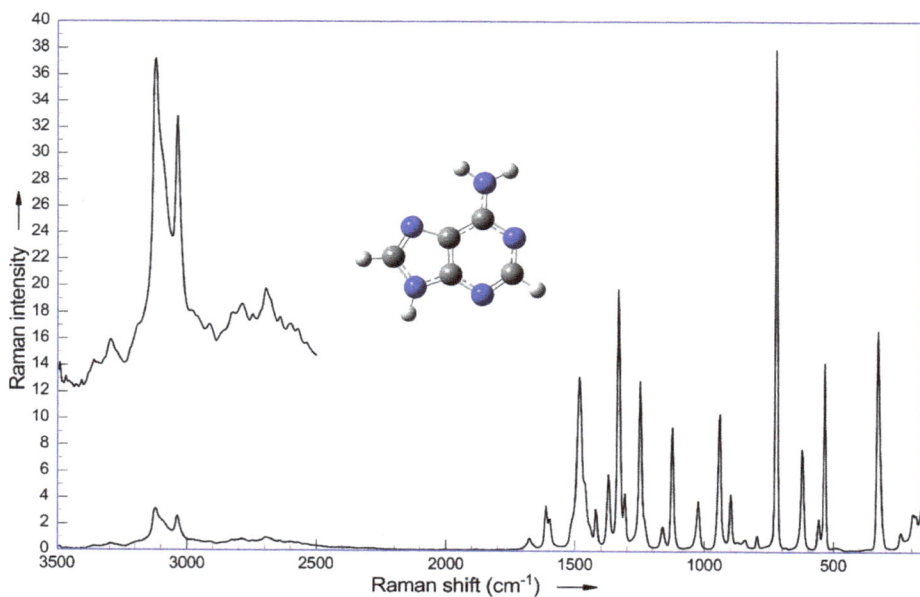

Figure 6.45: Raman spectrum of adenine (λ_{exc} = 1,064 nm, 270 mW; data courtesy of Bruker Optik GmbH, Ettlingen).

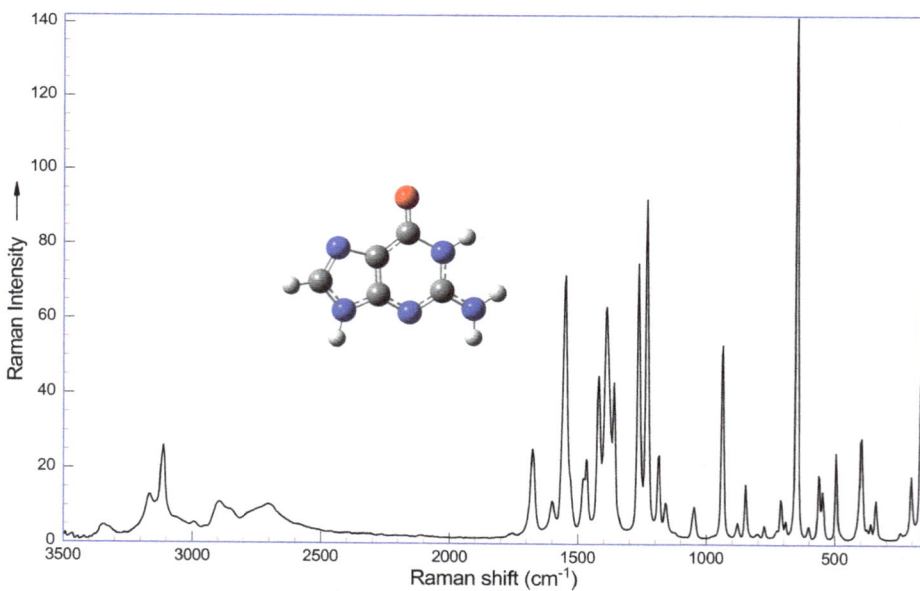

Figure 6.46: Raman spectrum of guanine (λ_{exc} = 1,064 nm, 270 mW; data courtesy of Bruker Optik GmbH, Ettlingen).

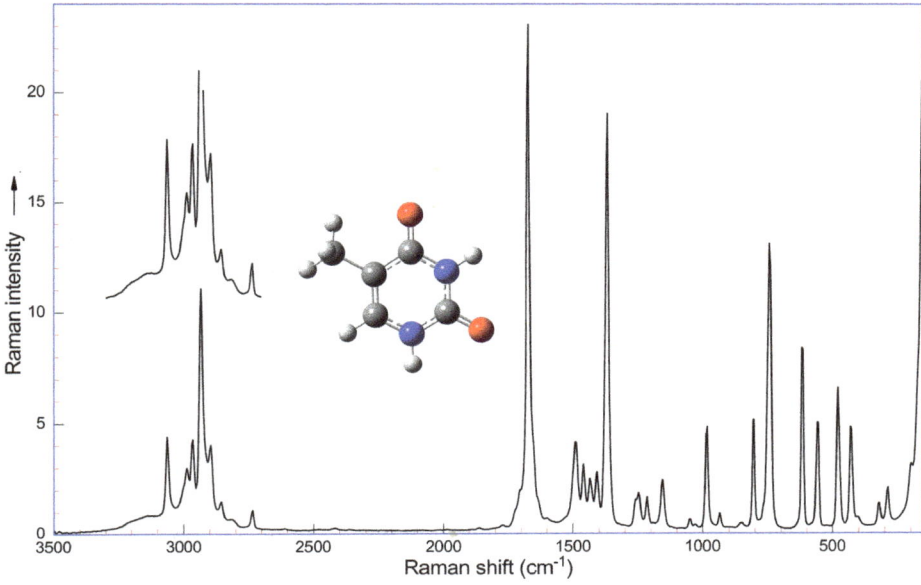

Figure 6.47: Raman spectrum of thymine (λ_{exc} = 1,064 nm, 270 mW; data courtesy of Bruker Optik GmbH, Ettlingen).

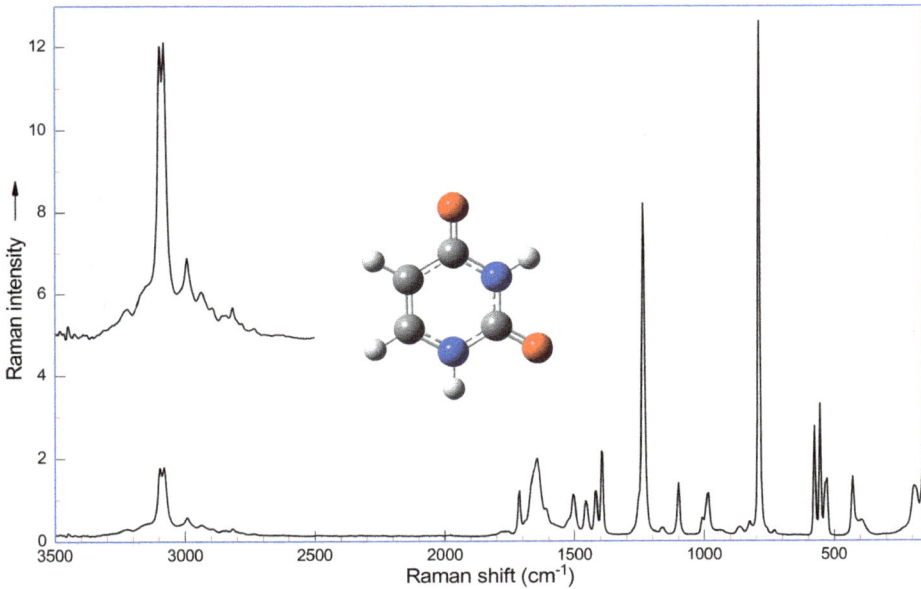

Figure 6.48: Raman spectrum of uracil (λ_{exc} = 1,064 nm, 270 mW; data courtesy of Bruker Optik GmbH, Ettlingen).

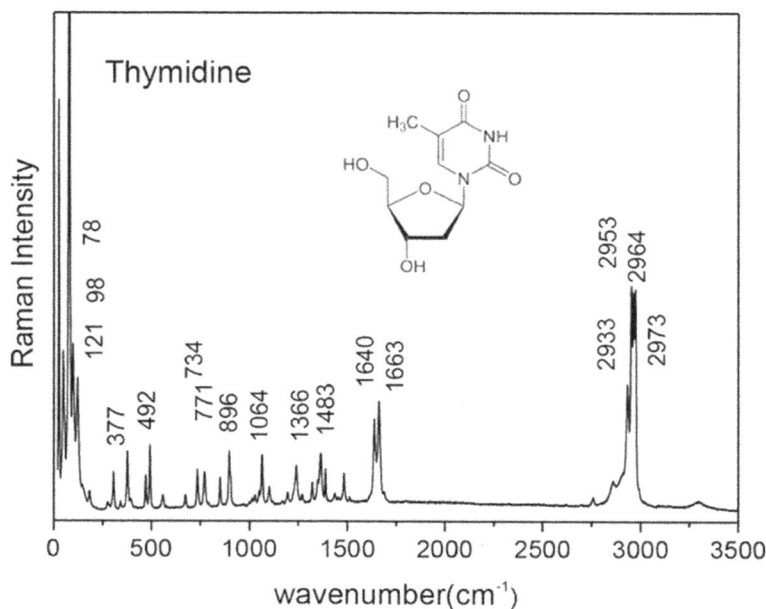

Figure 6.49: Raman spectrum of thymidine (Freire *et al.* 2017). Copyright ©2017 by the Author(s). Licensed by Creative Commons (CC 3.0).

6.11 Vitamins

Vitamins are a group of substances that the body cannot produce itself; there are 13 essential vitamins. Vitamins are needed for normal cell function, growth, and development. They can be grouped into fat-soluble vitamins (vitamins A, D, E, and K) and water-soluble vitamins (vitamin C and all the B vitamins).

In their report on industrial uses of Raman spectroscopy, Pinzaru *et al.* (2004, see Chapter 19) mention the water-soluble vitamins B_1, B_3, B_6, and C, and the water-insoluble vitamins A, B_{12}, K_1, and K_3.

6.11.1 Vitamin A

For all mammals, vitamin A, a group of carotenoids, is essential. It is needed for vision (forming rhodopsin), for embryo development and growth, and for maintenance of the immune system. A density functional and vibrational spectroscopic analysis of β-carotene is reported by Schlücker *et al.* (2003). The optimized geometry together with IR intensities and Raman activities, calculated at the BPW91/6-31G* level, is presented. The centrosymmetric structure of β-carotene is verified both theoretically and experimentally, by identifying a stable calculated structure with C_i symmetry and the mutually exclusive occurrence of bands in the experimental FT-IR and Raman spectra (Figure 6.50). In addition to the natural abundant (C_6–C_7) s-*cis* form, the all-*trans* form was found as a second conformer. s-*cis*-β-Carotene was found to be 8.8 kJ/mol more stable than the all-*trans* form.

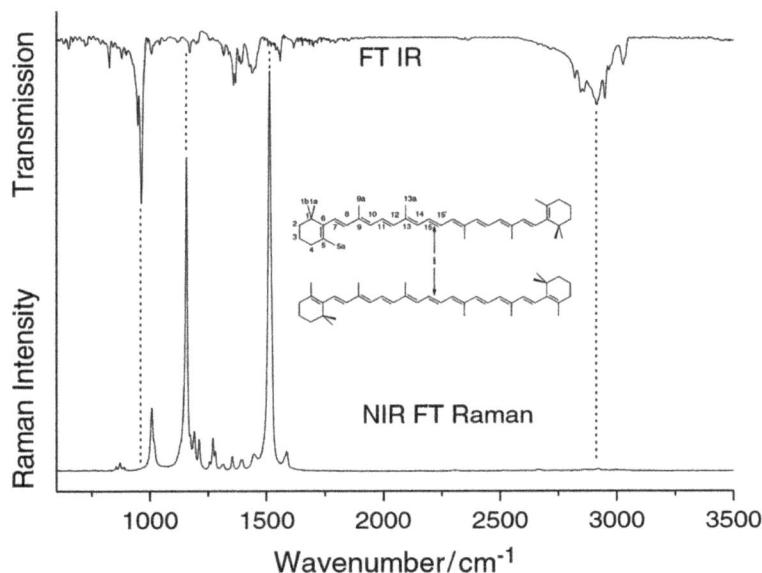

Figure 6.50: Experimental FT-IR and NIR-FT-Raman spectra of β-carotene (Schlücker *et al.* 2003). Copyright ©2003 John Wiley & Sons, Ltd., with permission.

6.11.2 The B vitamins

Vitamin B is the designation for a group of vitamins, comprising eight water-soluble vitamins, all serving as precursors of coenzymes. The Raman spectra of vitamins B_1, B_2, and B_{12} are shown in Figures 6.51–6.53.

Vibrational studies on vitamin B_6 (pyridoxine hydrochloride) are published by Cinta *et al.* (1999). The authors publish experimental spectra (IR, FT-Raman at different pH values, and SERS). With the aid of semiempirical and *ab initio* theoretical calculations, they give a complete assignment of the vibrational bands.

The vibrational spectrum of pantothenic acid is investigated by Srivasta *et al.* (2014) using experimental IR and Raman spectroscopies and DFT methods available. Vibrational assignments of the observed IR and Raman bands are proposed in light of the results obtained from computations. In order to assign the observed IR and Raman frequencies, the PEDs are also computed. Optimized geometrical parameters suggest that the overall symmetry of the molecule is C_1. The molecule is found to possess eight conformations. Conformational analysis is carried out to obtain the most stable configuration of the molecule, and the vibrational features of the lowest energy conformer are studied. The two methyl groups have slightly distorted symmetries from C_{3v}. The acidic O–H bond is found to be the smallest one. To investigate molecular stability and bond strength, the authors use natural bond orbital analysis.

Srivastava *et al.* (2014) study the vibrational spectrum of pantothenic acid (vitamin B_5, responsible for the metabolism of carbohydrates and fats, cholesterol as well) using experimental IR and Raman spectroscopies and DFT methods.

"What is the effect of SERS-active metals used in surface-enhanced vibrational spectroscopy of B vitamins?" Kokaislová and Matějka (2012) ask. Studying the SERS spectra of riboflavin, nicotinic acid, folic acid, and pyridoxine adsorbed on silver substrates and on gold substrates, the authors propose that the B-vitamins examined interact with the different metals via a different mechanism of adsorption.

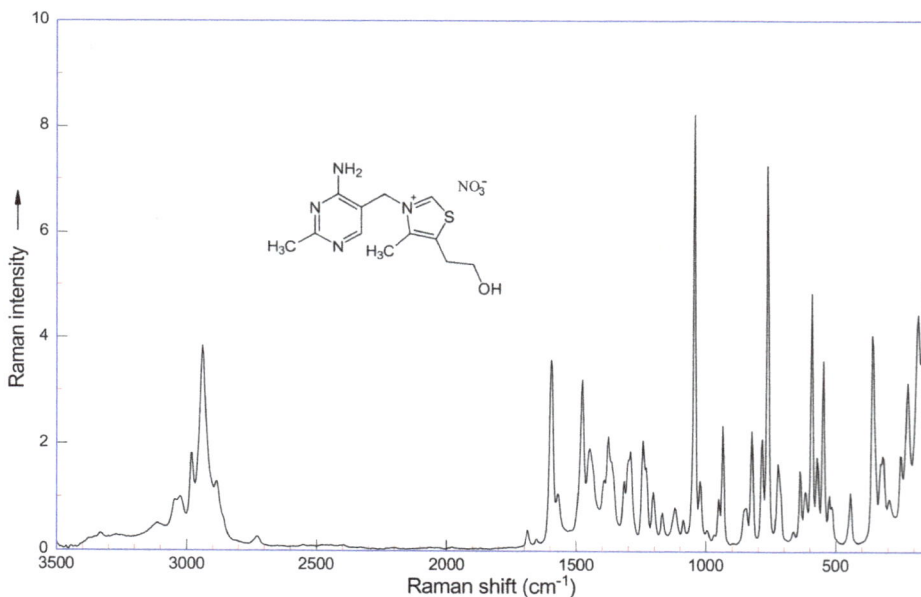

Figure 6.51: Raman spectrum of thiamine (vitamin B_1) nitrate (λ_{exc} = 1,064 nm, 270 mW; data courtesy of Bruker Optik GmbH, Ettlingen).

SERS is used by Radu *et al.* (2016) for the detection of vitamins B_2 and B_{12} in cereals. Despite the complexity of the matrix, the authors report the simultaneous detection of two B vitamins (riboflavin, vitamin B_2; and cyanocobalamin, vitamin B_{12}) by means of SERS. Radu *et al.* praise SERS as featuring molecular fingerprint identification and high analytical sensitivity together with low processing time and cost.

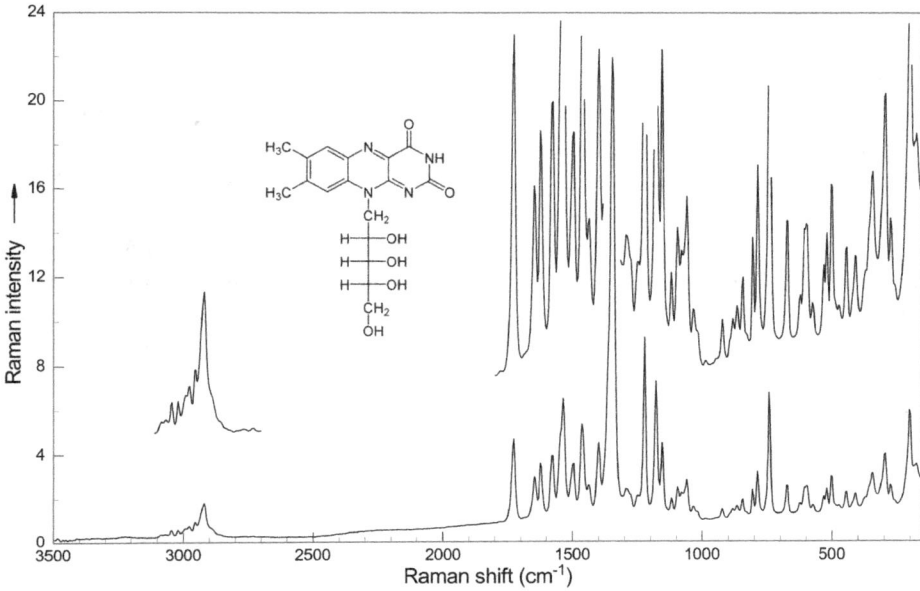

Figure 6.52: Raman spectrum of riboflavin (vitamin B$_2$) (λ_{exc} = 1,064 nm, 270 mW; data courtesy of Bruker Optik GmbH, Ettlingen).

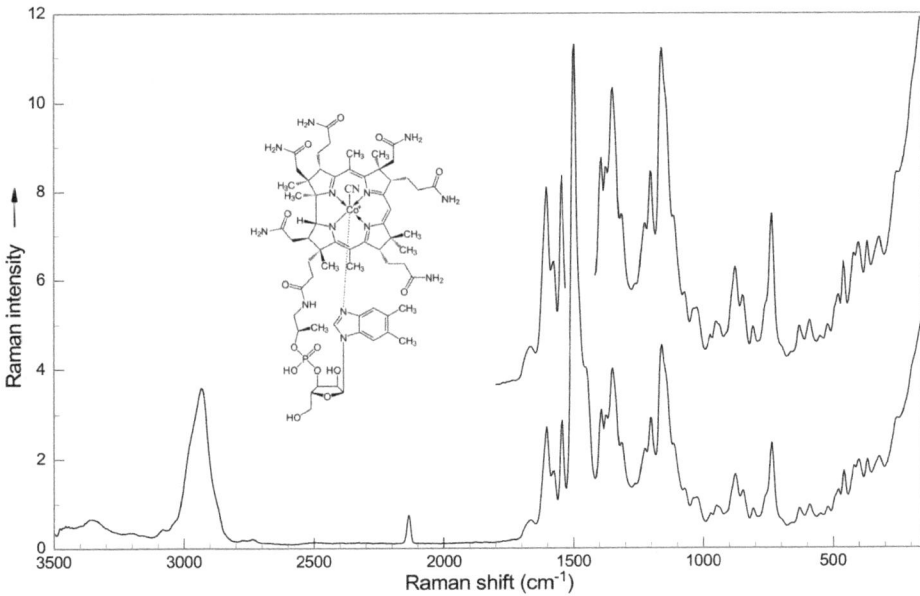

Figure 6.53: Raman spectrum of cyanocobalamin (vitamin B$_{12}$) (λ_{exc} = 1,064 nm, 270 mW, data courtesy of Bruker Optik GmbH, Ettlingen).

6.11.3 Vitamin C

Ascorbic acid, better known as vitamin C, is needed to avoid a haemorrhagic condition (scurvy) of a person's skin and gums with attack on collagen in binding tissues.

Vibrational spectroscopic studies on crystalline L-(+)-ascorbic acid and sodium ascorbate are reported by Hvoslef *et al.* (1971). IR and Raman spectra of the acid and its anion are shown and discussed together with those of the deuterated compound.

A review of the studies on L-(+)-ascorbic acid (vitamin C, AH_2, Figure 6.54) and its mono- and di-deprotonated anions (AH^- and A^{2-}) with Raman spectroscopy (Figure 6.55) in visible and UV light is given by Berg (2015). To complement the many previous calculations, the author found it necessary to perform a comparative study on the AH^- and A^{2-} anions, as well as similar recalculations on the AH_2 molecule at the DFT/B3LYP/6-311^{++}G(d,p) level. Assignments are given.

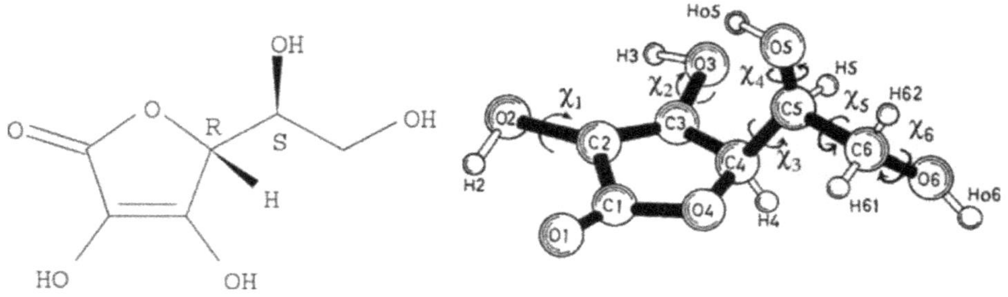

Figure 6.54: L-Ascorbic acid, $C_6H_8O_6$ (systematic name (5*R*)-5-((1*S*)-1,2-dihydroxyethyl)-3,4-dihydroxyfuran-2(5H)-one), or here just AH_2. Copyright ©2015 Rolf W. Berg. Licensed by Creative Commons Attribution 3.0.

6.11.4 Vitamin D

Vitamin D is a group of fat-soluble secosteroids responsible for the build-up of bone and thereby preventing rickets, a childhood disease. A Raman spectrum of vitamin D_2 is shown in Figure 6.56.

To establish non-invasive optical sensors for ergocalciferol (vitamin D_2) and cholecalciferol (vitamin D_3), operating in the tens of nmol/L region, Balcers *et al.* (2022) model optical properties and optical detection using absorption and Raman spectroscopy (Figure 6.57).

6.11.5 Vitamin E

Vitamin E – the tocopherols – is a potent antioxidant and therefore an important cell protective property against free radicals. Besides, vitamin E can mediate inflammations

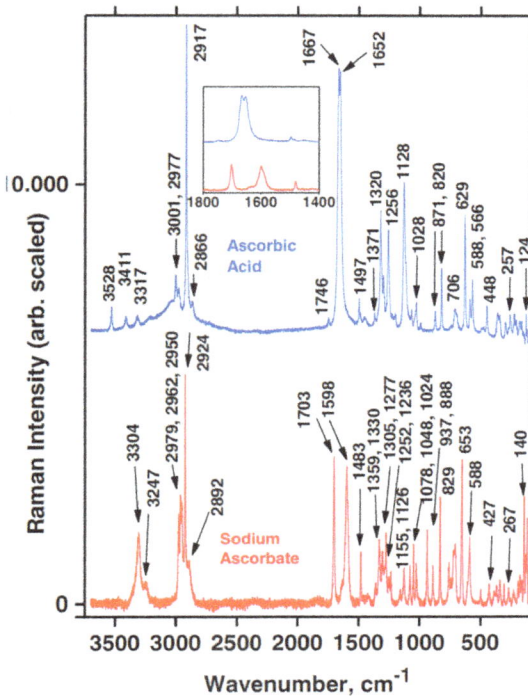

Figure 6.55: Reference Raman spectra for AH$_2$ and NaAH solids. Measurement details: laser, 532 nm; power level, ~200 mW; slit width, ~8 cm^{-1}; accuracy ±1 cm^{-1}. Insert shows details of the range from 1,800 to 1,400 cm^{-1}. Copyright ©2015 Rolf W. Berg. Licensed by Creative Commons Attribution 3.0.

and protect arteries against calcification (arteriosclerosis). A DFT-based computational study by Das *et al.* (2014) investigates the structures, potential energy surface, transition states and vibrational frequencies of a vitamin E precursor – chroman – in S$_0$ and S$_1$ states. The study calculates the barrier to planarity to be 3,500 ± 300 cm^{-1} in S$_0$ with the twisting angle as 30.5°, and the corresponding values of barrier as 3,800 cm^{-1} and twisting angle as 29° in S$_1$.

6.11.6 Vitamin K

The K-vitamins K$_1$–K$_3$ (*K* from the German *Koagulation* for coagulation) are fat-soluble compounds with the common 2-methyl-1,4-naphthochinon structure (menadione). They are important for the clotting of blood, stability of bones, and cardiovascular health. MIR (Figure 6.58) and NIR spectra of crystalline menadione (vitamin K$_3$) are studied by Beć *et al.* (2021). Detailed band assignments based on DFT calculations are presented. The anharmonicity calculations of the authors also supply overtone and combination bands.

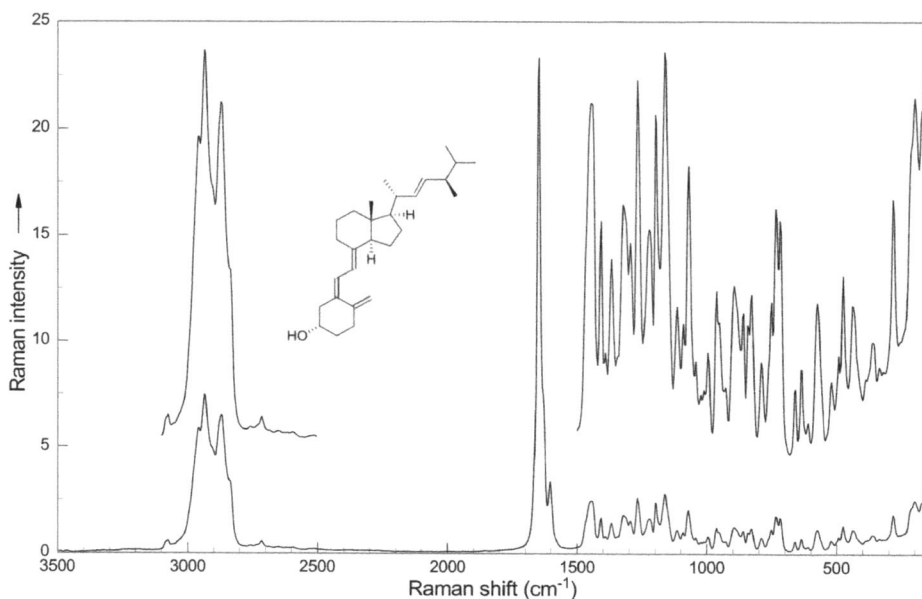

Figure 6.56: Raman spectrum of vitamin D_2 (λ_{exc} = 1,064 nm, 270 mW; data courtesy of Bruker Optik GmbH, Ettlingen).

6.12 Antibiotics

The use of Raman spectroscopy for the quantitative analysis of penicillin in fermentation broths is investigated by Clarke *et al.* (2005). Normal Raman spectroscopy as well as SERS is applied to penicillin G, penicillin V, ampicillin, carbenicillin, and 6-aminopenicillanic acid. With the aid of PCA, SERS could clearly be used to identify unequivocally the different penicillins. As modern fermentation processes yield high concentration of penicillins, Raman spectroscopy can be applied to the quantitative analysis of penicillins in fermentation broths.

The antibiotic chloramphenicol was first isolated from the bacterium *Streptomyces venezuelae* EHRLICH. Sajan *et al.* (2008) measure NIR-FT Raman (Figure 6.59), FT-IR, and SERS spectra, supplemented by density functional computations on the B3LYP/6-31G(d,p) level. The absence of a C–H stretching vibration and the observed C–H out-of-plane bending modes in the SERS spectrum suggests that the chloramphenicol molecule may be adsorbed in a flat orientation with respect to the silver surface.

For SERS of D-penicillamine on silver colloids, the paper of Lopez *et al.* (2004) may be consulted. Assignments of the spectral bands are proposed. As expected, a stable chemical bond is established between the sulfur atom and silver. Two of the conformers

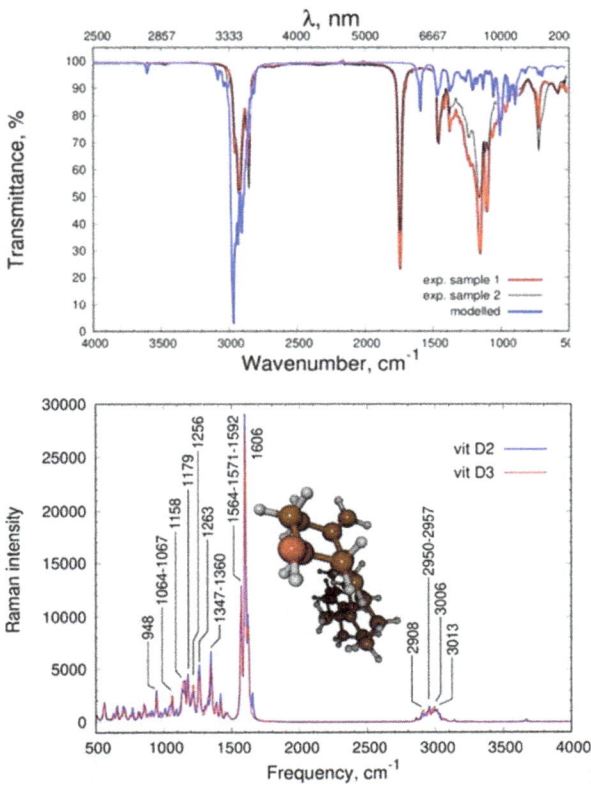

Figure 6.57: IR and Raman spectra of vitamins D_2 and D_3 (Balcers *et al.* 2022). Copyright ©2022 Elsevier B.V., with permission.

of D-penicillamine are adsorbed in their dianionic forms, carboxylate and amino groups interact with the metal as well.

Miyaoka *et al.* (2014) successfully detect the antibiotic amphotericin B produced in *Streptomyces nodosus* TREJO using Raman microspectroscopy *in situ*. The Raman spectrum of the pure compound in dimethyl sulfoxide with excitation at 532 nm (Figure 6.60) shows the band of the hydrophobic polyene domain at 1,559 cm^{-1}, corresponding to the C=C stretch, and at 1,157 cm^{-1}, corresponding to the C–C stretch enhanced by resonance.

6.13 Miscellaneous

Shikimic acid, a precursor to an antiviral drug, is isolated from the spice star aniseed (*Illicium verum* HOOK, Schisandraceae), as noted by Berger and Sicker (2009). Unusual

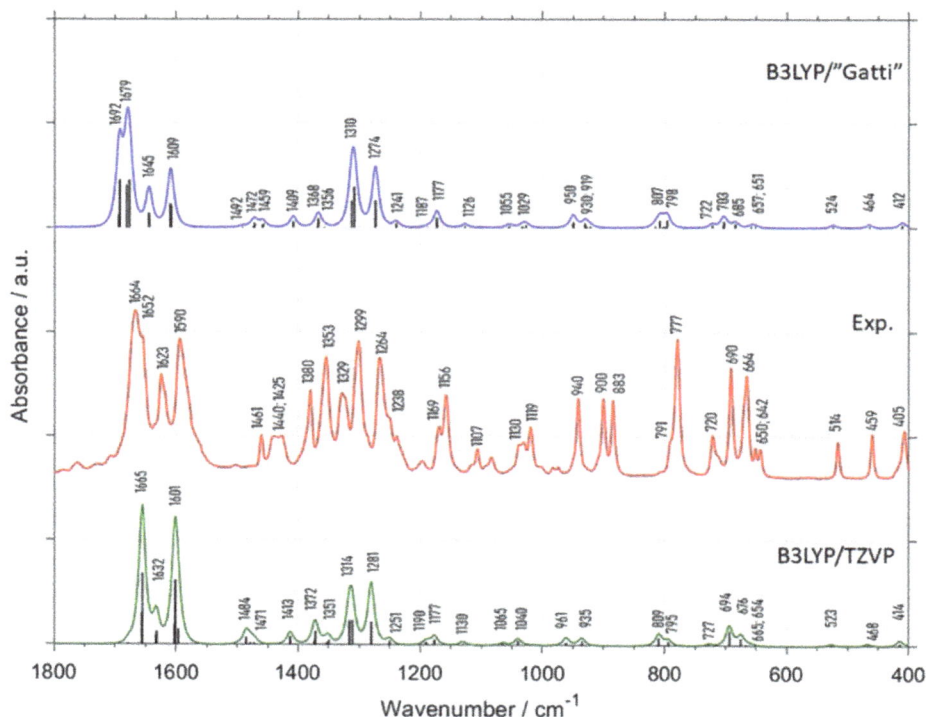

Figure 6.58: Spectrum of crystalline menadione in the 1,800–400 cm^{-1} range together with the harmonic spectra calculated by periodic DFT (B3LYP/"Gatti" and B3LYP/TZVP) (Beć et al. 2021). Copyright ©2021 by the authors, licensed by Creative Commons (CC 4.0).

sharp O–H stretching bands in the region around 3,240–3,500 cm^{-1} are observed in its IR spectrum. Broad aliphatic C–H bands are observed in the usual region, and the sharp bands at 1,690 cm^{-1} for C=O and at 1,650 cm^{-1} for C=C are remarkably close together.

Aleuritic acid is obtained from shellac, the secretion of the lac insect (*Kerria lacca* KERR, Kerriidae), by Berger and Sicker (2009). A broad O–H stretching band at 3,250 cm^{-1}, a split C–H band down to 2,900 cm^{-1}, and a C=O band at 1,700 cm^{-1} can be found in its IR spectrum.

6.14 Biological samples

See also Chapters 11 and 17.

Hernandez *et al.* (2014) reconsider disulfide linkage Raman markers. The authors state that during the last decades, Raman spectroscopy has been routinely used for probing the conformational features of disulfide linkages in peptides and proteins. The simple rule used consists of analysing the Raman bands in 550–500 cm^{-1} (ν(S–S)), assigning

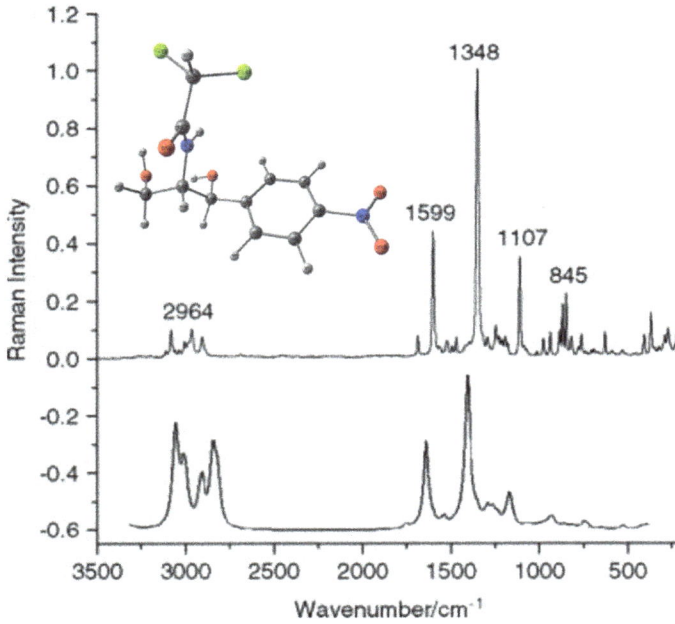

Figure 6.59: NIR-FT-Raman spectrum (top) of chloramphenicol and the calculated (bottom) Raman spectrum at B3LYP/6-31 (d,p) (Sajan *et al.* 2008). Copyright ©2008 John Wiley & Sons, Ltd., with permission.

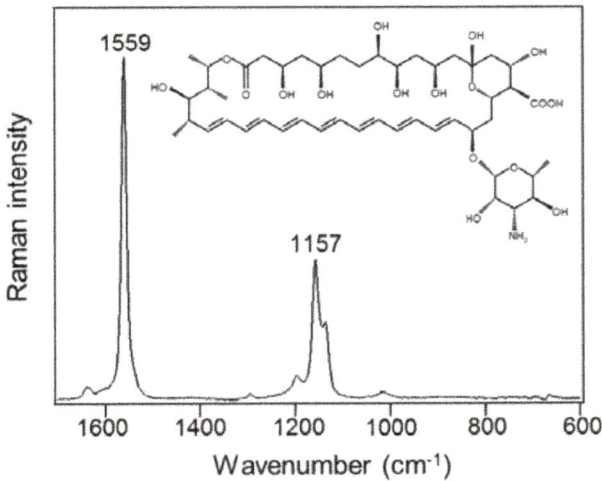

Figure 6.60: Raman spectrum of standard amphotericin B in dimethyl sulfoxide (30 mg/mL) (Miyaoka *et al.* 2014). Copyright ©2014 by the authors, licensed by Creative Commons (CC 3.0).

Raman markers near 500 cm^{-1}, near 520 cm^{-1}, and near 540 cm^{-1} (Figure 6.61) to three families of rotamers defined along the three successive bonds of the −C–S–S–C moiety, referred to as ggg, ggt, and tgt. Hernandez *et al.* advocate an accurate analysis of disulfide vibrational features using the five torsion angles (χ_1, χ_2, χ_3, $\chi_{2'}$, and $\chi_{1'}$) along the five successive bonds joining the two α-carbon atoms in the cysteine (Cys–Cys) unit. The authors conclude that the combined use of the old and recent conformational notations allows a more accurate structural and vibrational analysis of disulfide linkage.

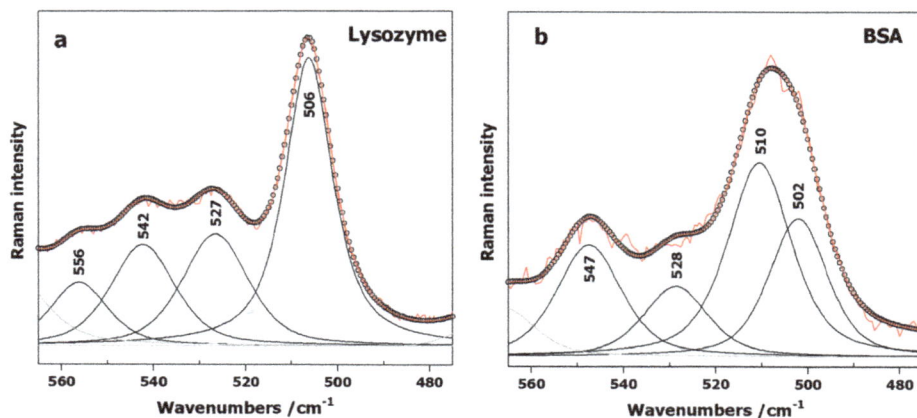

Figure 6.61: Raman spectra of chicken white egg lysozyme and bovine serum albumin (BSA) displayed in the 560–480 cm^{-1} spectral region corresponding to their v(S-S) markers. Band decomposition of the shown spectral region has been performed by pseudo-Voigt functions with a width at half height not exceeding ~ 15 cm^{-1}. Circles correspond to the sum of components (Hernandez *et al.* 2014). Copyright ©2014 John Wiley & Sons, Ltd., with permission.

FTN (ferritin) is a ubiquitous intracellular iron storage protein found in animals, plants, and bacteria. It has a spherical shape with an outer and inner diameter of about 12 and 8 nm, respectively. Inside the protein shell, a core of iron hydroxide can be found. Hartmann *et al.* (2019) report non-destructive chemical analysis of FTN using Raman microspectroscopy. Hartmann *et al.* compare FTN with apoferritin and reference minerals. Their results reveal that the iron core of natural FTN is composed of iron(III) hydroxide ferrihydrite ($Fe_2O_3 \cdot 0.5\ H_2O$).

Zhao *et al.* (2017) consider Raman microscopy as a superb tool to visualize and study the spatial–temporal distribution of chemicals. They review the development and applications of vibrational/Raman tags, particularly coupled with stimulated Raman scattering (SRS) microscopy. While label-free imaging has been the prevailing strategy in Raman microscopy, SRS imaging of vibrational tags has enabled researchers to study a wide range of small biomolecules. The technique made possible the imaging distribution and dynamics of small molecules such as glucose, lipids, amino acids, nucleic acids, and drugs that are otherwise difficult to monitor with other means.

In another paper, Zhao *et al.* (2017) review the application of vibrational tags in biological imaging by Raman microscopy. Label-free imaging is normally preferred to vibrational/Raman tags, but they are still used with SRS microscopy. As the label-free technique is often disturbed by overlapping spectra of cell components, introducing chemical bonds that vibrate in the cell-silent Raman window (1,800–2,600 cm^{-1}) serves as a solution. Labels such as nitrile and alkyne are examples of vibrational tags that can be used. A scheme (Figure 6.62), explaining the tags used to examine the distribution of cell components, is given by the authors.

Carrying out SRS under electronic pre-resonance conditions (epr-SRS) enables imaging with exquisite vibrational selectivity and sensitivity (down to 250 nM with ATTO740, 1 ms time constant) in living cells, as reported by Wei *et al.* (2017). The authors create a palette of triple-bond-conjugated near-IR dyes that each displays a single peak in the cell-silent Raman spectral window. Wei *et al.* report experiments on neuronal co-cultures and brain tissues that reveal cell-type-dependent heterogeneities in DNA and protein metabolism under physiological and pathological conditions.

The review of Talari *et al.* (2015) about Raman spectroscopy of biological tissues starts with a long and very useful list of peak frequencies and their assignments. The authors state that they want to develop a database of molecular fingerprints. Talari *et al.* then discuss the use of Raman spectroscopy to harvest biochemical information, to analyse breast, brain, cervical, gastrointestinal, lung, oral, and skin cancer, and the multivariate methods used to improve the amount of information obtainable from the spectra.

Vibrational spectroscopy has become very important in the analytics of food. The following paragraphs try to give a recent picture of the subject.

Pudney *et al.* (2010) describe how a confocal Raman system works and review the role of confocal Raman spectroscopy in food science. The authors advise the use of the system and how to avoid some of the common pitfalls, especially with regard to depth resolution. Strategies to obtain the desired analytical results are discussed.

Dahlenborg *et al.* (2012) investigate the local distribution of different components in white chocolate. Confocal Raman microscopy suggests that specific surface imperfections on chocolate could be part of a network of pore structures at and beneath the chocolate surface, which could be related to oil migration, and thus, to fat bloom formation.

Huen *et al.* (2014) image frozen bread dough by confocal Raman microscopy to observe ice crystals and their interaction with the other components of the frozen product: starch, gluten, and yeast. Monitoring the spatial distribution of the components, images of the frozen dough microstructure are generated. Huen *et al.* obtain an image of a continuous network of ice, not isolated crystals as one would expect.

Motoyama *et al.* (2016) demonstrate Raman microspectroscopy for simultaneously imaging the crystallinity and polymorphic types of fats, using porcine adipose tissue as a model. The crystalline states of fats strongly influence their physical properties,

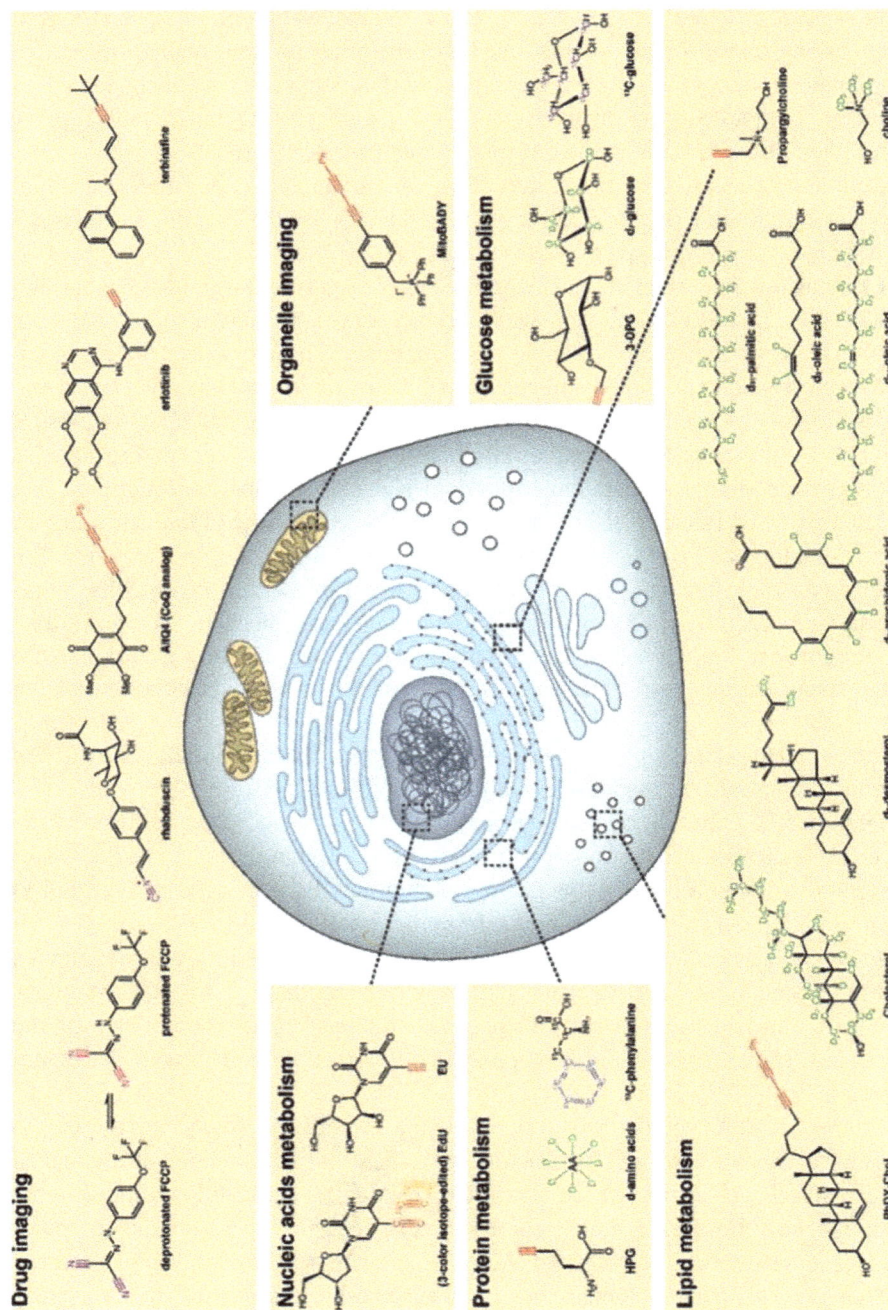

Figure 6.62: Raman/vibrational tags and their uses (Zhao *et al.* 2017). Copyright ©2017 Royal Society of Chemistry, with permission.

for example, the emergence of the β-crystal polymorph results in food product deterioration. Motoyama *et al.* analysed the whole fingerprint regions of Raman spectra by multivariate techniques to image the crystalline states of fats. The authors found out that β′-crystal polymorphs form a colloidal-gel-like network.

Smith *et al.* (2017) investigates the use of confocal Raman microscopy for observing the microstructure of processed cheese including the distribution of components and additives therein, valuable for the control of the quality of cheese. Commonly present additives, such as trisodium citrate, sorbic acid, paprika, and corn starch were analysed using reference Raman spectra. The study of Smith *et al.* shows that fat, protein, water, and the additives' distribution can be imaged effectively using high spatial resolution Raman microscopy.

Fat spreads are food emulsions such as butter, margarine, mayonnaise, and salad dressings. van Dalen *et al.* (2017) use confocal Raman microscopy for the spatial mapping of the microstructure of fat spreads, which is important to their sensorial properties such as texture, mouthfeel, and spreadability. van Dalen *et al.* are able to distinguish the solid and liquid phases of the lipids, and the spatial distribution of sunflower oil, water, emulsifier, and solid fat in various investigated spreads.

7 Vibrational spectroscopy of inorganic compounds

Vibrational spectra of inorganic compounds follow the same rules as organic compounds, but of course the C–H bands are missing. That means that you can recognize the spectrum of an inorganic compound at once. Most important for the interpretation of the spectra are symmetry considerations. If the symmetry of two compounds is the same, their spectra are very similar, for example, this is the case for $NaBrO_3$, $NaClO_3$, Na_2CO_3, $NaNO_3$, and Na_2SO_3. On the contrary, if two compounds are chemically related, but of different symmetry, their spectra look very different. Take, for example, Na_3PO_4, Na_2HPO_4, and NaH_2PO_4.

Spectroscopical data have been taken from the work of Socrates (2001) and references cited therein, if not given otherwise. In Socrates' book, much more data can be found, and are not given here due to lack of space.

7.1 Boron compounds

In boron hydrides, normal bonds of boron to hydrogen can be found together with special bridge bonds from two boron atoms to one hydrogen, also called "banana bonds" (Figure 7.1) after their curved shape. The symmetry of B_5H_9 is C_{4v}, whereas the symmetry of $B_{10}H_{14}$ is C_{2v}.

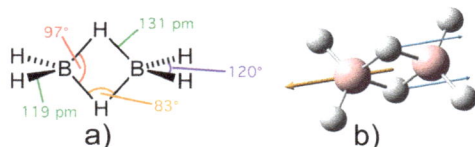

Figure 7.1: (a) Geometry of the diborane molecule (licensed by Wikimedia commons) and (b) asymmetric in-phase vibration calculated (density functional theory, 6-311^{++}G(d,p)) at 1,701 cm^{-1} (Hoffmann 2022).

The stretching vibration of the boron–hydrogen bond can be found at 2,565–2,480 cm^{-1}. This band is medium to strong in the infrared (IR) and weak to medium in the Raman spectrum. The in-plane B–H deformation vibration shows a band at 1,180–1,110 cm^{-1}; it is strong in the IR, and the out-of-plane B–H bending band at 920–900 cm^{-1} is medium to strong in the IR and weak to medium in the Raman spectrum.

The four possible stretching vibrations of the hydride bridge bond are shown in Figure 7.2. The symmetric in-phase vibration occurs at 2,140–2,080 cm^{-1} with weak-to-medium intensity in the IR, the symmetric out-of-phase vibration can be found at 1,990–1,850 cm^{-1} with weak intensity and more than one band, whereas the asymmetric in-phase vibration shows up at 1,800–1,700 cm^{-1} with weak-to-medium intensity in the IR; finally, the asymmetric out-of-phase vibration produces a very strong band at 1,610–1,525 cm^{-1}.

https://doi.org/10.1515/9783110717556-007

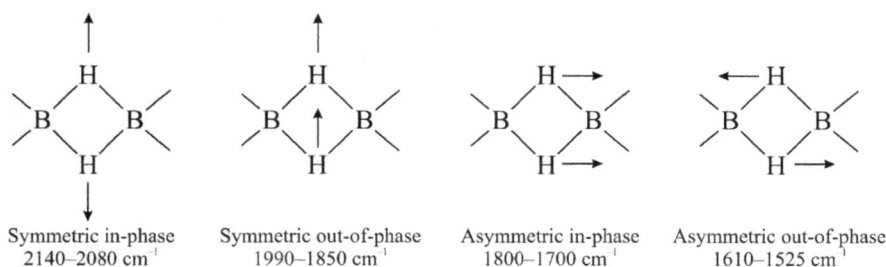

| Symmetric in-phase 2140–2080 cm^{-1} | Symmetric out-of-phase 1990–1850 cm^{-1} | Asymmetric in-phase 1800–1700 cm^{-1} | Asymmetric out-of-phase 1610–1525 cm^{-1} |

Figure 7.2: B–H stretching of two boron atoms bridged by two hydrogens.

For IR absorption spectra of B_2H_6 and B_2D_6 dispersed in neon at 4 K, see Chapter 13.

Alkyl diboranes have their symmetric BH_2 stretching frequency at 2,640–2,570 cm^{-1} (IR m-s, Raman m-w), whereas the asymmetric BH_2 stretch can be found at 2,535–2,485 cm^{-1} (IR m-s, Raman m-w). The BH_2 deformational vibration shows up at 1,205–1,140 cm^{-1} (IR m-s, Raman m-w) and the BH_2 wagging vibration at 975–920 cm^{-1} (IR m-s, Raman m-w).

In the IR spectrum of boronic acid anhydrides, the B–O stretch occurs as a strong band at 1,390–1,355 cm^{-1}.

The N–H stretch in borazines and borazoles is found as a medium band in the IR as well as in the Raman spectra at 3,500–3,600 cm^{-1}.

Natural boron contains isotopes [10]B (19.8%) and [11]B (80.2%), so in some boron compounds isotopic splitting of the vibrational bands can be found.

IR spectra of boron compounds are studied by Bellamy *et al.* (1958). The 56 borates, boronates, and boronites examined have the B–O stretching frequency at 1,350–1,310 cm^{-1}, and 51 of them in the narrower range of 1,346–1,316 cm^{-1}.

Becher and Thévenot (1974) publish a review of the IR spectra of some boron compounds. They study boron carbide (B_4C) [=$B_{11}C(CBC)$] and its isotypic derivatives $B_{12}O_2$, $B_{12}P_2$, and $B_{12}As_2$.

In a series of papers, Ford (2013) studies the vibrational spectra of boron halides and their molecular complexes. As an example, the paper *"Ab initio* studies of the boron trifluoride-nitrous acid complex" may be mentioned.

7.2 Inorganic carbon compounds

The IR spectra of rare earth carbonates are reported and discussed by Caro *et al.* (1972). The normal hydrated carbonates $Ln_2(CO_3)_3 \cdot n\,H_2O$ (n = 8 from La to Nd; 3 > n > 2 from Sm to Tm) have an IR spectrum similar to that of carbonate complexes, and the CO_3 group appears to be strongly coordinated to the rare earth atoms. The spectrum of the CO_3 functionality in the hydroxycarbonates $Ln_2(CO_3)_x(OH)_{3-x} \cdot n\,H_2O$ (x near 2.0 and n near 0.5) corresponds to that of a hydrogen-bonded CO_3 group.

Using a home-built low-cost Raman system, Wang *et al.* (Wang, Mittauer *et al.* 2009) study the temperature effect on Raman scattering of carbon disulfide. The v_1 band at 655 cm^{-1} (symmetric stretching mode) and the v_3 band at 1525 cm^{-1} (asymmetric stretching mode) are identified based on the frequencies calculated by normal mode analysis. Boltzmann distribution of the molecules among the vibrational energy levels and the temperature of the sample can be calculated using the ratio of the intensities of the *anti*-Stokes and Stokes bands. The asymmetric stretching and bending modes are only infrared active, but that is not true for molecules with two different sulfur isotopes.

Liquid carbon disulfide is studied by Gong *et al.* (Gong, He *et al.* 2017). The paper concentrates on the effects of Fermi resonance of v_1 and $2v_2$ on the Raman scattering of the fundamental mode v_2 (Figure 7.3). While modes v_1 at 656 cm^{-1} and overtone $2v_2$ at 796 cm^{-1} are allowed bands of the compound, the stretching vibration v_2 at 396 cm^{-1} is forbidden by the selection rules. However, a weak band has been observed along with the strong allowed totally symmetrical stretching mode v_1. The observation is ascribed to the effects of the vibrational anharmonicity of the free molecule and intermolecular interaction potential.

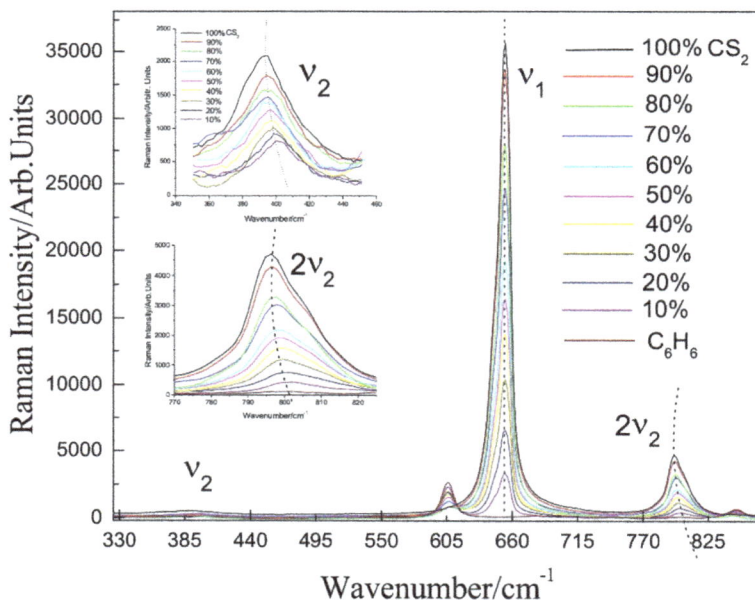

Figure 7.3: Raman spectra of CS$_2$ for different concentrations. The insets show an enlarged view of the v_2 and $2v_2$ bands from 340 to 460 cm^{-1} and 770–825 cm^{-1} (Gong, He *et al.* 2017). Copyright ©2017 Elsevier Ltd., with permission.

7.3 Phosphorus compounds

For phosphorus compounds, the P–H stretching can be found at 2,500–2,225 cm^{-1} as a sharp band in the IR of medium intensity. The P–C stretching shows up at 795–650 cm^{-1} as a sharp band with medium-to-strong intensity. The P=O stretching band occurs at 1,350–1,150 cm^{-1} (strong), and Fermi resonance or other effects can give rise to a doublet (Socrates 2001). These and other bands may be found in Tab. 7.1.

Table 7.1: Characteristic infrared bands of phosphorus compounds (Corbridge 1969; Thomas 1974; Günzler and Gremlich 2003).

Wavenumber (cm^{-1})	Assignment	IR	Raman
2,425–2,325	Phosphorus acid and ester P–H stretching	m-s	m-w
2,320–2,270	Phosphine P–H stretching		
1,090–1,080	Phosphine PH_2 bending	m	m-w
990–910	Phosphine P–H wagging		
2,700–2,100	Phosphorus acid and ester O–H stretching		
1,040–930	Phosphorus ester P–OH stretching		
1,050–970	Asymmetric P–OC$_{aliphatic}$ stretching	s, br	
830–750	Symmetric P–OC$_{aliphatic}$ stretching		
1,250–1,160	P–OC$_{aromatic}$ stretching	s	
1,050–870	P–OC$_{aliphatic}$ stretching	w	
1,320–1,280	P–C$_{aliphatic}$ stretching	m	
1,450–1,430	P–C$_{aromatic}$ stretching	m-s	
1,260–1,240	$(C_{aliphatic})_3$P=O stretching		
1,350–1,150	P=O stretchings		
770–760	P^{III}–F stretching	m-s	w
930–805	P^{V}–F stretching	m-s	w
865–655	P=S stretching	m-s	s, p
610–435	P–Cl stretching	m-s	
485–400	P–Br stretching	m-s	
550–525	P–S stretching	s	

7.4 Silicon compounds

For silicon compounds, we find the following characteristic IR bands (Hediger 1971; Socrates 2001; Günzler and Gremlich 2003):

The Si–H stretching vibration may be found at 2,250–2,100 cm^{-1} with medium-to-strong intensity in the IR, Raman intensity medium to strong, polarized.

The Si–H deformation can be found at 985–800 cm^{-1} with medium to strong intensity in the IR, weak in the Raman spectrum.

Assignment	Wavenumber (cm^{-1})		IR	Raman	IR	Raman
R$_2$Si–H	2,157–2,117 cm^{-1}				s	m-s, p
–OSi–H	2,230–2,120 cm^{-1}					
>SiH$_2$	965–923 cm^{-1}	920–843 cm^{-1}	m-s	w	m-s	w
C–SiCH$_3$	760–620 cm^{-1}				s-m	s
O–Si–CH$_3$	800–770 cm^{-1}					
Si–N–Si	950–830 cm^{-1}				s	
Si–NH$_2$	1,250–1,100 cm^{-1}				m	w
R$_3$Si–OH	900–810 cm^{-1}					
Si–OC	1,110–1,000 cm^{-1}	850–800 cm^{-1}	vs	w	s-m	w
Si–O–Si	1,090–1,030 cm^{-1}				vs	w

7.5 Sulfur compounds

For sulfur compounds we find the following characteristic IR bands (Hediger 1971; Socrates 2001; Günzler and Gremlich 2003):

Assignment	Wavenumber (cm^{-1})
R–S–H	2,590–2,530 cm^{-1} (IR w, Ra s, p)
R–(C=O)–S–H	≈2,550 cm^{-1} (IR w, Ra s, p)
R–CH$_2$–SH	730–570 cm^{-1} (IR w)
R–CH$_2$–S–CH$_2$–R	730–570 cm^{-1} (IR w-m, Ra s, p)
RSO–OH	870–810 cm^{-1} (IR m-s)
R–O–SO–OR	740–720 cm^{-1} (IR m-s, Ra m)
	710–690 cm^{-1} (IR m-s, Ra m)
R$_2$C=S	1,075–1,030 cm^{-1} (IR s, Ra s)
(RS)$_2$C=S	1,080–1,050 cm^{-1} (IR s, Ra s)
(RO)$_2$C=S	1,117–1,075 cm^{-1} (IR s, Ra s)
RS(R)C=S	1,210–1,080 cm^{-1} (IR s, Ra s)
R$_2$S=O	1,060–1,015 cm^{-1} (IR s, Ra m-w)
R–(S=O)–OH	≈1,100 cm^{-1} (IR vs, Ra m-s)
(RO)$_2$S=O	1,225–1,195 cm^{-1} (IR s, Ra s, p)

	ν_{as}	ν_s
R–SO$_2^-$M$^+$	1,030 cm^{-1} (IR s)	980 cm^{-1} (IR s)
R$_2$SO$_2$	1,370–1,290 cm^{-1} (IR vs, Ra w)	1,170–1,110 cm^{-1} (IR vs, Ra s)
R–SO$_2$–OR	1,375–1,350 cm^{-1} (IR s, Ra s-m)	1,185–1,165 cm^{-1} (IR s, Ra s)
(RO)$_2$SO$_2$	1,415–1,390 cm^{-1} (IR s, Ra s-m)	1,200–1,185 cm^{-1} (IR vs, Ra s)
R–SO$_2$–NR$_2$	1,365–1,315 cm^{-1} (IR vs, Ra m)	1,180–1,150 cm^{-1} (IR vs, Ra s)
R–SO$_2$–Hal	1,385–1,375 cm^{-1} (IR vs, Ra m-s)	1,180–1,170 cm^{-1} (IR vs, Ra s)
Aryl–SO$_2$–Hal	1,410–1,385 cm^{-1} (IR m-s, Ra m-w)	1,205–1,175 cm^{-1} (IR s, Ra s)

(continued)

	ν_{as}	ν_s
R–SO$_2$–OH	1,355–1,340 cm^{-1} (IR s, Ra s-m)	1,165–1,150 cm^{-1} (IR s)
R–SO$_3^-$M$^+$	1,250–1,140 cm^{-1} (IR vs)	1,070–1,030 cm^{-1} (IR s-m)
RO–SO$_3^-$M$^+$	1,315–1,220 cm^{-1} (IR vs, Ra s-m)	1,140–1,050 cm^{-1} (IR m, Ra s)
R–SO$_2$–SO$_2$–R	1,360–1,300 cm^{-1} (IR vs, Ra w)	1,150–1,110 cm^{-1} (IR vs, Ra s)

Wavenumber (cm^{-1})	Assignment
700–600	C–S stretching (IR vs, Ra s)
550–450	S–S stretching (IR w, Ra s-m, p)
2,500	S–H stretching (IR w, Ra s)
1,360–1,290	SO$_2$ asymmetric stretching (IR vs, Ra v)
1,190–1,120	SO$_2$ symmetric stretching (IR vs, Ra s, p)
1,060–1,020	S=O stretching (IR vs, Ra m-w)

Steinfeld *et al.* (1970) study IR double resonance in sulfur hexafluoride. High power irradiation of SF$_6$ by a CO$_2$ laser provides enough ground state depletion to be monitored by a conventional IR absorption spectrometer. The ν_3 band (965 cm^{-1}) is irradiated; and ν_4 (617 cm^{-1}) and ν_3 bands are monitored. Integrated absorption intensities provide information on ground state populations, which were determined as a function of laser power. The level population information coupled with contour analysis of the monitored bands is interpreted in terms of the vibrational and rotational equilibration processes in SF$_6$.

7.6 Noble gas compounds

Noble gases were supposed to be completely inert. But then Bartlett, in the spring of 1962, produced, by reacting the red gas PtF$_6$ with the colourless xenon, a mustard-yellow compound, supposed to be Xe$^+$ PtF$_6^-$. Later, the substance was identified as a mixture of fluoroxenyl compounds. Inspired by Bartlett's seminal work, shortly after this breakthrough, the German chemist Hoppe (1962) in Münster succeeded in synthesizing xenon difluoride (XeF$_2$). Other papers about xenon compounds by direct synthesis from xenon and fluorine were published the same year. Hargittai (2009) gives a nice account about Neil Bartlett and the first noble gas compound.

Smith (1963) studies xenon fluorides. Using IR spectroscopy with silver fluoride windows, for gaseous xenon difluoride, Smith finds ν_3 with its R-branch maximum at 566 cm^{-1} and its P-branch maximum at 550 cm^{-1}. He concludes that for the fact that ν_3 had no Q-branch; it should be a linear molecule (Figure 7.4). A combination band,

$v_1 + v_3$, can be found with its centre of the P- and R-branches at 1,070 cm^{-1}. The band of v_1 should therefore be at 515 cm^{-1} and v_1 is the IR-inactive Xe–F symmetrical stretching.

Nabiev *et al.* (2014) study molecular and crystal structures of compounds of the noble gases krypton and xenon. IR spectra are shown and discussed as well.

For the XeF$_2$ bending mode, bands are found at 213 cm^{-1} for v_2 and at 547 cm^{-1} for v_3. For krypton difluoride (KrF$_2$), v_2 is found at 236 cm^{-1} and v_3 at 580 cm^{-1}.

Frontera (2020) reviews noble gas bonding interactions involving xenon oxides and fluorides (e.g. XeF$_2$, XeF$_6$ (Figure 7.4 and 7.5), and XeO$_3$ (Figure 7.8)). The author outlines noble gas (or aerogen) bonding (NgB) as the attractive interaction between an electron-rich atom or group of atoms and any element of Group-18 acting as an electron acceptor.

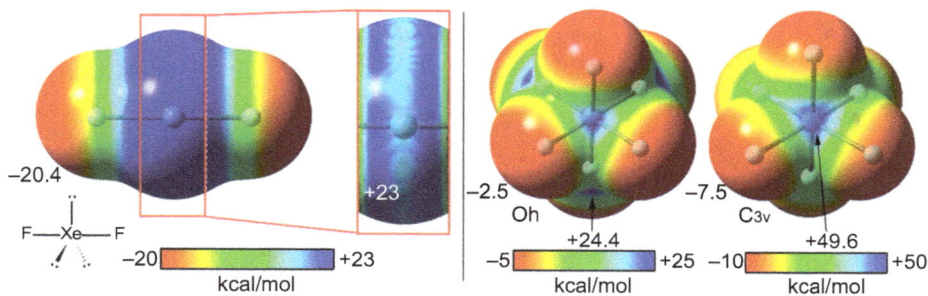

Figure 7.4: Left: MEP surface of XeF$_2$ showing the location of the positive belts. Isosurface 0.01 a.u. Right: MEP surfaces of the octahedral and C$_{3v}$ forms of XeF$_6$ (Frontera 2020). Copyright ©2020 by the author, licensed by Creative Commons (CC 4.0).

The vibrations of XeCl$_2$ occur at below 200 cm^{-1} for v_2 and at 313 cm^{-1} for v_3.

XeF$_4$ shows v_2 at 290 cm^{-1}, v_6 at 568 cm^{-1}, and v_7 at 123 cm^{-1}, the multiplet of a linear molecule.

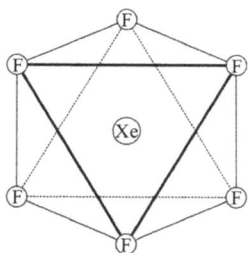

Figure 7.5: The octahedral isomer of xenon hexafluoride, and Xe–F stretching at 619 cm^{-1}.

Huston and Claassen (1970) present Raman spectra for XeO$_4$ vapour and solid (Figure 7.6). Observed fundamental frequencies (in cm^{-1}) for XeO$_4$ are 773 (A$_1$) for the vapour, and 767.1 (A_1), 277 (E), 867 (F_2), and 303 (F_2) for the solid. The Urey–Bradley constant F is −0.18 mdyn/Å for XeO_4. Claassen and Huston (1971) report matrix-isolation Raman and IR spectra, molecular structure, and force constants for XeO_3F_2.

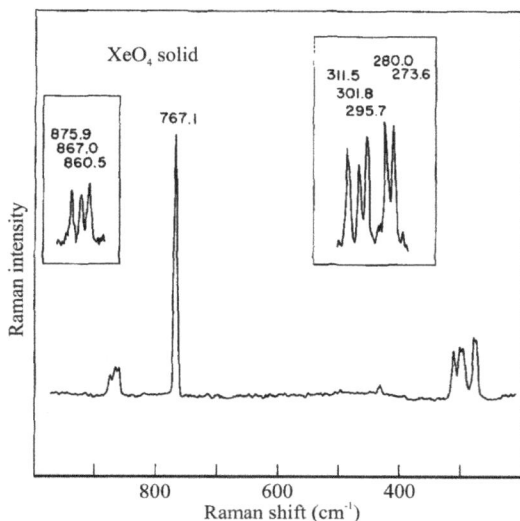

Figure 7.6: Raman spectrum of solid XeO_4 at approximately −195 °C (Huston and Claassen 1970). Copyright ©1970 AIP Publishing, with permission.

The ions $MoOCl_4Br_2^-$ and $WOCl_4Br_2^-$ are studied by Brunette and Leroy (1974), and they arrive at the symmetry C_{4v} for those ions. They report that, on the contrary, $XeF_5O^-Cs^+$ is distorted by the electron pair at the xenon atom and thus reduced symmetry of C_s by its splitting of the E modes.

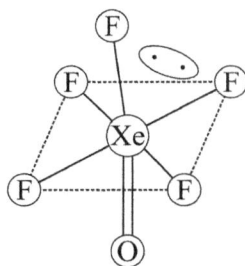

Figure 7.7: The geometry of the ion XeF_5O^- as a caesium salt (Schrobilgen *et al.* 1980). Copyright ©1980 Royal Society of Chemistry, with permission.

Schrobilgen *et al.* (1980) synthesize $XeOF_5^-$ and $[(XeOF_4)_3\,F]^-$ anions as their caesium salts and give a complete assignment of the bands of their Raman spectra. Based on these data and the X-ray powder diffraction data, they derive the structure of anions (Figure 7.7).

$[XeOXeOXe]^{2+}$, the "missing oxide of xenon(II)", is reported by Ivanova *et al.* (2015). They publish the synthesis, the Raman spectrum, and the X-ray crystal structure of $[XeOXeOXe][\mu\text{-}F(ReO_2F_3)_2]_2$.

The experimental Raman bands are: $\nu_1(A_1)$ at 883 cm^{-1}, $\nu_2(A_1)$ at 524 cm^{-1}, $\nu_3(A_1)$ at 420 cm^{-1}, $\nu_4(A_1)$ at 361 cm^{-1}, $\nu_5(B_1)$ at 544 cm^{-1}, $\nu_6(B_1)$ at 177 cm^{-1}, $\nu_7(B_2)$ at 390 cm^{-1}, $\nu_8(E)$ at 473, 468, 435 cm^{-1}, $\nu_9(E)$ at 410, 396 cm^{-1}, $\nu_{10}(E)$ at 384, 365 cm^{-1}, and $\nu_{11}(E)$ at 293, 374 cm^{-1}.

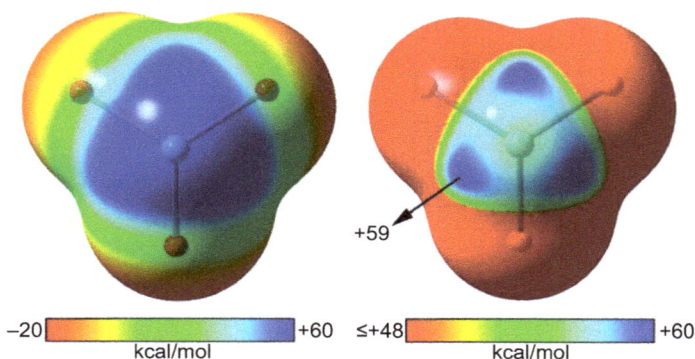

Figure 7.8: Molecular electrostatic potential (MEP) surfaces of XeO_3 using two different scales. Isosurface 0.01 a.u. (Frontera 2020). Copyright ©2020 by the author, licensed by Creative Commons CC (CC 4.0).

Tavčar and Žemva (2013) publish the Raman spectra of the first xenon(II) compounds of lithium together with synthesis and crystal structure (Figure 7.9). The sum formulae of the compounds reported are $[Li(XeF_2)_3](AsF_6)$ and $[Li(XeF_2)_n](AF_6)$ (A = P, As, Ru, Ir).

7.7 Halogen compounds

Crossover from hydrogen to chemical bonding is detected by Dereka *et al.* (2021). The $[F\text{–}H\text{–}F]^{(-)}$ ion represents a bare short H-bond, whose distinctive vibrational potential in water is revealed with femtosecond two-dimensional IR spectroscopy.

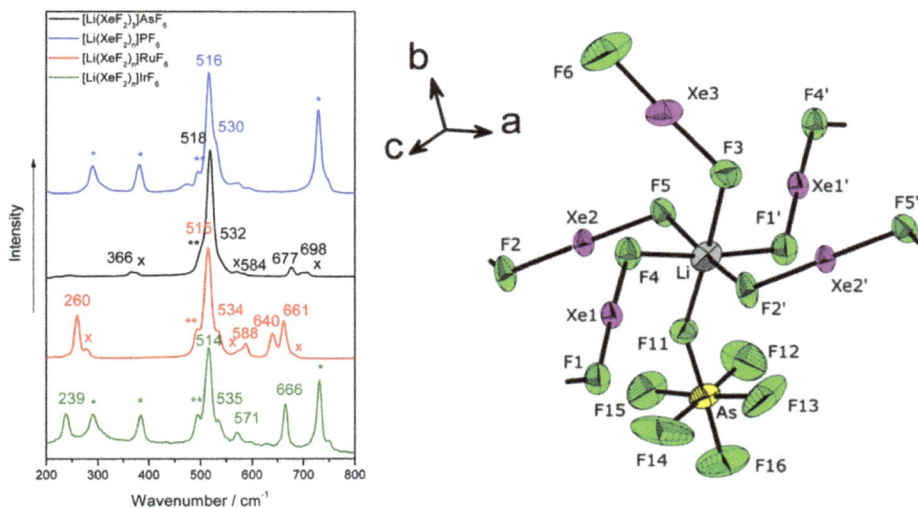

Figure 7.9: Comparison of the Raman spectra of $[Li(XeF_2)_3]AsF_6$, $[Li(XeF_2)_n]PF_6$, $[Li(XeF_2)_n]RuF_6$, and $[Li(XeF_2)_n]IrF_6$: *, FEP reaction vessel; **, free XeF_2; ×, $LiAF_6$.(left); coordination sphere of Li in $[Li(XeF_2)_3]AsF_6$. Thermal ellipsoids are drawn at the 50% probability level (right) (Tavčar and Žemva 2013). Copyright ©2013 American Chemical Society, with permission.

7.8 Inorganic anions and cations

The spectra of inorganic anions and cations can be found in the two-volume work about IR and Raman spectra of Nakamoto (1997).

Inorganic anions		
cm^{-1}	cm^{-1}	cm^{-1}
AsO_4^{3-} 800		
BrO_3^- 810–790		
CO_3^{2-} 1,450–1,410	880–800	
HCO_3^- 1,420–1,400	1,000–990	840–830
		705–695
HSO_4^- 1,180–1,160	1,080–1,000	880–840
ClO_3^- 980–930		
ClO_4^- 1,140–1,060		
CrO_4^- 950–800		
$Cr_2O_7^{2-}$ 950–900		

(continued)

	Inorganic anions		
	cm^{-1}	cm^{-1}	cm^{-1}
IO_3^-	800–700		
MnO_4^-	920–890	850–840	
N_3^-	2,170–2,080	1,375–1,175	
NO_2^-	1,400–1,300	1,250–1,230	840–800
NO_3^-	1,410–1,340	860–800	
PO_4^{3-}	1,100–950		
$S_2O_3^{2-}$	1,660–1,620	1,000–990	
SO_4^{2-}	1,130–1,080	680–610	
SO_3^{2-}	≈1,100		
SiF_6^{2-}	≈725		
SiO_4^{2-}	1,100–900		

	Inorganic cations		
	cm^{-1}	cm^{-1}	cm^{-1}
NH_4^+	3,335–3,030	1,485–1,390	
NO_2^+	1,410–1,370		
NO^+	2,370–2,230		

7.9 Carbon allotropes

Diamond shows a single sharp line in its Raman spectrum at 1,330 cm^{-1}. As in its pure crystalline form, it shows no IR absorption, and can be used as a window in IR spectroscopy.

Graphite also shows no IR bands. In the Raman spectrum, the relative intensities of the band at 1,575 cm^{-1} and 1,350 cm^{-1} can be used to study its purity.

Raman spectra of C_{60} and C_{70} are published by Dennis et al. (1991). C_{60} is one of the rare examples of molecules with I_h symmetry.

Compton and Hammer (2014) publish the Raman under liquid nitrogen (RUN) spectrum of C_{60} dissolved in carbon disulfide (Figure 7.10). The CS_2 Raman spectrum has been removed.

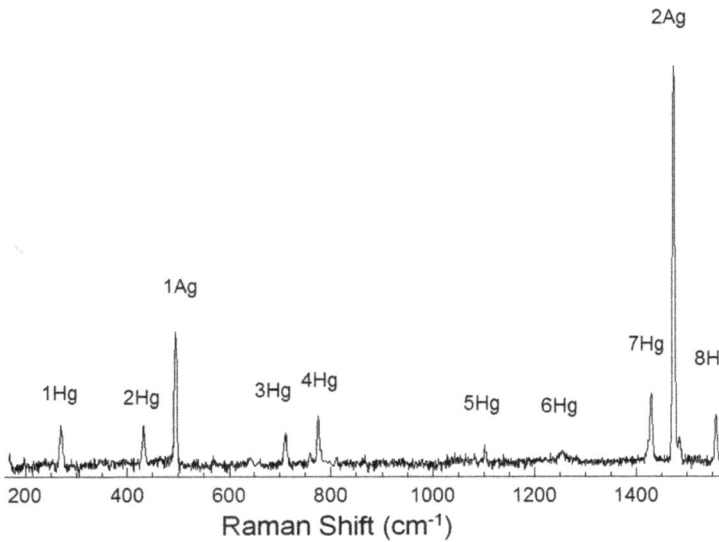

Figure 7.10: RUN spectrum of C_{60} dissolved in carbon disulfide. The CS_2 Raman spectrum has been removed (Compton and Hammer 2014). Copyright ©2014 by the authors, licensed by Creative Commons (CC 4.0).

By molecular mechanics, Adhikari *et al.* (2011) calculate the first natural vibrational frequencies of the fullerene family. Eight spherical fullerenes, including C_{60},

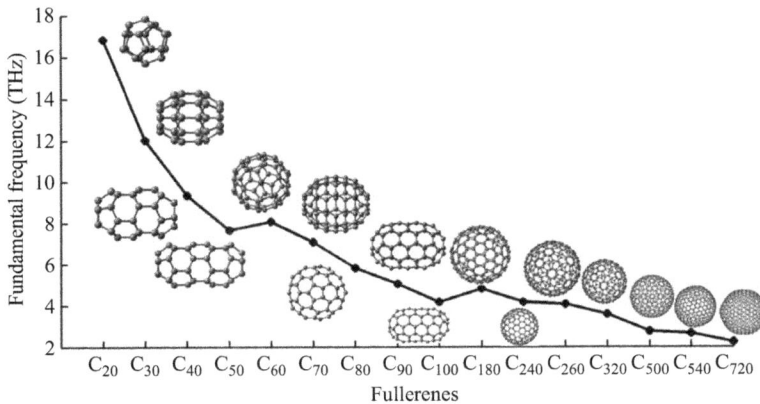

Figure 7.11: The variation of the first natural frequency across the complete range of fullerenes calculated using the molecular mechanics method. Both spherical and ellipsoidal types of fullerenes are considered. The eight spherical types of fullerenes include C_{60}, C_{80}, C_{180}, C_{240}, C_{260}, C_{320}, C_{500}, and C_{720} and the eight ellipsoidal (or non-spherical) types of fullerenes include C_{20}, C_{30}, C_{40}, C_{50}, C_{70}, C_{90}, C_{100}, and C_{540} (fullerenes are not drawn to scale). The frequencies (f) decrease approximately with an increase in the fullerene mass (M) as $f \propto M^{-1/2}$ (Adhikari and Chowdhury 2011). Copyright ©2011 Elsevier B.V., with permission.

C_{80}, C_{180}, C_{240}, C_{260}, C_{320}, C_{500}, and C_{720}, and eight ellipsoidal (or non-spherical) fullerenes, including C_{20}, C_{30}, C_{40}, C_{50}, C_{70}, C_{90}, C_{100}, and C_{540}, are considered (Figure 7.11).

C_{70} has D_{5h} symmetry as has dicyclopentadienyl ruthenium.

Dicyclopentadienyl iron (ferrocene) has D_{5d} symmetry as found out by the X-ray analyses of Seiler and Dunitz (1979).

Raman spectroscopy of carbon nanotubes (CNTs) is detailed by Dresselhaus *et al.* (2005). The theoretical foundations for the various types of Raman scattering processes relevant to CNTs are reviewed. The authors report experimental techniques and novel experimental findings for first-order and second-order Raman spectra for single-walled CNTs (SWCNTs).

Because CNTs show absorption in every part of the optical spectrum, Raman spectra of CNTs are always resonant Raman spectra. Therefore, a calculation of the normal vibrational spectra (Figure 7.12) of a small model SWCNT (Figure 7.13) has been performed by Hoffmann (2021) on the B3LYP/6-31G(d) level.

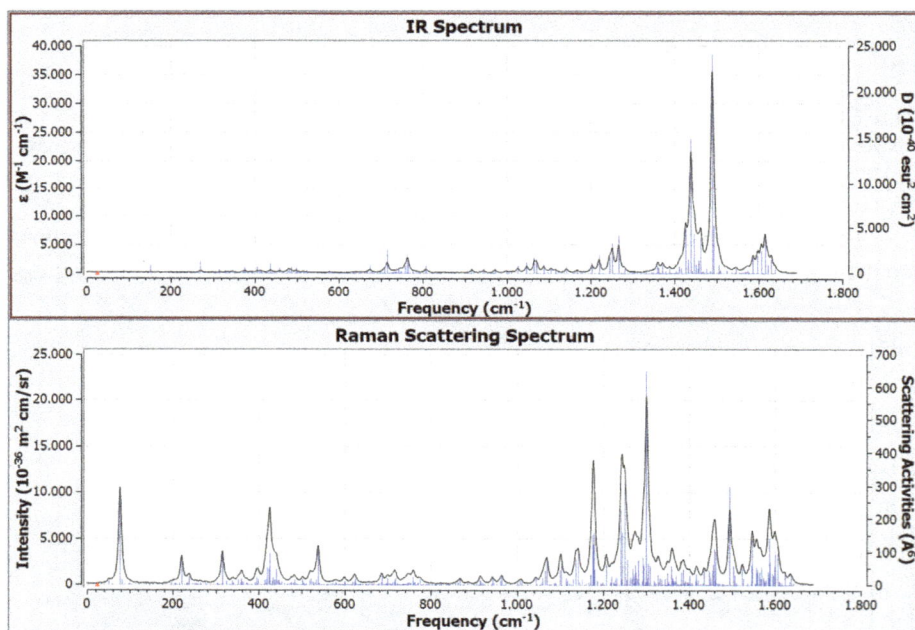

Figure 7.12: Calculated spectra of a small model SWCNT using density functional theory with the B3LYP functional and the 6-31G(d) basis set, $f = 0.9648$ (Hoffmann 2021).

Quite a few authors show spectra of CNTs taken under different conditions. Hoffmann *et al.* (2017) show the calculation of the tube diameters using the radial breathing modes in the Raman spectrum (Figure 7.14).

Weisman and Kono (2019) in their review paper show the resonance Raman spectrum (Figure 7.15) from a bundle of SWCNTs.

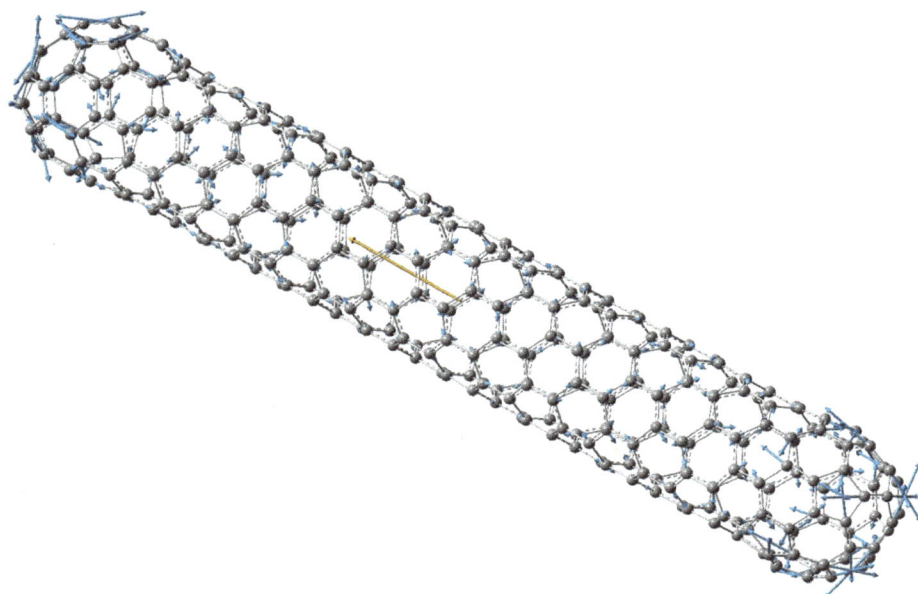

Figure 7.13: Infrared-active vibration (calc. 924 cm^{-1}) of a small model single-walled carbon nanotube of D$_2$ symmetry (model, density functional theory, B3LYP/6-31G(d)). Displacement vectors are shown in blue and dipole moment in red (Hoffmann 2021).

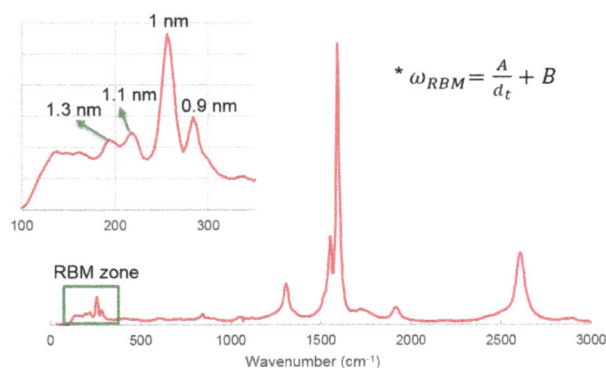

Figure 7.14: Calculation of the tube diameters using the radial breathing modes (RBMs) in the Raman spectrum (Hoffmann *et al.* 2017). Copyright ©2017 John Wiley and Sons, with permission.

Dresselhaus *et al.* (2007) discuss the differences between metallic and semiconducting SWCNTs and show spectra (Figure 7.16).

A review about the characterization of graphene, graphite, and CNTs by Raman spectroscopy is published by Dresselhaus *et al.* (2010). The comprehensive book chapter is a good introduction into the field with many well-explained figures and spectra.

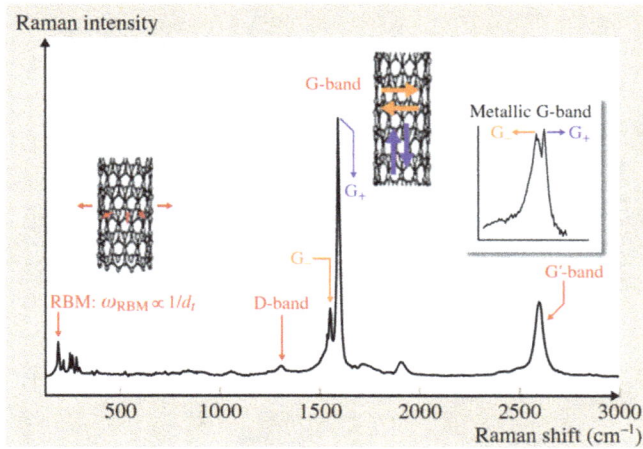

Figure 7.15: A typical resonance Raman spectrum of a bulk SWCNT sample, taken by E. Hároz with 785 nm excitation, for SWCNTs produced at Rice University. The spectrum illustrates the primary vibrational modes observed for SWCNTs: RBM, D-band, G-band, and G' (or 2D)-band. The right inset shows the lineshape of the G-band for metallic SWCNTs in resonance with E_{laser}. Also featured is the G'-band, which is the second order of the D-band, but by symmetry, the G'-band is Raman-allowed and not defect-induced (Weisman and Kono 2019). Copyright ©2019 by the authors, licensed by Creative Commons (CC 4.0).

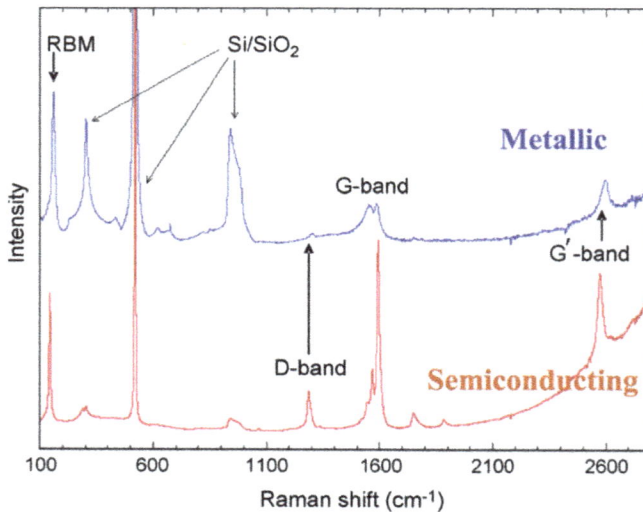

Figure 7.16: Raman spectrum from an individual metallic SWCNT (upper trace) and a semiconducting SWCNT (lower trace) showing a signal comparable to the Raman intensity from the substrate Si atoms (Dresselhaus *et al.* 2007). Copyright ©2007 Elsevier B.V., with permission.

A recent tutorial review by Jorio *et al.* (2021) describes Raman spectroscopy for CNT applications. The authors first introduce the relevant basic aspects of Raman spectroscopy of graphene-related materials, and then explain the use the Raman spectral features for the control and characterization of nanotube properties relevant for applications. Advanced techniques in the CNT field are also introduced.

7.10 Binary metal and metal organic compounds

7.10.1 Binary metal compounds

Huston and Claassen (1970) present Raman spectra for OsO_4 vapour. Observed fundamental frequencies (in cm^{-1}) for gaseous OsO_4 are 965.2(A_1), 333.1(E), 960.1(F_2), and 322.7(F_2). The Urey–Bradley constant F is 0.32 for OsO_4.

7.10.2 Metal organic compounds

The gas-phase IR spectrum of bis(benzene)chromium(0) (Figure 7.17) is recorded by Ngai *et al.* (1969). The authors find the spectrum consistent with D_{6h} symmetry. Cyvin *et al.* (1971) publish a normal coordinate analysis of light and heavy bis(benzene) chromium.

Figure 7.17: German stamp celebrating the 100th birthday of Ernst Otto Fischer. The stamp shows a portrait of the chemist who won the Nobel Prize for synthesizing the metal organic compound: bis(benzene) chromium.

Brunvoll *et al.* (1971) report the results of a total normal coordinate analysis of ferrocene, $Fe(C_5H_5)_2$. As in the previously published total vibrational analysis of bis(benzene) chromium, interesting kinematic couplings between some ligand and some framework normal modes of the complex can be observed, and these account for some frequency shifts from free (ionic) to complexed cyclopentadienide without any charge in the harmonic force field of the ligand. Calculated mean amplitudes for ferrocene are compared with those determined by an electron diffraction study and with the corresponding values in the free cyclopentadienide anion.

Brunvoll *et al.* (1971) perform a normal coordinate analysis of bis(benzene)chromium, $(C_6H_6)_2Cr$, assuming D_{6h} symmetry. Kinematic couplings explain some frequency shifts from free to complexed benzene.

Wang *et al.* (2021) combine Fourier-transform IR spectra of ferrocene (Fc) (Figure 7.18) and density functional theory-based quantum mechanical calculations, which confirm the dominance of the eclipsed Fc conformer in the fingerprint region of 400–500 cm^{-1} in solutions. Solution IR spectra of Fc measured in acetonitrile ($\varepsilon = 35.69$), dichloromethane ($\varepsilon = 8.93$), tetrahydrofuran ($\varepsilon = 7.43$), and dioxane ($\varepsilon = 2.21$) show two well-defined bands in the 480–500 cm^{-1} region with the higher wavenumber band higher in intensity. The

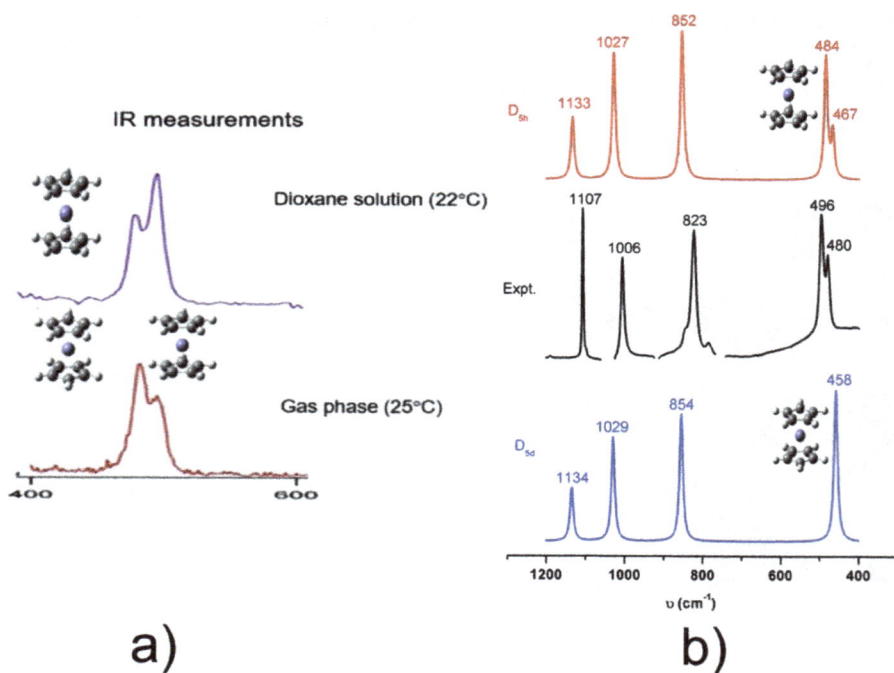

Figure 7.18: (a) IR spectra of ferrocene in dioxane solution and in the gas phase, and (b) experimental and calculated IR spectra for ferrocene conformers (Wang *et al.* 2021). Copyright ©2021 Elsevier B.V., with permission.

present study confirms that solvent effects enhanced the dominance of the D_{5h} Fc conformer which resulted in a switch of the IR profile patterns in the 400–500 cm^{-1} range, from low-wavenumber band higher in intensity pattern in the gas phase to higher wavenumber band higher in intensity in solutions. It further suggests that the effects of solvents on the IR spectra of Fc in this region are small, and the solvent model effects are also small for the IR spectrum in the region of 400–600 cm^{-1} of Fc.

7.11 Radioactive samples

A novel technique for Raman analysis of highly radioactive samples is introduced by Colle *et al.* (2017). Their hermetically closed sample compartments can be used by any standard micro-Raman spectrometer. As an example, it was employed to take Raman spectra of AmO_2, NpO_2, and a sample from Chernobyl lava.

8 Near-infrared spectroscopy

Near-infrared (NIR) spectroscopy (NIRS) is an important and extremely useful method of analysis for some research areas and applications, ranging from materials science via chemistry to life sciences. A good introduction to all aspects of NIRS is given by Siesler *et al.* (2002). The book features fundamental theory and instrumentation of NIRS and practical applications such as sample preparation and investigations of polymers, textiles, drugs, food, and animal feed. Special topics, such as two-dimensional correlation analysis, are covered in separate chapters.

Other good books on the topic are published by Workman (2000) and Workman and Weyer (2012). The former is a multi-volume work covering all spectroscopies, and the latter focuses on interpretation of NIRS.

Still the oral route for the delivery of drugs is preferred. NIRS has been adapted by researchers as a fast and non-destructive analytical method for the evaluation of tablets. Patel (2017), in his review, after looking at the basics, focuses on evaluation of tablet hardness, content uniformity, and dissolution by NIRS. Additionally, advantages and limitations of NIRS are discussed.

The review of Pasquini (2003) features the basic theory of NIRS and its applications in the field of analytical science. It also discusses modern instrument design, practical aspects, and applications in a number of different fields. As an example, Pasquini shows overtone and combination bands of three hydrocarbons in the NIR absorbance spectrum (Figure 8.1). The review also depicts the state of the art of NIRS in Brazil.

Assignment of near-infrared bands of common organic functionalities (Socrates 2001) is given in Table 8.1. The NIR analysis of polymers has an advantage: NIR absorptions are quite low so that a lot of polymer samples, especially foils and sheets, can be analysed without sample preparation. The usable bands are, for example, 5,950 and 4,610 cm^{-1} for polystyrene and 3,640–3,600 cm^{-1} for phenolic polymers.

For the analysis of the structural isomers of polyisoprene the sharp peak at 4,060 cm^{-1} of *cis*-1,4-polyisoprene can be used, as it becomes very weak for *trans*-1,4-polyisoprene.

The composition of copolymers can also be determined advantageously by NIR analysis. For example, the polystyrene–butadiene copolymer can be analysed using the 4,250 cm^{-1} peak of aliphatic C–H and the 4,580 cm^{-1} peak for aromatic C–H vibrations.

Nishinari *et al.* (1989) are studying the hydrolysis of starch. For the monitoring of the reaction they are using the NIR band at 4,760 cm^{-1}.

The analysis of food components by NIR is quite common. Water in foods can be measured by using the band at 5,160 cm^{-1}, for the analysis of unsaturated vegetable oil bands at 4,340 and 4,265 cm^{-1} can be applied, and proteins show up with a band at 4,590 cm^{-1}.

https://doi.org/10.1515/9783110717556-008

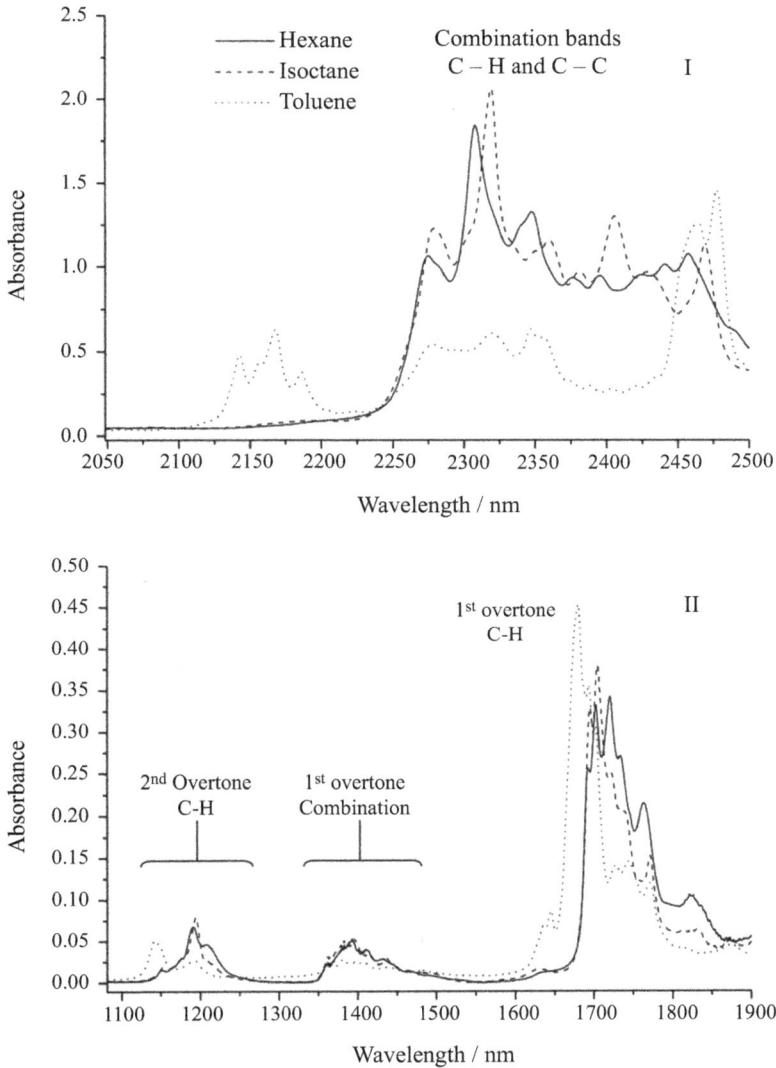

Figure 8.1: NIR absorbance spectra of some hydrocarbons. I region of combination bands II region of overtone bands. Optical path = 1 mm (Pasquini 2003). Copyright ©2003 by the author, licensed by Creative Commons (CC 3.0).

The peaks at 5,815 cm^{-1} (CH$_2$) and 4,675 cm^{-1} (–CH=CH–) can be used to replace the iodine number procedure for oils and fats.

Lodder *et al.* (1989) test the feasibility of determination of cholesterol and other blood constituents (e.g. lipoproteins and triglycerides) by NIR reflectance analysis. They report characteristic bands of lipoproteins at 5,715 and 4,330 cm^{-1}. As the different

Table 8.1: Assignment of near-infrared bands of common organic functionalities (Socrates 2001).

Wavelength (nm)	Wavenumber (cm^{-1})	Assignment
2,200–2,500	4,550–4,000	Combination C–H stretching
2,000–2,200	5,000–4,550	Combination N–H stretching, combination O–H stretching
1,650–1,800	6,250–5,550	First overtone C–H stretching
1,400–1,500	7,000–6,500	First overtone N–H stretching, first overtone O–H stretching
1,300–1,420	7,700–7,050	Combination C–H stretching
1,100–1,225	9,090–8,200	Second overtone C–H stretching
950–1,100	10,500–9,090	Second overtone N–H stretching, second overtone O–H stretching
850–950	12,000–10,500	Third overtone C–H stretching
775–850	12,900–12,000	Third overtone N–H stretching

composition of sera and small changes at the analytical wavelengths disturb the measurements, the determination of cholesterol or triglyceride by near-infrared reflectance spectroscopy is judged by the authors to be very unreliable.

Sugar content is analysed by Kawano *et al.* (1993) in fruits like mandarins. The authors used transmittance spectra of intact fruits acquired by fibre optics.

A recent book edited by Ozaki *et al.* (2021) about NIRS reviews theory, spectral analysis, instrumentation, and applications of the method up to 2021.

In a recent review, Barton *et al.* (2022) address the problem of using Raman data from different instruments and sources. The authors highlight the necessity for Raman spectroscopy harmonization and advocate multivariate data analysis for this purpose. The methods used are discussed in detail.

9 Vibrational spectroscopy of gases

In the liquid and solid states, the rotation of molecules is hindered, but in the gaseous state they can rotate freely. That means in the vibrational spectra, one can also observe the rotation of molecules. The same selection rules are obeyed: transitions show up in the infrared spectra if the dipole moment changes during the transition, and they show up in the Raman spectrum if the polarizability is changing. Two different clusters of transitions can be observed: pure rotational spectra and rovibrational spectra. Pure rotational spectra are centred about zero wavenumbers (Figure 9.3). As vibrations are excited at higher energy as rotations, in gaseous molecules, rotations are always excited together with vibrations (Figures 9.2, 9.5, and 9.6). The rovibrational branches are centred around the wavenumber of the vibration:

$$\Delta J = -1 \qquad \text{P branch}$$

$$\Delta J = 0 \qquad \text{Q branch}$$

$$\Delta J = +1 \qquad \text{R branch}$$

For linear molecules $\Delta J = \pm 1$, which means the Q branch is missing.

Symmetric top molecules may be classified as either prolate top or oblate top. A symmetric top is a rotor in which two moments of inertia are the same:

$$I_A \neq I_B = I_C$$

Examples of oblate symmetric tops are benzene (C_6H_6), cyclobutadiene (C_4H_4), and ammonia (NH_3). Prolate tops are chloroform ($CHCl_3$) and methylacetylene ($CH_3C{\equiv}CH$). A linear molecule counts as symmetric.

The different top molecules may be identified by their roto-vibrational spectra. The contours of such spectra are given by Weidlein *et al.* (1982).

Under temperatures of liquid nitrogen, single-line Raman spectra of liquid gases may be obtained (Figure 9.1), as rotations are subdued (Compton and Hammer 2014).

The pure rotational Raman spectrum of oxygen (Figure 9.2) is studied by Berard and Lallemand (1979) at room temperature and 1–15 atm with a spectral resolution of $0.2\ \text{cm}^{-1}$.

The rotational–vibrational Raman spectrum of oxygen and nitrogen is studied by Harney and Milanovich (1976). They directly determine α_e for $^{16}O^{16}O$, $^{14}N^{14}N$, and $^{18}O^{18}O$.

Bloom *et al.* (1976) measure stable isotope ratios in nitrogen and oxygen using Raman scattering. Their analysis shows that the measurement is possible with an accuracy of <±0.1%.

The Raman spectrum of the spherical rotator CD_4 is studied by Brodersen *et al.* (1977). The investigators determine B_0 from the ν_3 Raman band of CD_4.

https://doi.org/10.1515/9783110717556-009

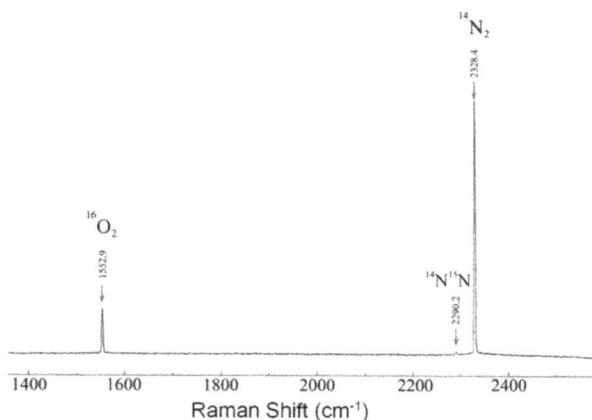

Figure 9.1: Raman spectrum of liquid nitrogen with the presence of some dissolved oxygen. Oxygen is generally not present in "fresh" samples of liquid nitrogen (Compton and Hammer 2014). Copyright ©2014 by the authors, licensed by Creative Commons (CC 4.0). Label of smallest peak corrected.

Newton *et al.* (2014) publish a comparative study of *in situ* N_2 rotational Raman spectroscopy (Figure 9.3) methods for probing energy thermalization processes during spin-exchange optical pumping.

Edwards *et al.* (1988) determine broadening coefficients, σ_p, for the pure rotational Raman lines of nitrogen and oxygen, perturbed by hydrogen, helium, and methane, and of hydrogen broadened by helium, argon, nitrogen, and carbon monoxide, over the pressure range of 2–100 bar with a host gas partial pressure of 1 bar. The authors discuss the broadening coefficients in relation to the dipole, quadrupole, and octopole moments and the polarizabilities of the perturbing species.

Schrötter describes the Raman spectroscopy of gases in the 2001 *Handbook of Raman Spectroscopy*, edited by Lewis *et al.* (2001).

Buric *et al.* (2017) review gas-phase Raman scattering. Their focus is on methods and applications in the energy industry.

Parrot *et al.* (2019) analyse mixtures of *para*-hydrogen (pH_2) and *ortho*-hydrogen (oH_2) by Raman spectroscopy. Spectra of normal hydrogen and para-enriched hydrogen at 4 bar are shown in Figure 9.4.

The optimization of a parabolic cell for gas Raman analysis is described by Yu *et al.* (2019). Testing their optimized cell, the authors find detection limits of 33 ppm-bar for H_2, 38 ppm-bar for CO, 19 ppm-bar for CH_4, 27 ppm-bar for C_2H_6, 23 ppm-bar for C_2H_4, and 20 ppm-bar for C_2H_2 with an exposure time of 200 s.

Rey *et al.* (2021) try a complete elucidation of the rotational–vibrational band structure in the SF_6 infrared spectrum at RT from full quantum-mechanical (CCSD(T)/ CVQZ) calculations. First, rotationally resolved spectra are reported for the three most abundant isotopologues ($^{32}SF_6$, $^{33}SF_6$, and $^{34}SF_6$) of the sulfur hexafluoride molecule.

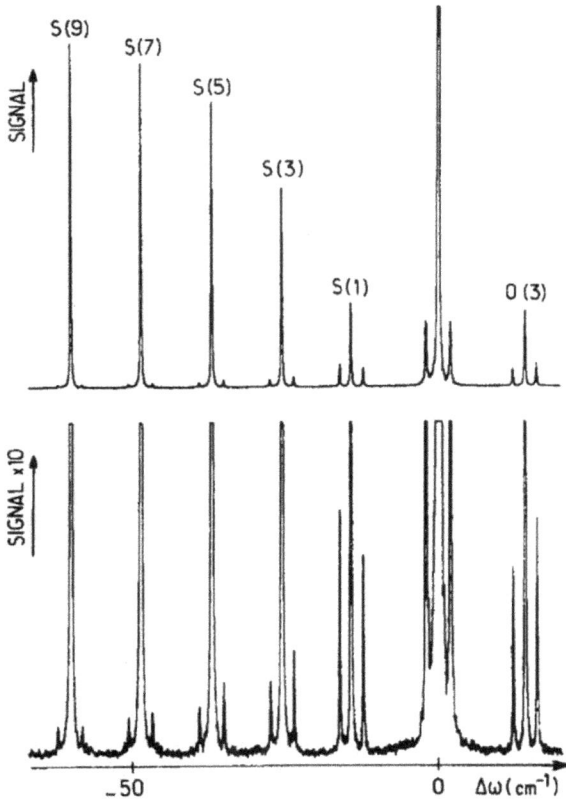

Figure 9.2: Rotational Raman spectra of O_2 at 1 atm and 290 K (Berard and Lallemand 1979). Copyright ©1979 Elsevier B.V., with permission.

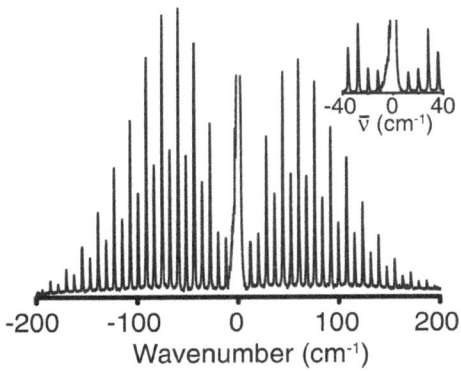

Figure 9.3: Pure rotational Raman spectrum of N_2 recorded at a resolution of 1.5 cm^{-1}. (Newton *et al.* 2014). Copyright ©2014 Springer Nature, with permission.

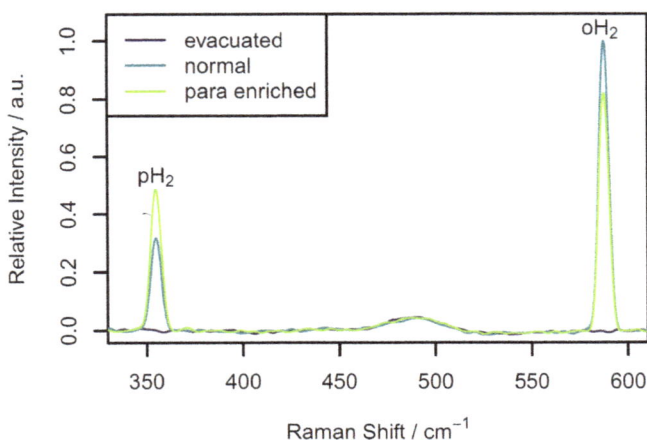

Figure 9.4: Raman spectra of an evacuated NMR tube, a tube filled with nH$_2$ at 4 bar, and a tube filled with pH$_2$-enriched hydrogen at 4 bar (Parrott *et al.* 2019). Copyright ©2019 by the authors, licensed by Creative Commons (CC 4.0).

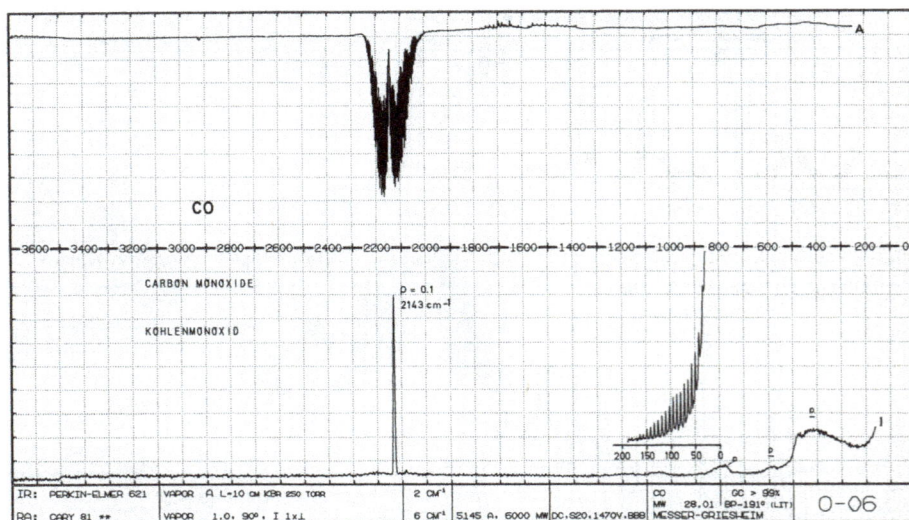

Figure 9.5: Infrared and Raman spectra of gaseous CO (from Schrader, B. (Ed.) (1996), Infrared Atlas of Organic Compounds, 2nd Ed., page O – 06. Copyright 1996 Wiley-VCH Verlag GmbH & Co. KGaA. Reproduced with permission).

Exploiting symmetry, highly excited rotational states were calculated up to $J = 121$, resulting in 6 billion transitions distributed over more than 500 cold and hot bands.

Andersen *et al.* (2019) publish the Raman and infrared spectra of 1,1-dicyclopropyl-2,2-dimethylethene (c-C$_3$H$_5$)$_2$C=C(CH$_3$)$_2$, DCPDME). The compound consists of a single

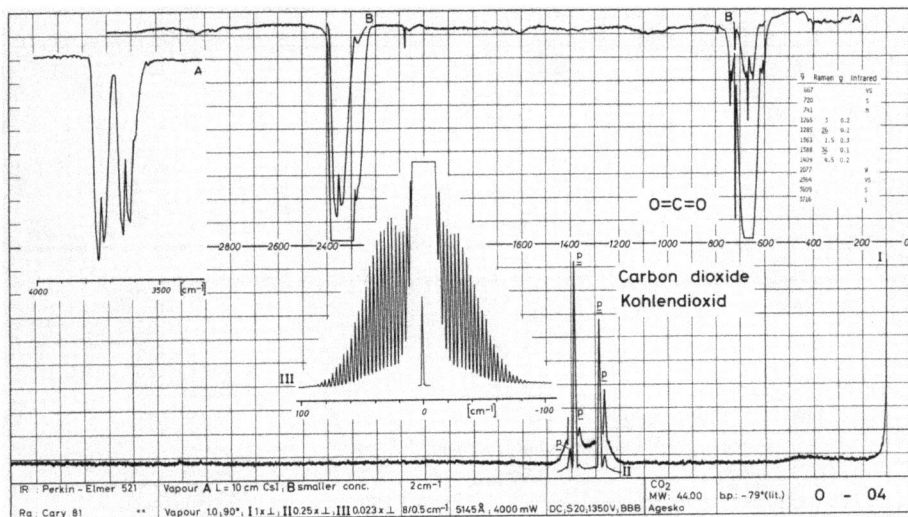

Figure 9.6: Infrared and Raman spectra of gaseous CO_2 (from Schrader, B. (Ed.) (1996), Infrared Atlas of Organic Compounds, 2nd Ed., page O − 04. Copyright 1996 Wiley-VCH Verlag GmbH & Co. KGaA. Reproduced with permission).

conformer in the crystalline solid, but different conformers (see also Figures 5.21 and 5.22) exist in the melt and in the gas phase (Figure 9.7). The measurements are interpreted using DFT calculations for spectral assignments with the functionals B3LYP, B3LYPD3, M06L, M062X, ωB97XD, and MP2 calculations employing the ccPVDZ and cc-PVTZ basis sets.

Okajima and Hamaguchi (2015) use N_2 to calibrate Raman spectra (Figure 9.8). The ratio of bands having the same rotational quantum number J was taken and used to calibrate intensities.

Jones (1972) uses a Fabry–Perot interferometer for the study of Raman spectra of gases under high resolution. The author shows photographically recorded spectra of N_2, O_2, H_2, and N_2O.

Studying gases, the amount of substance in the laser focus is quite low. That is why special sample arrangements have been designed to increase the signal-to-noise ratio of gases. One of these is the use of capillaries.

Boutilier (1981) uses computer-controlled absorption, fluorescence, and coherent *anti*-Stokes Raman spectrometry for liquid-chromatographic detection. The application of a silicon-intensified target vidicon detector allows for the spectra to be recorded and stored every 3 s without stopping flow.

Rodgers and Strommen (1981) construct a multifunctional spinning cell for obtaining Raman spectra of microsamples in the backscattering geometry. The cell is very effective in reducing the power density on the sample, what is desirable if the sample is photochemically or thermally unstable or thermal noise is generated.

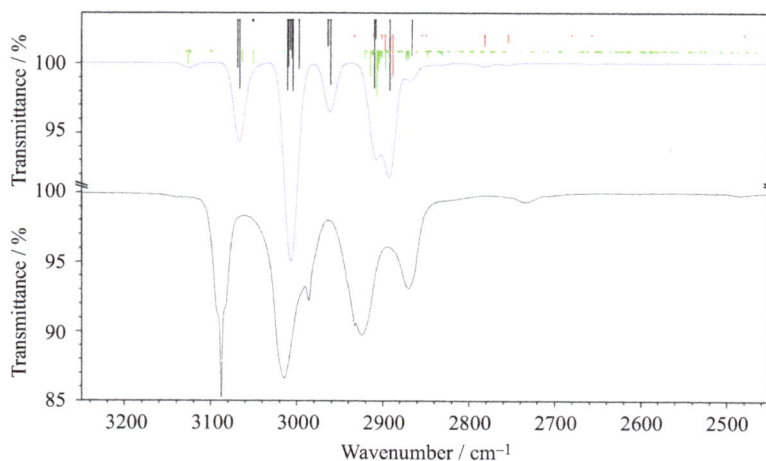

Figure 9.7: Gas-phase infrared spectrum of DCPDME in the 3,250–2,450 cm^{-1} range (black curve), simulated anharmonic spectrum (blue curve), and anharmonic intensities; fundamentals (black sticks), overtones (red sticks), and combination bands (green sticks) (Andersen *et al.* 2019). Copyright ©2019 Elsevier B.V., with permission.

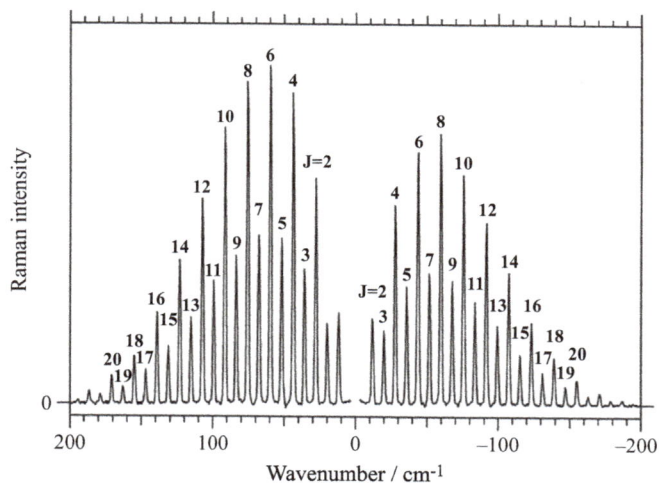

Figure 9.8: The rotational Raman spectrum of N$_2$ with rotational quantum number *J* was shown. Spectrum acquired in 30 min with a 532 nm Nd-YVO$_4$-laser and 20 mW laser power (Okajima and Hamaguchi 2015). Copyright ©2015 John Wiley and Sons Ltd., with permission.

Abad *et al.* (1994) record stimulated Raman spectra of the central region of the v_1 band of ethylene, both in a jet at temperatures down to ≈ 25 K. Only Q branches are recorded in this experiment. A vib-rotational analysis was performed, and improved values were obtained for the band origin and the rotational constants of the $v_1 = 1$ state of ethylene.

The stimulated Raman spectrum of cyanogen $^{12}C_2^{14}N_2$ (Figure 9.9) is investigated by Bermejo *et al.* (1997) at nearly Doppler resolution. For this purpose, the authors used the stimulated Raman technique and recorded the regions around the Q branches of the v_1 (2,330 cm^{-1}) and v_2 (845 cm^{-1}) vibrations. In addition to the fundamentals, hot bands arising from $v_5 = 1$, $v_5 = 2$, and $v_4 = 1$ have been observed. Two of cyanogens five fundamental modes, $v_3(\Sigma_u^+)$ and v_5 (Π_u), are infrared active, and the other three, v_1, $v_2(\Sigma_g^+)$, and $v_4(\Pi_g)$, are Raman active. Rotational constants for the excited states have been obtained, and the assignments have been tested by calculation of the Raman contours.

Müller and Zumbusch (2007) review coherent *anti*-Stokes Raman scattering microscopy. Some limitations of the techniques are discussed by the authors, but, as many problems cannot be solved with conventional techniques, they foresee a rapid growth of coherent *anti*-Stokes Raman spectroscopy (CARS) applications.

Intracavity phase-matched CARS is used by Zaitsu and Imasaka (2014) for trace gas detection, achieving a 6,000-fold improvement of the CARS signal under optimized

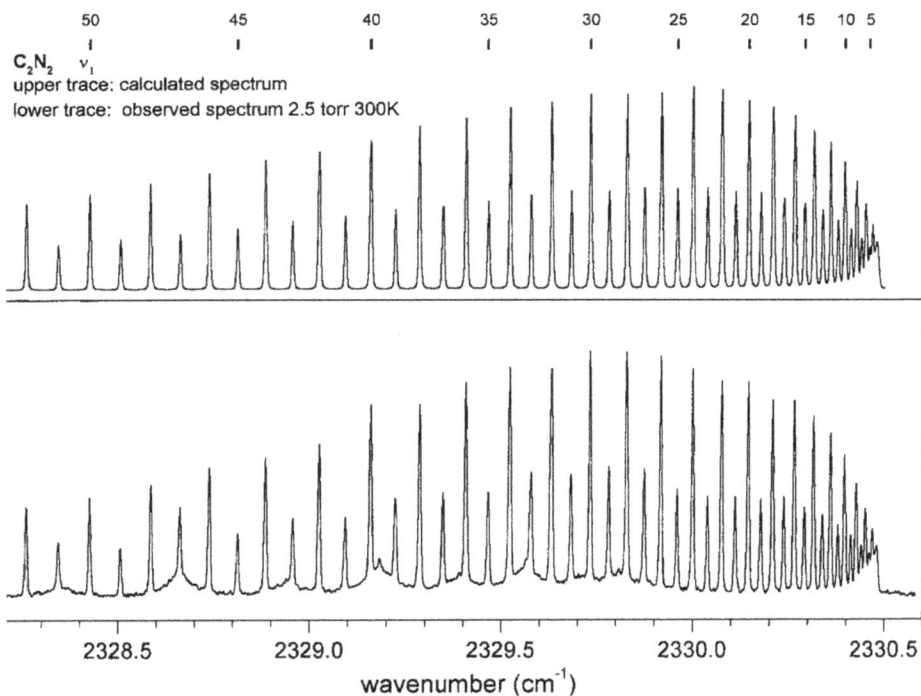

Figure 9.9: Stimulated Raman spectrum of the v_1 band of $^{12}C_2^{14}N_2$. The broad structure observed in the experimental contour is the spectrum of atmospheric nitrogen outside the cell (Bermejo *et al.* 1997). Copyright ©1997 John Wiley and Sons, with permission.

conditions. The authors apply careful control of cavity dispersion to satisfy the phase-matching condition of CARS in a high-finesse optical cavity.

High-resolution Raman spectroscopy of $^{12}C_2H_2$ and $^{13}C_2H_2$ is reported by Bermejo *et al.* (1998). The authors propose a generalization of a previously reported technique for investigating vibrationally excited states via high-resolution Raman spectroscopy. In the first step, a vibrational state, typically not accessible by a dipole moment transition from the ground state, is populated in a pulsed stimulated Raman process. After a delay of several nanoseconds, a high-resolution spectrum corresponding to transitions from this long-lived state is recorded following a quasi-cw-stimulated Raman spectroscopy scheme. The use of a pulsed dye laser for providing one of the pumping radiations overcomes some of the limitations of the previously proposed scheme, where this radiation was generated by a Raman shifter filled at relatively high pressure with the gas under study. Experimental aspects are discussed, and the spectra of the Q branches of $2\nu_2(\Sigma_g^+)-\nu_2(\Sigma_g^+)$ of $^{12}C_2H_2$ and $^{13}C_2H_2$ are presented. The analysis of the measured transitions yielded accurate values of the spectroscopic parameters for the $\nu_2 = 2$ vibrationally excited state for both isotopomers and improved values of the corresponding anharmonicity constants x^0_{22} are derived.

10 Time-resolved vibrational spectroscopy

Schrader *et al.* (1993) use the step-scan technique for time-resolved (TR) Raman measurements. This technique, previously developed for infrared (IR) spectroscopy, may be used in near-IR Fourier transform (FT)-Raman spectroscopy as well. It allows the study of photochemical and photophysical processes, the application of modulation techniques, and the investigation of "noisy" samples. Photo-isomerization of the dye merocyanine 540 has been observed with the step-scan technique upon periodic excitation with a flash lamp.

The review of Diem (2015) features the TR methods in vibrational spectroscopy. Diem notes that methods employing vibrational spectroscopy have the advantage that vibrational transitions occur at very fast time scales. That means that fast photochemical reactions can be studied in TR-FT-IR measurements where the reaction can be initiated by a laser pulse and can be repeated thousands of times. Diem describes the hydrolysis reaction of rat sarcoma virus protein (Ras)-bound guanosine-5′-triphosphate to Ras-bound guanosine-5′-diphosphate as an example, which details reaction the mechanism and dynamics revealed by TR-FT-IR spectroscopy.

Hamaguchi and Iwata (2017) describe the principle and practice of TR Raman spectroscopy (TRRS) in the subpicosecond to the hectosecond time regimes. The authors also give an overview of its historical developments. Prototype set-ups are illustrated for nanosecond and transform-limited picosecond TRRS. Intrinsic limit of time resolution due to the time-energy uncertainty principle is described. Application examples given include division of a fission yeast cell in the hectosecond regime, structure relaxation of transient deoxyhaemoglobin in the nano- and microsecond regimes, and excited electronic state internal conversion of β-carotene in the subpicosecond regime.

Solvent effects on the structure of the triplet excited state of xanthone are studied by Kumar and Umapathy (2016). TR resonance Raman spectroscopy of triplet excited states of xanthone is carried out in aprotic (acetonitrile) and protic solvents (methanol). A depolarized Raman peak at 614 cm^{-1} (an out-of-plane mode) appears in acetonitrile but is absent in methanol solvent; it is identified as a vibronic active mode coupling these two states. In strong contrast to the ground state of xanthone, a few Raman peaks of the $\pi\pi^*$ triplet state show a blue shift with increasing solvent polarity, which is corroborated by time-dependent density functional theory calculations.

Bakker *et al.* (2018) write an editorial to a virtual special issue about TR vibrational spectroscopy. A table of contents leads to various papers about the subject.

https://doi.org/10.1515/9783110717556-010

11 Vibrational spectroscopy of crystals

When a substance crystallizes, new bands appear in its vibrational spectra. Vibrational bands below 200 cm^{-1} show torsional and translational vibrations typical for molecules of crystalline substances on their lattice positions. They disappear when the substance melts or is dissolved.

Claassen *et al.* (1963) report the vibrational spectra of XeF$_4$. The Raman spectrum, taken from a single crystal, shows bands at 543 and 502 cm^{-1} (vs, stretching vibrations) and weaker ones at 235 and 442 cm^{-1}. Bands observed at 123, 291 (v_2, out-of-plane), and 586 cm^{-1} in the infrared spectrum are assigned to fundamentals. The races of all bands are: 586 E$_u$, 543 A$_{1g}$, 502 B$_{2g}$, 442 (overtone, 2 v_4, B$_{1u}$), 291 B$_{2u}$ (v_2), 235 B$_{1g}$ (v_3), 221 B$_{1u}$ (v_4), and 123 cm^{-1} E$_u$. Based on those bands, Claassen *et al.* describe XeF$_4$ as a planar molecule of symmetry D$_{4h}$.

Reference spectra of common minerals are published by Zeller and Juszli (1975). The spectroscopists show infrared spectra of 52 minerals in an application note, using spectral ranges from 4,000 to 240 cm^{-1} and 525 to 32 cm^{-1}. Some of the minerals (e.g. halite and fluorite) were studied in reflection from 4,000 to 240 cm^{-1} as well.

An article on Raman spectroscopy of amino acid crystals is published by Freire *et al.* (2017). The authors measured the Raman spectra of the (predominantly) α-form of glycine in the high wavenumber region for two scattering geometries, Z(YY)Z and Z(XX)Z. In Figure 11.1, it is possible to observe bands associated with symmetric stretching of CH$_2$, v_s(CH$_2$), at 2,971 cm^{-1} (Figure 11.2), antisymmetric stretching v_{as}(CH$_2$), at 3,006 cm^{-1}, and N–H stretching, v(N–H), at 3,146 cm^{-1}. Depending on the kind of polymorph, the stretching of N–H, in particular, is observed at different wavenumbers and with different intensities.

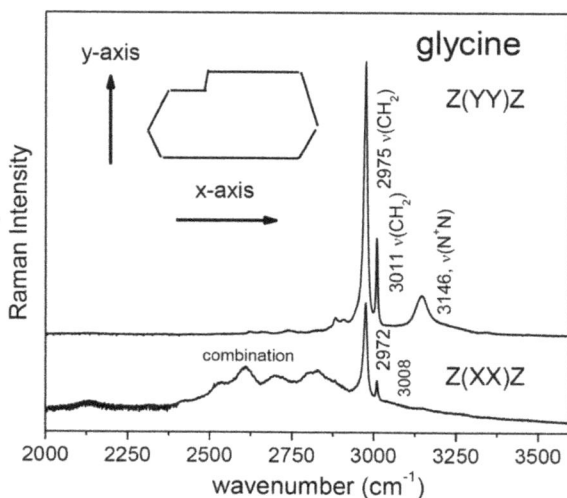

Figure 11.1: Polarized Raman spectra of glycine (α-form) for two scattering geometries (Freire 2017). Copyright ©2017 by the author, licensed by Creative Commons (CC).

https://doi.org/10.1515/9783110717556-011

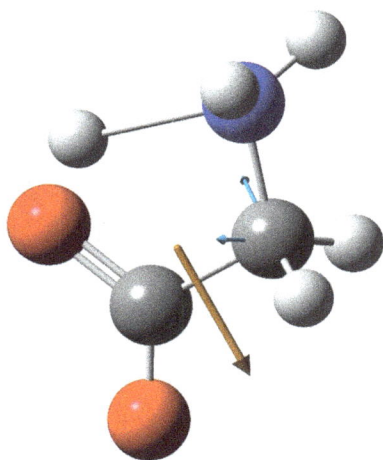

Figure 11.2: ν_s(CH$_2$) of glycine. Calculated 2,947 cm^{-1} (B3LYP, 6–311$^+$G(d,p), f = 0.9648), experimental 2,975 cm^{-1} (Hoffmann 2022).

Larkin *et al.* (2014) characterize polymorphs of active pharmaceutical ingredients using low-frequency Raman spectroscopy. As an example, the authors compare the conventional and the low-frequency Raman spectra of carbamazepine (CBZ). While the Raman spectra of CBZ forms II and III from 3,800 to 200 cm^{-1} are very similar, the low-frequency spectra show a clear fingerprint of the polymorph under study.

First-order Raman spectrum of fluorite contains only one single band as a result of the O$_{5h}$ symmetry and selection rules for the Raman bands. This Raman band is located at 322 cm^{-1} and is due to the T$_{2g}$ mode as found out by Srivastava *et al.* (1971). The frequency of the Raman active mode in CaF$_2$, EuF$_2$, PbF$_2$, SrCl$_2$, BaF$_2$, and BaCl$_2$ has been measured at 300 and 77°K.

FT-Raman spectroscopy is very valuable in the determination of polymorphs according to the opinion of Ayala *et al.* (2006). The authors choose olanzapine as an example and compare density functional theory (DFT) calculations (B3LYP, 6–31 G(d)) with experimental spectra.

Heyler *et al.* (2013) analyse molecular and polymorph structure with THz-Raman spectroscopy. This spectroscopic technique can yield rich information regarding the vibrational energies associated with molecular and intermolecular structures and features. These modes correspond to very low frequency (~5 to 200 cm^{-1}, or 150 GHz to 6 THz) emissions, which have been traditionally difficult and/or expensive to access, as most Raman spectrometers use filters that remove lines up to 150 cm^{-1} from the Rayleigh line. Heyler *et al.* reduced this blocked spectral region to 5 cm^{-1} by the use of ultra-narrowband volume holographic grating filters. As examples, they showed the spectra of CBZ polymorphs, sulfur allotropes, and ammonium nitrate for home-made explosive detection.

Compton and Hammer (2014) publish the Raman under liquid nitrogen (RUN) spectrum of benzene and naphthalene in the low energy region (Figures 11.3 and 11.4). Lattice modes are visible in benzene's Raman spectrum.

The discrimination of the amorphous and crystalline forms of a substance can be done by Raman spectroscopy, as shown by Larkin *et al.* (2015). The authors used low- and mid-frequency Raman spectroscopy to monitor the amorphous-crystalline trans- formation of indomethacin.

A database of Raman spectra of precious gemstones and minerals (Figures 11.5–11.7) used as cut gems is set up by Culka and Jehlička (2019). They obtain the spectra using a portable sequentially shifted excitation Raman spectrometer.

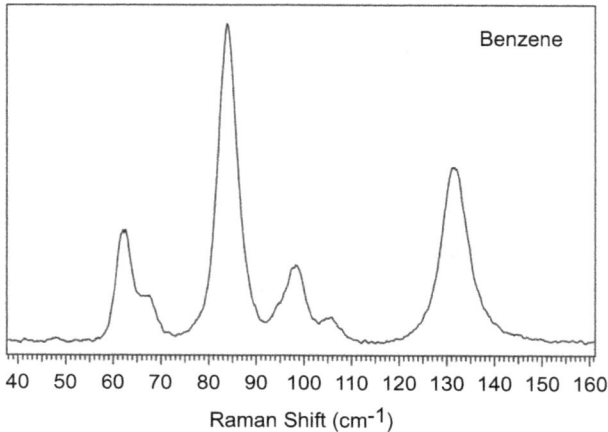

Figure 11.3: RUN (Raman under liquid nitrogen) spectrum of benzene in the low energy region of the spectrum recorded at UM. Lattice modes involving motions between benzene molecules are evident (Compton and Hammer 2014). Copyright ©2014 by the authors, permission by Creative Commons (CC 4.0): http://creativecommons.org/licenses/by/4.0/.

Porto *et al.* (1967) measure first-order Raman spectra in five materials belonging to the D_{4h} point group: TiO_2, MgF_2, ZnF_2, FeF_2, and MnF_2. As expected, four Raman-active phonon frequencies were observed: each spectrum exhibited strong lines of A_{1g} and E_g symmetries, a weak high-frequency line of B_{2g} symmetry, and a very sharp B_{1g} line at quite low fre- quency (<100 cm^{-1} in all materials except TiO_2). As an example, the Raman spectrum of MgF_2 is shown in Figure 11.8. Spectrum (a): α_{zz} component showing A_{1g} phonon at 410 cm^{-1}; spectrum (b): α_{zx} component showing E_g phonon at 295 cm^{-1}; spectrum (c): α_{xx} component showing B_{1g} and A_{1g} phonons at 92 and 410 cm^{-1}, respectively. Gain increase by a factor of 10 shows clearly the B_{1g} phonon. Spectrum (d): α_{yy} and α_{xy} components together showing B_{1g}, and A_{1g} again. The small peak at 515 cm^{-1} should be noticed, which is due to the B_{2g} phonon.

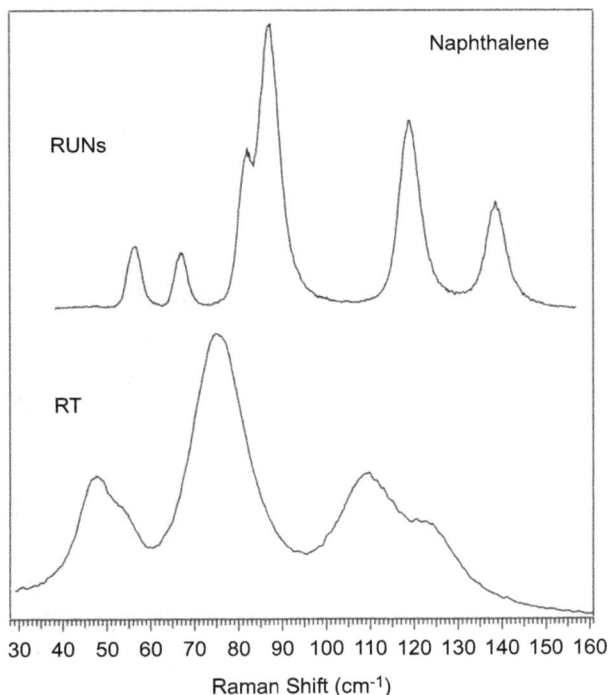

Figure 11.4: Room-temperature Raman spectrum versus RUN (*R*aman *u*nder liquid *n*itrogen) spectrum of naphthalene in the low energy region. Normal modes are observed to blue shift and sharpen under liquid nitrogen (Compton and Hammer 2014). Copyright ©2014 by the authors, licensed by Creative Commons (CC 4.0).

Isolation of few-layer black phosphorus by mechanical exfoliation down to two layers is described by Castellanos-Gomez *et al.* (2014). The authors characterize the exfoliated flakes by Raman spectroscopy and TEM measurements. They found that the flakes are highly crystalline and even in free-standing form. A strong thickness dependence of the band structure is found by DFT calculations.

In a review, Ferrari and Basko (2013) praise Raman spectroscopy as a versatile tool for studying the properties of graphene. It proves to be useful to determine the number and orientation of layers, the quality and types of edge, and the effects of perturbations, such as electric and magnetic fields, strain, doping, and disorder and functional groups; it also provides insight into all sp^2-bonded carbon allotropes, having graphene as their fundamental building block. Furthermore, the authors point out the potential of Raman spectroscopy for layered materials other than graphene.

Figure 11.5: Analysed cut gemstones and minerals (the scale of white lines is 5 mm): 1, green beryl (emerald); 2, colourless beryl; 3, bluish beryl (aquamarine); 4, colourless synthetic corundum; 5, blue synthetic corundum; 6, synthetic ruby (red synthetic corundum); 7, orange synthetic corundum; 8, bluish zircon; 9, pyrope almandine (red-brown garnet); 10, green olivine; 11, yellowish titanite; 12, smoky grey obsidian; 13, grossular yellow garnet; 14, grossular green garnet; 15, grossular orange garnet; 16, yellow scapolite; 17, violet scapolite; 18, colourless quartz; 19, violet quartz; 20, red-brown quartz; 21, yellow tourmaline; 22, brown tourmaline; 23, yellowish apatite; 24, colourless fluorite; 25, milky fluorite; 26, yellow fluorite; 27, pink fluorite; 28, blue kyanite; 29, colourless pollucite; 30, colourless sanidine; 31, yellowish labradorite; 32, colourless petalite; 33, bluish zircon; and 34, colourless cubic zirconia (Culka and Jehlička 2019). Copyright ©2019 John Wiley and Sons Ltd., with permission.

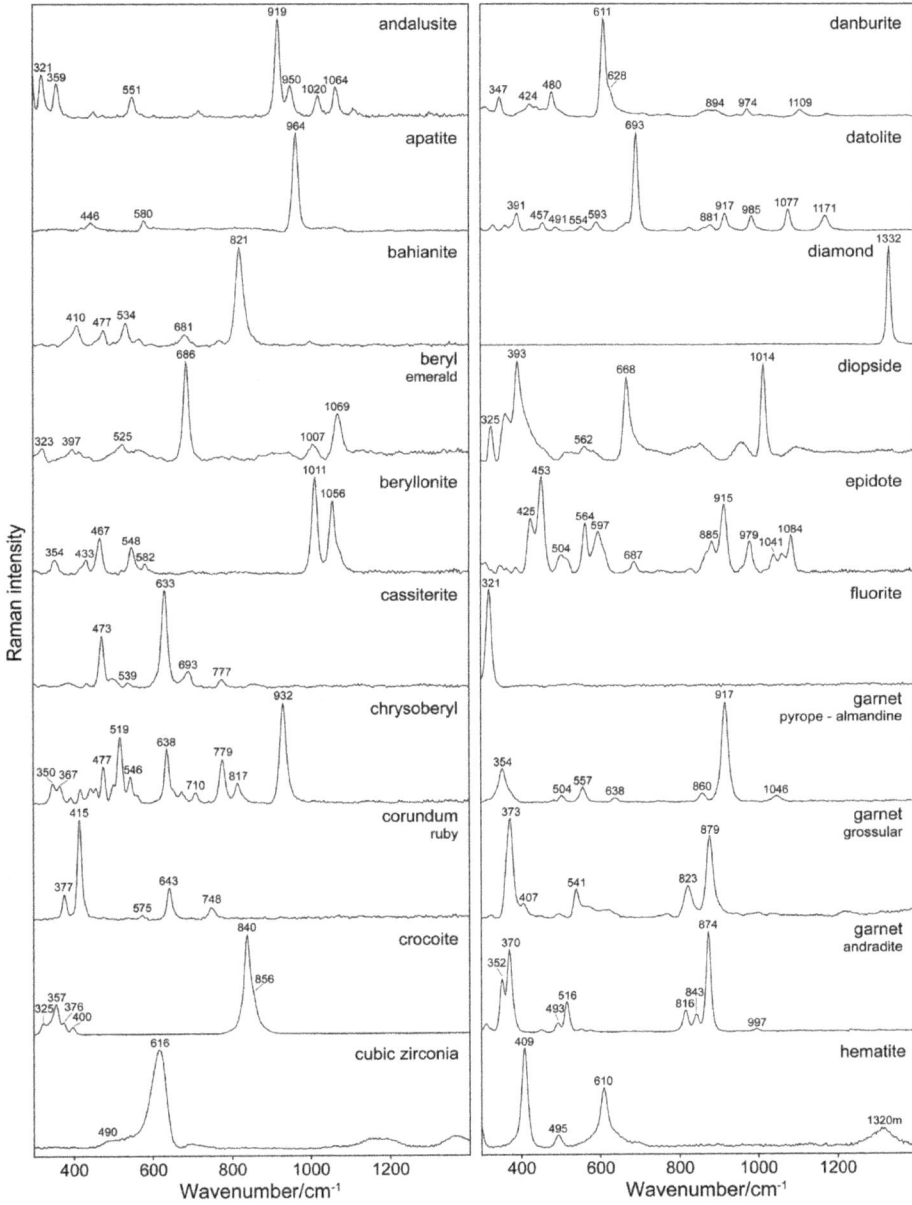

Figure 11.6: Raman spectra of gemstones from Figure 11.5 (Culka and Jehlička 2019). Copyright ©2019 John Wiley and Sons Ltd., with permission.

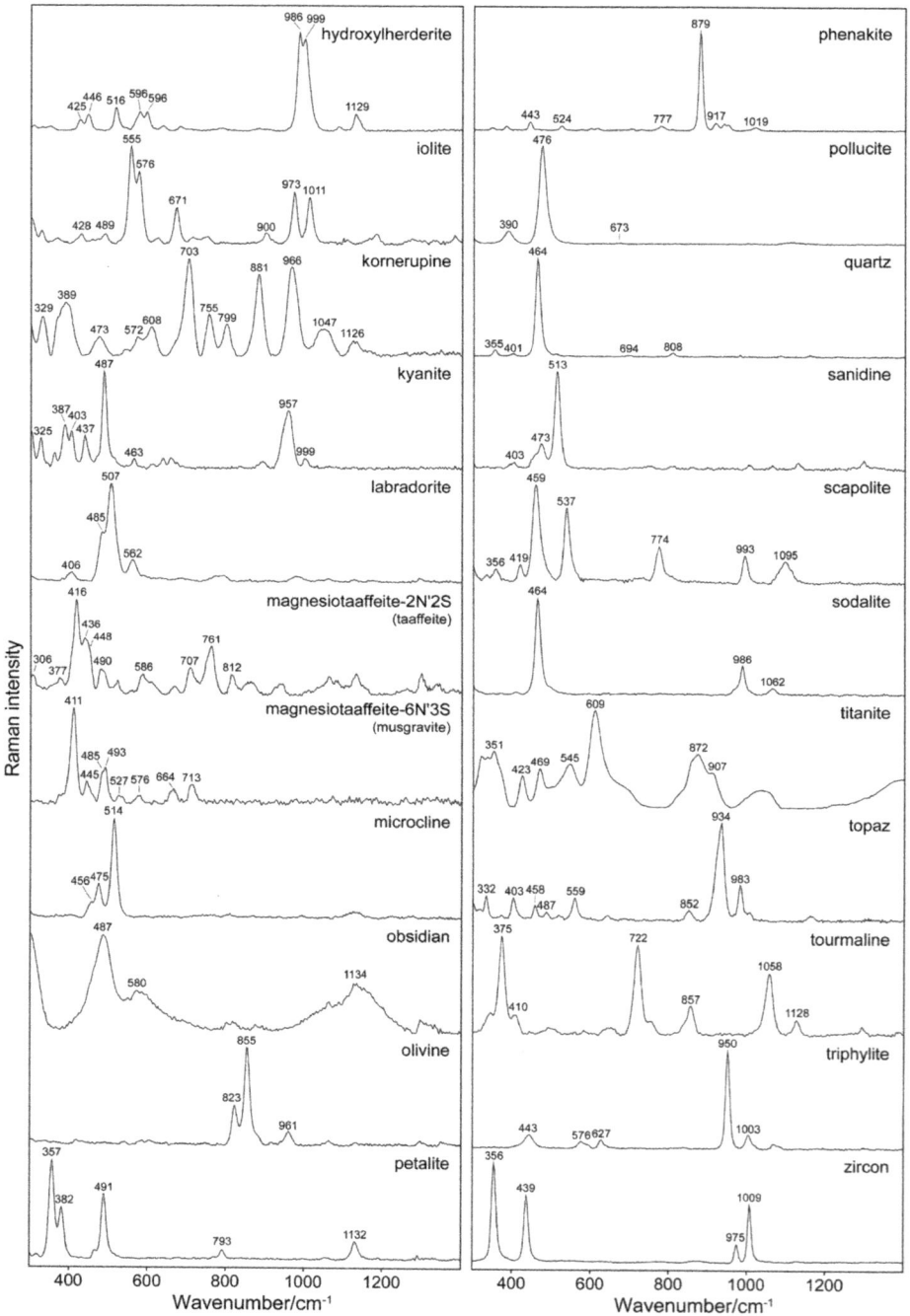

Figure 11.7: Raman spectra of gemstones from Figure 11.5 (Culka and Jehlička 2019). Copyright ©2019 John Wiley and Sons Ltd., with permission.

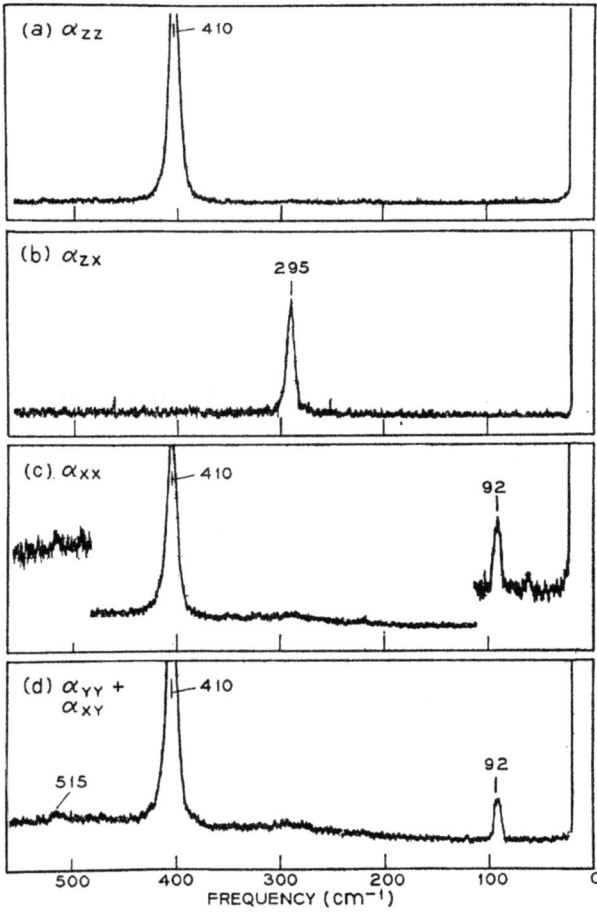

Figure 11.8: Raman spectrum of MgF_2 (Porto *et al.* 1967). Copyright ©1966 American Physical Society, with permission.

12 Vibrational spectroscopy of polymers

Polymeric compounds consist of very long macromolecules. This could convince you to believe that it is impossible to calculate all vibrational modes. Fortunately, this is not necessary: For a polymeric chain of infinite length, it could be shown that it is sufficient to calculate the mode of the repetition unit. The polymer chain can then be regarded as a one-dimensional crystal, due to the fact that chain-to-chain interactions can be neglected compared to the much larger intramolecular interactions. The whole vibrational problem can be treated as a problem of classical solid-state physics.

Besides group frequencies also specific polymer bands occur: vibrations of the chain, mirroring the chain structure.

Hummel (1974) provides a list of the problems to be solved in the analysis of polymers:
a) Chemical structure: structural units, entanglement, end groups, branching
b) Steric order: *cis–trans*-isomerism, taxis
c) Conformational order: "one-dimensional crystallinity" like planarity or helicity
d) Three-dimensional order: crystalline and mesomorphous phases

A simplified model of the change of morphology during processing of high polymers is shown by Tadokoro and Kobayashi (Figure 12.1).

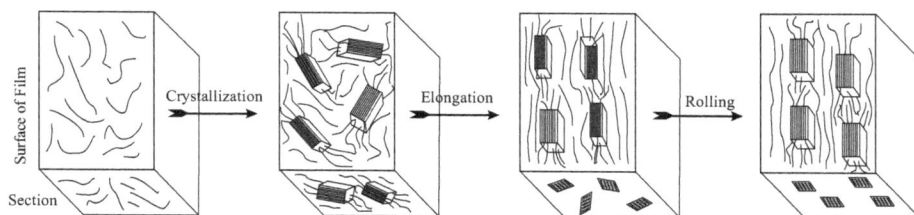

Figure 12.1: Simplified models for the morphology of high polymers (redrawn after Tadokoro and Kobayashi in Dieter O. Hummel (Ed.) "Polymer Spectroscopy", Verlag Chemie 1974).

12.1 Interpretation of polymer spectra

Sloane *et al.* (1973) report Raman spectra of some polymers and copolymers of styrene, butadiene, and methylmethacrylate.

Smith writes a regular column: "IR Spectral Interpretation Workshop". The last issues of this very readable column lately explain "The Infrared spectra of Polymers" with nine papers so far (Smith 2021, 2022).

Care has to be taken as the crystallinity and orientation affect the spectra of polymers; this is especially true for polyethylene terephthalate.

https://doi.org/10.1515/9783110717556-012

12.1.1 Hydrocarbon polymers

The simplest polymers are polyethylenes. They consist of long chains of methylene groups: $-(CH_2)_n-$. Methylene groups on adjacent molecules can be rocking in-phase and out-of-phase with each other.

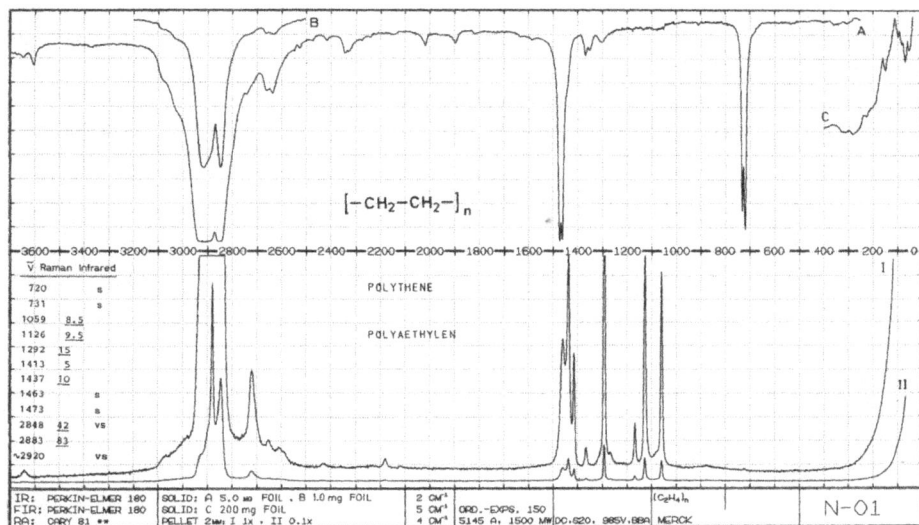

Figure 12.2: Infrared and Raman spectra of solid polyethylene (from Schrader, B. "*Infrared Atlas of Organic Compounds*", 2nd Ed., page N–01. Copyright 1996 Wiley-VCH Verlag GmbH & Co. KGaA. Reproduced with permission).

The thermoplast polyethylene comes with different structures depending on the synthesis. Whereas high-density polyethylene (HDPE) contains a high percentage of crystalline regions, the low-density polyethylene (LDPE) has more branched structures. This can be seen in their Raman spectra: the LDPE spectrum is more complicated than the HDPE spectrum (Figure 12.2).

The less stable LDPE is used for the manufacture of foils for carrier bags and sacks, pipes, and cable insulation sheaths.

HDPE is processed to bottles for cleansing agents, petrol canisters, garbage cans, bottle crates, and foils for landfill protection layers. Ultra-high-molecular-weight polyethylene (UHMWPE) can be processed to artificial joints and ropes for ships.

Schachtschneider and Snyder (1963) carry out normal co-ordinate calculations on the extended n-paraffins C_2H_6, through n-$C_{14}H_{30}$, and polyethylene using a perturbation method.

Tasumi *et al.* (1964) publish an infrared (IR) study, enhanced by an analysis of normal vibrations, of stereoregulated polydideuteroethylene.

Gall *et al.* (1972) publish the laser Raman spectra of several different samples of polyethylene and gave complete vibrational assignments of polyethylene (Table 12.1). The authors found out that the spectra are sensitive to crystallinity and to the presence of impurities such as vinyl groups. The possible effects of the presence of gauche CH_2 groups are discussed.

Table 12.1: Complete vibrational assignments of polyethylene (Gall *et al.* 1972).

Infrared	Raman	Assignment
2,920 cm^{-1}		$\nu_{as}(CH_2)$
2,851 cm^{-1}		$\nu_s(CH_2)$
	2,882 cm^{-1}	$\nu_{as}(CH_2)$
	2,847 cm^{-1}	$\nu_s(CH_2)$
1,473 cm^{-1}		
1,465 cm^{-1}		$\delta(CH_2)$
	1,463 cm^{-1}	$\delta(CH_2)$
	1,453 cm^{-1}	Fermi resonance
	1,441 cm^{-1}	and overtones
	1,416 cm^{-1}	
	1,372 cm^{-1}	
	1,295 cm^{-1}	
1,176 cm^{-1}		(CH_2)
	1,170 cm^{-1}	(CH_2)
	1,131 cm^{-1}	$\nu_s(C-C)$
	1,064 cm^{-1}	
1,050 cm^{-1}		
730 cm^{-1}		CH_2 rocking
720 cm^{-1}		
	104 cm^{-1}	
77 cm^{-1}		Lattice mode
	34 cm^{-1}	LAM ($m = 3$)
	12 cm^{-1}	LAM ($m = 1$)

The intensities are as follows:

2,920 cm^{-1} (IR vs, br, Ra sh) C–H stretching
2,883 cm^{-1} (IR vs, sh, Ra vs) C–H stretching
2,848 cm^{-1} (IR vs, br, Ra vs) C–H stretching
1,463 cm^{-1} (IR s, Ra m) C–H deformation
1,229 cm^{-1} (IR vw, Ra s) skeletal vibrations
1,126 cm^{-1} (IR vw, Ra s) skeletal vibrations
1,059 cm^{-1} (IR vw, Ra s) skeletal vibrations
725 cm^{-1} (IR m, Ra vw) doublet, CH_2 rocking

Lu *et al.* (1998) use Raman microscopy to analyse a series of high-modulus polyethylene fibres. A high degree of orientation within the samples was detected using a polarization analyser. Measurements of the 1,131 cm^{-1}:1,064 cm^{-1} band intensity ratio of the fibres can be related to their Young's modulus.

Real-time micro-Raman measurements on stressed polyethylene fibres are reported by Tarantili *et al.* (1998). Studying single fibres of UHMWPE, they evaluated strain rate effects and molecular stress redistribution.

Parker *et al.* (2000) test the validity of the use of the *n*-alkanes as model compounds for polyethylene. They found an excellent agreement when comparing the experimental data for the ν_5 mode of perdeuterated polyethylene and the results for perdeuterated *n*-hexadecane. Observed and calculated (DFT, B3LYP/6-31G**) frequencies for the longitudinal acoustic modes of $C_{16}D_{34}$ were also compared and agreed well. The authors stated to have found the first experimental validation of the assumption that the force fields derived from the *n*-alkanes are transferable to polyethylene.

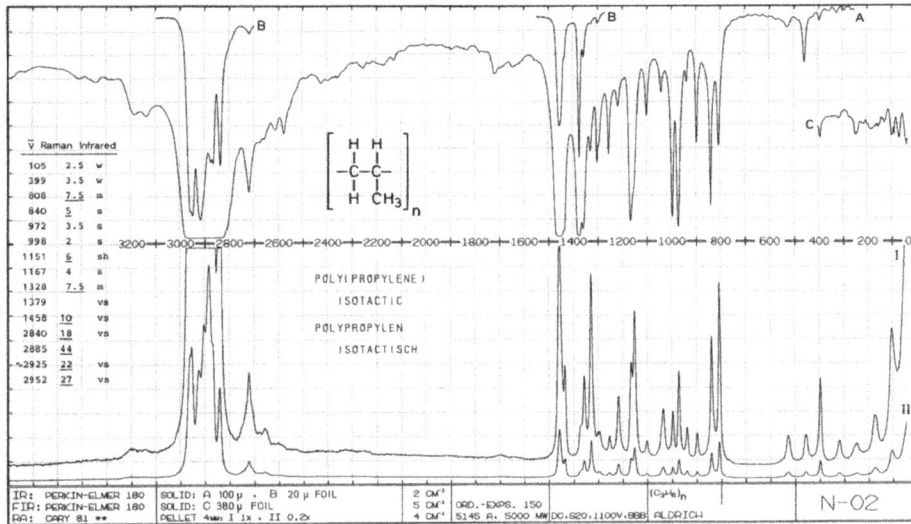

Figure 12.3: Infrared and Raman spectra of solid isotactic polypropylene (from Schrader, B. *"Infrared Atlas of Organic Compounds"*, 2nd Ed., page N–02. Copyright 1996 Wiley-VCH Verlag GmbH & Co. KGaA. Reproduced with permission).

Isotactic polypropylene is used in injection moulding for parts for electrical engineering, automobile parts, and domestic appliances. It is also extruded to fibres and foils. As a foamed material, it features high energy absorption, so it is used in vehicle bumpers. Its vibrational spectrum (Figure 12.3) shows bands at 2,950 cm^{-1} (Ra s IR s), 1,460 cm^{-1} (Ra s IR s), 1,380 cm^{-1} (Ra s IR s, δ_s CH$_3$), 1,155 cm^{-1} (Ra s IR m), and 970 cm^{-1} (Ra s IR m).

The isotactic form of polypropylene has sharp bands from 1,250 to 835 cm^{-1} (Ra m-w, IR m). Most of those bands disappear in atactic polypropylene, and only the bands at 1,155 and 970 cm^{-1} remain.

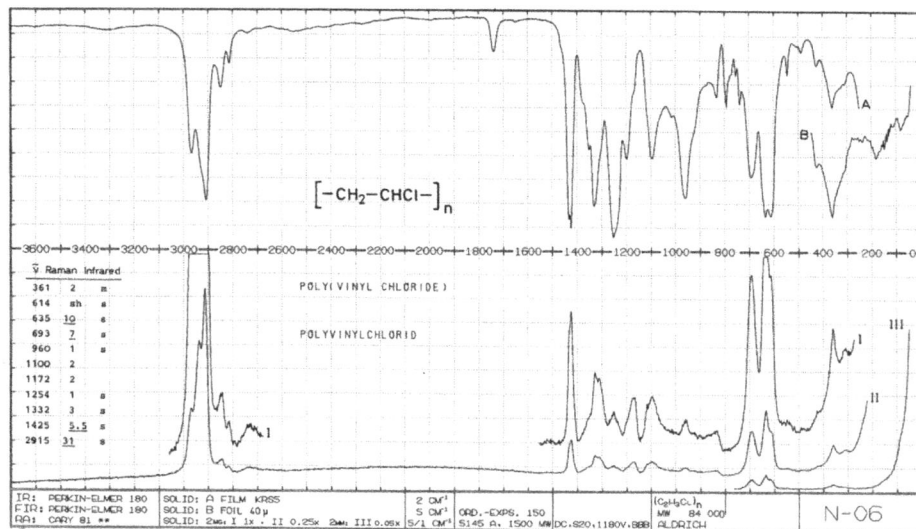

Figure 12.4: Infrared and Raman spectra of solid isotactic polyvinylchloride (from Schrader, B. "Infrared Atlas of Organic Compounds", 2nd Ed., page N–06. Copyright 1996 Wiley-VCH Verlag GmbH & Co. KGaA. Reproduced with permission).

Polyvinylchloride (PVC) is processed into tubes for drainage systems and floor coverings. It is one of the most frequently used polymers. In its vibrational spectrum (Figure 12.4), one observes the CH$_2$ deformational band at 1,425 cm^{-1} (IR s, shifted 30 cm^{-1} compared to other aliphatic CH$_2$ groups). The C–Cl stretching is observed as broad strong bands in the region of 710 to 590 cm^{-1}, where the bands at 693 cm^{-1} (IR s, Ra s) and 635 cm^{-1} doublet (IR s, Ra s) can be discriminated.

Quite often a band at 1,720 cm^{-1} is found, originating from the C=O stretching of a plasticizer.

PVC has a distorted planar zig-zag structure of in-phase and out-of-phase adjacent CCl$_2$ groups, as found out by Hendra *et al.* (1969).

Prokhorov *et al.* (2016) investigate Raman spectra of a number of industrial grades of PVC powder and films.

Poly(methacrylic acid methylester) is an optically clear and weatherproof thermoplast. It is used as a replacement for glass, as its impact resistance is six times higher.

In the vibrational spectra of polyacrylates, we find two lines at 1,260 cm^{-1} and 1,150 cm^{-1} from C–O–C stretching modes, and polymethacrylates have even two doublets (Figure 12.5). Other bands can be found at 1,200 and 835 cm^{-1}.

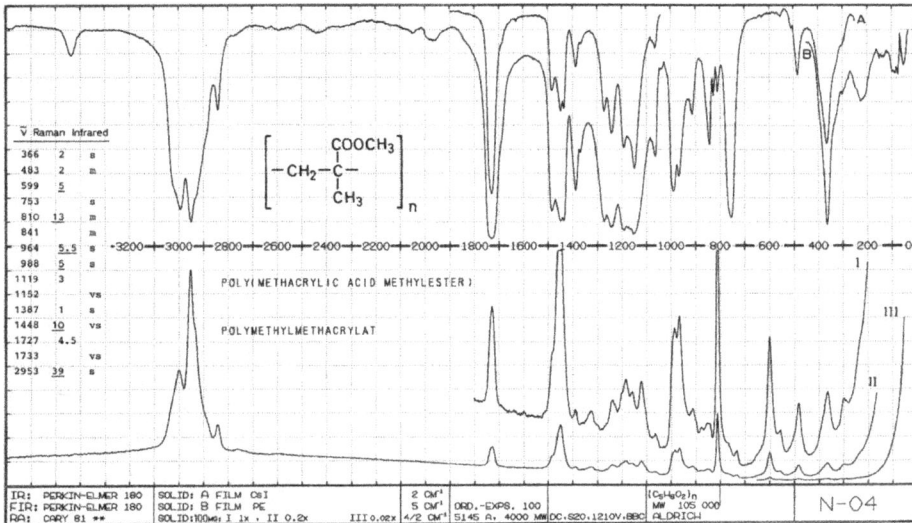

Figure 12.5: Infrared and Raman spectra of solid poly(methacrylic acid methylester) (from Schrader, B. *"Infrared Atlas of Organic Compounds"*, 2nd Ed., page N–04. Copyright 1996 Wiley-VCH Verlag GmbH & Co. KGaA. Reproduced with permission).

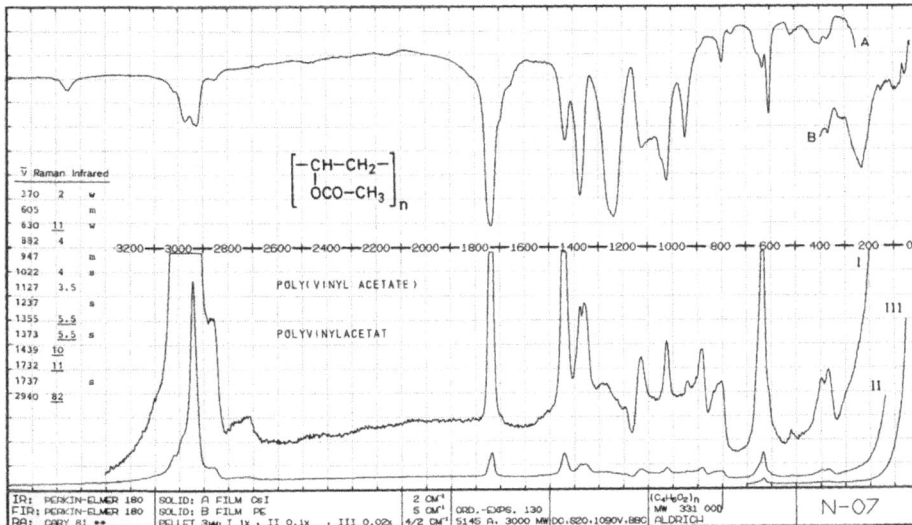

Figure 12.6: Infrared and Raman spectra of solid poly(vinylacetate) (from Schrader, B. *"Infrared Atlas of Organic Compounds"*, 2nd Ed., page N–07. Copyright 1996 Wiley-VCH Verlag GmbH & Co. KGaA. Reproduced with permission).

Poly(vinylacetate) shows good outdoor durability (high light and weather resistance). It is used as a bonding agent in emulsion paints and adhesives. At 1,737 cm^{-1}, we find the C=O stretching vibration of the ester as a band strong in Raman and IR (Figure 12.6). The acetate C–O–C groups asymmetric stretching can be found at 1,237 cm^{-1} as a strong IR band, which is only weak in the Raman spectrum. The band at 1,022 cm^{-1} is of medium intensity both in the Raman and the IR.

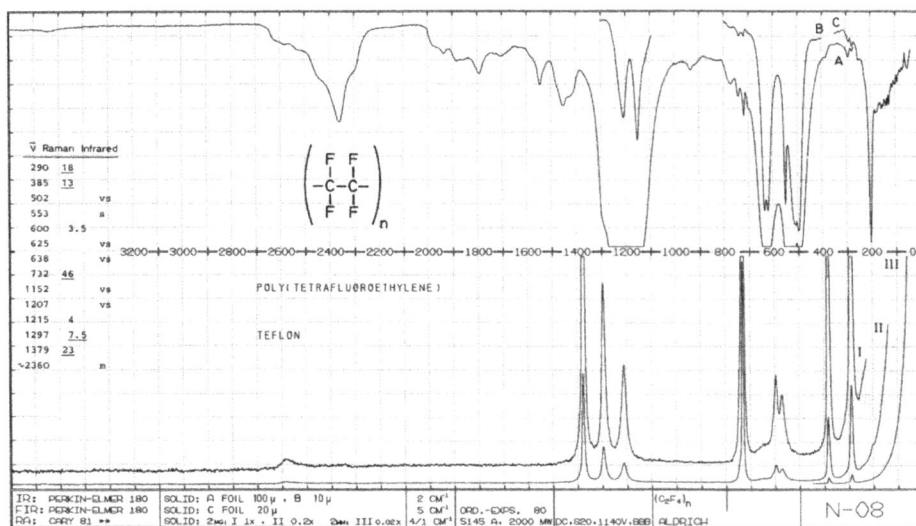

Figure 12.7: Infrared and Raman spectra of solid poly(tetrafluoroethylene) (from Schrader, B. *"Infrared Atlas of Organic Compounds"*, 2nd Ed., page N–08. Copyright 1996 Wiley-VCH Verlag GmbH & Co. KGaA. Reproduced with permission).

Poly(tetrafluoroethylene) shows low wettability and good sliding properties. The Raman spectrum in Hendra (1974) shows only one weak band above 1,379 cm^{-1}, the CF$_2$ stretching overtone at 2,360 cm^{-1} (IR m), whereas many overtone and combination bands may be observed in the IR (Figure 12.7). Other vibrational bands are 1,381 cm^{-1} (s, ν_{sym} CF$_2$), 1,217 cm^{-1} (Ra s, ν_{asym} CF$_2$), 1,207 cm^{-1} (IR vs), 1,152 (IR vs), 732 cm^{-1} (Ra vs), 576 cm^{-1} (Ra m, CF$_2$ rock), 551 cm^{-1} (Ra vw, IR m, CF$_2$ wagging), 385 cm^{-1} (Ra s, IR sh, δ CF$_2$), 290 cm^{-1} (Ra s, IR vw, CF$_2$ twisting), and 93 cm^{-1} (Ra vw, δ_{C-C} skeletal bend).

The thermoplastic poly(vinylfluoride) (PVF) finds multiple uses as it is resistant to sunlight degradation and solvents; it is chemical inert; and it has high tensile strength, toughness, and dimensional stability. Alaaeddin *et al.* (2019) review properties, applications, and manufacturing prospects of the polymer. Its vibrational spectrum (Boerio and Koenig 1971) shows a 2,940 cm^{-1} C–H stretching band, the C–H deformation band at 1,430 cm^{-1}, and the C–F stretching band at 1,085 cm^{-1} (IR). Additionally, Raman bands not observed in the IR spectra are found at 2,973, 1,437, 1,327, 1,198, and 1,059 cm^{-1}.

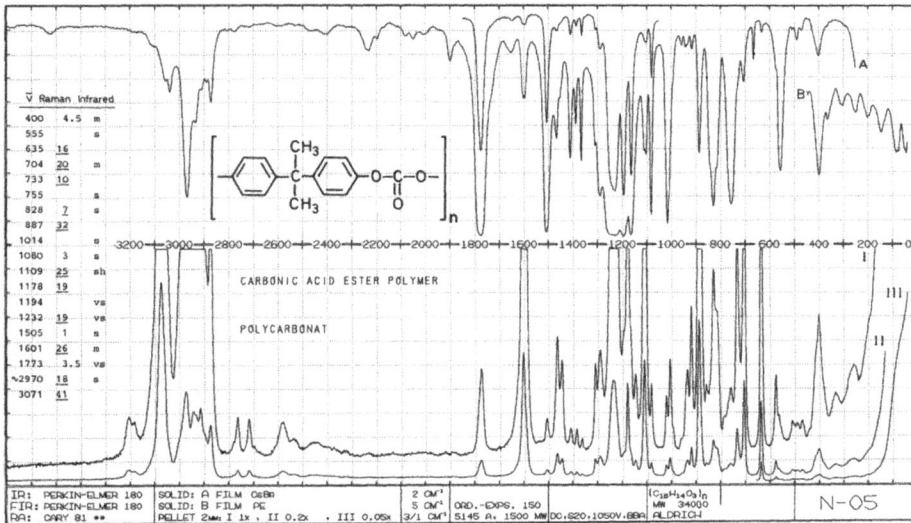

Figure 12.8: Infrared and Raman spectra of solid polycarbonate (from Schrader, B. "*Infrared Atlas of Organic Compounds*", 2nd Ed., page N–05. Copyright 1996 Wiley-VCH Verlag GmbH & Co. KGaA. Reproduced with permission).

The transparent polycarbonate is processed glass fibre reinforced for housings, for helmets, glazing for windows and doors, and for traffic signs. It is also used as a substrate of compact discs.

Near 1,785 cm^{-1} in the vibrational spectra of polycarbonates, we find the C=O stretching vibration (Ra m, IR vs), and near 1,250 cm^{-1} the C–O–C vibration (Ra s, IR vs br). Thus, the spectrum of the aromatic polycarbonate shown in Figure 12.8 has the vibration of the *p*-substituted aromatic ring near 860 cm^{-1} (Ra vs, IR m).

Polystyrene, as a transparent material, is used as a lightweight and nearly unbreakable substitute for glass. Its vibrational spectra show the characteristic bands of aliphatic and aromatic hydrocarbons. In the IR spectrum of polystyrene (Figure 12.9), strong bands from =C–H stretching at 3,100–3,000 cm^{-1} are prominent. In the region from 2,000 to 1,600 cm^{-1}, we find a series of characteristic combination bands (w, only in IR): 1,940, 1,870, 1,800, 1,740, and 1,670 cm^{-1}. Bands from the monosubstituted benzene ring can be found at 1,603 cm^{-1} (IR m, aromatic C=C stretching) and at 1,600–1,430 cm^{-1} (aromatic ring stretching). The bands at 758 cm^{-1} (Ra w, IR s) and at 699 cm^{-1} (Ra w, IR vs) belong to the C–H out-of-plane vibration of a monosubstituted aromatic group. At 620 cm^{-1}, the Raman spectrum shows a sharp, strong line (IR w), assigned to C–C *trans*-gauche (Menezes *et al.* 2017) vibrations within the phenyl ring.

Polystyrene can be fractionated by liquid chromatography. The purified compounds can then be identified by spectroscopy, as reported by Sackett *et al.* (1978).

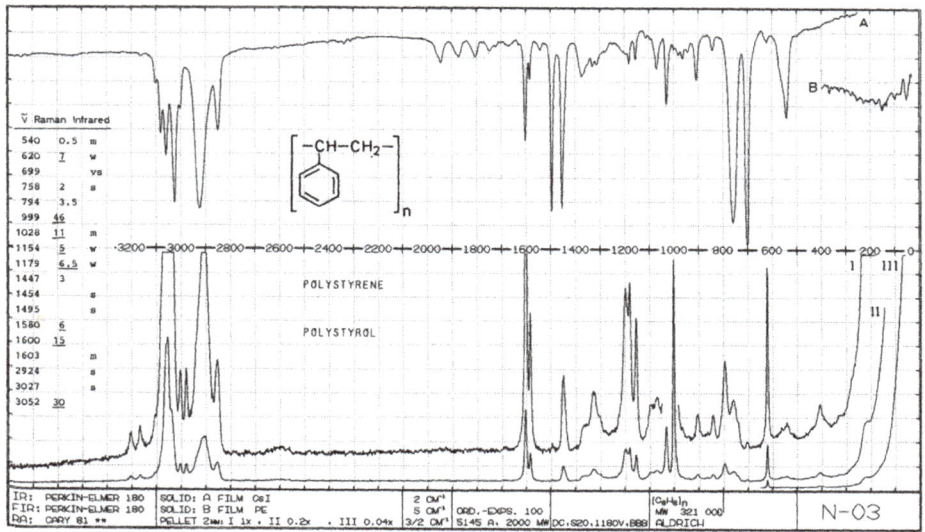

v̄ Raman	Infrared	
540	0.5	m
620	7	w
699		vs
758	2	s
794	3.5	
999	46	
1028	11	m
1154	5	w
1179	6.5	w
1447	3	
1454		s
1495		s
1580	6	
1600	15	
1603		m
2924		s
3027		s
3052	30	

POLYSTYRENE

POLYSTYROL

IR: PERKIN-ELMER 180   SOLID: A FILM CsI
FIR: PERKIN-ELMER 180   SOLID: B FILM PE
RA: CARY 81 **   PELLET 2mm I 1x , II 0.2x   . III 0.04x

2 CM⁻¹
5 CM⁻¹
3/2 CM⁻¹

ORD.-EXPS. 100
5145 A: 2000 MW DC.S20.1180V.BBB

[C₈H₈]n
MW 321 000
ALDRICH

N-03

Figure 12.9: Infrared and Raman spectra of solid polystyrene (from Schrader, B. *"Infrared Atlas of Organic Compounds"*, 2nd Ed., page N–03. Copyright 1996 Wiley-VCH Verlag GmbH & Co. KGaA. Reproduced with permission).

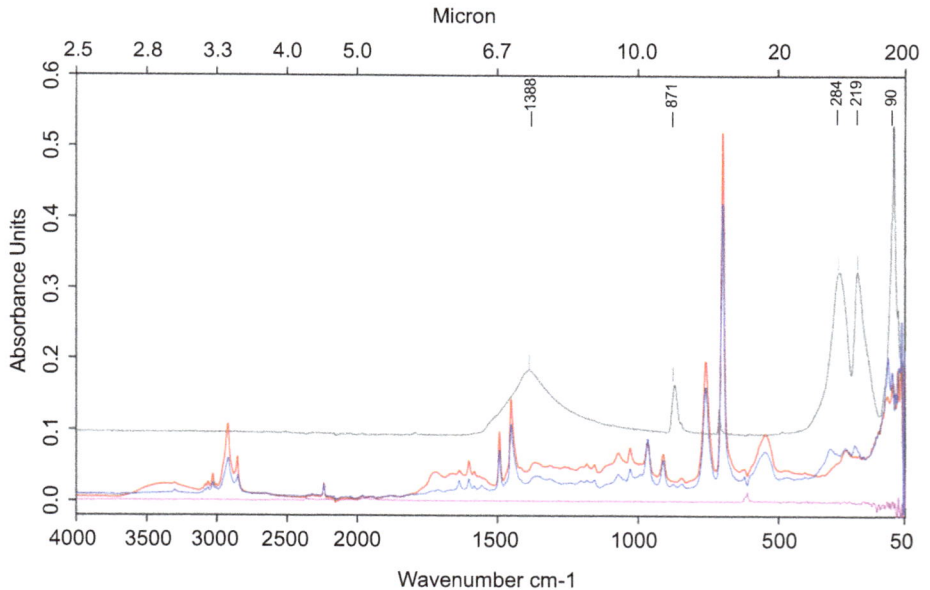

Figure 12.10: The IR spectrum of an acrylonitrile-butadiene-styrene copolymer (ABS) and its filler, $CaCO_3$, may be obtained with a single measurement from 6,000 cm^{-1} to 80 cm^{-1} (www.bruker.com/optics).

A polystyrene film was often used in grating IR spectrometers for wavelength calibration. For Fourier-transform (FT) spectrometers, this is obsolete, as these are self-calibrating using their laser.

Polymers can be made from a single monomer or several monomers. The former ones are called homopolymers, and the latter are called copolymers. The IR spectrum of the acrylonitrile-butadiene-styrene copolymer (ABS) together with the spectrum of its filler is shown in Figure 12.10 as an example.

ABS features high impact resistance together with high stiffness. This is useful in automobile construction, housing, domestic appliances, electrical devices, and safety helmets. The copolymer's vibrational spectrum contains the bands described for polystyrene and the bands of the acrylonitrile and butadiene components. The $C{\equiv}N$ stretching band is found at 2,240 cm^{-1} (Ra s, IR w).

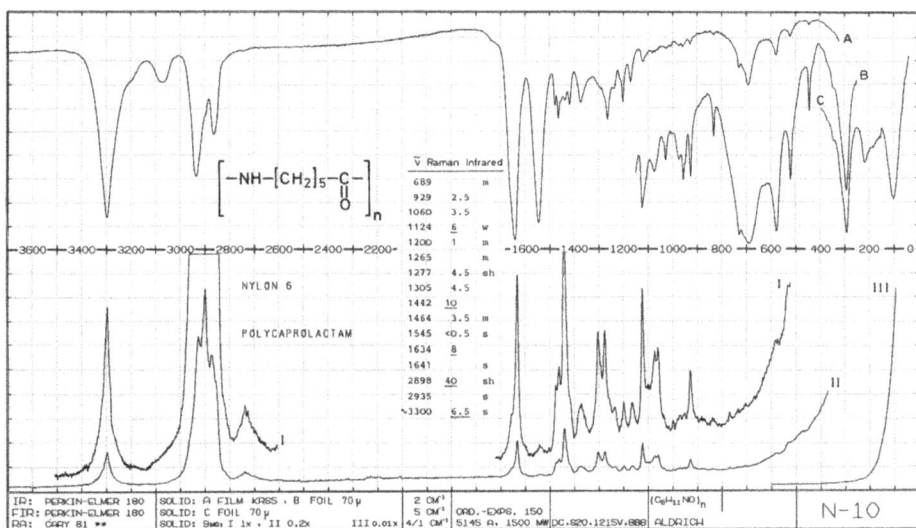

Figure 12.11: Infrared and Raman spectra of solid polycaprolactam (from Schrader, B. "*Infrared Atlas of Organic Compounds*", 2nd Ed., page N–10. Copyright 1996 Wiley-VCH Verlag GmbH & Co. KGaA. Reproduced with permission).

Polyamides are best known by their trademarks such as Nylon™ and Kevlar™. These compounds are used to make fibres of extraordinary strength. As an example, the vibrational spectra of polycaprolactam (Nylon 6) are shown in Figure 12.11. As other amides, polyamides show amide I, II, and III bands in their IR spectra at approx. 1,640 cm^{-1} (Ra m, IR s), 1,540 cm^{-1} (Ra w, IR s), and 1,280 cm^{-1} (Ra m, IR sh), respectively.

The amorphous polysiloxanes ("silicones") are characterized by the disubstituted siloxane group as a repeating unit. They are used as isolating filling gaps, hydraulic pump oils, and thermosetting high-temperature varnishes. Their vibrational spectra contain the bands of $(-Si-O-)_n$ chain and its substituents.

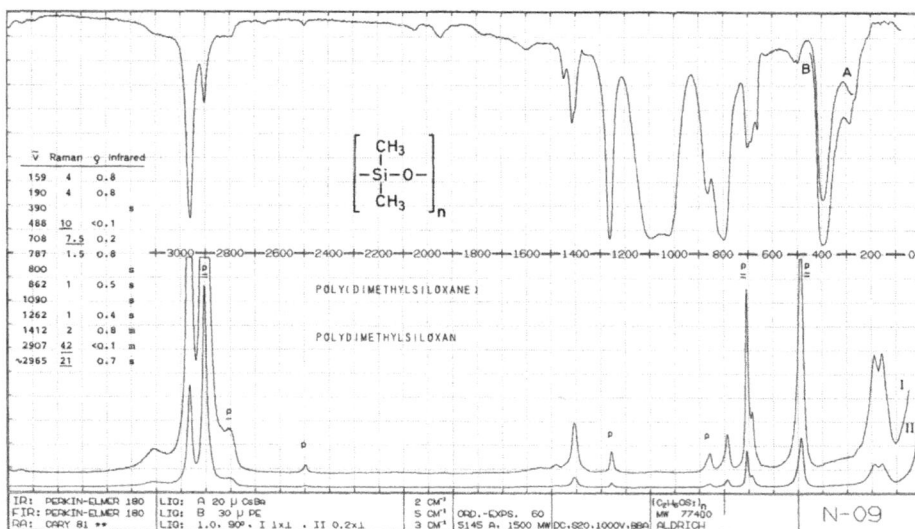

Figure 12.12: Infrared and Raman spectra of solid poly(dimethylsiloxane) (from Schrader, B. *"Infrared Atlas of Organic Compounds"*, 2nd Ed., page N–09. Copyright 1996 Wiley-VCH Verlag GmbH & Co. KGaA. Reproduced with permission).

For example, in the spectra of poly(dimethylsiloxane) (PDMS, Figure 12.12), we find the bands of methyl groups at 2,965 cm^{-1} (Ra s, IR s, ν_{as} CH$_3$), 2,907 cm^{-1} (Ra vs, IR m, ν_s CH$_3$), and 1,261 cm^{-1} (Ra w, IR m, δ_s CH$_3$). At 1,100–1,000 cm^{-1}, a very broad and strong two-peak IR band, which does not appear in the Raman spectrum, can be detected. It originates from the Si–O–Si stretching vibration.

The band of the symmetric CH$_3$Si deformation can be found at 1,262 cm^{-1} (Ra w, IR s). At 800 cm^{-1}, the band of the Si–C (Ra –, IR s br) shows up, whereas the Si–O can be found at 862 cm^{-1} (Ra w, IR m br).

Epoxy resins are all the resins that are produced by an epoxy compound and an alcohol, a carbonic acid, or an amine. The curing of the mixture can be followed by monitoring the epoxy bands. After curing, the resulting product is a network of high chemical resistance. An example is the condensation product of bis(4-hydroxyphenyl) propane-2 (bisphenol A) and epichlorohydrin. This resin contributes 75% of all epoxy resins. Vašková and Křesálek (2011) monitor epoxy resin cross-linking via Raman microscopy in nearly real time.

Hexagonal and orthorhombic polyoxymethylene and the deuterated compound are studied by Zerbi and Hendra (1968). Laser-excited Raman spectra of the compound have been measured, and the experimental frequencies have been compared with the calculated ones. The ν_{as} (COC) of both forms shows at 1,097 cm^{-1} and 1,091 cm^{-1}, respectively (Ra w, IR wws), the ν_s (COC) at 932 cm^{-1} (Ra w, IR wws).

Natural rubber consists mainly of polymers of the organic compound isoprene. Rubber, an elastomer, is a general term for natural or synthetic polymers that show rubber-like elasticity at room temperature. Most of the synthetic rubbers are polymers made from 1,3-dienes like butadiene or isoprene (methylbutadiene). Cornell and Koenig (1970) present the Raman spectra of cis-1,4-, *trans*-1,4-, and 3,4-polyisoprene, and Vašková and Křesálek (2011) show the IR and Raman spectra of styrene–butadiene rubber.

Polyurethanes are used as rigid and flexible foams, varnishes, and coatings. They are manufactured from isocyanates and diols or polyols. The urethane structure – NH–C(=O)–O– shows the following IR absorptions:

Polyurethanes from aliphatic isocyanates: 1,695, 1,538, 1,250, 781 cm^{-1}.

Polyurethanes from aromatic isocyanates: 1,695, 1,600, 1,538, 775–763 cm^{-1}.

The transparent thermoplastic polysulfones are known for their toughness and stability at high temperatures. Technically used polysulfones contain an aryl–SO_2–aryl subunit. In the IR region, they show strong bands at 1,170–1,120 cm^{-1} and at 1,360–1,300 cm^{-1}. Due to better mechanical properties, often polyethersulfones are used.

Polyolefins as packing materials cause a great problem for the environment because they can only be degraded very slowly. Polylactic acid (PLA) has been proposed as a replacement because it is produced from biological raw materials and can be composted and decomposed easily in humid and warm conditions. Kobayashi *et al.* (2022) study the relationship of amorphous structure to the tear strength of PLA films.

12.2 Rheo-optical spectroscopy of polymers

The behaviour of polymers during stretching can be advantageously studied by rheo-optical spectroscopy. A miniature device was designed by Siesler (Figure 12.14), which is mainly used with IR and near-IR spectrometers using linearly polarized radiation, but has also been used with Raman spectroscopy.

Stress/strain diagrams provide insight to the mechanical properties of a polymer. A polymer film sample is periodically stretched and relaxed (Figure 12.13). During these cycles, stress is plotted against strain. In a dedicated stretching machine, spectra of the sample can be measured by IR and Raman spectrometers. Using linear polarized radiation, the orientation of functional groups in the polymer and crystallization phenomena can be detected.

From the experimental spectra, the orientation of functional groups can be calculated as follows:

The structural absorbance A_0 calculated as

$$A_0 = \left(A_\parallel + 2A_\perp\right)/3$$

and the oriental function f of absorption bands whose transition moment directions have been assumed to be parallel or perpendicular to the polymer axis chain:

$$f_{\parallel} = (R-1)/(R+2)$$

and

$$f_{\perp} = -2\,(R-1)/(R+2).$$

by a procedure appropriately correlate the successively measured integral or peak-maximum absorbance values $A_{\parallel}$ and $A_{\perp}$ of selected absorption bands. The dichroic ratio R is defined as $R = A_{\parallel}/A_{\perp}$.

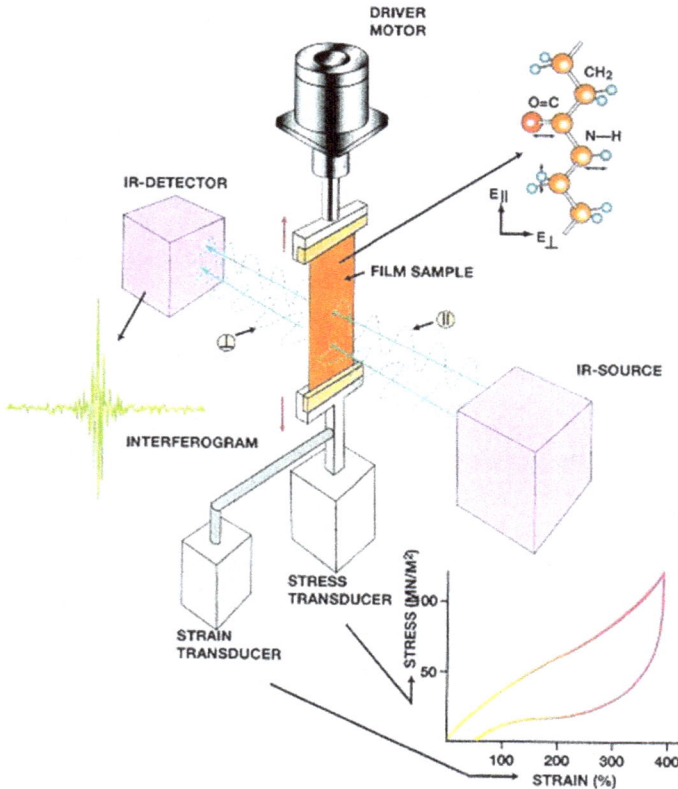

Figure 12.13: Measurement principle of rheo-optical measurements. Courtesy of Prof. H. W. Siesler, Essen.

Lu *et al.* (1984) study the electric-field-induced microstructural changes in PVF$_2$ at high temperature (65 °C) by spectroscopy. The authors found that the reduction of the coercive field is directly related to faster dipolar orientation at 65 °C, compared to the measurements at room temperature. The α- to β- or δ-phase transformation can take place at a much faster rate and lower field strength.

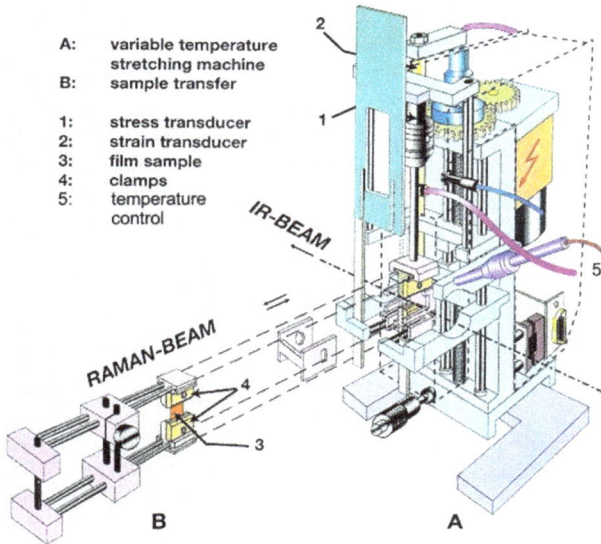

Figure 12.14: Experimental set-up of the miniature stretching machine. Courtesy of Prof. H. W. Siesler, Essen.

Hoffmann *et al.* (1993) review rheo-optical FT IR and Raman spectroscopy of polymers. As examples of the technique, bidirectional drawing of poly(ethylene terephthalate) and the strain-induced conformational changes of poly(vinylidene fluoride) are detailed.

Rodríguez-Cabello *et al.* (1995) study the rheo-optical behaviour of uniaxially stretched poly(vinylidene fluoride) by FT-Raman spectroscopy. Poly(vinylidene fluoride) exists in four different polymorphs named I(β), II(α), III(γ), and $IV_p(\delta)$ or $IV_p(\alpha_p)$.

A low-temperature FT near-IR spectroscopic study of strain-induced orientation and crystallization in a PDMS network is published by Klimov *et al.* (2005). The mechanical properties, morphology, and orientation of a PDMS network have been studied during cyclic elongation and recovery (Figure 12.15) by simultaneous FT near-IR polarization spectroscopy at temperatures ranging from room temperature to –40 °C. Completely different orientation/recovery mechanisms and changes in the state of order of PDMS were detected as a consequence of cyclic loading/unloading with decreasing temperature. The differences observed at –20 °C compared to room temperature are explained in terms of conformationally regular chain segments, whereas the cooling to –40 °C leads to the formation of lamellar crystals.

Huang *et al.* (2021) report enhanced piezoelectricity from highly polarizable oriented amorphous fractions in biaxially oriented poly(vinylidene fluoride) (BOPVDF) with pure β crystals. Figure 12.16 shows the FT-IR spectra of a sample film before and after extensive unipolar poling at 650 MV/m, from which all α-absorption bands have been disappeared. The spectrum shows that pure β-phase crystals were successfully prepared for the BOPVDF.

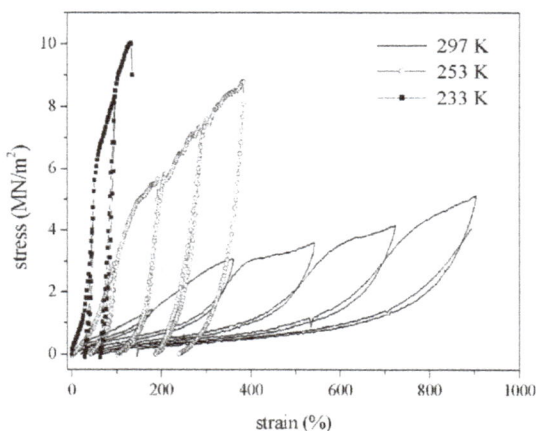

Figure 12.15: Stress/strain diagrams of the elongation/recovery cycles of PDMS films measured at +24 °C (297 K), –20 °C (253 K), and –40 °C (233 K) (Klimov *et al.* 2005). Copyright ©2005 John Wiley and Sons, with permission.

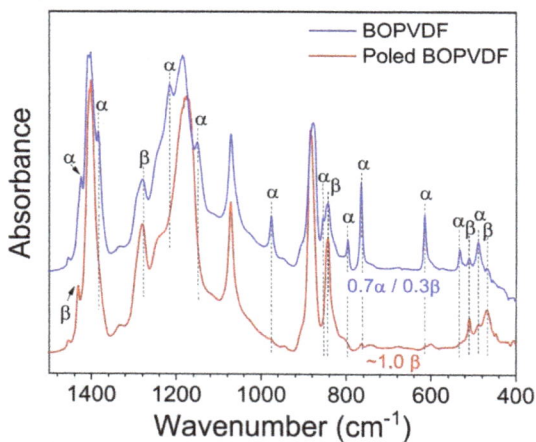

Figure 12.16: FT-IR spectra for the fresh and poled BOPVDF films in the transmission mode. Absorption bands of α- and β-crystals are labelled (Huang *et al.* 2021). Copyright ©2021 by the authors, licensed by Creative Commons (CC 4.0).

Kida *et al.* (2015) study HDPE during tensile deformation using Raman spectroscopy (Figure 12.17). Observing the bands of the amorphous and crystalline chains of the polymer during stretching, they observe peak shifts in both the C–C stretching region (around 1,100 cm^{-1}) and the CH$_2$ bending modes (around 1,400 cm^{-1}).

Vogel *et al.* (2009) observe the orientation of poly(3-hydroxybutyrate) (PHB) and PLA segments in PHB/PLA blend films (PHB < PLA) during uniaxial elongation up to 250% strain at 50 °C by *in situ* rheo-optical FT-IR spectroscopy. From the orientation

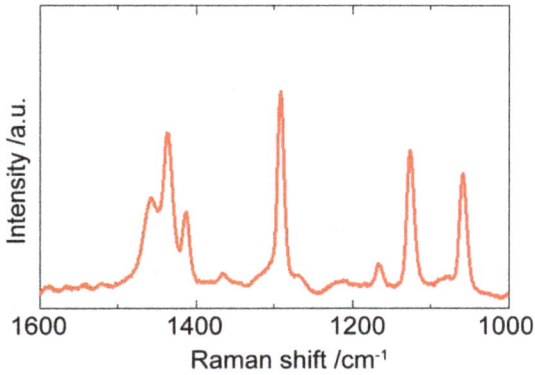

Figure 12.17: Part of the Raman spectrum of undrawn HDPE specimen (Kida *et al.* 2016). Copyright ©2016 by the authors, licensed by Creative Commons (CC 3.0).

functions of the ν(C=O) bands of the blend components, it was derived that the PLA chains orient in the direction of elongation while the PHB chains orient perpendicular to the drawing direction. Blends with PHB > PLA or pure PHB could only be drawn in ice water, resulting in chains aligned parallel to the drawing direction for both blend components.

13 Matrix isolation spectroscopy

The thermal motion of molecules at room temperature is complicating their spectra, so there are a couple of advantages of taking vibrational spectra of molecules in a matrix of condensed noble gases. This technique, called matrix isolation spectroscopy, was first intended to enable the observation of unstable molecules. The spectra obtained are not disturbed by neighbouring molecules and only very subtly changed by surrounding rare gas atoms, thus resembling gas phase spectra without rotational lines. The temperature of the matrices is normally about 10–20 K, but can be as low as 4.2 K using liquid helium. Thanks to the low temperatures, vibrational bands are extremely sharp, preventing partial overlap and allowing for increased accuracy in the measurement of band position. Contrary to gas phase spectra, in these matrices rotation of molecules (with the exception of a few small hydrides) is not observed, also sharpening the lines.

Barletta *et al.* (1971) succeed in measuring the Raman spectrum of S_2 in an argon matrix at 7 K. Using a krypton laser for excitation, they measure a band at 716 cm^{-1}, which is close to 725 cm^{-1} given by other authors for the IR absorption.

The Raman spectrum of SF_6 in argon at 4.2 K is shown by Shirk and Claassen (1971). All three Raman-active fundamental vibrations of the molecule could be detected at various excitation frequencies. The authors also studied $CHCl_3$ in an argon matrix (see Section 5.1.3).

The infrared spectrum of matrix-isolated molecular HOF is reported by Goleb *et al.* (1972). The authors observed fundamental frequencies at 886.0, 1,359.0, and 3,537.1 cm^{-1} in N_2 at ca. 8 K.

Peng *et al.* (2016) study the infrared and ultraviolet spectra of diborane (B_2H_6, Figure 13.1) and deuterated diborane (B_2D_6) in a neon matrix. Complete vibrational assignments (B3LYP/ 6-311^{++}G(d,p)) are given.

Bava *et al.* (2021) prepare FC(S)SF, FC(S)SeF, and FC(Se)SeF (Figure 13.2) through matrix (Ar) photochemical reactions of F_2 with CS_2, SCSe, and CSe_2. FTIR data and calculated frequencies, computed with the NBO method at the B3LYP/6-311$^+$G* level of approximation, of the fluorine complexes are given. For the $CS_2\cdots F_2$ complex (Figure 13.3) $\nu_{as}(CS_2)$ is given as 1,524.0 cm^{-1} and for $\nu(F_2)$ the experimental frequency is 867.7 cm^{-1}.

https://doi.org/10.1515/9783110717556-013

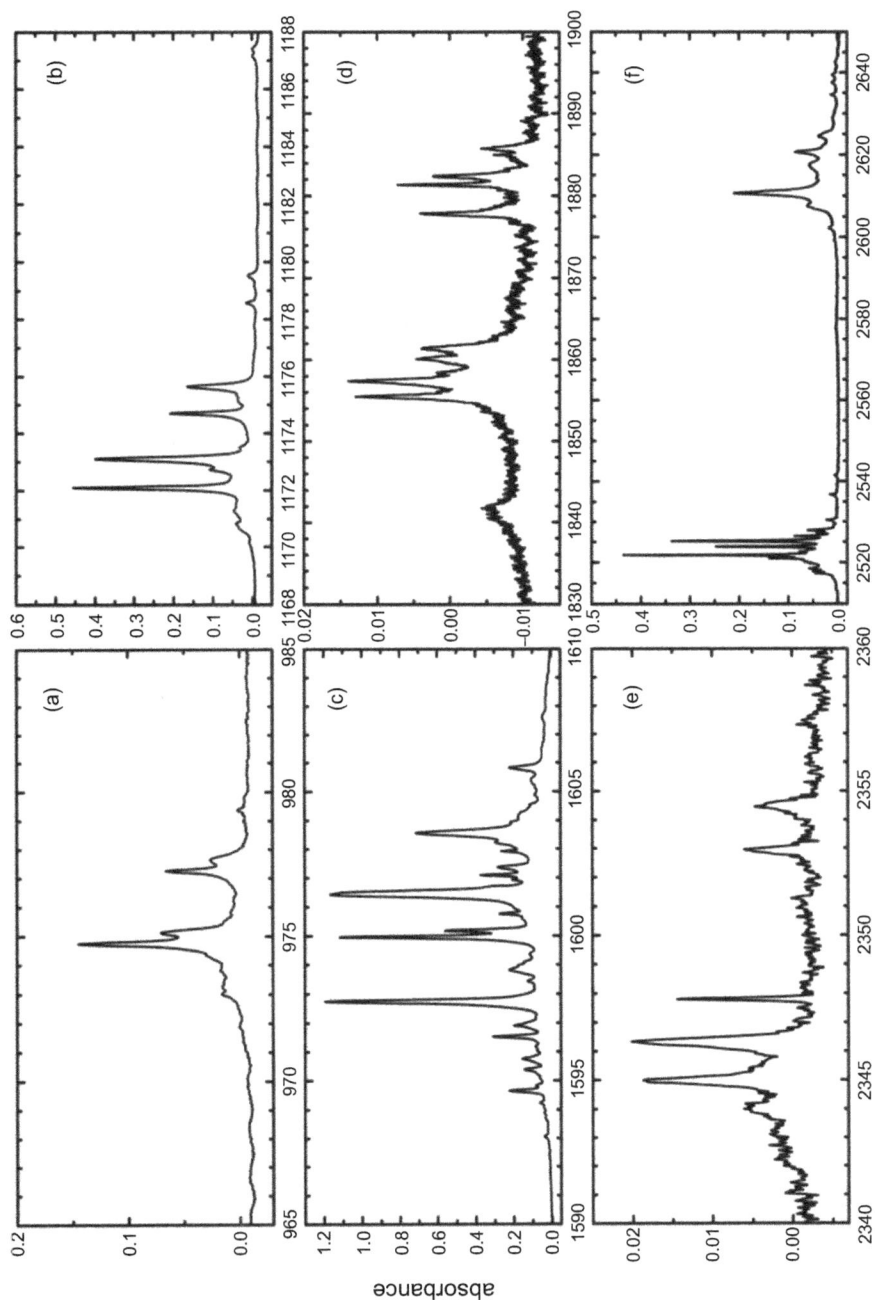

Figure 13.1: Infrared absorption spectra of B_2H_6 dispersed in neon (1:1,000) at 4 K: (a) 965–985, (b) 1,168–1,188, (c) 1,590–1,610, (d) 1,830–1,900, (e) 2,340–2,360, and (f) 2,510–2,650 cm^{-1} (Peng *et al.* 2016). Copyright ©2016 American Chemical Society, with permission.

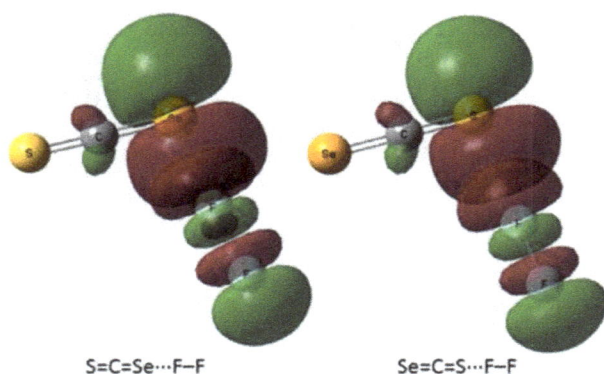

$$S=C=Se\cdots F{-}F \qquad Se=C=S\cdots F{-}F$$

Figure 13.2: Schematic representation of the orbital interactions between the lone electron pair located at a selenium or sulphur atom of the SCSe molecule and the unoccupied σ *anti*-bonding orbital of the F_2 molecule in the SCSe$\cdots F_2$ and SeCS$\cdots F_2$ complexes (Bava *et al.* 2021). Copyright ©2021 Royal Society of Chemistry, with permission.

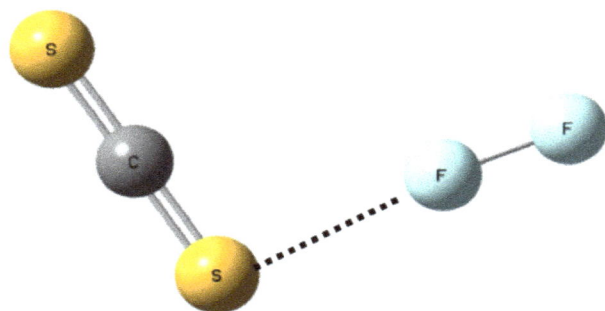

Figure 13.3: Optimized geometry of the $CS_2\cdots F_2$ complex (Bava *et al.* 2021). Copyright ©2021 Royal Society of Chemistry, with permission.

14 Nonlinear Raman spectroscopy

Borman (1982) recapitulates two laser conference papers by Eckbreth about nonlinear Raman spectroscopy. He publishes quantum diagrams of the various techniques (Figure 14.1), drawn by Eckbreth. In these quantum diagrams of the nonlinear Raman processes, spontaneous Raman is included as a reference. Levels shown are virtual states (--------) and ground and excited vibrational levels in the ground electronic state of the molecule. Ground vibrational levels are marked "G". For each Raman process shown, arrows entering at left are incident photons, up arrows represent photon absorption, down arrows represent photon generation, and arrows at right are output photons. The figure is redrawn after Alan Eckbreth in Borman (1982).

In a series of review papers, Kiefer 2007–2009 and later Nafie 2010–2020 report "Recent advances in linear and non-linear Raman spectroscopy".

Spontaneous Raman, or "normal" Raman, spectroscopy is characterized by an incoming laser beam promoting the molecules to a higher "virtual" level. Coming back to normal, most molecules emit light of the laser frequency: Rayleigh scattering. Very few molecules emit light of a lower frequency: Raman scattering, making this a very inefficient process. Therefore, some complicated experiments have been designed, improving the efficiency of Raman scattering.

In stimulated Raman scattering (SRS), the virtual state is first populated by a second light source; then, the exciting laser depopulates this state, resulting in an enormous enhancement of the Raman efficiency. This process is similar to the amplification process in a laser.

In SRS microscopy, like coherent *anti*-Stokes Raman scattering (CARS) microscopy, both the pump and Stokes photons are incident on the sample. If the frequency difference SRS = pump − Stokes matches a molecular vibration (vib), stimulated excitation of the vibrational transition occurs. Unlike CARS, in SRS, there is no signal at a wavelength that is different from the laser excitation wavelengths. Instead, the intensity of the scattered light at the pump wavelength experiences a stimulated Raman loss, with the intensity of the scattered light at the Stokes wavelength experiencing a stimulated Raman gain. The key advantage of SRS microscopy over CARS microscopy is that it provides background-free chemical imaging with improved image contrast, both of which are important for biomedical imaging applications where water represents the predominant source of non-resonant background signal in the sample.

In hyper-Raman spectroscopy, two pump photons are converted into one photon of Raman scattered light and one phonon. The scattered light occurs at optical frequencies somewhat lower than *twice* the frequency of the pump light.

https://doi.org/10.1515/9783110717556-014

14.1 Stimulated Raman gain

If a sample is irradiated by a pulsed laser of frequency ω_0 and a continuous laser covering the vibrational frequency region $\omega_0 + 3{,}500$ cm^{-1}, absorptions by Raman-active vibrations ω of the sample may be observed at $\omega_0 + \omega$. An emission at ω_0 may also be observed. This process, called inverse Raman effect, extends the general applicability of Raman spectroscopy to unstable species, fluorescing compounds, and low-pressure gases.

14.2 Coherent *anti*-Stokes Raman

CARS is a nonlinear four-wave mixing process that is used to enhance the weak (spontaneous) Raman signal. In the CARS process, a pump laser beam (at frequency pump) and a Stokes laser beam interact, producing an *anti*-Stokes signal at a frequency two times the pump frequency minus the Stokes frequency.

14.3 Coherent Stokes Raman

Coherent Stokes Raman spectroscopy (CSRS) is closely related to but less commonly used as CARS. Unlike CARS, CSRS is nonparametric; the final state is not the same as the initial state. Therefore, CSRS spectra may exhibit extra peaks due to coherence dephasing. Furthermore, the CSRS output beam is generated to the Stokes (lower frequency) side of ω_3.

14.4 Resonance Raman spectroscopy

In 1946, Harrand and Lennuier (1946) – when exciting some yellow crystals with yellow light – notice an intimate connection between the absorption and an enhancement of certain vibrational transition bands related to the chromophore. The normal Raman effect is a very weak one. It is excited by radiation in a region where the substance under study is not absorbing. Excitation in an absorbing region results in a huge enhancement of the vibrations belonging to the atoms of the absorbing chromophore.

A complete assignment of the resonance Raman (RR) spectra of cytochrome c is given by Berezhna *et al.* (2003). The results illustrate that cytochrome c has an altered vibrational spectrum in solution, in intact, and in swollen mitochondria. The authors found out that cytochrome c has a different vibrational spectrum in solution, in intact, and in swollen mitochondria.

Robert (2004) publishes a review of RR studies in photosynthesis. He focuses on (bacterio)chlorophyll and carotenoid molecules within photosynthetic systems. Marker

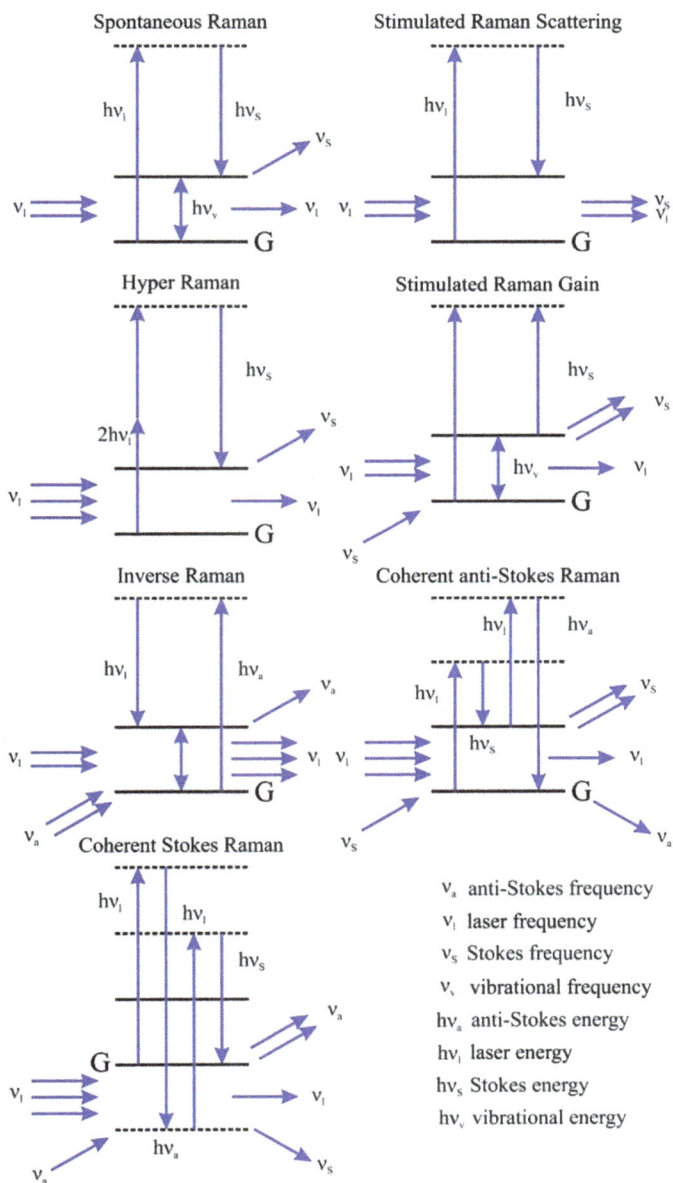

Figure 14.1: Quantum diagram of the various nonlinear Raman techniques (redrawn after Eckbreth).

bands for the coordination state of the central Mg atom and for the interaction state of the various carbonyl functionalities of (bacterio)chlorophyll molecules are identified and some examples are given.

When Berg (2015) studies fruit juice, beer, or wine at 229 nm excitation, only the Raman spectrum of ascorbic acid, present or added for preservation, can be detected. Raman spectroscopy investigations of L-(+)-ascorbic acid and its mono- and di-deprotonated anions (AH⁻, Figure 14.2, and A^{2-}) are reviewed, and new measurements are reported with several wavelengths. Results are interpreted by new DFT/B3LYP quantum chemical calculations with the $6–311^{++}G(d,p)$ basis set.

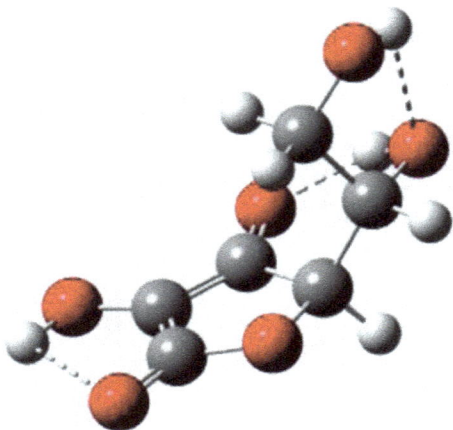

Figure 14.2: Ascorbic acid: Optimized structure of the AH⁻ ion in the PCM water model as obtained after Gaussian minimization based on Berg's guess. This global minimum conformer of AH⁻ has three internal hydrogen bonds (Berg 2015). Copyright ©2015 Rolf W. Berg. Licensed by Creative Commons (CC 4.0).

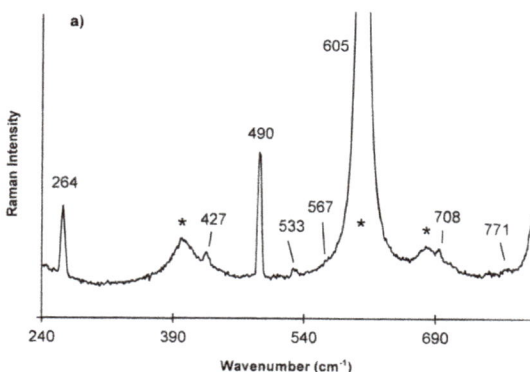

Figure 14.3: Resonance Raman spectrum of C_{60} in benzene at 413.1 nm excitation for the wavenumber region 240–750 cm⁻¹. Solvent peaks are asterisked (Gallagher *et al.* 1997). Copyright ©1997 American Chemical Society, with permission.

Galagher *et al.* (1997) study the RR spectrum of C_{60} (Figures 14.3 and 14.4). The authors report that the I_h point group is a good descriptor for the 10 fundamental Raman modes of C_{60}, the slight reduction in the high symmetry of the molecule, due to solvent and ^{13}C effects, activates at least 6 of the remaining 36 Raman-silent modes.

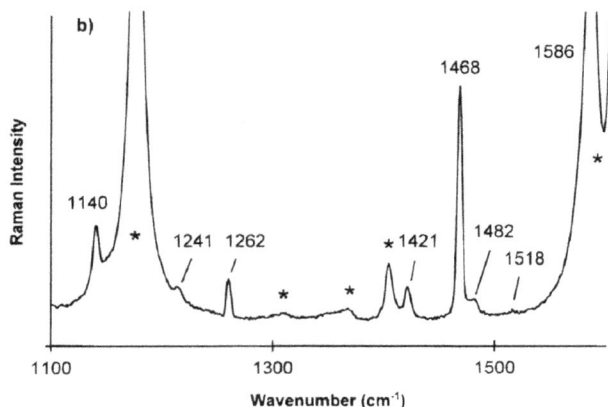

Figure 14.4: Resonance Raman spectrum of C_{60} in benzene at 413.1 nm excitation for the wavenumber region 1,000–1,650 cm^{-1}. The band at 1,262 cm^{-1} is a mercury emission line from fluorescent lights for calibration. Solvent peaks are asterisked (Gallagher *et al.* 1997). Copyright ©1997 American Chemical Society, with permission.

Baiardi *et al.* (2014) perform an accurate yet feasible computations of RR spectra (RRS) for metal complexes in solution: $[Ru(bpy)_3]^{2+}$ as a case study (Figure 14.5).

Optical activity (OA) can also be measured in RRS. Raman OA (ROA) of a photoreceptor protein under pre-resonance and resonance conditions is studied by Haraguchi *et al.* (2015). The photoactive yellow protein is isolated from *Salinibacter ruber*. When the RRS is excited with 532 nm, most of the RROA bands are negative (Figure 14.6). They can be assigned to the *p*-coumaric acid chromophore, except for a small, non-resonant protein amide I band at 1,667 cm^{-1} (C=O stretching).

Zając and Bour (2022) review the measurement and theory of RROA for gases, liquids, and aggregates. They point out the advantages for RR methods compared to the normal Raman: the intensity of the part of the Raman and the ROA spectrum related to the chromophore is greatly enhanced, which means that the necessary amounts of sample are greatly reduced. This is especially important in biochemistry, where samples are often only available in minor amounts.

14.5 Coherent *anti*-Stokes Raman spectroscopy (CARS)

In the mid-1960s, a signal from a new vibrational technique, called the "Three wave mixing experiment", was recorded for the first time. This technique, later called *coherent anti-Stokes Raman scattering* (Terhune 1963), uses two lasers to create high imaging contrast without labelling. Additionally, the objects, even living, remain intact.

Coherent *anti*-Stokes Raman spectroscopy is reviewed by Begley *et al.* (1974)

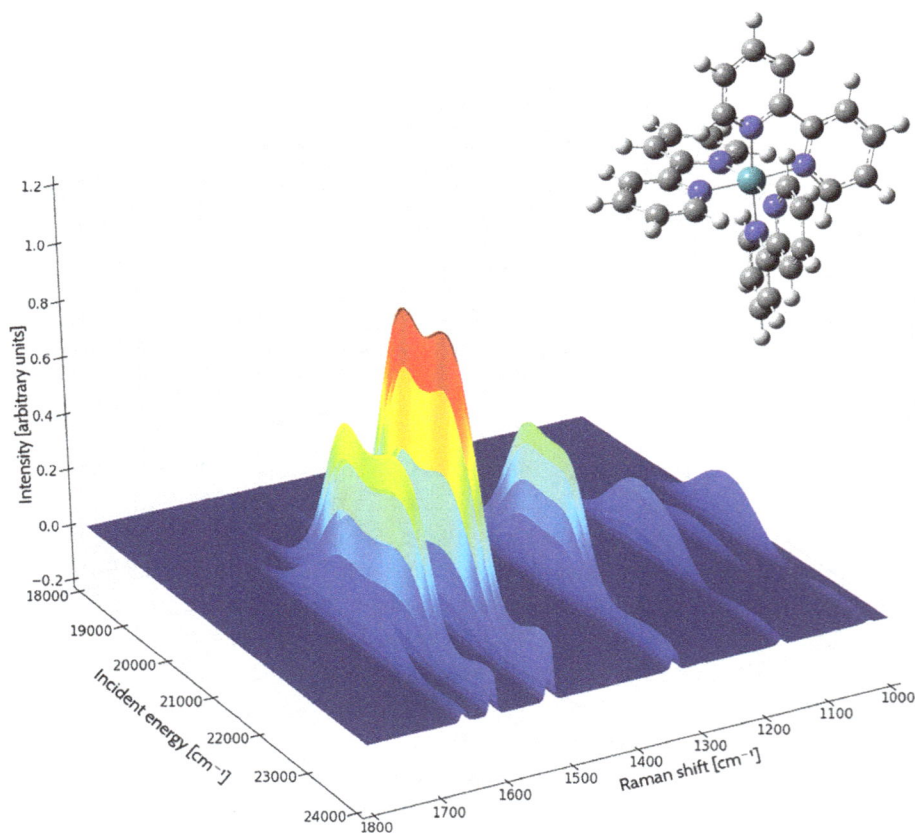

Figure 14.5: Resonance Raman spectrum of the tris(2,2'-bipyridyl)-ruthenium(II) complex. Each curve shows the intensity of the Raman shift produced by various frequencies of incident light. The area under each curve is filled, coloured by the magnitude of the intensity (blue to red as the intensity increases). The spectrum was computed in acetonitrile solution (Baiardi *et al.* 2014). Copyright ©2014 Royal Society of Chemistry, with permission.

A new technique known as CARS microscopy was first reported as a vibrational imaging tool in 1982 by Duncan *et al.* (1982). This technique allows non-invasive characterization and imaging of chemical species and biological systems without preparation or labelling with natural or artificial fluorophores that are prone to photobleaching.

Frunder *et al.* (1986) describe the construction of a CARS spectrometer with continuous wave (CW) intracavity excitation for high-resolution Raman spectroscopy and the stabilization of the argon laser to a hyperfine component in the Doppler-free polarization spectrum of iodine. The authors evaluated the CARS spectrum of the Q-branch of the rovibrational band of nitrogen, and obtained the vibration–rotation interaction constant as $\alpha_e = 0.0173850 \pm 0.0000035$ cm^{-1}.

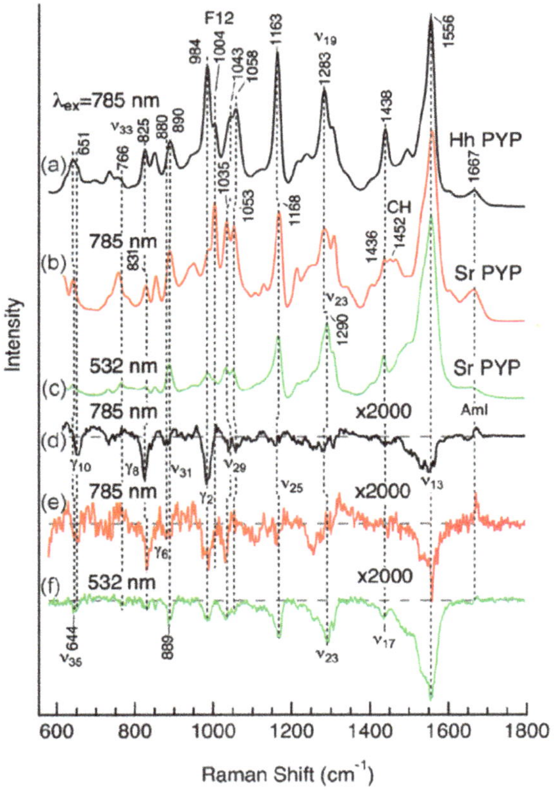

Figure 14.6: Resonance Raman and RROA spectra of a photoactive yellow protein obtained with 532 and 785 nm excitations (Haraguchi *et al.* 2015). Reprinted with permission. Copyright ©2015 John Wiley and Sons.

Masiello *et al.* (2004) use high-resolution infrared (IR) and coherent *anti*-Stokes Raman spectroscopies as a part of a series of investigations of isotopic forms of sulfur trioxide. The fundamental modes and several hot bands of $^{32}S^{16}O_3$ are described in detail.

Djaker, Lenne *et al.* (2007) present an application of CARS microscopy with high optical resolution and spectral selectivity, in resolving structures in surface ex vivo *stratum corneum* by looking at the CH_2 stretching vibrational band. They use a CARS microscope (Figure 14.7) with a pump laser at 730 nm and a Stokes laser tunable from 750 to

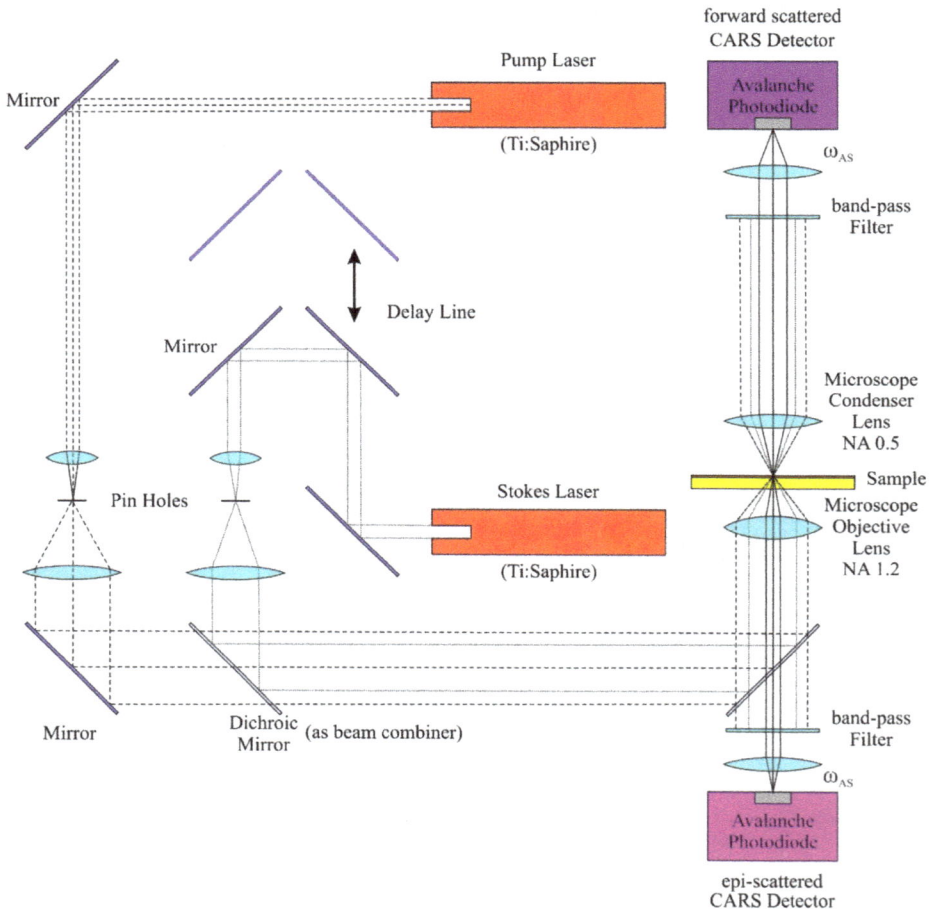

Figure 14.7: Typical set-up for CARS microscopy (drawn after Djaker *et al.* (2007) Two laser sources provide the pump and Stokes light and are synchronized through a picosecond path difference mirror set-up. In this set-up, the incident light is focused through an optically transmissive sample substrate. Both the forward scattered light (F-CARS) and epi-scattered light (E-CARS) are spectrally filtered by band-pass filters and are subsequently detected by two avalanche photodiodes. CARS images are obtained by raster scanning the sample. Copyright ©2007 Elsevier B.V., with permission.

950 nm (repetition rate of 80 MHz and pulse duration of 3 ps) to prevent two-photon electronic resonances. The spectral resolution of the instrument is about 2.5 cm^{-1}. The sample is scanned by a piezo stage with a spatial resolution below 1 μm.

An IR–CARS spectrometer is built by Volkov *et al.* (1981), featuring a high resolution of ~10^{-3} cm^{-1} with an automated data acquisition and processing system, which can be used to record simultaneously both IR and Raman spectra of a given substance in the frequency range 1,900–5,000 cm^{-1}. Results of a study of the rotational structure of the vibrational states $v_1(A_1)$ and $v_3(F_2)$ of the $^{28}SiH_4$ molecule are given.

Coherent *anti*-Stokes Raman scattering: spectroscopy and microscopy are reviewed by El-Diasty (2011).

Coherent *anti*-Stokes Raman scattering with single-molecule sensitivity using a plasmonic Fano resonance is claimed by Zhang *et al.* (2014).

Wang *et al.* (2016) contribute a review: A new synchronized source solution for coherent Raman scattering microscopy. The authors show applications at the microscopy of mouse skin.

The initial decomposition micromechanism of energetic materials is the interest of Peng *et al.* (2018). For this purpose, they use coherent *anti*-Stokes Raman scattering of liquid nitromethane. The dephasing time of the C–H stretching vibration located at 3,000 cm^{-1} is measured as 0.18 ps, which is far less than the dephasing time 6.2 ps of the C–N stretching vibration located at 917 cm^{-1}. The vibrational dephasing time is closely related to thermal collision, which means the C–H bond of the nitromethane molecule is excited first, causing an initial chemical reaction under thermal loading.

An effective tool for visualization of single-walled carbon nanotubes (CNTs) is coherent *anti*-Stokes Raman microscopy for Paddubskaya *et al.* (2018). Large well-registrable CARS signals have been obtained at μW-range excitation in CNTs with 0.8 and 1.1 nm diameters. Even volume images of SWCNTs in polyvinylalcohol could be obtained.

Even ROA can be detected by CARS. Hiramatsu *et al.* (2012) observe ROA by heterodyne-detected polarization-resolved coherent *anti*-Stokes Raman scattering. In their experiments, the contrast ratio of the chiral signal to the achiral background has been improved markedly. For (−)-β-pinene, it is 2 orders of magnitude better than that in the reported spontaneous ROA measurement.

Coherent Raman microscopy is a mighty means to detect biomolecules in unlabelled specimens by their Raman fingerprints. As high radiation power is needed for those experiments, photodamage may occur. This can partly overcome by quantum-enhanced nonlinear microscopy, as Casacio *et al.* (2021) show experimentally.

14.6 Stimulated Raman spectroscopy

Raman spectra of near Doppler resolution have only been achieved so far by means of nonlinear techniques, based on third-order terms of the susceptibility tensor χ (1). Among these techniques, the so-called inverse Raman or stimulated Raman has the advantage that the intensity depends linearly on the Raman scattering cross section, and therefore the spectra are free from the characteristic distortions associated with coherent *anti*-Stokes Raman spectroscopy (CARS). The selection rules are also the same; thus, this technique provides a unique tool to the high-resolution study of molecular vibrations which are not allowed in electric dipole spectroscopy, like those belonging to totally symmetric species in nonpolar molecules. The spectrum of cyanogen, NC–CN, is shown (Figure 14.8) as an example by Bermejo *et al.* (1997).

Gómez (1999) describes the theory of nonlinear Raman spectroscopy. The author advocates the semi-classical approach, in which the electric field operator is replaced by its expectation value and considered as a function.

Quick *et al.* (2015) revisit β-carotene by transient absorption and stimulated Raman spectroscopy. The spectra from states S_2, S_1, and S_0 of β-carotene in *n*-hexane are obtained and discussed. An early band around 1,770 cm^{-1} is seen having either broad (Stokes of λ_R = 573, 621 nm) or slightly dispersive (*anti*-Stokes of λ_R = 776 nm) line-shape. Internal conversion $S_2 \rightarrow S_1$ (200 fs) turns it into the band for symmetric C=C stretching in S_1.

Fluorescence spectroscopy and Raman spectroscopy are two major classes of analytical techniques. The former features superb sensitivity down to the single-molecule level, whereas the latter provides vibrational specificity. Combining both spectroscopies, Xiong *et al.* (2019) demonstrate stimulated Raman excited fluorescence (SREF) as a new hybrid spectroscopy on Rhodamine 800.

Xiong *et al.* (2019) achieve vibrational specificity and nanomolar sensitivity of 11 visible dyes by SREF spectroscopy. The authors study the C=C skeletal mode (1,602–1,652 cm^{-1}) of the red dyes.

Hu *et al.* (2020) measure the polarization- and orientation-dependent scalar, vector, and tensor light shifts in ^{87}Rb using near-resonant, stimulated Raman spectroscopy. Using an atomic fountain clock, the authors attack the light shift-related systemic errors in the atomic optic-based quantum sensors, including the atomic magnetometer, atomic interferometer, atomic clock, and quantum register.

Xiong *et al.* (2021) point out that the advent of far-field super-resolution fluorescence microscopy has greatly sharpened our vision in the microscopic world over the past two decades. While Raman spectroscopy offers significantly higher chemical specificity than fluorescence spectroscopy, the fluorescence spectrum is some powers of ten stronger than the Raman spectrum. Combining the advantages of both spectroscopies, the authors have developed an ultrasensitive vibrational spectroscopic technique called SREF. This technique is outlined in Figure 14.9.

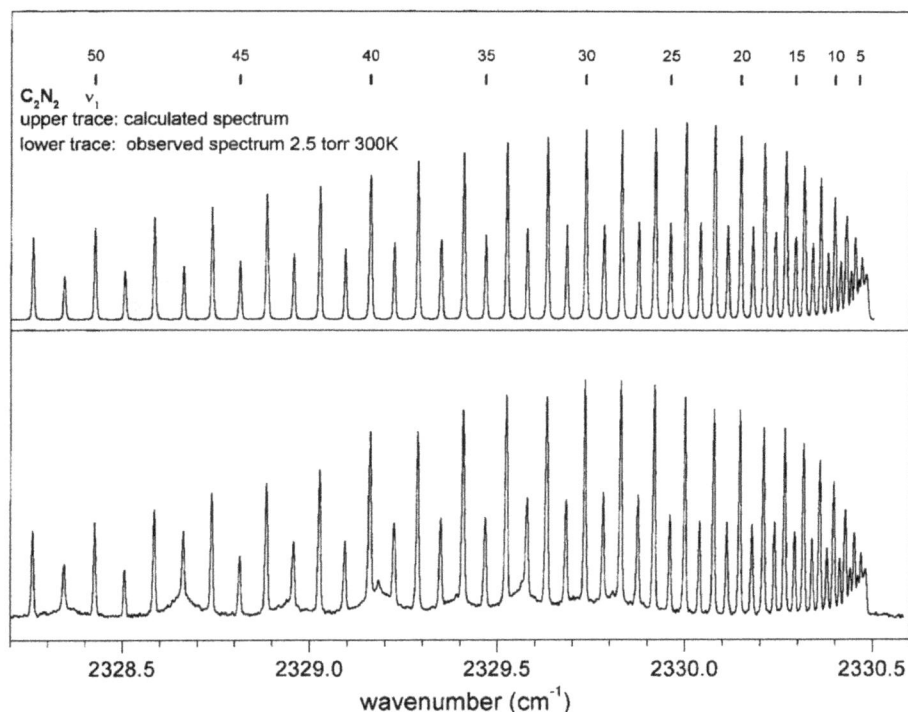

Figure 14.8: Stimulated Raman spectrum of the v_1 band of $^{12}C_2^{14}N_2$. The broad structure observed in the experimental contour is the spectrum of atmospheric N_2 outside the cell (Bermejo *et al.* 1997). Copyright ©1997 Academic Press, with permission.

A review by Xiong *et al.* (2022) discusses SREF microscopy. It outlines the history of this new technique, the underlying physical principles, and the remaining technical challenges and future opportunities.

14.7 The hyper-Raman effect

Hyper-Raman (HR) spectra can provide vibrational information on molecules where ordinary Raman scattering is suppressed due to symmetry issues. Reviews by Ziegler (1990, 2001) highlight the technique and applications. The technique allows one to observe not Raman-active modes (the so-called silent modes). As the HR effect is a three-photon process, high-energy lasers are needed to observe it. Two photons are necessary to reach the virtual state, then one photon is emitted. Alongside the hyper-Rayleigh line ($2v_0$) also lines at $2v_0 \pm v_i$ (Stokes and *anti*-Stokes) are observed in the spectrum. The Raman spectrum is similar to a normal Raman spectrum but it appears doubly frequency-shifted.

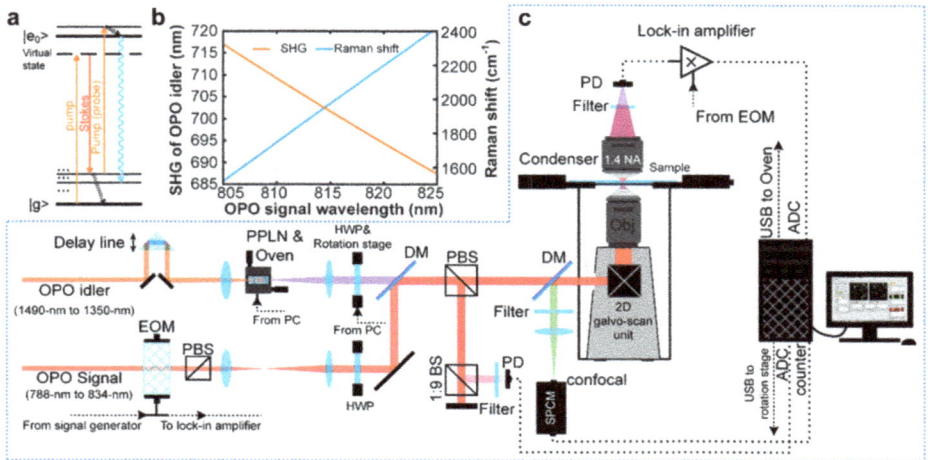

Figure 14.9: New stimulated Raman excited fluorescence (SREF) laser microscopy system opening excitation band near 600 nm. (a) Energy diagram for SREF process. (b) Wavelength of SHG of OPO idler (red) and corresponding resonance Raman shift (blue) as a function of OPO signal wavelength when OPO signal is set as the Stokes beam and the SHG is set as the pump beam. (c) System set-up for SREF excitation band around 600 nm. PPLN, periodically poled lithium niobate; QWP, quarter-wave plate; DM, dichroic mirror; PBS, polarization beam splitter; BS, beam splitter; SPCM, single-photon counting module; DAC, digital-to-analogue converter; ADC, analogue-to-digital converter; EOM, electro-optical modulator; APD, avalanched photodiode; PD, photodiode (Xiong *et al.* 2021). Copyright ©2021 by the authors. Licensed by Creative Commons (CC 4.0).

The selection rules are different from those for the normal Raman effect: all IR-active vibrations are allowed, additionally the phenomenon can be used to identify modes of vibration that are forbidden in normal Raman *and* IR spectra. Totally symmetric vibrations are forbidden, but vibrations of low symmetry result in strong signals.

"Hyper-Raman scattering has also attracted considerable attention as a tool for the study of crystals, owing primarily to the HRS symmetry selection rules. In centrosymmetric materials *ungerade* modes are HR active. This is also the case in IR, but in contrast to conventional linear Raman spectroscopy. In crystals and other solid-state materials, it is often low-frequency dipole-allowed (ungerade) modes, such as transverse optic (TO), LO and mixed material–radiation character oscillations, i.e. polaritons, that are of great interest. Some of these degrees of freedom, for both centrosymmetric and noncentrosymmetric materials, may be "silent modes", that is, neither Raman nor IR allowed, but permitted by the three-photon selection rules that govern HRS. Of considerable practical significance, the observation of low-frequency modes is easier in spontaneous HRS than in RS spectra since the incident excitation radiation is far remote from the scattered frequencies and, in centrosymmetric crystals, hyper-Rayleigh scattering is formally forbidden aside from the effects of defects and quadrupole polarization contributions, further minimizing the interference from broad baseline effects in the low frequency regime. Furthermore, since HR excitation is nonresonant at the incident laser frequency, HRS studies probe bulk solid-state properties unambiguously, in contrast to IR reflectance measurements or even resonance Raman studies, which may be affected by crystal surface effects. Crystals are inherently concentrated samples, which further contribute to the high quality of the reported HRS spectra. In addition, the orientational dependence of these ordered materials allows the polarization dependence of the observed spontaneous HRS to be an important label of vibrational symmetry.

Denisov *et al.* (1987) have given an excellent and extensive review of many applications of HRS for the study of structure and dynamics in condensed-phase materials."

Text reproduced with permission from Ziegler (2001). Copyright ©2002 John Wiley & Sons, Ltd.

Okuno (2021) reports HR spectra of the alcohols methanol, ethanol, 1-propanol, and 2-propanol. Okuno compares the HR spectra, excited at 532 nm, with IR and Raman spectra, and finds similarities between HR and IR spectra both in the fingerprint and in the C–H stretching regions.

15 Spatially offset Raman spectroscopy (SORS)

Normal Raman spectroscopy of turbid samples has only a limited penetration depth. If, for example, a white powder shall be characterized in a plastic container, chances are that fluorescence and Raman from the container will obscure the weak spectrum from the powder. But if laser spot and spot of detection are spatially offset from each other, the substance gives a clear Raman spectrum. Matousek *et al.* (2005) report their feasibility study using *trans*-stilbene powder beneath a 1 mm thick diffusely scattering over-layer of poly(methyl methacrylate) powder. The improvement (in contrast) of the spatially offset Raman spectroscopy (SORS) technique compared to normal Raman spectroscopy was 19 times high.

Eliasson and Matousek (2007) use SORS in the identification of counterfeit pharmaceutical tablets and capsules. Unlike normal Raman spectroscopy with its low penetration depth, the technique works through different types of packaging that give bad results otherwise.

In a review about SORS, Matousek *et al.* (2016) nicely illustrate the method at the example of a drinking cup made from a bilayer of polytetrafluoroethylene (PTFE) and polypropylene (PP) (Figures 15.2 and 15.3). The principle of the measurement is illustrated in Figure 15.1.

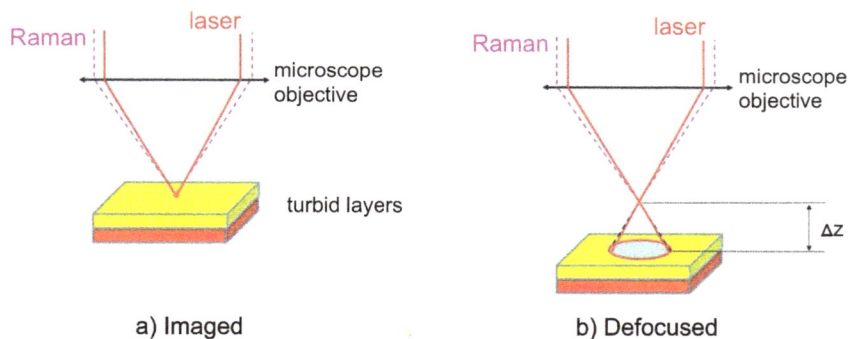

Figure 15.1: Schematic diagram of micro-SORS analysis. The first measurement is taken in a standard "imaged" position (a) and the second in a "defocused" position (b) implemented by displacing (Δz) the sample away from the microscopic objective (Matousek *et al.* 2016). Copyright ©2016 RSC, with permission.

https://doi.org/10.1515/9783110717556-015

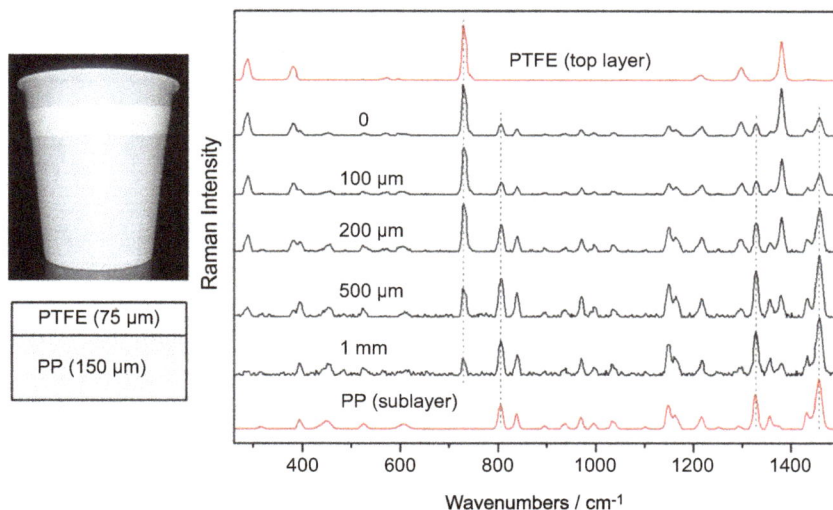

Figure 15.2: Micro-SORS measurement of a two layer diffusely scattering polymer system (the top layer was a polytetrafluoroethylene (PTFE) tape placed over a sublayer of a plastic drinking cup made of polypropylene (PP)); image and stratigraphic scheme of the sample (left); defocused Raman spectra shown for different distances Δz from the "imaged" plane ($\Delta z = 0$) indicated next to each spectrum (right). The spectra are offset for clarity. Note that the line markers are for guidance to emphasize the changing relative intensity of PTFE and PP with the defocusing distance; the reference spectra (the top and bottom) are acquired on the sample using conventional Raman spectroscopy (Matousek *et al.* 2016). Copyright ©2016 The Royal Society of Chemistry, with permission.

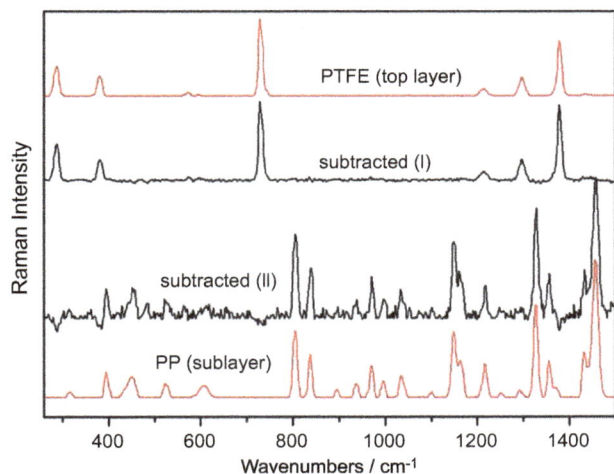

Figure 15.3: Numerical recovery of pure Raman spectra of individual layers from the micro-SORS measurements carried out on the two-layer polymeric system depicted in Figure 15.2: (I) recovered top layer spectrum and (II) recovered sublayer spectrum. Reference spectra are shown in top and bottom positions (Matousek *et al.* 2016). Copyright ©2016 The Royal Society of Chemistry, with permission.

16 Surface-enhanced Raman spectroscopy (SERS)

A huge enhancement effect of the Raman spectrum was discovered accidentally by Fleischmann, Hendra, and McQuillan (1974) in a work to study the role of absorption in electrochemistry. The enhancement was found while taking the Raman spectra of pyridine absorbed at a silver electrode.

Yet, it took several years until this effect was (re)discovered as surface-enhanced Raman scattering (SERS). Jeanmaire and Van Duyne (1977) studied the Raman spectrum of heterocyclic, aromatic, and aliphatic amines that were absorbed on an anodized silver electrode, and found a high enhancement of the signals. That was the first time that SERS was recognized as such.

Sharma *et al.* (2012) publish a review focused on SERS. They describe its history, materials, and applications, and speculate about future prospects.

Dick *et al.* (2000) prepare self-assembled monolayers from carboxylic acid-terminated alkanethiols, $HS(CH_2)_x COOH$, with variable chain length x. To these layers on a silver film over nanosphere (AgFON) electrode, cytochrome c (Cc) is bound and SERS is measured. The electrochemistry detected this way demonstrates that $Fe^{2+}Cc$ electrostatically bound to the $x = 5$ AgFON/S$(CH_2)_x$COOH surface exhibits reversible oxidation to ferricytochrome c, whereas $Fe^{2+}Cc$ electrostatically bound to the $x = 10$ surface exhibits irreversible oxidation, which means heterogeneous electron transfer is distance and orientation dependent. In comparison, $Fe^{2+}Cc$ covalently bound to both $x = 5$ and $x = 10$ surfaces exhibit oxidation with an intermediate degree of reversibility.

Ricci *et al.* (2018) study the Raman and SERS spectra of the blue pigment indigo and the indigo–Ag_2 complex (Figure 16.1). Using time-dependent density functional theory (DFT) in conjunction with the B3LYP functional and LANL2DZ/6-31$^+$G(d,p) basis sets, static and pre-resonance Raman spectra of the indigo–Ag_2 complex are calculated.

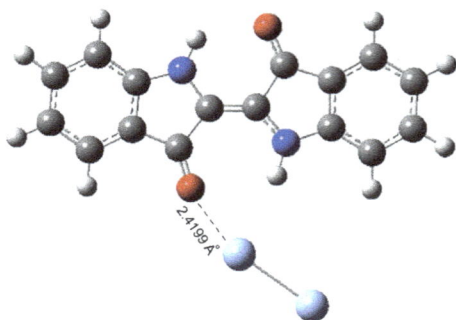

Figure 16.1: The optimized structure of the indigo–Ag_2 complex (Ricci *et al.* 2018). Copyright ©2018 Elsevier B.V., with permission.

https://doi.org/10.1515/9783110717556-016

The Raman spectra of D-penicillamine in aqueous solution and on silver colloids are recorded by López-Ramírez *et al.* (2004). Vibrational assignments of the spectra obtained were proposed. The analysis of the SERS spectra shows that a stable chemical bond is established between the sulfur atom and silver, and the two conformers of the molecule are adsorbed in their dianionic forms.

Near-infrared Fourier-transform SERS spectroscopy of 1,4-benzodiazepine drugs (Figure 16.2) is reported by Trachta *et al.* (2004). The authors employed gold films over nanospheres as the SERS substrate. The Raman spectra of crystalline reference samples and the calculation of normal modes by DFT (B3LYP/6–31G*) were compared with the SERS spectra. Most of the bands observed in the SERS spectra belong to vibrations located on rings a and b of the benzodiazepine part of the drugs. It was concluded that the latter is adsorbed with the molecular plane nearly orthogonal to the metal surface.

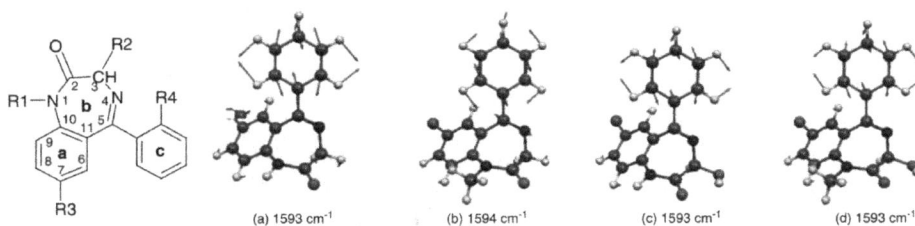

Figure 16.2: Structure of four of the 1,4-benzodiazepines investigated and numbering of atoms and selection of molecule-specific calculated normal modes (B3LYP/6–31G*; scaling factor 0.9613) of the 1,4-benzodiazepines investigated: (a) nordiazepam; (b) diazepam; (c) oxazepam; and (d) temazepam (Trachta *et al.* 2004). Copyright ©2004 John Wiley and Sons Ltd., with permission.

You *et al.* (2016) publish a computational study on SERS of *para*-substituted benzene-thiol derivatives adsorbed on gold nanoclusters. With the substituents H, F, Cl, Br, OH, SH, SeH, NH_2, and CH_3, DFT calculations of SERS were performed on a series of bridge-type and vertex-type x-BT/Au_{13} complexes (Figure 16.3) for geometric, electronic, and excitation properties to determine the key factor in spectral enhancement.

An *et al.* (2016) publish an experimental and DFT study on SERS of melamine on silver substrate (see Chapter 5, Figure 5.82).

You *et al.* (2018) report a theoretical study on surface-enhanced Raman spectroscopy of aromatic dithiol derivatives adsorbed on gold nanojunctions (Figure 16.4). DFT calculations were performed on dithiols HS-Ar-X-Ar-SH (X=O, S, Se, NH, CH_2, N=N, CH=CH, C≡C) and their single-end-linked or double-end-linked Au_{13} complexes. The calculated geometric, spectral, electronic, and excitation properties were used to discuss the dominant factor of enhancement.

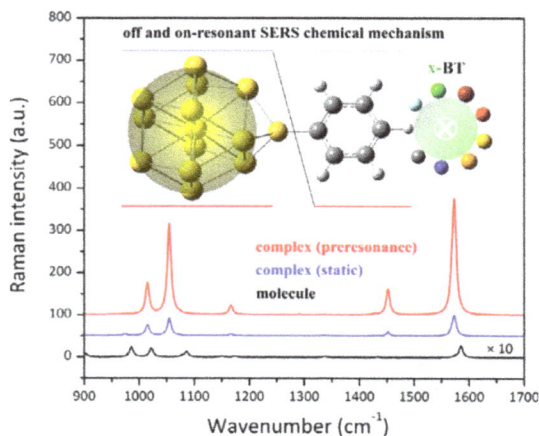

Figure 16.3: Calculated normal Raman spectra of bridge-type OH-BT/Au$_{13}$ complex (You *et al.* 2016). Copyright ©2016 Elsevier B.V., with permission.

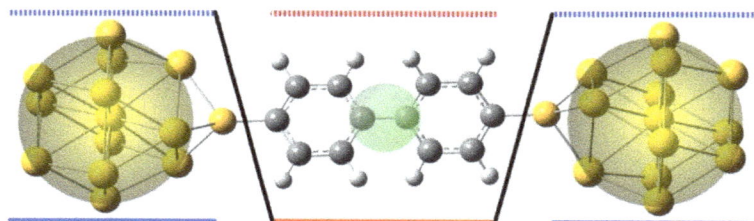

Figure 16.4: Aromatic dithiol derivatives interacting with a gold junction (You *et al.* 2018). Copyright ©2018 Elsevier B.V., with permission.

According to López-Tobar *et al.* (2017), large size citrate-reduced gold colloids appear to be optimal SERS substrates for cationic peptides. Surface-enhanced Raman spectra of two cyclic cationic peptides were simultaneously analysed as a function of concentration and colloidal size. Large size (~95 nm) citrate-reduced gold nanoparticles are shown to be the most adequate substrates for providing an optimal Raman signal enhancement within the 10^{-6}–10^{-7} m peptide concentration range. Ultraviolet–visible absorption, z-potential, transmission electron microscopy, and Raman data are consistent with a permanent evolution of the colloidal coverage, assignable to the gradual transformation of citrate anions and their substitution by oxidized products.

17 Tip-enhanced Raman spectroscopy (TERS) and nano-IR

Since the invention of scanning probe microscopy by Binnig and Rohrer (1982), there has been a constant improvement in techniques, and a variety of sophisticated techniques for the analysis of surface properties such as electrical conductivity, thermal properties, and Young's modulus have been developed. The resolution of the techniques has reached a single molecule and even an atom level. As the resolution of the surface structures is quite satisfying, there has been a demand for the analysis of chemical composition as well. Vibrational spectroscopy is the means of choice to fulfil this demand, but the special resolution of these techniques has been limited to microscopic resolution, which means limited by Abbe's law and therefore well below the resolution of scanning probe microscopy.

17.1 Tip-enhanced Raman spectroscopy (TERS)

Raman spectroscopy has always provided the composition of the sample at hand, and even strain and temperature and alignment of molecules can be analysed. The spatial resolution of the technique was limited. This was changed when the electromagnetic field at the surface of a laser-irradiated sample could be hugely enhanced by using gold or silver as the tip of the scanning element in atomic force microscope (AFM).

The first to propose a microscopic method to overcome the resolution limit of light microscopy was E. H. Synge. In a letter to Albert Einstein, he advocated the use of a tiny gold particle to shine light on a sample (**Figure 17.1**). Synge (1928) later published a paper where he shows a different arrangement.

Already some excellent books on the similar subject of nanophotonics have appeared, which treat tip-enhanced Raman spectroscopy (TERS) in a few chapters, for example Kawata *et al.* (2001, 2002), Kawata and Shalaev (2007), and Novotny and Hecht (2012). They all are more theoretically based, and none of them focused on the application of TERS. Additionally, they are all quite outdated, with the exception of the Novotny/Hecht book (2012), which is a (slightly) updated version of the 2006 edition.

As SERS is an enhancement of the Raman spectrum by metal nanoparticles (NPs), it is not very surprising that the enhancement can also be reached by a metal tip whose apex is the size of one of those NPs. The technique is called tip-enhanced Raman spectroscopy. The first paper (submitted 11/1999) on the technique by Stöckle *et al.* (2000) studies brilliant cresyl blue and C_{60}. In the same year, Pettinger *et al.* (2000) publish a paper "Surface Enhanced Raman Spectroscopy: Towards Single Molecule Spectroscopy". Three years later, Pettinger *et al.* (2003) describe enhancement on the CN^- band at 2,131 cm^{-1}, typical for the CN^- stretch vibration at Au. In contrast to a typical SERS experiment, the sample can be scanned resulting in an image with high spatial resolution.

https://doi.org/10.1515/9783110717556-017

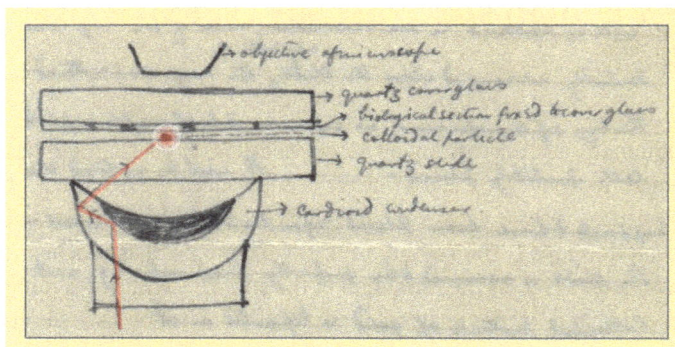

Figure 17.1: In an April 1928 sketch sent to Albert Einstein, Edward Hutchinson Synge proposed a new microscopic method: using a tiny gold particle (red) between two quartz slides to scatter incident light from below onto a sample. Light that did not strike the particle would be totally internally reflected, and an objective lens of a microscope could be positioned to accept some of the gold-scattered light. That arrangement, Synge wrote, could be used to image a biological specimen fixed to the top cover slip at a resolution below the diffraction limit (courtesy of the Albert Einstein Archives, Hebrew University of Jerusalem, Israel.).

Two principal experimental set-ups for a TERS experiment are in use: The first uses an inverted microscope for the illumination of the tip by a laser from below (Figure 17.2), and the second uses side illumination at ≈60° by the laser (Figure 17.3).

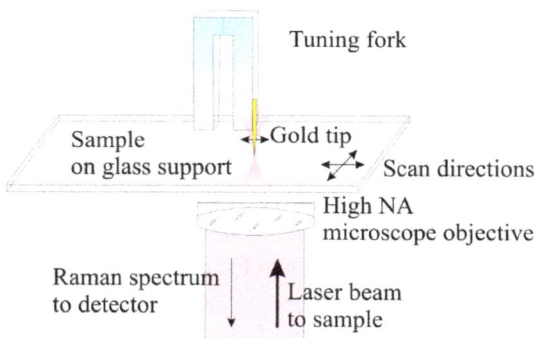

Figure 17.2: Sketch of a typical TERS experiment using a gold tip attached to a tuning fork. The sample on a glass support is illuminated by a laser beam focused with an oil-immersion microscope objective of high numerical aperture.

An article by Deckert-Gaudig *et al.* (2017) reviews the development of TERS from the beginning (in 2000) to the state of the art (in 2016). One of the authors' conclusion is that "Thanks to commercially available set-ups, TERS is no longer only for 'experts' but rather is open to everyone."

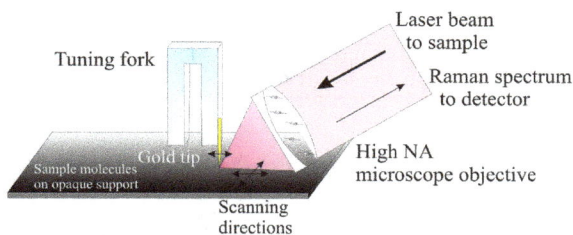

Figure 17.3: Sketch of a typical TERS experiment using a gold tip attached to a tuning fork using the side illuminating mode. The tip and the sample on an opaque support are illuminated by a laser beam focused by a long working-distance microscopic objective of high numerical aperture.

17.1.1 Inorganic samples

Kharintsev *et al.* (2007) study atomic force and shear force-based TERS and imaging. Using a scanning near-field optical spectrometer (SNOM) in inverted configuration (Figure 17.4), they compare the Raman spectra of single-walled carbon nanotube (SWCNT) bundles (Figure 17.5) when a gold-coated AFM tip and a self-made gold SFM tip are brought close to and withdrawn from a sample. The enhancements for different Raman lines are calculated and compared (Table 17.1). The comparison ended in favour of the gold tip. The paper also shows confocal far-field Raman images and tip-enhanced near-field Raman intensity distribution images (Figure 17.6) of an SWCNT bundle for three characteristic vibrational modes. The images were acquired by a self-made gold tip in SFM mode.

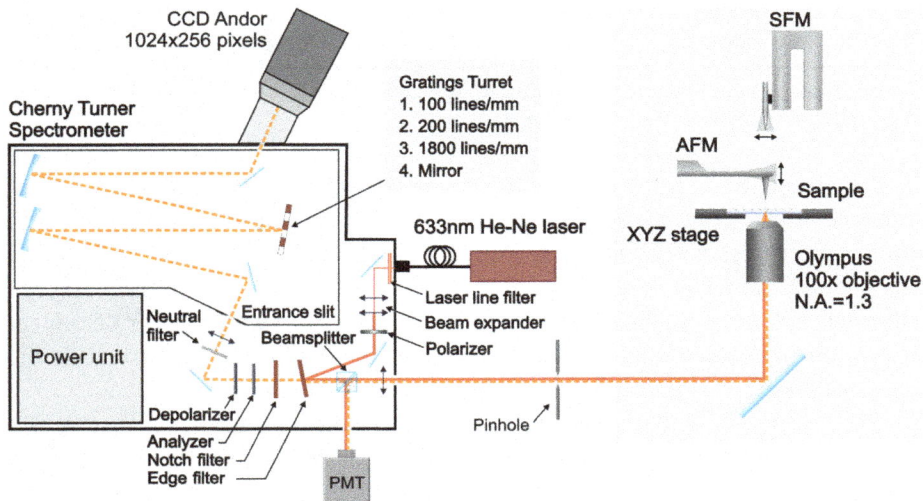

Figure 17.4: Layout of the scanning near-field optical spectrometer (inverted configuration) (Kharintsev *et al.* 2007). Copyright ©2007 IOP Publishing Ltd., with permission.

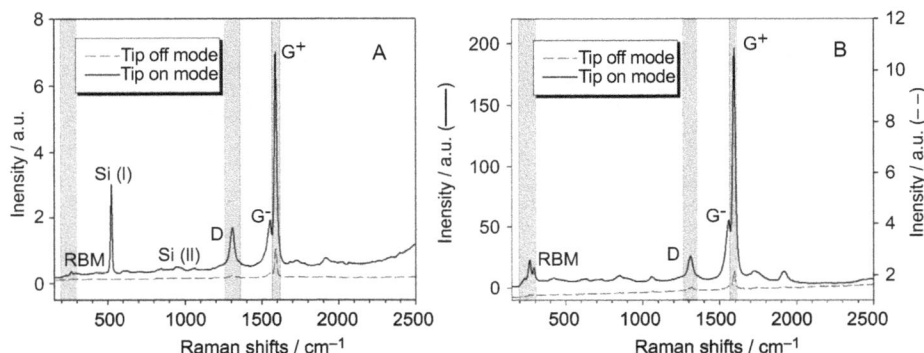

Figure 17.5: Raman spectra of SWCNT bundles when a gold-coated AFM tip (A) and a self-made gold SFM tip (B) are brought close to and withdrawn from a sample, respectively; data presented are original data without applying smoothing procedures (Kharintsev *et al.* 2007). Copyright ©2007 IOP Publishing Ltd., with permission.

Table 17.1: The field enhancement values for characteristic vibrational modes.

Mode	RBM	D	G⁻	G⁺	Correction factor
$\Delta v/\text{cm}^{-1}$	263	1,313	1,557	1,594	
AFM	4*	12*	9*	6*	None
	70	214	160	118	d_{spot}/d_{tip}
	144	432	324	216	A_{spot}/A_{tip}
SFM	457*	190*	263*	254*	None
	27,420	11,400	15,780	15,240	d_{spot}/d_{tip}
	182,800	76,000	105,200	101,600	A_{spot}/A_{tip}

An asterisk indicates that a correction factor d_{spot}/d_{tip} or A_{spot}/A_{tip} (from $R_{spot} = 150$ nm, R_{tip} (AFM) = 25 nm, and R_{tip} (SFM) = 8 nm) is not taken into account.

Self-assembled, in-plane Ge/Si nanowires are grown by Reparaz *et al.* (2013) combining molecular beam epitaxy and the metal-catalyst assisted-growth method. Local strain and composition of the nanowires are analysed by means of tip-enhanced Raman (TER) scattering. The authors find out that the Ge distribution in the (001) crystallographic direction is inhomogeneous, displaying a gradient with a higher Ge content close to the top surface, and in contrast, the (uncapped) wires exhibit essentially the same small residual compressive strain everywhere along the wire.

Kharintsev *et al.* (2013) design optical plasmonic antennas for implementing TER scattering measurements. By dc-pulsed low-voltage electrochemical etching of a gold wire, they engineer TERS-active metallic gold tips with diverse shapes (Figure 17.7) and sizes in a highly reproducible fashion. The underlying etching mechanism at a voltage-driven meniscus around a gold wire immersed into an electrolyte is discussed in detail. The authors show that the developed method is suitable to produce not only the

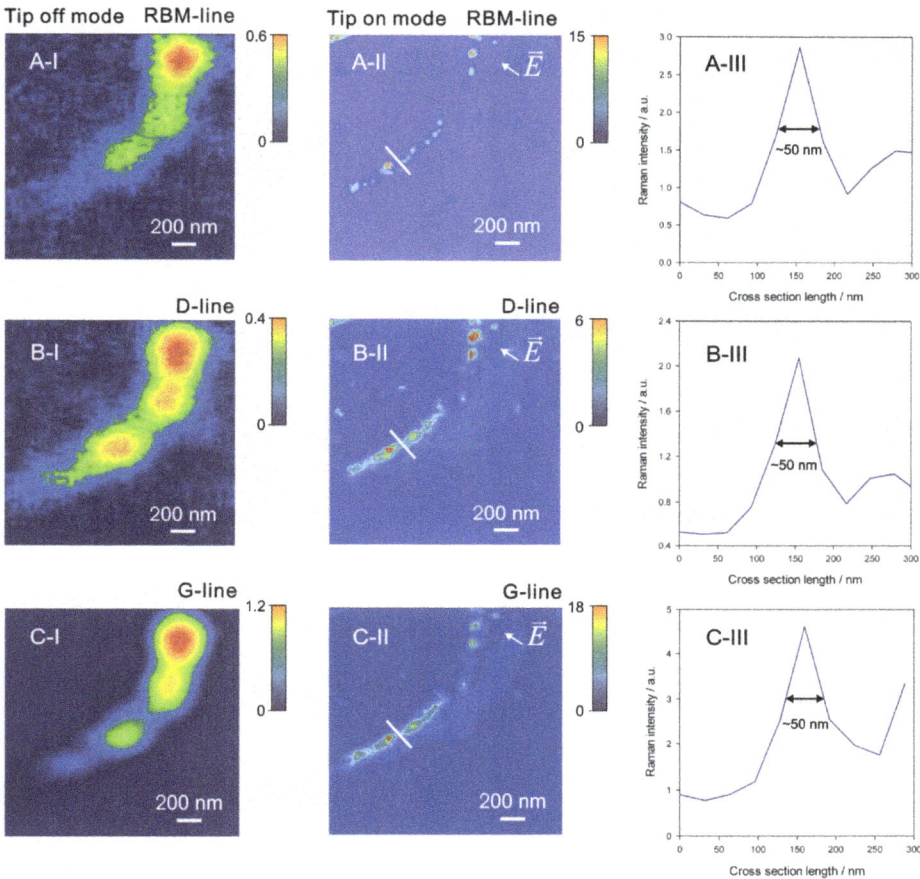

Figure 17.6: Confocal far-field Raman images (A-I, B-I, and C-I) and tip-enhanced near-field Raman intensity distribution images acquired by a self-made gold tip in SFM mode (A-II, B-II, and C-II) of an SWCNT bundle for characteristic vibrational modes: radial breathing mode at 290 cm^{-1} (A), D-line at 1,390 cm^{-1} (B), G^{+}-line at 1,594 cm^{-1} (C), and their corresponding cross sections (A-III, B-III, C-III, original data) (Kharintsev *et al.* 2007). Copyright ©2007 IOP Publishing Ltd., with permission.

simplest geometries such as cones and spheroids, but more complex designs. Attempts have been made to design plasmonic tapered antennas with quasi-uniformly spaced nano-sized bumps on the mesoscopic zone for the extra surface plasmon-light coupling. The capability of the patterned antenna to enhance and localize optical fields is demonstrated with near-field Raman microscopy and spectroscopy of SWCNT bundles.

A problem in TERS is the production of the right polarization. As we always need a strong polarization in the direction of the long axis of the gold tip, we have to use a high numerical aperture objective when using an inverted microscopic configuration for TERS. A strong longitudinal field is produced by Zhang *et al.* (2014) through the

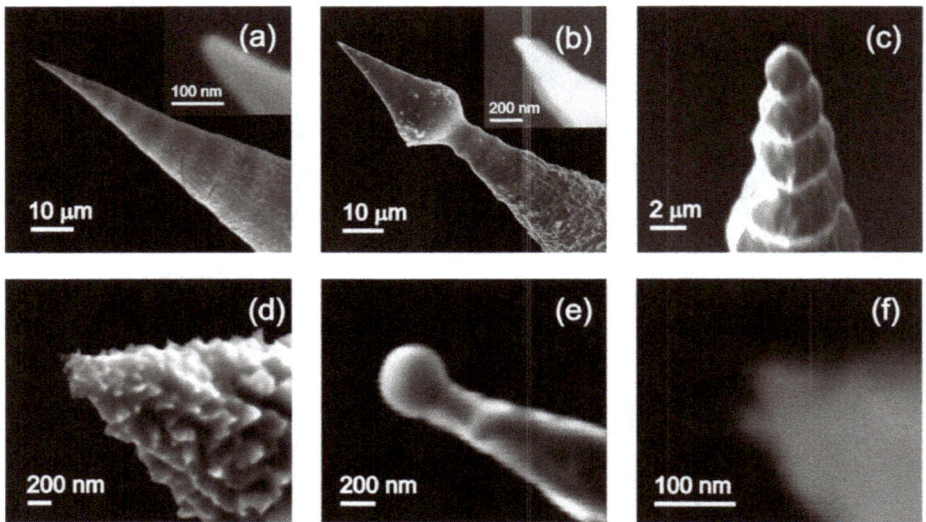

Figure 17.7: SEM images of electrochemically etched plasmonic optical antennas at different dc-pulsed voltage regimes: (a) smooth cone, (b) lance, (c) pyramid, (d) rough cone, (e) spheroid, and (f) dimer (Kharintsev *et al.* 2013). Copyright ©2013 IOP Publishing Ltd., with permission.

use of a radially polarized laser beam (Figure 17.8). With this field, a 1.7×10^4 times enhancement of the spectrum of SWCNTs is achieved, which is 6 times higher than that with focused linearly polarized excitation.

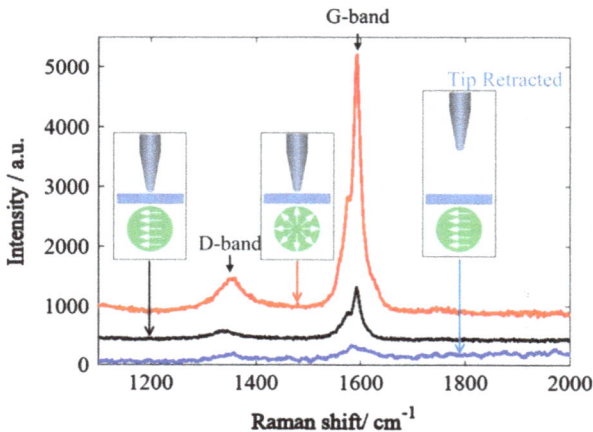

Figure 17.8: Tip-enhanced Raman spectra with radially polarized (RP) and linearly polarized (LP) excitation as well as far-field Raman spectrum of SWCNTs (Zhang *et al.* 2014). Copyright ©2014 Elsevier B.V., with permission.

A recent review by Mrđenović *et al.* (2023) focuses on nanoscale chemical analysis of 2D molecular materials using tip-enhanced Raman spectroscopy.

17.1.2 Organic samples

Bortchagovsky *et al.* (2013) cover their silver tip by thiophenol as an internal standard for TERS. Thiophenol shows sharp lines in the resulting TER spectra when chemisorbed onto freshly etched silver tips and does not contaminate the sample.

Rzeźnicka *et al.* (2013) report TERS of 4,4′-bipyridine and 4,4′-bipyridine *N,N′*-dioxide adsorbed on gold thin films. Among the signals was an Au−N stretching vibration at 185 cm^{-1}. By the aid of the theoretical calculations of the Raman scattering intensities for each vibrational normal mode, a standing-up, tilted orientation at monolayer coverage and a vertical orientation with the longitudinal molecular axis parallel to the surface at low surface coverage (Figure 17.9) are concluded.

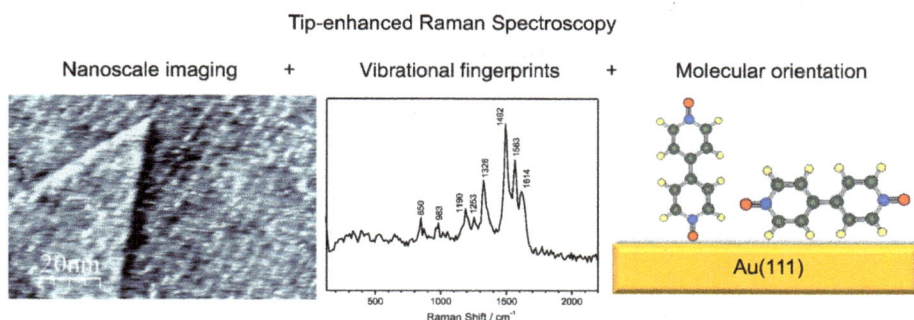

Figure 17.9: Molecular orientation of 4,4′-bipyridine N,N′-dioxide adsorbed on gold thin films (Rzeźnick *et al.* 2013). Copyright ©2013 Elsevier B.V., with permission.

Zhang *et al.* (2013) study TERS mapping of *meso*-tetrakis(3,5-di-tertiarybutylphenyl)-porphyrin (Figure 17.11). The authors are able to show a TERS image of a single porphyrin ring (H$_2$TBPP, Figure 17.10). Using scanning tunnelling microscopy (STM) with a silver tip on a silver substrate under high-vacuum (HV) conditions, they achieve a resolution below 1 nm. Using a DFT calculation (174 atoms, B3LYP/6-31G(d)), the authors also provide a video showing important vibrational modes.

Zhang *et al.* (2014) review HV-TERS. First, introducing the principles of normal TERS, the authors proceed presenting applications, arriving finally at recent progress on HV-TERS from three groups.

Titanium nitride (TiN) is mechanically, chemically, and thermally more robust than gold. Scherger *et al.* (2016) propose the use of TiN-covered probes for TERS, as it can provide a plasmon resonance very similar to that of gold. Measurements on thin films of poly(3,4-ethylenedioxythiophene)/poly(styrenesulfonate) show an increase in observed intensity up to 3.1 times, but the TiN has to be coated with Al$_2$O$_3$ for use in air. Scherger *et al.* (2017) also demonstrate that TiN-covered probes can be used for measurements in liquid environments.

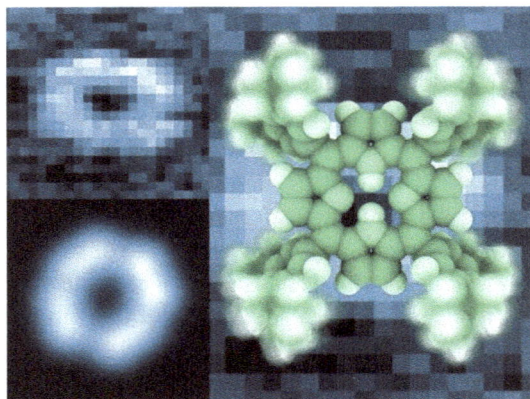

Figure 17.10: TERS image of a single porphyrin ring (Zhang *et al.* 2013). Copyright ©2013 Springer Nature, with permission.

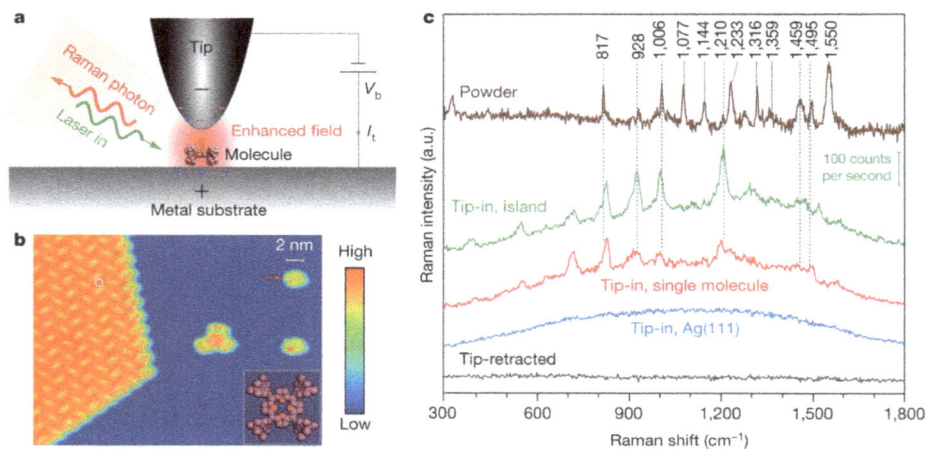

Figure 17.11: (a) Schematic tunnelling-controlled TERS in a confocal-type side-illumination configuration, in which Vb is the sample bias and it is the tunnelling current. (b) STM topograph of sub-monolayered H2TBPP molecules on Ag(111) (1.5 V, 30 pA, 35 nm × 27 nm). The inset shows the chemical structure of H2TBPP, and the white circle indicates one representative site for TERS measurements on the molecular islands. (c) TERS spectra for different conditions. The tip-in spectra were acquired at 120 mV, 0.5 nA, and 3 s. The green spectrum is taken on top of the molecular island (the green scale bar shows the signal level detected by charge-coupled device (CCD)). The red spectrum is taken on top of a single molecule (marked by the red arrow in (b)). The blue spectrum is taken on bare Ag(111). The black spectrum is taken on top of the molecular island but with the tip retracted 5 nm from the surface (120 mV, 3 s). For comparison, a standard Raman spectrum (brown) is shown on the top for a powder sample of H2TBPP molecules (Zhang *et al.* 2013). Copyright ©2013 Springer Nature, with permission.

Foti *et al.* (2022) compare commercial silver-coated AFM tips to home-made bulk gold tips for TERS. Using their TERS equipment, the authors study the multi-walled carbon nanotube (MWCNT)/polymer interface of FFUR-14-functionalized MWCNTs.

17.1.3 Biochemical samples

Rasmussen and Deckert (2006) perform TERS measurements on nucleotides and DNA pyrimidine bases. The latter results were compared with standard SERS and Raman measurements of the nucleotides and the pure bases. The nucleotides show a weaker scattering efficiency than the pure bases, but the spectral features of the base patterns dominate.

Liu *et al.* (2009) investigate adsorbed non-resonant molecules on single-crystal gold surfaces by TERS. They study TERS of benzenedithiol, 4,4′-bipyridine, and a series of oligo(phenylene ethynylene)s of different lengths. Finite-difference time-domain simulations suggest an explanation of the enhancements. A method of tip regeneration is also detailed.

Amyloid plaques, associated with Alzheimer's disease, contain the Aβ(16–22) sequence. This peptide can self-assemble into nanotapes. Paulite *et al.* (2013) image these nanotapes by STM and gap-mode STM-TERS. Gap-mode STM-TERS uses a solid silver tip in feedback with a gold substrate. Comparing both techniques (Figure 17.12), Paulite *et al.* find that the areas weakly observed as a feature in the STM image can be identified as nanotape/peptide structures using TERS imaging of the aromatic ring breathing marker band (1,004 cm^{-1}).

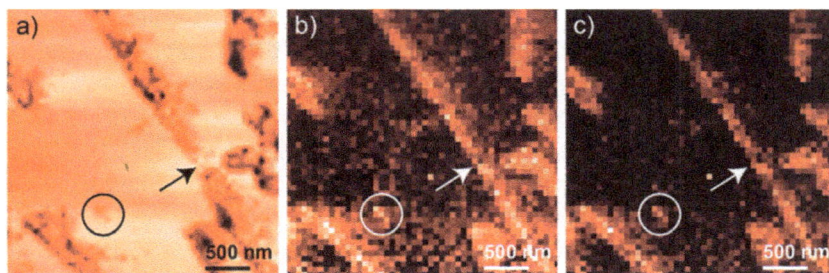

Figure 17.12: Simultaneously acquired (a) STM and (b,c) TERS (acquisition time 1 s/pixel, 2 mW incident power) images of individual nanotapes with 3 × 3 µm^2 scan size and 50 × 50 pixels. The colour-coded TERS images display the intensity of the aromatic ring breathing marker band (1,004 cm^{-1}), (b) value of the peak integral, and (c) peak maximum. The arrow and circle illustrate that areas weakly observed as a feature in the STM image can be identified as nanotape/peptide structures using TERS imaging (Paulite *et al.* 2013). Copyright ©2013 American Chemical Society, with permission.

A 1.7 nm resolution chemical analysis of carbon nanotubes (CNTs) by TER imaging at ambient temperature is reported by Chen *et al.* (2014). This was achieved by using a gap-mode configuration with a nanogap formed by a clean gold tip and gold substrate with an extremely small gap distance of ≈1 nm. Different types of CNTs, local defects, diameters, and bundling effects could all be clearly analysed at high resolution.

Balois *et al.* (2015) use tip-enhanced THz Raman spectroscopy for local temperature determination at the nanoscale. The authors are able to determine the local sample temperature of semiconducting CNTs (radial breathing mode (RBM) frequency of 173.83 cm^{-1}) and metallic CNTs (RBM frequency of 187.70 cm^{-1}) through the determination of the ratio between *anti*-Stokes and Stokes Raman peaks of the RBM.

Insulin amyloid (misfolded proteins) fibrils are studied by Moretti *et al.* (2013). Using a top-illuminated TERS set-up, they achieved a Raman signal enhancement of ~10^5, allowing them to display a TERS map visualizing the supramolecular organization of the fibrils and the orientation of a few molecules in different points along the fibrils.

Using a biomembrane model, Böhme *et al.* (2010) test biochemical TERS imaging below the diffraction limit. The model (Figure 17.13) consists of a supported phospholipid film, which is labelled by a protein (streptavidin). A silver-coated TERS tip is placed at 40 positions of the sample, separated by 75 nm. The spectra obtained can be easily classified as typical for lipids, proteins, or a mixture of both.

Najjar *et al.* (2014) study combed double-stranded (ds) DNA of a λ-phage virus on octadecyltrichlorosilane (OTS)-modified borosilicate glass substrates. The authors resolve cross sections of nanowire-shaped thin DNA bundles by TERS. They reach an enhancement factor (EF) higher than 6 × 10^2 and a lateral spatial resolution better than 9 nm, using tips coated with an Ag/Au bilayer. TER spectra at single points show very intense spectral features which can be assigned to typical vibrational modes of adenine (A), guanine (G), cytosine (C), thymine (T), and the DNA backbone composed of sugar and phosphodiester groups. The assignments are given in a table. While the TER signature of nucleobases is congruent with observations in single-stranded (ss) DNA, additional modes tied to the DNA backbone can be discerned in ds DNA. TERS enables ss and ds DNA samples to be distinguished from each other and hence can be exploited for the detection of DNA hybridization. Moreover, no TER contribution of the OTS layer appears, suggesting that functionalized DNA strands could be studied without spectral perturbation from the substrate.

Double strand breaks (DSBs) are damages in DNA that can lead to genetic defects and cell apoptosis. Lipiec *et al.* (2014) study DSBs using TER scattering. The authors find out that the P–O–H in-plane bending mode at 1,160 cm^{-1} and the CH$_2$/CH$_3$ scissoring mode at 1,368 cm^{-1} along with the CH$_2$ deformation mode at 1,439 cm^{-1} are enhanced in the TER spectra. This hints to cleavage by free radicals at the 3- and 5-bonds of deoxyribose upon exposure to UVC radiation.

Ramanauskaite *et al.* (2018) compare the different procedures for the fabrication of Ag probes for TERS. The authors focus on technologically simple methods allowing

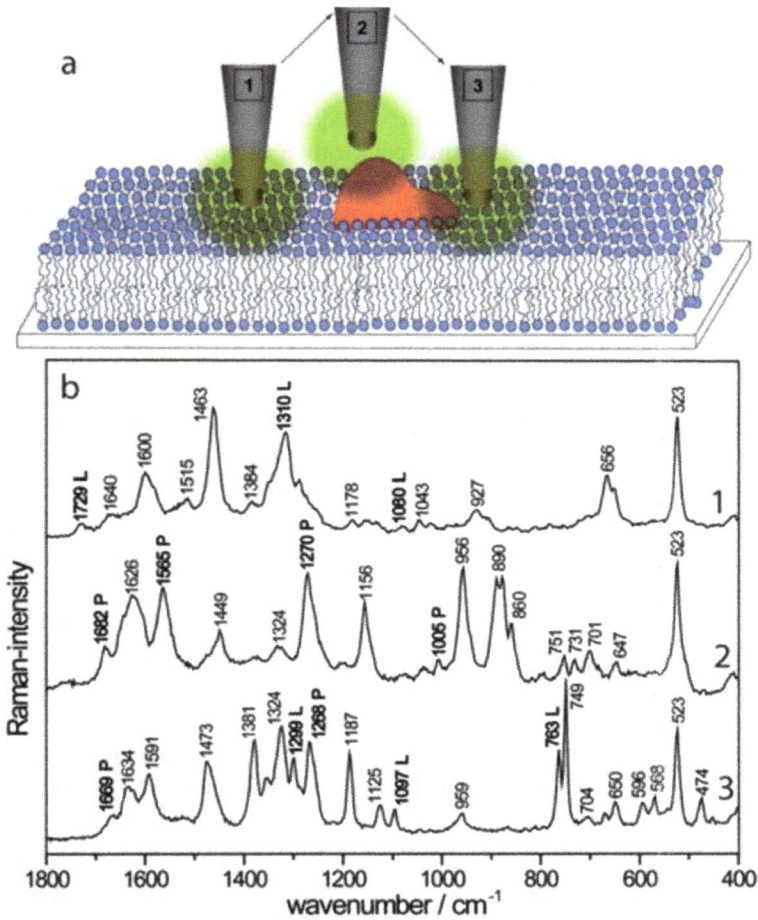

Figure 17.13: (a) Schematic tip–sample interactions for TERS measurements on biomembrane mimic structures (not to scale). For three different tip positions, the field interacts with lipid domains (position one), protein domains (position two), or a mixture of both biomolecules simultaneously (position 3). (b) Raman spectra recorded on such membrane structures can be related to the tip position indicated in Figure 17.13(a). Specific marker bands for each class are marked with L (lipid domain) or P (protein domain) (Böhme *et al.* 2010). Copyright ©2010 John Wiley and Sons, with permission.

Si cantilevers coated with plasmonic silver nanostructures and bulk metal Ag tips with good shape reproducibility to be produced. The EFs of the produced tips were measured using graphene oxide (GO) as the sample in a top-illumination TERS set-up.

A TERS system based on an AFM and radially polarized laser beam is developed by Fan *et al.* (2020), and a procedure for facilitating hotspot alignment via the intrinsic photoluminescence (PL) from the silver-coated TERS probe is described. Therefore, Fan *et al.* use the intensity change of Ag PL to evaluate the stability of the Ag-coated probe during TERS experiments. Tracking the Ag PL of the TERS probe is also helpful to detect damage of the precious probe and hotspot alignment.

Zhou *et al.* (2021) choose the vector beam to form a tightly focused vertical electric field. The altered laser incidence along normal direction proved several unexpected advantages, such as nearly 100% purity of vertical polarization component and precisely symmetrical shape of point spread function. For the bottom illuminated method, taking the opaqueness and absorption characteristics of the beam by the metal substrate into consideration, the substrate thickness becomes the critical factor concerning gap plasmons coupling effect and incident light transmittance. Here, the spatial resolution lines of the vector beam excited from the bottom are compared with that of side method excited by linearly polarized beam, which shows that the vector beam presents greater symmetric characteristics.

Advanced Raman spectroscopic methods are used by Malard *et al.* (2021) to study 2D materials. The techniques of coherent *anti*-Stokes spectroscopy, stimulated Raman scattering, and TERS have overcome the limitations of the parent technique. Malard *et al.* discuss the physical information on graphene, hexagonal boron nitride (hBN), and transition metal di- and mono-chalcogenides, which can be obtained from Raman spectroscopic investigations.

Yang *et al.* (2014) describe the controlled fabrication of "Tip-On-Tip" TERS probes. The novel TERS probes exhibit enhanced near-field Raman signals compared to an Ag-coated SPM probe, which are attributed to the intense electromagnetic field around the apexes of the Ag nanoneedles and the periodic structure of the Ag nanoneedles.

Legge *et al.* (2020) characterize a commercially available powder containing few-layer graphene flakes before and after plasma or chemical functionalization with either nitrogen or oxygen species. The authors use TERS imaging of the submicron-sized flakes, revealing the location of the defects (edge versus basal plane) and variations in the level of functionalization. The effect of functionalization on the mechanical properties of nanocomposites is also studied.

Zeng *et al.* (2015) study interfacial properties by electrochemical TERS with the example of (4′-(pyridin-4-yl)biphenyl-4-yl)methanethiol (4-PBT). The EC-TERS set-up used (Figure 17.14) focuses the laser light horizontally to the EC-STM cell to minimize the optical distortion. It consists of an insulated gold tip approached on a single-crystal gold substrate until the tunnel current reaches 500 pA. Protonation and deprotonation of the molecule can be observed as a subtle change in the molecular configuration with potential, as shown by the spectra supported by DFT calculations.

Su *et al.* (2018) succeed in mapping of functional groups on a carboxyl-modified graphene oxide (GO–COOH) surface with a spatial resolution of ≈10 nm using TERS. The scientists extended the capability of TERS by measuring local electronic properties *in situ*, in addition to the surface topography and chemical composition.

A side-illuminated TER study of edge phonons in graphene at the electrical breakdown limit is reported by Okuno *et al.* (2016). As nanoscale integration of graphene into a circuit requires a stable performance under high current density, the authors

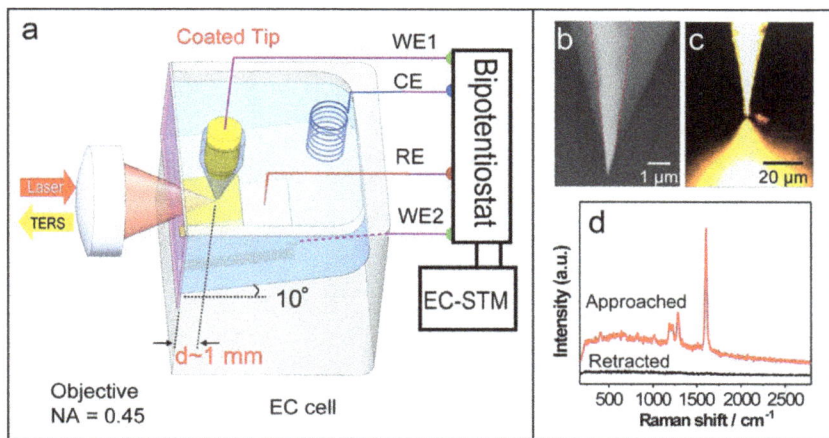

Figure 17.14: (a) Schematic illustration of the EC-TERS set-up. (b) SEM image of an insulated gold tip. (c) Microscopic image of the tip, single-crystal substrate, and laser spot in an EC-TERS system. (d) TERS of 4-PBT adsorbed on the Au(111) surface obtained while the tip was approached (top) and retracted (bottom). The acquisition time was 1 s. The bias voltage was 500 mV, and the tunnelling current was 500 pA (Zeng *et al.* 2015). Copyright ©2015 American Chemical Society, with permission.

explore the effects of a high current density on graphene, using TERS. They find that the high current density induces Raman bands at 1,456 and 1,530 cm^{-1}, which were assigned to edge-phonon modes originating from zigzag and armchair edges. This leads the scientists to conclude that C–C bonds are cleaved due to the high current density, leaving edge structures behind, which are detected through the observation of localized phonons.

"TERS at work" is observed by Krayev *et al.* (2016). The authors study GO and the 2D semiconductors, MoS$_2$ and WS$_2$. Measured by the gap mode, the TERS signal of these 2D materials becomes strongly enhanced over wrinkles and creases. The resonant Raman signal of MoS$_2$ contains additional peaks normally forbidden by selection rules. The spatial resolution of the measurements is 10–20 nm.

A four-component sample made of several carbon allotropes (SWCNTs, GO, C$_{60}$ fullerene, and an organic residue; Figure 17.15) is studied by Sheremet *et al.* (2016) using TERS. For the organic residue, the authors suggest a decreased van der Waals interaction with GO in contrast to graphene. A spatial resolution of 15 nm and below is realized.

Xue *et al.* (2011) study the phase separation behaviour of a partially miscible poly (methyl methacrylate)/poly(styrene-co-acrylonitrile) (PMMA/SAN) (70/30 wt%) thin film using TERS. For the first time, non-destructive chemical mapping is carried out using TERM as well (Figure 17.16). Spin-coated uniform PMMA/SAN films with a thickness of about 500 nm are annealed above the lower critical solution temperature to induce phase separation. The treated films show visible domains of different sizes

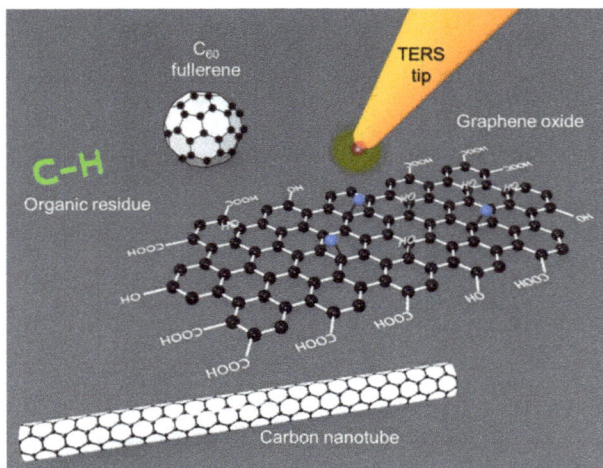

Figure 17.15: Sketch (not to scale) of the four-component sample with graphene oxide, carbon nanotubes, C_{60} fullerene, and an organic residue deposited on a gold substrate. The AFM TERS tip was made of solid gold. The laser excitation was 638 nm of a solid-state laser focused in side illumination by a long working distance of 100× objective, N.A. 0.7 (Sheremet *et al.* 2016). Copyright ©2016 Elsevier B.V., with permission.

under optical microscopy. The phase separation behaviour of films is then studied using high-resolution TERM. The TERM images at different stages are compared to obtain new insights into the phase separation behaviour of the PMMA/SAN blend films. The interface width (~200 nm, Figure 17.17) at the early stage of phase evolution is visualized. The comparison of TERM images at different stages of the phase separation process reveals an unexpected transition of PMMA from the dispersed phase to the continuous phase (Figure 17.18).

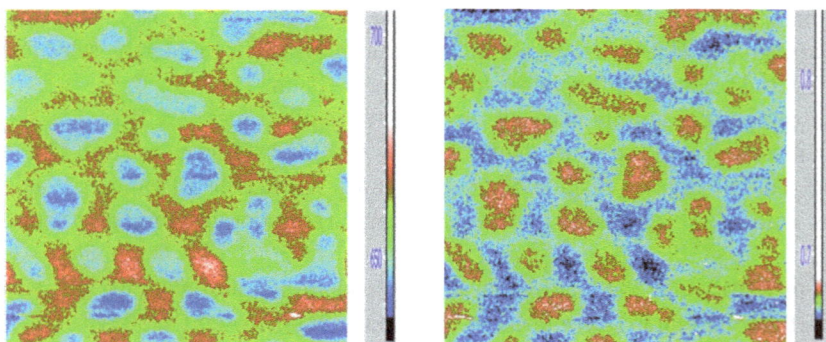

Figure 17.16: TERM in terms of the maximum Raman intensity of the Raman bands (a) at 1,002 cm^{-1} corresponding to SAN and (b) at 800 cm^{-1} corresponding to PMMA on a sample annealed at 250 °C for 2 min (Film I) (Xue *et al.* 2011). Copyright ©2011 American Chemical Society, with permission.

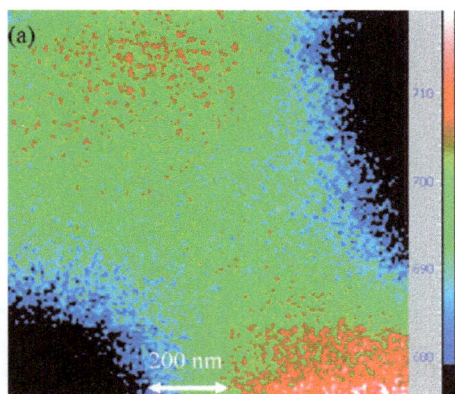

Figure 17.17: TERM at the boundary region between two phases as indicated in Figure 17.16(a) in terms of the maximum Raman intensity of the Raman bands at 1,002 cm^{-1}, corresponding to SAN. The arrow indicates the interface width between SAN and PMMA phases (Xue *et al.* 2011). Copyright ©2011 American Chemical Society, with permission.

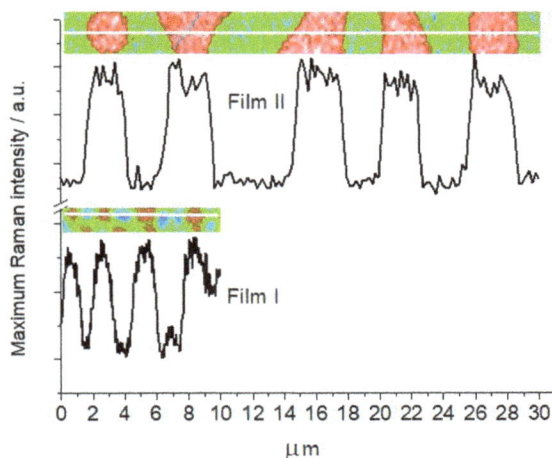

Figure 17.18: Raman intensity profile at 1,002 cm^{-1} corresponding to SAN on a sample annealed at 250 °C for 2 min (Film I) and for 5 min (Film II) (Xue *et al.* 2011). Copyright ©2011 American Chemical Society, with permission.

Chemical imaging of patterned phthalocyanine films (Figure 17.19) using tip-enhanced near-field optical microscopy below the diffraction limit is published by Hermann and Gordon (2016). The experimenters build a TERS instrument around a tuning fork-based AFM attached to an inverted optical microscope. The Au tip is used in shear force mode to reach a spatial resolution of <50 nm. Patterned organic thin

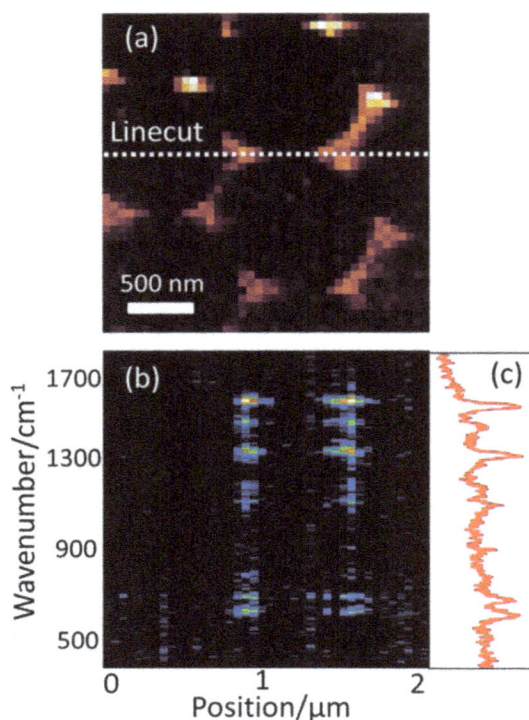

Figure 17.19: A tip-enhanced near-field optical microscope has been developed and used to chemically image patterned Cu(II) phthalocyanine thin films with spatial resolution better than 50 nm (<λ/10). Near-field spectra of coumarin-6 films and the distance scaling (Hermann and Gordon 2016). Copyright ©2016 John Wiley and Sons Ltd., with permission.

films (coumarin-6 and Cu(II) phthalocyanine) are produced by evaporation using a colloidal crystal masking process (1 μm silica spheres).

The stress on CNTs embedded in a peeled-off polymer film is mapped by Hoffmann *et al.* (2017) using TERS. An SWCNT network embedded in an epoxy/amine polymer is mapped using the G$^+$-line at 1,592 cm^{-1}, the G'-line at 2,690 cm^{-1}, and the D-(defect) line at 1,350 cm^{-1} (Figure 17.20). The areas of maximal intensity (coded in white) are similar in the upper half of all three mappings, but show large differences in the lower half of the D-line mapping. Here the D-line intensity is enlarged from defects and stress in the nanotubes caused by peeling off the film from a glass substrate. The experiments show that TERM can be used to successfully identify and characterize the structural integrity of SWCNTs even underneath a layer of pure polymer and provides valuable information about homogeneity and stress effects for a polymer composite.

TERS blinking measurements are used by Agapov *et al.* (2015) to identify the individual isotopes of non-Raman-resonant polystyrene in a miscible binary blend. Thin

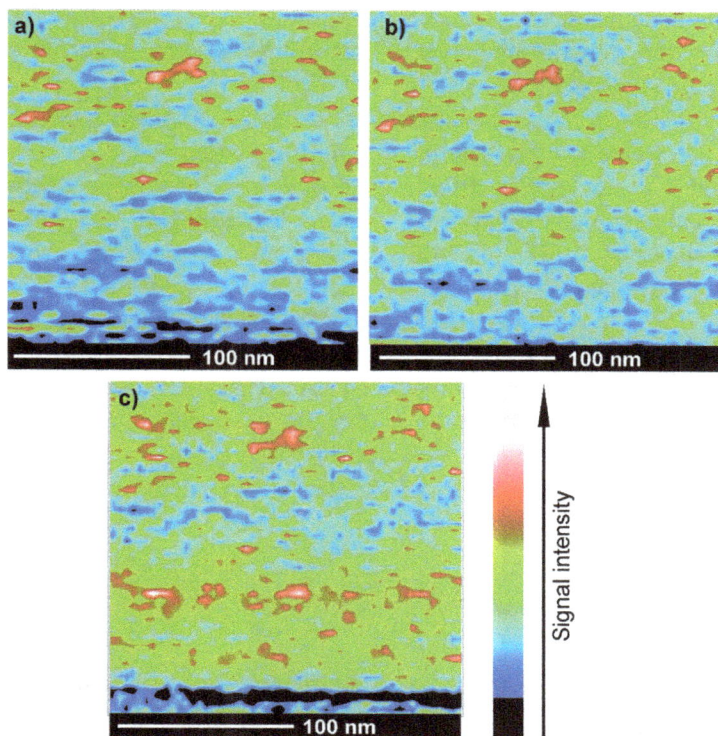

Figure 17.20: Tip-enhanced Raman mapping of an SWCNT network embedded in an epoxy/amine polymer. (a) G^+-line, (b) G'-line, and (c) D-line. The areas of maximal intensity (coded in white) correspond to (a) 4,200 counts, (b) 1,300 counts, (c) 830 counts, respectively (Hoffmann *et al.* 2017). Copyright ©2017 John Wiley and Sons Ltd., with permission.

films of hydrogenous polystyrene (h-PS) and perdeuterated polystyrene containing 20 wt% h-PS, having strong characteristic peaks at 976 cm^{-1} and 1,003 cm^{-1}, respectively, due to the benzene ring breathing mode are studied.

TERS can be used to image plasmon-enhanced local electric field variations with extremely high spatial resolution under ambient conditions, as reported by Bhattarai *et al.* (2017). Using a silver AFM tip coated with 4-mercaptobenzonitrile molecules, they image step edges on an Au(111) surface. A resolution (full width at half-maximum) of only ~3.5 nm is obtained.

Poliani *et al.* (2017) present an experimental study on the near-field light matter interaction by TER scattering with polarized light in three different materials: germanium-doped gallium nitride (GaN), graphene, and CNTs. The authors explain the experimental data with a tentative quantum mechanical interpretation, which takes into account the role of plasmon polaritons and the associated evanescent field.

Su *et al.* (2018) describe mapping of functional groups on a carboxyl-modified GO (GO–COOH) surface with a spatial resolution of ≈10 nm using TERS. The authors also visualize structural defects on the sample.

Su *et al.* (2021) visualize structural modification of patterned graphene nanoribbons (GNRs) using TERS with a spatial resolution of 5 nm. The measurements reveal a structurally modified 5–10 nm strip of disordered graphene at the edge of GNRs. Nanoscale organic contaminants on GNRs are detected by hyperspectral TERS imaging. Su *et al.* speculate on the future use of TERS for nanoscale chemical and structural characterization of graphene-based devices.

Meyer *et al.* (2017) demonstrate polarization-dependent TERS on double perovskite La$_2$CoMnO$_6$ thin films, grown on SrTiO$_3$ (100) substrates, by using an STM based set-up in the side-illumination geometry. The obtained TERS (Figure 17.21) spectra are analysed for different polarization configurations. They are set in relation to the corresponding far-field spectra and are compared with theoretical selection rule calculations, considering a tip amplification and depolarization model for the tip-induced near field.

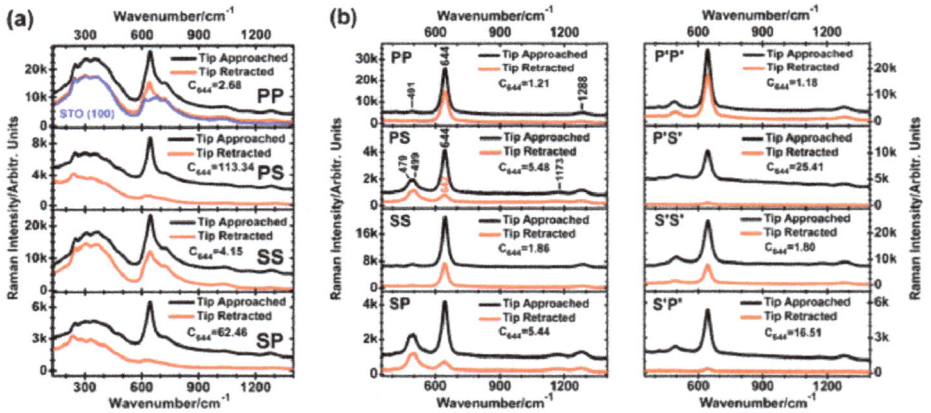

Figure 17.21: Polarized far-field and tip-enhanced Raman spectra of LCMO films with (a) d = 25 nm and (b) d = 120 nm film thickness on STO (100) at room temperature. The spectra are taken in the parallel pp, ss, and the crossed scattering ps and sp configurations. The STO (100) background is displayed for an orientation of the substrate contribution on the spectra of the 25 nm LCMO film. For the 120 nm LCMO film, additional far-field and tip-enhanced spectra in the p′p′, s′s′, p′s′, and s′p′ configurations with an in-plane rotation of ϕ = 45° are displayed (Meyer *et al.* 2017). Copyright ©2017 John Wiley and Sons, Ltd., with permission.

Bonhommeau and Lecomte (2018) review TERS as a tool for nanoscale chemical and structural characterization of biomolecules. They compare it with optical microscopy techniques of which the relatively new super-resolved fluorescence microscopy techniques, such as stimulated emission depletion, reach a value as low as about 5 nm. Drawbacks of the techniques are discussed as well.

Low-cost tips for TERS are fabricated by two-step electrochemical etching of 125 μm diameter gold wires, as reported by Foti *et al.* (2018). The tips are tested on dyes (Figure 17.22), pigments, and biomolecules and feature EFs higher than 10^5 and a spatial resolution of 5 nm in TERS mapping.

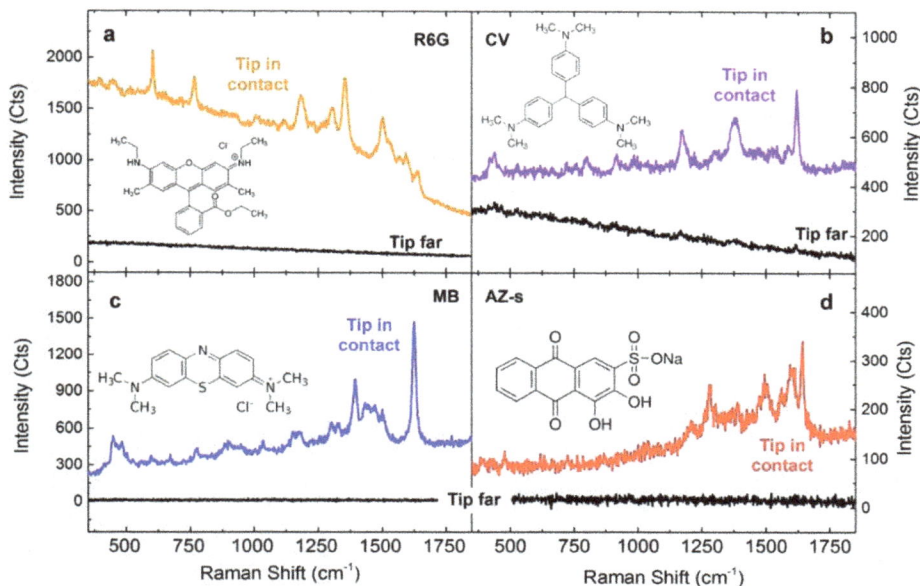

Figure 17.22: TERS spectra (coloured lines) of different molecules acquired with the tip in contact with the surface: (a) rhodamine 6G (R6G, $P = 1$ mW, $t = 5$ s), (b) crystal violet (CV, $P = 1$ mW, $t = 3$ s), (c) methylene blue (MB, $P = 0.1$ mW, $t = 1$ s), and (d) alizarin-s (AZ-s, $P = 1$ mW, $t = 5$ s). Black lines represent signal intensity acquired in the same conditions when the tip is far from the sample (Foti *et al.* 2018). Copyright ©2018 by the authors, licensed by Creative Commons (CC 4.0).

Multiple vibrational modes in ultra-HV-TERS combined with molecular-resolution STM are observed by Jiang *et al.* (2012).

Zhang *et al.* (2013) report chemical mapping of a single molecule by plasmon-enhanced Raman scattering. The authors map single *meso*-tetrakis(3,5-di-*t*-butylphenyl) -porphyrin (H2TBPP) molecules on an Ag(111) surface.

The technique of TERM is now sufficiently advanced to allow for spatially re-solved analysis of single fibres. Deckert's group in Jena studies insulin protofilaments and fibrils. Kurouski *et al.* (2014) characterize the surface of insulin protofilaments and fibril polymorphs using TERS.

Van den Akker *et al.* (2015) study the heterogeneity of the molecular structure of individual hIAPP amyloid fibrils on the nanoscale with TERS. With these experiments, the authors can identify and spectrally map alternating hydrophobic and hydrophilic (or a mixture of both) nanometre-sized domains on insulin fibrils for the first time.

Deckert-Gaudig *et al.* (2016) spatially resolve hydrophilic and hydrophobic domains on individual insulin amyloid fibrils by TERS.

Chiang *et al.* (2016) achieve conformational contrast of surface-mediated molecular switches of Ångstrom-scale spatial resolution in ultra-HV-TERS. The authors interrogate the conformational switch between two metastable surface-mediated isomers of H_2TBPP on a Cu(111) surface.

Lee *et al.* (2019), using TER-STM at the precisely controllable junction of a cryogenic ultra-HV-STM, show that Ångstrom-scale resolution is attained at subatomic separation between the tip atom and a molecule in the quantum tunnelling regime of plasmons. Lee *et al.* succeed in recording vibrational spectra within a single molecule (Co(II)–tetraphenyl porphyrin immobilized on Cu(100) at 6 K) and obtain images of normal modes (Figure 17.23).

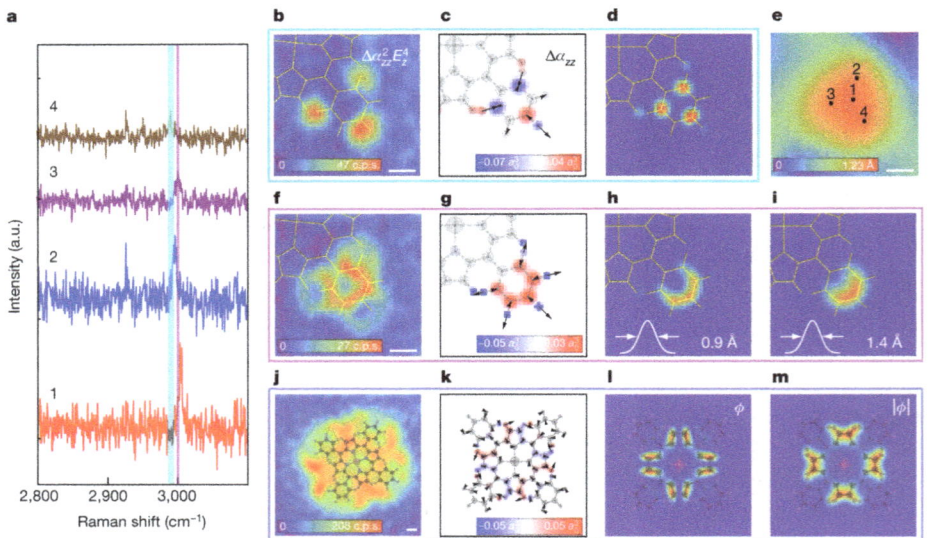

Figure 17.23: Close-ups of TER-STM images and their simulations. (a) Variation of the C–H stretching frequency within a phenyl group is shown by four representative spectra recorded on the points indicated in the scanning tunnelling microscope (STM) topography in (e). The spectra are extracted from a hyperspectral image recorded on a 40 × 40 grid with 4-s integration time per grid point. (b–d) TERS map of the asymmetric C–H stretching mode, integrated over the blue band (2,986–2,995 cm^{-1}) in (b); colour-coded atomic partitioning of the polarizability differences $\Delta\alpha_{zz}$ in atomic units a_0^3 (c); and simulated Raman image (d). (e) STM topography (11 × 11 Å^2) of the lower-right phenyl group. (f–i) TERS map integrated over the purple band (2,999–3,003 cm^{-1} (f)), $\Delta\alpha_{zz}$ (g), and images obtained using a Gaussian with FWHM of 0.9 Å (h) and 1.4 Å (i). (j–m) 29 × 29 Å^2 Raman map (80 × 80 grid) integrated over a band (1,538–1,572 cm^{-1} (j)), $\Delta\alpha_{zz}$ map (20 × 20 Å^2 (k)) and images obtained using signed polarizability differences (phase ϕ preserved) versus their magnitudes (l, m). Scale bars are 2 Å. Experimental Raman maps in (b–m) are low-pass-filtered for clarity (Lee *et al.* 2019). Copyright ©2019 Springer Nature, with permission.

Vibrational modes of a single molecule can be visualized by TERS under STM conditions with atomic resolution. The origin of the spatially varying bright spots still lacks theoretical interpretation. Liu *et al.* (2019) systematically study the Raman scattering images of a single Co(II)–tetraphenylporphyrin molecule and employ a locally integrated Raman polarizability density. An intuitive explanation of the origin of the experimental Raman scattering images is given.

Chen *et al.* (2014) analyse CNTs by TER imaging, achieving a resolution of 1.7 nm. Using a gold STM tip on a gold substrate at ambient temperature, they can discriminate different types of CNTs, locate defects, and measure diameters and bundling effects (Figure 17.24).

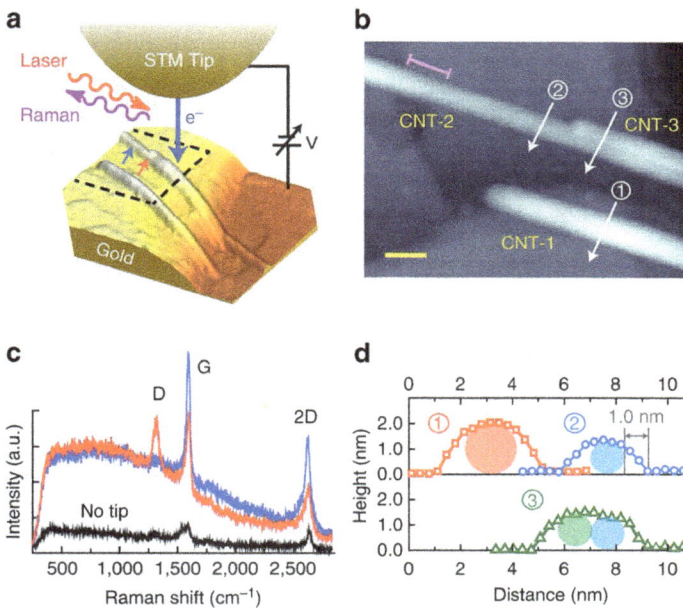

Figure 17.24: TERS experiment and the structure of CNT bundle. (a) Schematic illustration of TERS experiment with laser excitation at the STM tunnelling junction. (b) High-resolution STM topography of three CNTs (dashed rectangular area in (a), 39 × 30 nm^2). The pink line segment on the upper left indicates the diameter transition region of CNT-2. Scale bar, 5 nm. (c) Comparison of far-field Raman (black, no tip) and TERS (red and blue) spectra taken at two locations on CNT-2 as indicated in (a). Accumulation time of each spectrum is 1 s. (d) Topographic profiles of CNTs along three-arrowed lines in (b). Circular shades represent the cross section of each CNT. The 10–90% height transition indicated in profile B defines the spatial resolution (1.0 nm) of our STM. Note that profile C includes the bundling structure of CNT-2 and CNT-3 (Chen *et al.* 2014). Copyright ©2014 Springer Nature, with permission.

Olschewski *et al.* (2015) introduce a manual and an automatic TERS-based virus discrimination. Two pathogenic viruses, varicella-zoster virus and porcine teschovirus, can be discriminated using TERS and chemometrics. TERS spectra of single virus particles can also be used with a classification accuracy of 91%.

A review by Wang *et al.* (2017) focuses on TERS for surfaces and interfaces. The group of authors concludes that TERS is an ideal tool for achieving an in-depth understanding of the surface and interfacial processes. TERS has been able to monitor the dynamic interactions of the molecules on the surface, and identify surface sites of different chemical and electronic properties, with spatial resolution of about 3 nm.

17.2 Nano-IR

An alternative method to TERS with high spatial resolution is nano-IR spectroscopy using a tunable quantum cascade laser (QCL). Dazzi *et al.* (2012) review the combination of AFM and IR spectroscopy (nano-IR) for nanoscale chemical characterization (Figure 17.25). Five years later, an update to the review is published by Dazzi *et al.* (2017) and another review with Dazzi as co-author in 2020.

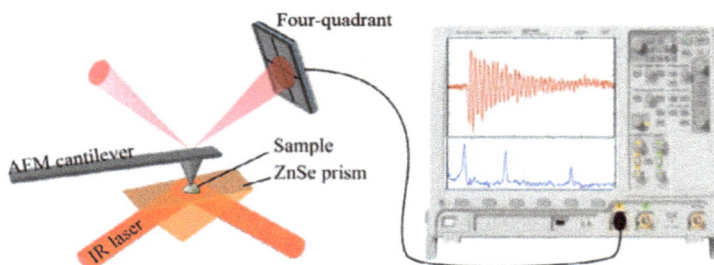

Figure 17.25: Scheme of the AFM–IR set-up. The AFM cantilever ring-down amplitude plotted as a function of laser excitation wavelength produces the IR spectrum (Dazzi *et al.* 2012). Copyright ©2012 by the authors, licensed by Creative Commons (CC 4.0).

AFM in contact mode is unsuitable for soft or loosely adhesive samples such as polymeric NPs. To solve this issue, tapping-mode AFM-IR is advocated by Mathurin *et al.* (2018). Mathurin *et al.* test PLGA NPs morphology and composition, they also visualize drug location and core–shell structures. The authors prove preferential accumulation of the incorporated drug, pipemidic acid, in the NPs' top layers, even at its low concentration of <1 wt%.

Paclitaxel is an effective chemotherapeutic agent against a broad range of cancers and is used in drug-eluting stents against restenosis. To understand the distribution of the drug in a polymer matrix, nanoscale partitioning of paclitaxel in hybrid lipid–polymer membranes is studied by Tuteja *et al.* (2018), using FT-IR spectra of the samples on a diamond ATR crystal (Figure 17.26).

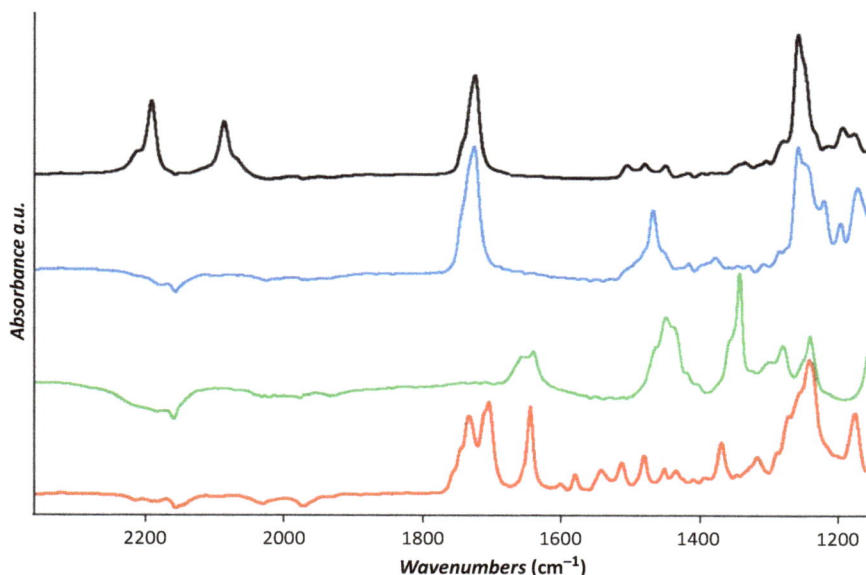

Figure 17.26: FT-IR spectra of paclitaxel (red), PBD-b-PEO (green), DPPC (blue), and d-DPPC (black). The spectral intensity has been normalized to the strongest peak in each spectrum, and the spectra are displayed with an offset for clarity. The negative bands between 1,900 and 2,200 cm^{-1} observed in the spectra are due to imperfect background compensation due to the absorption of the diamond ATR crystal (Tuteja *et al.* 2018). Copyright ©2018 Royal Society of Chemistry, with permission.

A single cotton fibre is studied by Igarashi *et al.* (2020), using AFM-IR (Figure 17.27, left). Looking for water bound to the fibre, they find two stretching modes of OH groups (Figure 17.27, right), clearly decoupled from each other, which arise from the effects of the air–water (hydrophobic) and water–cellulose (hydrophilic) interfaces.

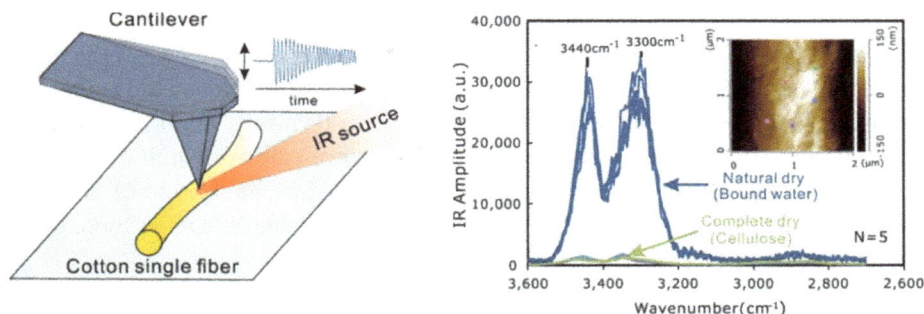

Figure 17.27: AFM–IR spectra of a cotton single fibre under the naturally dried condition (50% RH) and a completely dry condition, respectively. Inset shows an AFM image of a cotton single fibre (Igarashi *et al.* 2020). Copyright ©2020 American Chemical Society, with permission.

IR absorption nanospectroscopy is applied by Ruggeri *et al.* (2020) to achieve single-molecule secondary structure determination of proteins. The method offers high throughput on a second timescale and with a high signal-to-noise ratio of ~10–20. The test molecules chosen are apoferritin and thyroglobulin.

Strongly absorbing nanoscale IR domains within strained bubbles at hBN–graphene interfaces are studied by Vincent *et al.* (2019).

Substrate-enhanced photothermal nano-imaging of surface polaritons in monolayer graphene is reported by Menges *et al.* (2021).

Phonon-polaritons arise from the strong coupling of photons to optical phonons. Future applications, the authors note, will include sensing, imaging, superlensing, and photonics-based communications by light confinement and harnessing below the diffraction limit. Yang *et al.* (2021) investigate phonon-polaritonic crystals that incorporate hyperbolic hBN to a silicon-based photonic crystal (Figure 17.28) by nano-IR. Near-field excited Archimedean-like tiling patterns are observed by SNOM.

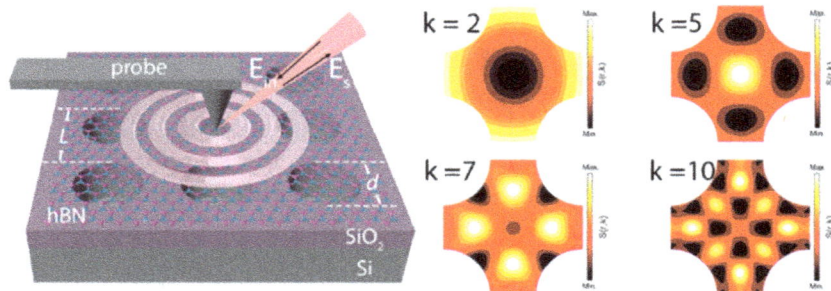

Figure 17.28: Phonon-polaritonic crystals (PPCs) based on hyperbolic hBN and the real-space phonon-polariton propagation. Archimedean-like tiling patterns were calculated from the wave interference of a Hankel function of the first kind with $k = 2$, 5, 7, and 10 cm^{-1} (Yang *et al.* 2021). Copyright ©2021. American Chemical Society, with permission.

Stani *et al.* (2022) are interested in the secrets of Stradivari's violins. One of the questions is if Stradivari used proteins on the wood. IR s-SNOM is used by the authors to acquire spectra of different spots on their tiny sample (Figure 17.29). It is possible to conclude that, for the analysed Stradivari's violins, proteins are too diluted in the wooden matrix to clearly stand out from the background at the microscale, whereas the nanoresolved approach clearly highlights the local chemical composition, overcoming the dilution effects of the analytes of interest.

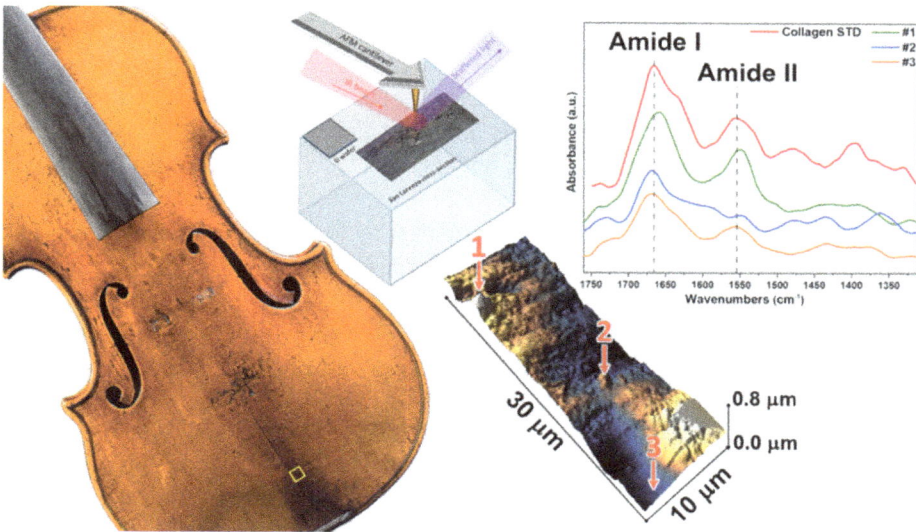

Figure 17.29: The tested Stradivari violin, a sketch of the s-SNOM method, AFM of the sample, and spectra of the collagen standard together with spectra of the three points on the sample (Stani *et al.* 2022). Copyright ©2022 American Chemical Society, with permission.

18 Vibrational optical activity (VOA)

Chemical compounds that are chiral, which means that the molecules exist in two forms which are mirror images of each other, interact differently with circularly polarized light. As natural products normally exist only as one of these chiral forms, and synthetic medical compounds show different actions depending upon the enantiomer used, certain methods are needed to discriminate between the different forms. One of the few methods to achieve this aim is the measurement of vibrational optical activity, mainly in the form of vibrational circular dichroism (VCD), as reviewed by Hoffmann (1995).

Polavarapu and Santoro (2020) write a review on applications of VCD. The very elaborate paper describes work on a large palette of natural product families. Another method is the measurement of Raman optical activity, mainly using circularly polarized light to measure Raman circular intensity difference (CID). A review of the field is given by Hoffmann (2010).

18.1 Vibrational circular dichroism (VCD)

Due to experimental difficulties, circular dichroism in the infrared (IR) region was measured much later than the "normal" CD in the visible region. As the first step towards the measurement of single-molecule effects, Schrader and Korte (1972) report the measurement of the IR rotatory dispersion of carvone in liquid crystalline solution. They use a modified commercial infrared spectrometer and observe a huge effect, which is not the result of the carvone itself but of the liquid crystal in which a helical arrangement (cholesteric state) is induced by the chiral solute. In this case, the liquid crystal acts as a kind of molecular amplifier, which allows the absolute configuration (AC) of tiny amounts of solutes to be determined reliably. At about the same time, Dudley *et al.* (1972) measure the IR circular dichroism of (−)-menthol in a liquid crystal. Their equipment consists of a normal IR spectrometer supplemented by a Fresnel rhomb made from sodium chloride. In the years to follow, the key to the measurement of VCD is the development of photoelastic modulators suitable for work in the IR spectral region. The first successful measurements of circular dichroism originating from vibrational transitions in the IR are done by Hsu and Holzwarth (1973) on thin slices of monocrystalline α-$NiSO_4 \cdot 6\ H_2O$ and α-$ZnSeO_4 \cdot 6\ H_2O$. For this measurement, the authors use a normal dispersive IR spectrometer supplemented by a linear polarizer and a photoelastic modulator made from germanium.

Soon after the first measurements of Holzwarth, Nafie *et al.* (1975) report an improved measurement of VCD of 2,2,2-trifluoro-1-phenylethanol.

The measurement of VCD is greatly simplified when Nafie *et al.* (1979) report the use of a Fourier-transform (FT)-IR instrument and publish a comparison of a dispersive and the FT-IR-VCD spectrum of (+)-camphor in CCl_4 solution in the C–H stretching region.

https://doi.org/10.1515/9783110717556-018

The measurement of VCD was confined to the region below 15 µm (700 cm^{-1}) resulting from the transmittance characteristics of the photoelastic modulator made from zinc selenide (ZnSe). This is changed by the construction of a photoelastic modulator made from cadmium telluride (CdTe) with sufficient transparency and modulation down to about 300 cm^{-1} by Hoffmann *et al.* (1987).

Commercial FT instruments use many optical elements which may produce artefacts. Therefore, a dedicated VCD interferometer is built by Hoffmann and Hochkamp (1992), which consists only of the minimal amount of optical elements necessary for a photoelastic modulator-based VCD instrument (Figure 18.1). For the measurement of interferograms, a simultaneous technique (Hoffmann and Hochkamp 1992) is used. Instead of recording the "AC" and the "DC" part one after the other, the two signals are sampled nearly simultaneously using an input multiplexer in front of the analogue-to-digital converter. As the "AC" part is always delayed by the passage through the lock-in amplifier, this can be done without introducing an error. The result of this procedure is that each slight change in the baseline of the interferometer has the same effect on both interferograms and should be compensated this way.

A very promising technique is the measurement of circular dichroism (and linear dichroism) by means of a polarizing interferometer proposed by Martin and Puplett (1970) in 1969. The Martin–Puplett interferometer uses a linear polarizer as a beam splitter. If the interferogram produced at the detector is Fourier transformed, the sine FT gives directly the circular dichroism, whereas the cosine FT directly produces the linear dichroism of the sample. Bomem Inc. (1987) in 1987 offers a commercial instrument, but it is not before 1990, that Ragunathan *et al.* (1990) report the first real VCD measurements with a Martin–Puplett interferometer on (–)-α-pinene.

Using the CdTe modulator in a dedicated VCD interferometer, the VCD spectrum of (+)-(*R*)-3-methylcyclohexanone in the spectral region of 700 to 350 cm^{-1} can be measured for the first time by Hoffmann and Hochkamp (1992) (Figure 18.2). As a detector, a long-wavelength-optimized HgCdTe detector has to be used, which has a lower peak detectivity (D$^{*}_{peak}$ 2 × 10^9 cm Hz$^{1/2}$ W^{-1}, cut-off$_{50\%}$ 350 cm^{-1}) than the normal detector (D$^{*}_{peak}$ 10^{10} cm Hz$^{1/2}$ W^{-1}, cut-off$_{50\%}$ 460 cm^{-1}). Only a bolometer has a higher detectivity in this range, but it will not operate at the high modulating frequency of 60 kHz. The sign of the bands corresponds well to the rotatory strengths calculated with distributed origin, and the sign of the 435 cm^{-1} band, however, is buried in noise. The comparison of spectral bands measured at the low resolution of 32 cm^{-1} to the rotatory strengths calculated with Gaussian at the 6-311^{+}G(p,d) level is shown in Figure 18.3. No scaling factor has to be used for the frequencies.

The first successful VCD measurements with a polarizing Michelson interferometer are done on pinene (Ragunathan *et al.* 1990). They use a wire grid polarizer on a BaF$_2$ substrate as the beam splitter, limiting the measurement to 11 µm on the long wavelength side and an optical long-pass filter (to prevent saturation of the detector) putting a limit of 6 µm to the short wavelength side. The spectrum of (–)-α-pinene as a 50 µm film of neat liquid using a Bomem spectrometer with a polarizing interferometer attachment is

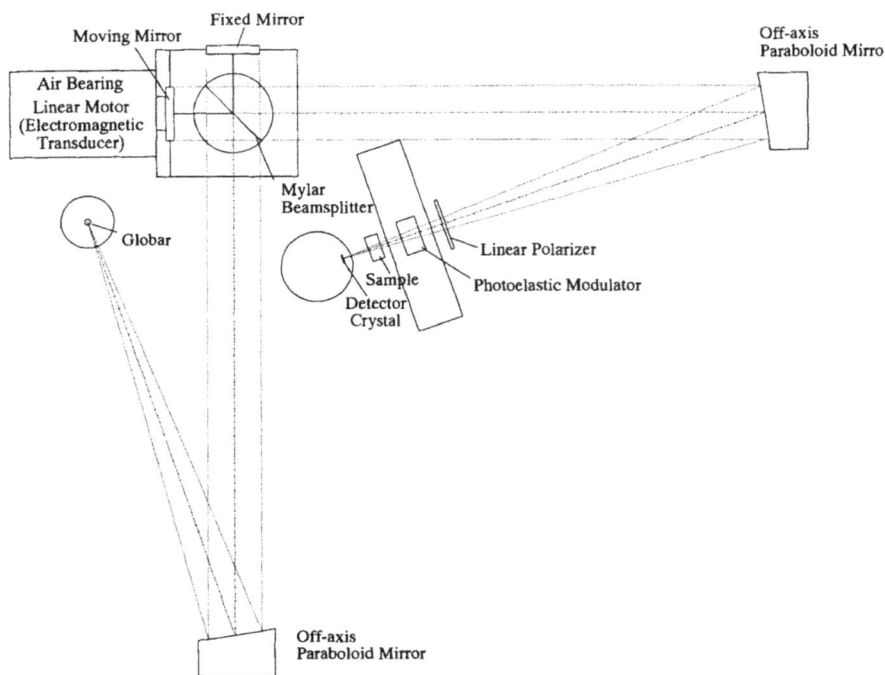

Figure 18.1: Optical layout of a dedicated VCD interferometer (Hoffmann 1995). Copyright ©1995 VCH/John Wiley and Sons Ltd., with permission.

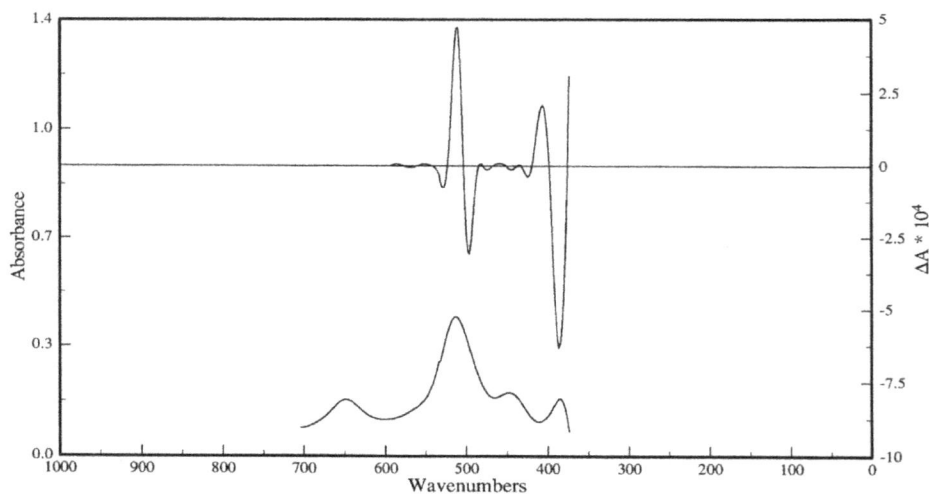

Figure 18.2: Experimental VCD of (*R*)-(+)-3-methylcyclohexanone belcw 650 cm^{-1} (Hoffmann and Hochkamp 1992).

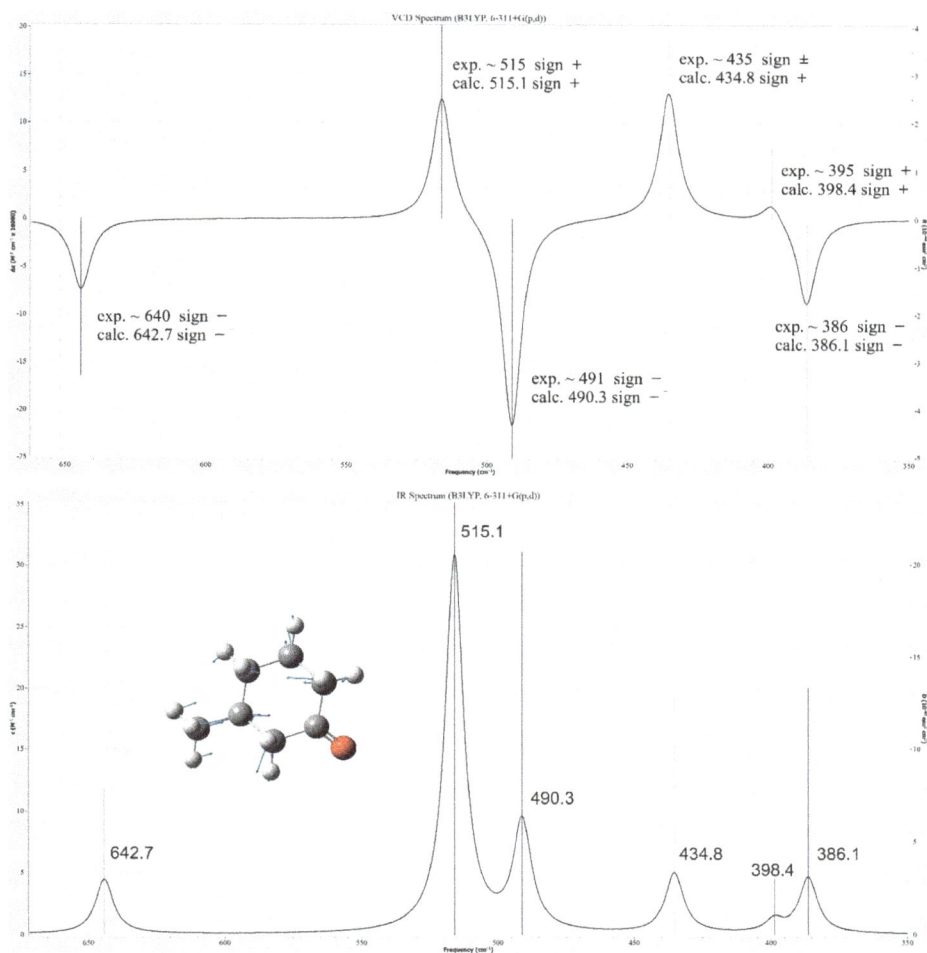

Figure 18.3: Calculated (*R*)-(+)-3-methylcyclohexanone IR (top) and VCD (bottom) spectrum (B3LYP 6-311⁺G (p,d)) in cm⁻¹. Hoffmann, unpublished). Inset shows the vibration of the molecule at 398.4 cm⁻¹.

a composite of 60,000 VCD scans and 12,000 transmission scans and takes 19.5 h to be measured. A spectrum with comparable signal-to-noise ratio is shown in Figure 18.5. This is measured in our laboratory in 1992 with a ZnSe PEM-based VCD spectrometer and is a composite of only 5,000 scans (5 blocks of 1,024 scans each). The noise estimate shown as the upper trace is the difference between two 5,000 scan VCD spectra.

The spectrum of crystalline α-NiSO₄ · 6 H₂O is reinvestigated by Hoffmann (1995). It is shown in Figure 18.4. Unlike the original spectrum of Hsu and Holzwarth (1973), it is measured on a modified Nicolet 7199 FT-IR spectrometer using a ZnSe photoelastic modulator, resulting in a wider wavelength range and higher resolution.

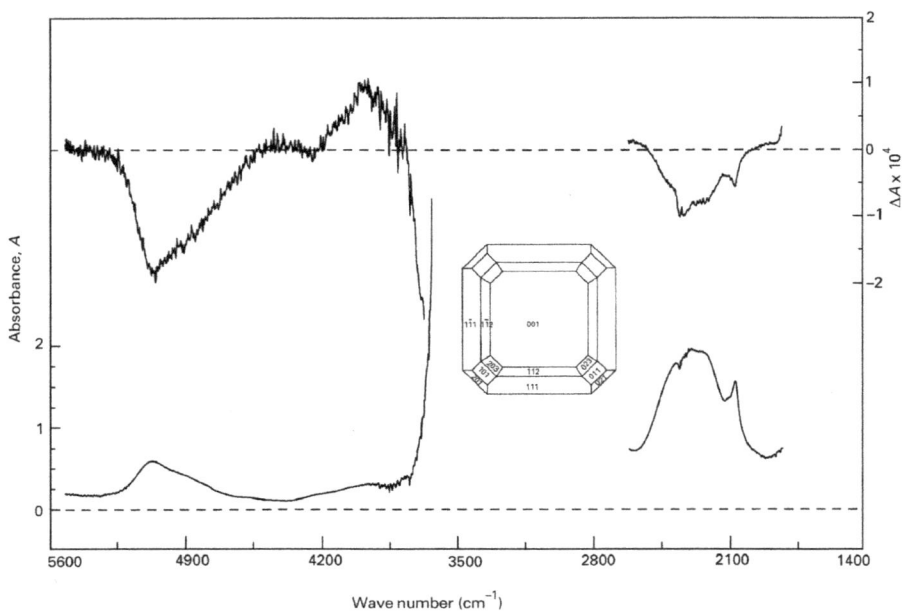

Figure 18.4: VCD spectrum of a α-NiSO$_4$ · 6 H$_2$O single-crystal slice (d = 63 μm, cleaved parallel (001) as shown in inset) (Hoffmann 1995). Copyright ©1995 VCH/John Wiley and Sons Ltd., with permission.

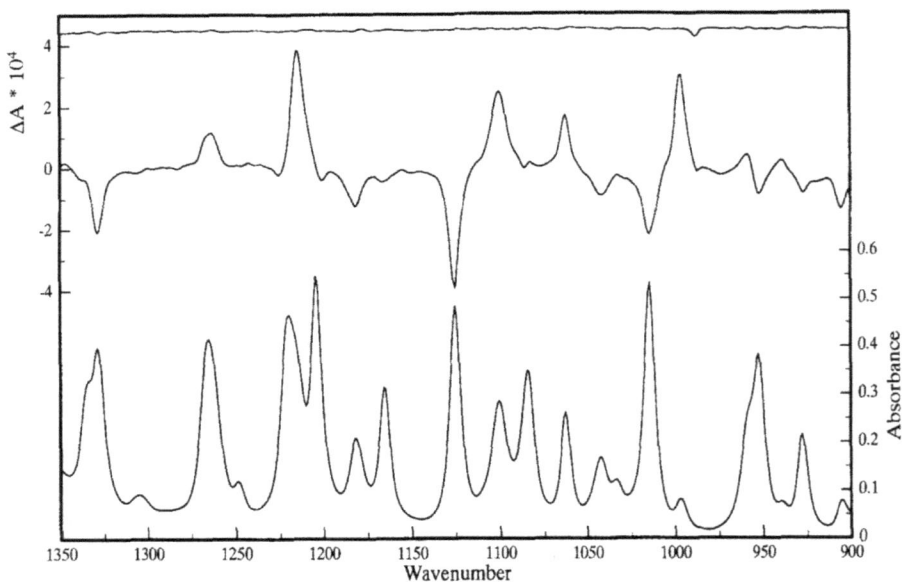

Figure 18.5: VCD spectrum of (−)-α-pinene (neat, 50 μm) (Hoffmann 1995). Copyright ©1995 VCH / John Wiley and Sons Ltd., with permission.

The Raman, IR, and VCD spectra of the steroid precursor (+)-5,6,7,8-tetrahydro-8-methylindan-l,5-dione are calculated on the 6-31G level of theory by Hoffmann (1993) and compared with the experimental spectra shown in the same paper.

The aim of the review by del Rio and Joseph-Nathan (2021) is to update VCD results for the assignment of the AC of natural products published from 2015 to 2019. During this period, VCD is reported in some 126 publications involving almost 300 molecules.

Paoloni *et al.* (2020) measure the spectra of two diols from the near-IR to mid-IR. They supply VCD data together with quantum chemical calculations using density functional theory (DFT). The spectra of 2,3-butanediol and *trans*-1,2-cyclohexanediol (Figures 18.6 and 18.7) are calculated nearly unsupervised using the DFT/GVPT2 model, which also takes anharmonicity into account.

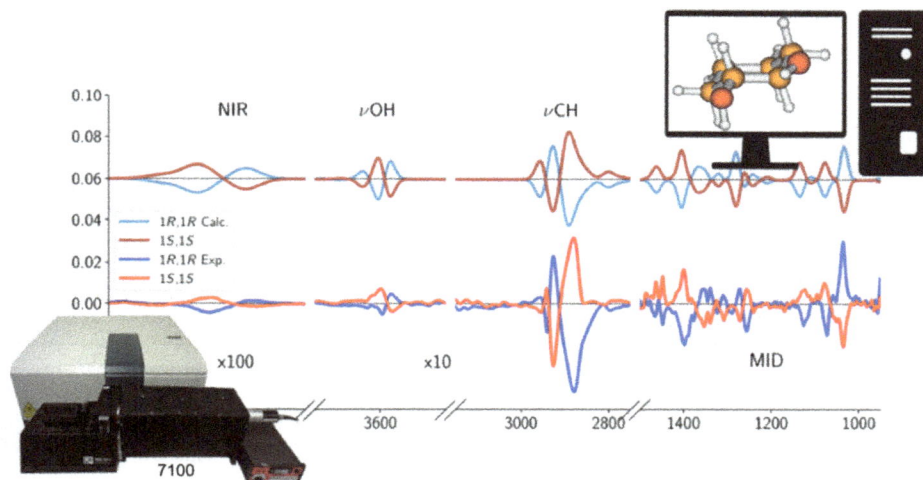

Figure 18.6: VCD of 1,2-cyclohexanediol in the NIR, ν_{OH}, ν_{CH}, and mid-IR region (Paoloni *et al.* 2020). Copyright ©2020 by the authors, licensed by Creative Commons (CC 4.0).

The spectra of each conformer are weighted with their respective Boltzmann population based on B2PLYP harmonic energy. The spectra are simulated, assigning Gaussian distribution functions of 10 cm^{-1} half-width at half-maximum.

Anisodamine, or 6β-hydroxyhyoscyamine, is a tropane alkaloid found in the roots and leaves of various species of *Physochlaina*, *Scopolia*, *Duboisia*, and *Datura* genera. It can be used, for example, against cardiovascular disorders, acute ischaemic renal failure, and snakebites. Munoz *et al.* (2006) report the AC of natural diastereoisomers of 6β-hydroxyhyoscyamine as shown by VCD. The predicted VCD and IR spectra of (3R,6R,2′S)-6β-hydroxyhyoscyamine and (3S,6S,2′S)-6β-hydroxyhyoscyamine are calculated using DFT with the B3LYP functional and 6-31G(d) basis set. Eight conformers are taken into account.

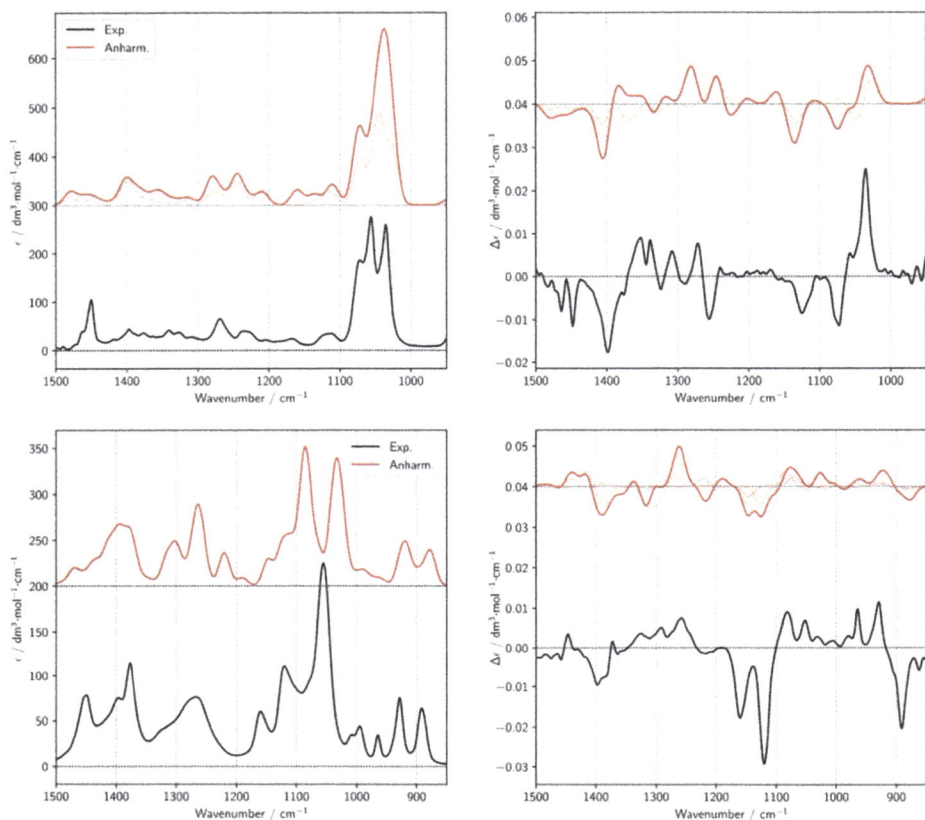

Figure 18.7: Comparison of experimental spectra of 1*R*,2*R*-cyclohexanediol (top left and top right images) and 2*R*,3*R*-butanediol (bottom left and bottom right images) with anharmonic calculations of IR and VCD spectra in the mid-region (Paoloni *et al.* 2020). Copyright ©2020 by the authors, licensed by Creative Commons (CC 4.0).

Cepharanthine, a compound isolated from the roots of *Stephania*, has been used for a wide variety of medicinal applications. Based on chiroptical spectroscopic studies, it is suggested that the previously stated AC of cepharanthine, (+)-(1*R*,10*S*), and also those of some of its analogues, have to be revised. While the experimental specific optical rotation (SOR) at sodium D-line is +230, the corresponding calculated values for (1*R*,10*S*) at B3LYP/6-311⁺G(d) level were +25 in gas phase and +60 using polarizable continuum model (PCM) representing the solvent. On the other hand, those calculated for (1*R*,10*R*) at B3LYP/6-311⁺G(d) level were +181 in gas phase and +206 using the PCM representing the solvent. The experimental SORs at other wavelengths, namely 365, 405, 436, and 546 nm, also match poorly with those calculated for (1*R*,10*S*), but matched better with those calculated for (1*R*,10*R*). These observations suggest that the originally assigned (+)-(1*R*,10*S*) has to be reassigned as (+)-(1*R*,10*R*), as stated by Ren *et al.* (2019).

The clerodane diterpenes anacolosins A–F and corymbulosins X and Y, isolated from the plant *Anacolosa clarkii* PIERRE (Olacaceae), display cytotoxic activities against paediatric solid tumour cell lines. Their AC is established by Cai *et al.* (2019) using the comparison of experimental and calculated VCD (Figure 18.8).

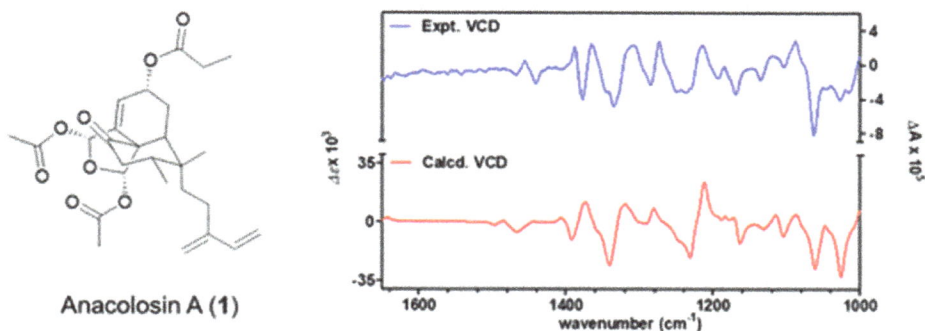

Figure 18.8: Experimental and calculated VCD of anacolosin A (Cai *et al.* 2019). Copyright ©2019 American Chemical Society, with permission.

Cao *et al.* (2019) isolate six new azaphilone derivatives from fungus *Pleosporales* sp. CF09-1. The new compounds are pleosporalones B and C and pleosporalones E–H and one known analogue pleosporalone D (Figure 18.10), but with unassigned AC. The ACs of C-2′ and C-3′ in 3 are assigned by comparison of experimental and calculated VCD (Figure 18.9).

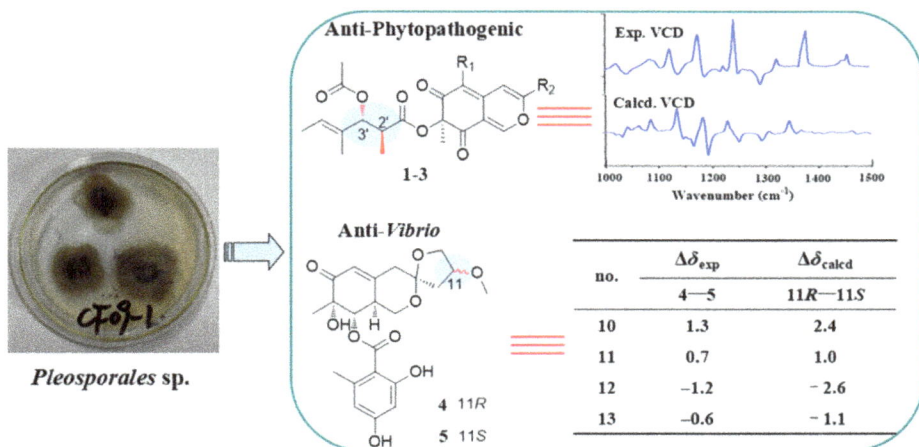

Figure 18.9: Absolute configurations of C-2′ and C-3′ in pleosporalone D (Cao *et al.* 2019). Copyright ©2019 American Chemical Society, with permission.

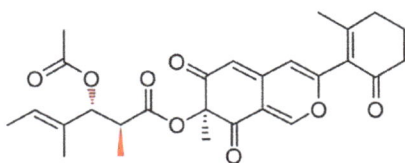

Figure 18.10: Structure of pleosporalone D (Cao *et al.* 2019). Copyright ©2019, American Chemical Society, with permission.

(1*R*,10b*R*)-10-((*R*)-1,2-Dihydroxyethyl)-1-hydroxy-8,9-dimethoxy1,5,6,10b-tetrahydropyrrolo [2,1-a]isoquinolin-3(2H)-one, an analogue of (−)-crispine A, with three stereogenic centres is synthesized from garcinia acid (a natural product isolated from tropical plant native to South Asia) by Johnson *et al.* (2019). Experimental chiroptical spectra (optical rotatory dispersion (ORD), electronic circular dichroism (ECD), and VCD) and the corresponding quantum chemical predicted spectra are used to establish the AC.

A long-standing debate around the AC of (+)-frondosin B (Figure 18.11) is ended by Joyce *et al.* (2018). When a total synthesis of a synthetic intermediate (originally extracted from marine sponge) was undertaken, the presence of a minor impurity (<7%) present in the late step of synthesis changed the sign of observed specific rotation. This sign reversal influenced the literature assignment of AC for this natural product. Experimental VCD and ECD measurements are used to determine the AC of the synthetic intermediate as (*R*), and this investigation led to the identification of minor impurities that resulted in conflicting assignments of AC in the earlier synthetic efforts.

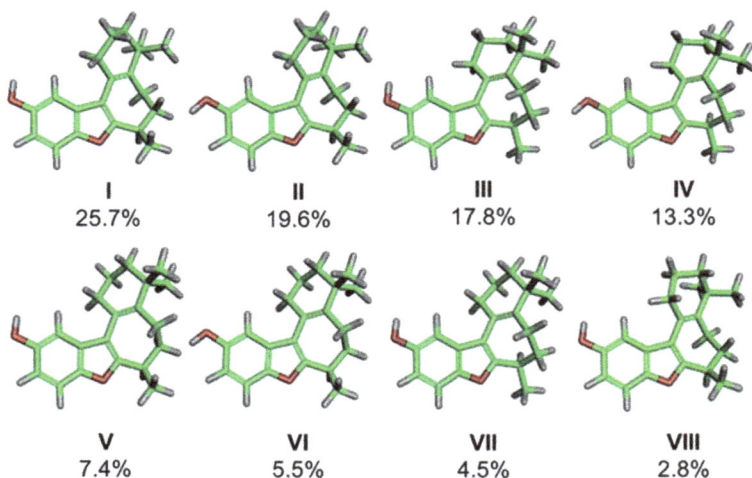

| I | II | III | IV |
| 25.7% | 19.6% | 17.8% | 13.3% |

| V | VI | VII | VIII |
| 7.4% | 5.5% | 4.5% | 2.8% |

Figure 18.11: Main conformers of frondosin B (Joyce *et al.* 2018). Copyright ©2018 by the authors, licensed by Creative Commons (CC 4.0).

C-8-substituted dibenzoylmethanes (DBMs) assume a di-keto form, which generates a chiral centre at position 8. The AC of the chiral centre generated in C-8-substituted DBMs was not known. To address this issue, 20-methoxy-8-(a-a-dimethylallyl)-[30,40:400,500]-furan-dibenzoylmethane, a C-8-substituted DBM, is isolated from the plant roots of *Dahlstedtia glaziovii*, and its AC is determined to be (*S*) based on the experimental and calculated VCD spectrum by Canzi *et al.* (2018).

Citrinin derivatives are a class of bioactive polyketide compounds isolated from the fungal genera *Penicillium*, *Aspergillus*, and *Monascus*. Penicitrinone A, a citrinin derivative, is obtained from the Bohai Sea fungus *Penicillium janthinellum*. Cao *et al.* (2019) address the problem of assigning the right diastereomer by comparing the experimental VCD of structure 1 with the corresponding DFT calculations for the two diastereomers 1a and 1b (Figure 18.12).

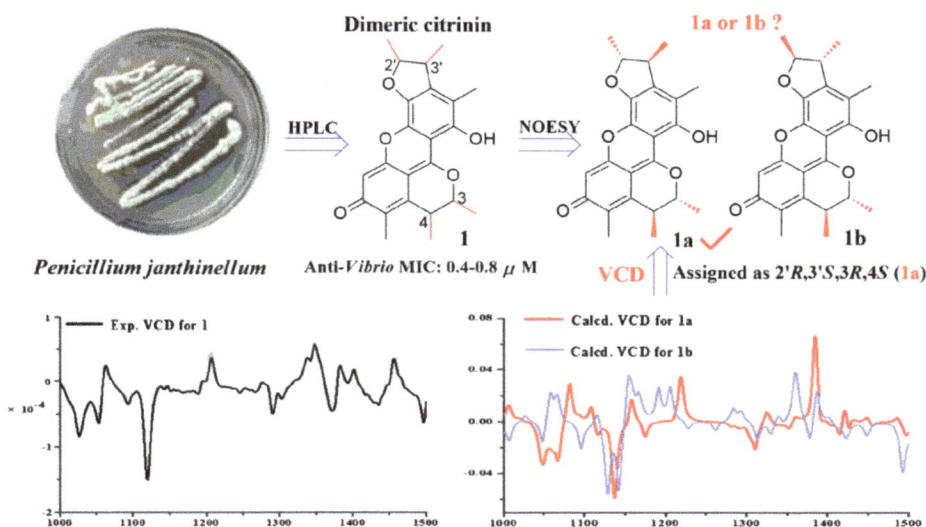

Figure 18.12: Structure and experimental and calculated VCD of citrinin (Cao *et al.* 2019). Copyright ©2019 Taylor & Francis, with permission.

Metabolites isolated from *Rutstroemia*, and their isolation and identification of three chiral compounds 9-*O*-methylfusarubin, 5-*O*-methylnectriafurone, and terpestacin, from a diseased *Bromus tectorum* plant, are reported. The terpestacin compound was previously known in the literature. The unknown absolute stereochemistry at C-3 of 9-*O*-methylfusarubin is determined from the analysis of experimental ECD and VCD (Masi *et al.* 2018).

Jonquailine, an alkaloid isolated from *Narcissus jonquilla quail* (an Amaryllidaceae species plant), shows significant activity against several malignant cancer cell types. Vergura *et al.* (2018) assign the previous unknown stereochemistry at C-8 of B ring. The task is accomplished by comparing DFT calculations of VCD to the experimental VCD spectrum of jonquailine. This method also assigns the AC of its analogues pretazettine and 8-*O*-methylpretazettine on the way.

Although two triterpenoids, icetexone and conacytone, were isolated four decades ago from the plant *Salvia ballotaeflora*, their characterization left several uncertainties. Esquivel *et al.* (2018) determine the AC of both compounds by VCD measurements on icetexone acetate and conacytone triacetate.

Five cascarosides, extracted from the bark of *Rhamnus purshiana* plant, are studied by Demarque *et al.* (2018). As the introduction of the C_{10}-stereocentre of (ox-) anthrones by plant organisms is not stereospecific, a VCD study can be used to characterize the C_{10}-stereocentre of (ox-)anthrones as (10S) or (10R). A detailed analysis of the underlying vibrational modes is also published.

Naturally occurring xanthocidin and its six derivatives are identified from the culture of *Streptomyces* sp. bacteria AcE210. The 1D and 2D NMR spectroscopy as well as high-resolution mass spectrometry are used to elucidate their planar structures. The experimental VCD spectra are measured for homoxanthocidin B in $CDCl_3$ at a concentration of 130 mM. Calculated VCD spectra are obtained using the B3LYP functional and 6-311^{++}G(d,p) basis set, as reported by Ortlieb *et al.* (2018).

The known antibiotic and cytotoxic compounds griseorhodin A and griseorhodin C are produced in solid culture by *Streptomyces puniceus* AB10, which is isolated from the leafcutter ant *Acromyrmex rugosus rugosus*. Using VCD and DFT (B3LYP /PCM (DMSO)/6-31G(d)) calculations, Ortega *et al.* (2017) establish their ACs as 6S,6aR,7S,8S and 6R,6aR,7S,8R, respectively (Figure 18.13).

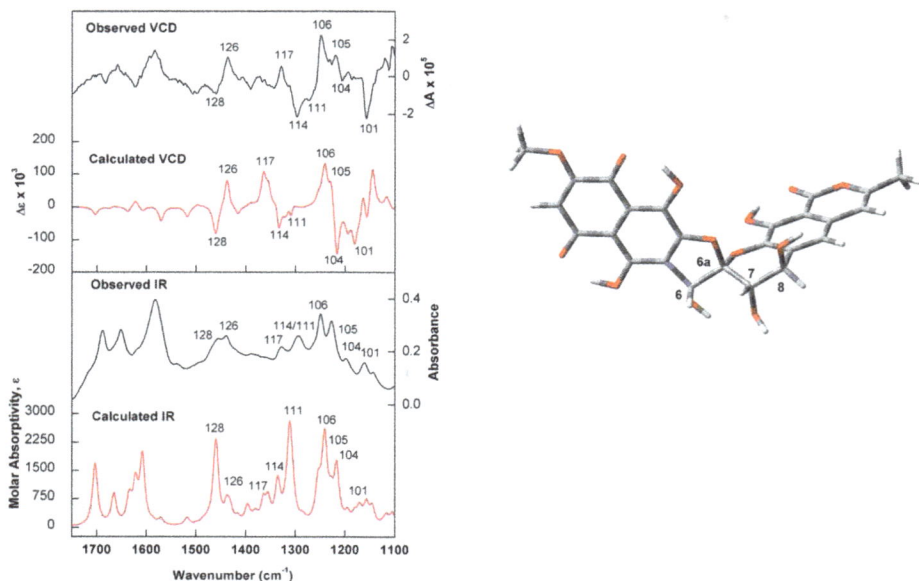

Figure 18.13: Comparison of the observed IR and VCD spectra of (−)-griseorhodin C with the calculated (B3LYP/PCM(DMSO)/6-31G(d)) IR and VCD spectra of the lowest energy conformer identified for (6R,6aR,7S,8R)-griseorhodin C (right) (Ortega *et al.* 2017). Copyright ©2017 Elsevier B.V., with permission.

The complicated task to differentiate between two epimeric compounds in the presence of an inherently dissymmetric chromophore, which normally dominates VCD and ECD spectra is tackled by Burgueno-Tapia *et al.* (2017). Diterpenoids from *Jatropha dioica* Sessé, jatropholone A and B, after comparison of their experimental and DFT calculated VCD spectra (Figure 18.14), allow the authors to conclude that although non-local (M/P) chirality generated by atropisomerism dominates over local chirality generated by an (*R/S*) change, the stereogenic centre can confidently be assigned. By contrasting their respective calculated and experimental IR and VCD spectra, the ACs of jatrophatrione and citlalitrione (a compound proposed as a taxonomic marker for the genus *Jatropha*) can also be established.

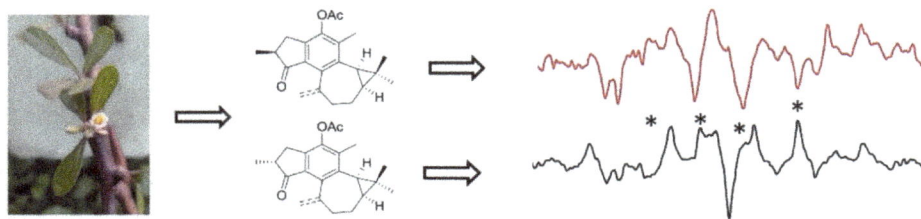

Figure 18.14: Diterpenoids from *Jatropha dioica* and their VCD (Burgueno-Tapia *et al.* 2017). Copyright ©2017 Elsevier B.V., with permission.

Naturally occurring plant-derived cyclic sotolon, (*R*)-3-hydroxy-4,5-dimethylfuran-2 (5H)-one, and maple furanone, (*R*)-5-ethyl-3-hydroxy-4-methylfuran-2(5H)-one, are used as food additives for wine and maple sugar. Their experimental SOR, ECD, and VCD spectra are studied by Yang *et al.* (2017). Calculated data are obtained with density functionals (B3LYP and MWP1MPW91) and basis sets (6-31$^+$G(d) and 6-311^{++}G(2d, p)), and are compared with the experimental results.

Melchor-Martinez *et al.* (2017) extract the roots of *Jatropha dioica* Sessé obtaining riolozatrione and a C-6 epimer of riolozatrione, 6-epi-riolozatrione, as a new structure and only the second reported riolozane diterpenoid. The two known diterpenoids jatrophatrione and citlalitrione are also isolated and characterized. The AC of the four compounds is established by means of VCD.

Caffeic acid ester derivatives, (2*S*,3*S*)-4-O-caffeoyl-2-C-methyl-D-erythronic acid, 3-O-caffeoyl-2-(2-propyl)-2-hydroxybutanedioic acid, and 3-O-caffeoyl-2-((*S*)-2-butyl)-2-hydroxy butanedioic acid, are isolated from *Tithonia diversifolia* (Hemsl.) A. Gray, the Mexican sunflower (Figure 18.15). Ccana-Ccapapinta *et al.* (2017) publish an ECD and VCD study of the compounds. The authors give a *caveat:* when deuterating hydroxyl groups in phenolic and saccharide moieties in methanol-d_4 to improve the agreement of calculated with experimental VCD data, a mirror image may be obtained. The deuteration status of the tertiary hydroxyl group at C-2 is critical for the correct reproduction of experimental VCD data in protic solvents. Ccana-Ccapapinta *et al.* recommend a combination of ECD and VCD in the case of stereochemical analysis of polar chiral natural product molecules.

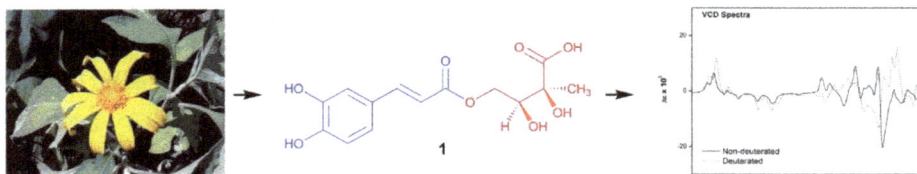

Figure 18.15: *Tithonia diversifolia* and VCD of caffeic ester (Ccana-Ccapatinta *et al.* 2017). Copyright ©2017 Elsevier B.V., with permission.

Four serrulatane-type diterpenoids, euplexaurenes A–C, are isolated together with an-thogorgiene P from Gorgonian samples of *Euplexaura* sp. (a sea fan) collected in the South China Sea. The relative configuration of the 5,3,6-tricyclic unit in euplexaurene A was deduced by NOESY experiments. The AC at C-8 is assigned as (*S*) based on Mosher's method using MTPA esters by Cao *et al.* (2017). Thus, the configuration of tricyclic nucleus of euplexaurene A is determined to be (1*S*,4*R*,5*R*,8*S*,9*R*,10*S*).

Correct assignment of the stereogenic centres of highly flexible linear diterpenes (LDs) is challenging. Merten *et al.* (2015) report the first application of VCD spectroscopy for the AC determination of LDs of algal origin and provide experimental and computational procedures, such as a fragmentation approach, which will facilitate the use of VCD spectroscopy for configuration assignments of LDs. The method is applied to elegandiol (Figure 18.16)

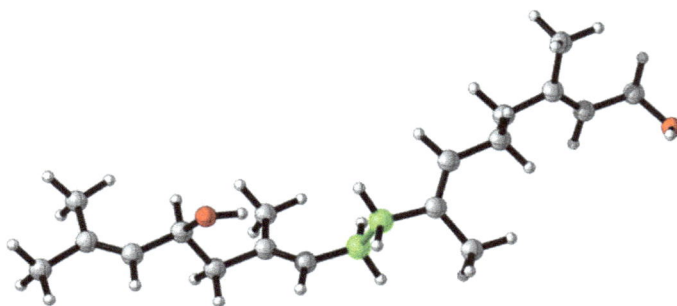

Figure 18.16: Lowest energy structure of elegandiol. The two highlighted carbon atoms, C8 and C9, mark the position of the cut for the fragment calculation (Merten *et al.* 2015). Copyright ©2015 Royal Society of Chemistry, with permission.

The AC of (−)-1-oxo-4,9-dihydroxy-8-methoxy-6-methyl-1,2,3,4-tetrahydroanthracene, or aloe-saponol III 8-methyl ether, extracted from the plant roots of *Eremurus persicus* (Jaub and Spach) Boiss, is determined to be (*R*) by Rossi *et al.* (2017), based on the visual comparison of experimental and calculated VCD and ECD spectra in CDCl₃. Predicted spectra were obtained using the B3LYP functional and TZVP basis set with PCM representing the solvent.

The antiprotozoal acyclic diterpene, bifurcatriol, 3,7,11,15-tetramethylhexadeca-2,10,14-triene-1,7,13-triol, a new LD with two stereogenic centres, is isolated from the Irish brown alga *Bifurcaria bifurcata*. Smyrniotopoulos *et al.* (2017) apply VCD spectroscopy for the correct assignment of the stereogenic centres of highly flexible LDs.

Diplodia corticola, a pathogen of cork oak, produces α-pyrones and furanones. Mazzeo *et al.* (2017) report ECD, VCD, and ORD spectra and compare them with DFT computations (B3LYP functional and TZVP basis set). Mazzeo *et al.* assign the AC of diplobifuranylones A–C as (2S,2′S,5′S,6′S), (2S,2′R,5′S,6′R), and (2S,2′S,5′R,6′R), while diplofuranone A is (4S,9R), and sapinofuranones B and C are (4S,5S).

Aparicio-Cuevas *et al.* (2017) isolate two new dioxomorpholines and three new derivatives from a marine-facultative *Aspergillus* species. The ACs of 1 and 2 (Figure 18.17) were elucidated by comparison of experimental and DFT-calculated (B3PW91/DGDZVP) VCD spectra.

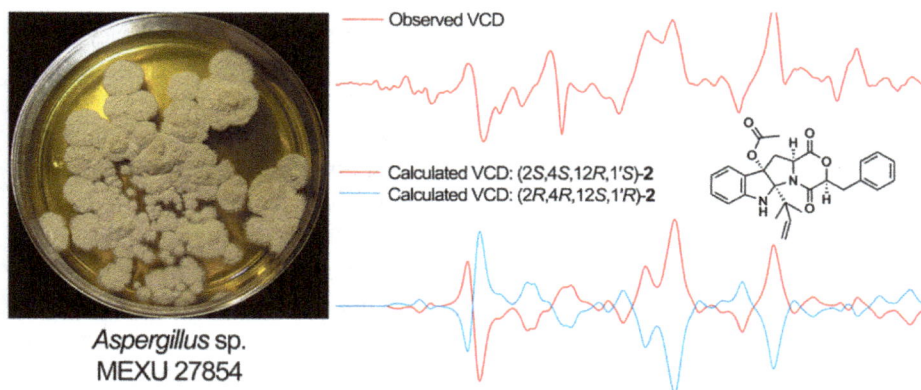

Figure 18.17: Calculated and experimental VCD of dioxomorpholines 2 (Aparicio-Cuevas *et al.* 2017). Copyright ©2017 American Chemical Society, with permission.

Callicapene M3, an isopimarane-type diterpenoid isolated from the *Callicarpa macrophylla* VAHL, is studied by Wang *et al.* (2017). Using ECD and VCD spectra with the aid of time-dependent DFT theoretical calculations, the AC of the compound is established as (4S, 5S, 9S, 10S, 13S, 14S)-14-α-hydroxy-7,15-isopimaradien-18-oic acid.

Trichothecenes, isolated from *Trichoderma albolutescens*, show antiviral activities against the pepper mottle virus. Ryu *et al.* (2017) establish the ACs of trichodermin, 1, and a new compound trichoderminol, 2 (Figure 18.18), by experimental and calculated VCD spectra (Figure 18.19).

In their review, Superchi *et al.* (2018) consider ECD still as the most common technique used in the context of fungal metabolites. But if the compound under study

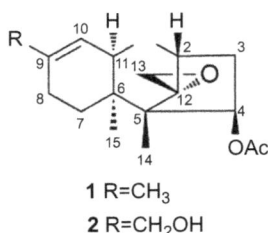

1 R=CH$_3$

2 R=CH$_2$OH

Figure 18.18: Structure of trichodermin, 1, and trichoderminol, 2 (Ryu *et al.* 2017). Copyright ©2017 American Chemical Society, with permission.

lacks chromophoric groups, OR and VCD will be very helpful, and in the presence of high conformational flexibility, the authors advise to use two or more chiroptical methods to deduce the AC.

Ocimene monoterpenoids from *Artemisia absinthium* are studied by Julio *et al.* (2017). The AC of the naturally occurring ocimenes (−)-(3S,5Z)-2,6-dimethyl-2,3-epoxyocta -5,7-diene and (−)-(3S,5Z)-2,6-dimethylocta-5,7-dien-2,3-diol, is established by comparison of experimental and calculated (B3LYP/DGDZVP) VCD spectra, the AC of the acetonide, and also the monoacetate of the diol.

Phomopsis amygdali is a plant-pathogenic fungus that produces the tricyclic hydrocarbon fusicocca diene. Merten *et al.* (2017) report the stereochemical assignment of fusicocca diene from NMR shielding constants and VCD spectroscopy.

Three chiral compounds such as *cis*-chrysanthenyl acetate, oxocyclonerolidol, and *cis*-acetyloxychrysanthenyl acetate are isolated from essential oils (EOs) of leaves and flowers of *Bubonium graveolens* (FORSK.) MAIRE (Compositae), using preparative high-performance liquid chromatography, by Said *et al.* (2017). The comparison of the experimental and calculated VCD spectra of pure isolated compounds provides their AC as being (1S, 5R, 6S)-(−)-2,7,7-trimethylbicyclo[3.1.1]hept-2-en-6-yl acetate, (2R, 6R)-(+)-6-ethenyl-2,6-dimethyl-2-(4-methylpent-3-en-1-yl) dihydro-2H-pyran-3(4H)-one), and (1S, 5R, 6R, 7S)-(−)-7-(acetyloxy)-2,6-dimethylbicyclo[3.1.1] hept-2-en-6-yl] methyl acetate.

Chiral compounds from *Talaromyces aculeatus* are studied by Ding *et al.* (2016). To determine the AC of 5′,6′-dihydroxy-2′,6′-dimethyl-3′-(2″-oxopentyl)cyclohex-2′-en-1′-yl-2,4-dihydroxy-6-methylbenzoate, it is converted to the acetonide (1a, Figure 18.20), and its experimental and predicted ECD and VCD spectra (B3LYP functional and 6-311⁺G(d) basis set) are analysed.

Plakinidone is found in the Caribbean sponge *Plakortis angulospiculatus*. The compound is a five-membered lactone, that is, a 3-methyl-4-hydroxy-2(5H)-furanone or tetronic acid ring. Predicted spectra of plakinidone are obtained using B3LYP functional and the aug-cc-pVDZ basis set using PCM for representing the acetonitrile solvent by Jimenez *et al.* (2016). By comparison of experimental and calculated spectra, the AC is assigned as (11S,17R).

The hard, white, crystalline alkaloid strychnine, isolated from the seeds of *Strychnos nux vomica* L. and similar plants, forms very bitter orthorhombic prisms from a solution in alcohol. Reinscheid *et al.* (2016) show calculated (MPW1PW91/cc-pvdz, IEFPCM: chloroform) IR and VCD spectra of the major conformer of (−)-strychnine base.

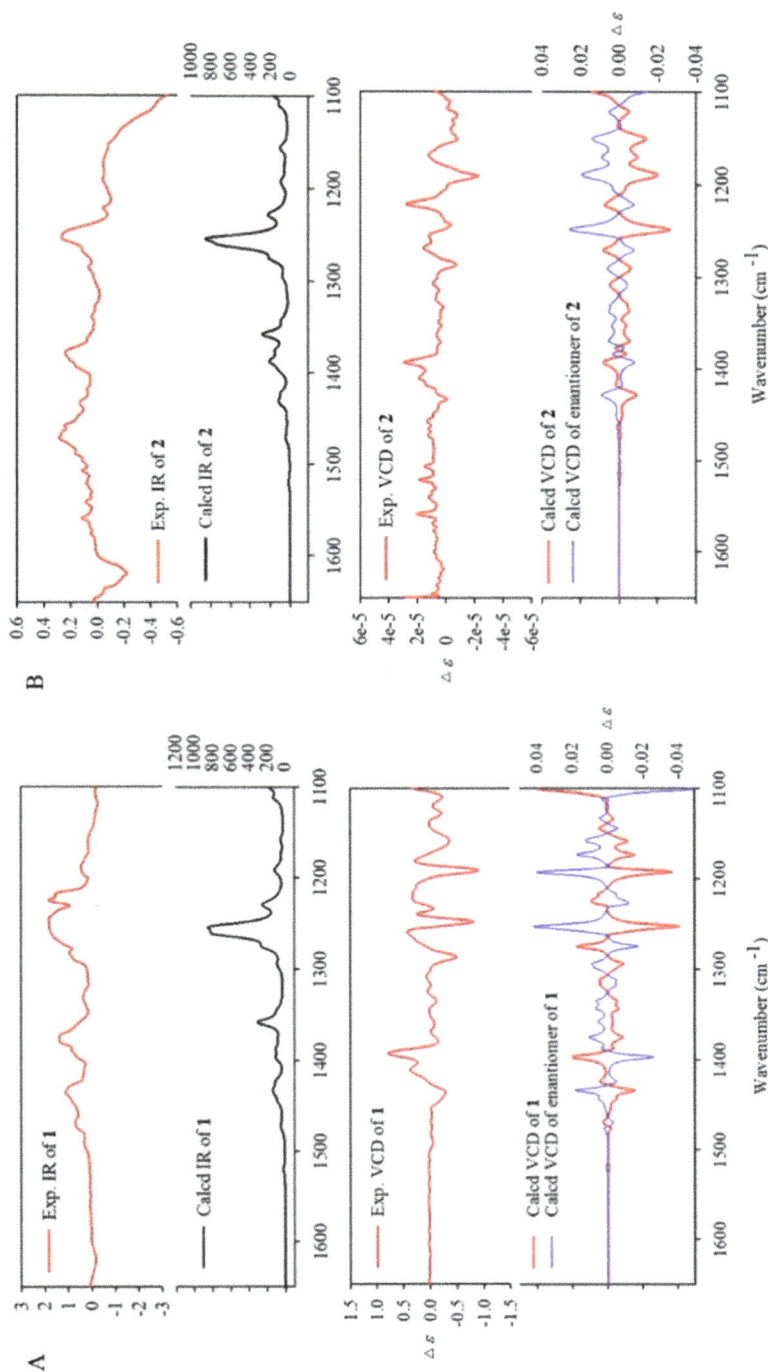

Figure 18.19: Comparison of experimental and calculated IR and VCD spectra of compounds (A) 1 and (B) 2 (Ryu et al. 2017). Copyright ©2017 American Chemical Society, with permission.

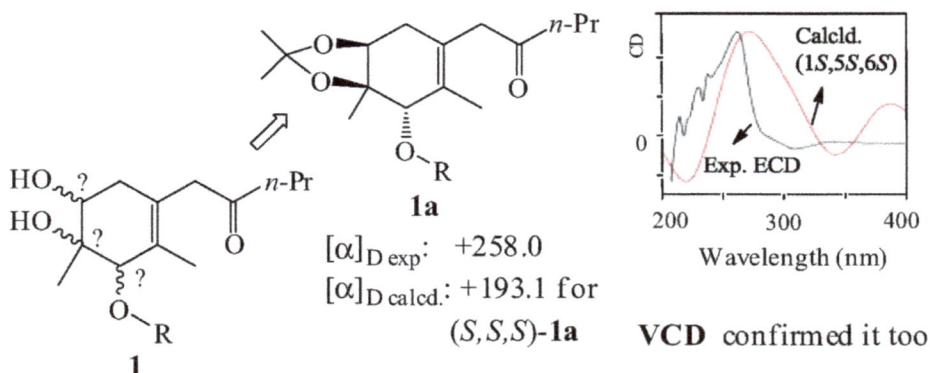

Figure 18.20: Structure, optical rotation, and ECD of the acetonide of 5′,6′-dihydroxy-2′,6′-dimethyl-3′-(2″-oxopentyl)cyclohex-2′-en-1′-yl-2,4-dihydroxy-6-methylbenzoate (Ding *et al.* 2016). Copyright ©2016 Elsevier B.V., with permission.

Four novel steganes such as (−)-(P,8R,8′R)-5′-desmethoxystegane, (−)-(P,7R,8R,8′R)-5′-desmethoxysteganol, (−)-(P,7R,8R,8′R)-5′-desmethoxysteganacin, (−)-(P,8R,8′R)-5′-desmethoxysteganone and (+)-(M,8R,8′R)-3,4-dihydroxy-3′,4′-dimethoxyisosteganolide are synthesized by Velázquez-Jiménez *et al.* (2016). The steganes' AC is assigned by VCD measurements (Figure 18.21) in combination with DFT calculations.

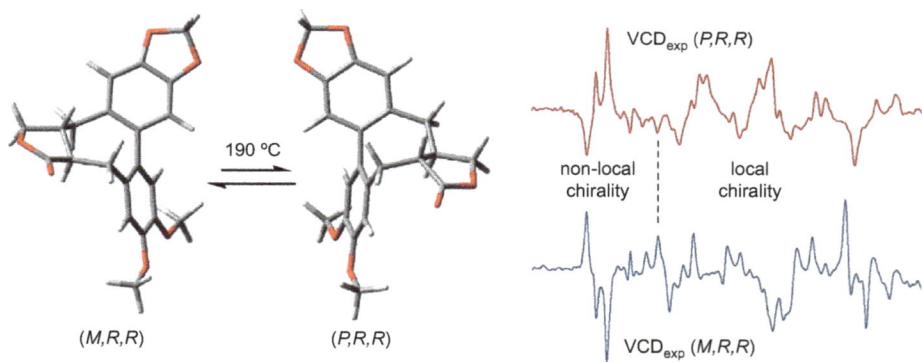

Figure 18.21: VCD spectrum of (+)-(M,8R,8′R)-5′-desmethoxyisostegane and the spectrum of its atropisomeric derivative, obtained by thermal inversion of the chiral axis (Velázquez-Jiménez *et al.* 2016). Copyright ©2016 Elsevier B.V., with permission.

Nakahashi *et al.* (2016) apply VCD to the stereochemical analysis of sphingosine, the base of the membrane components sphingolipids. All stereoisomers of sphingosine (D-*erythro*-sphingosine, L-*erythro*-sphingosine, D-*threo*-sphingosine, and L-*threo*-sphingosine) can be discriminated by VCD patterns in the mid-IR region. Derivatizing sphingosine with glutaraldehyde improves the solubility in non-polar solvents. The resulting rigid cyclized structure enhances the VCD intensities.

Isocorilagin, the α-anomer of the ellagitannin, is a phytochemical marker in *Phyllanthus* species. It can be isolated, for example, from *Pelargonium reniforme*. Sprenger *et al.* (2016) carefully investigate the corilagin structure in both methanol and DMSO solutions using NMR, electronic and vibrational CD, and DFT and MD (molecular dynamics) calculations, and find out that isocorilagin is the result of a solvent-induced conformational transition of corilagin (Figure 18.22), rather than its diastereoisomer. Corilagin changes from $B_{1,4}$ and $^\circ S_5$ conformations of the β-glucose core in DMSO-d_6 to an inverted $^1C_4$ conformation in methanol-d_4.

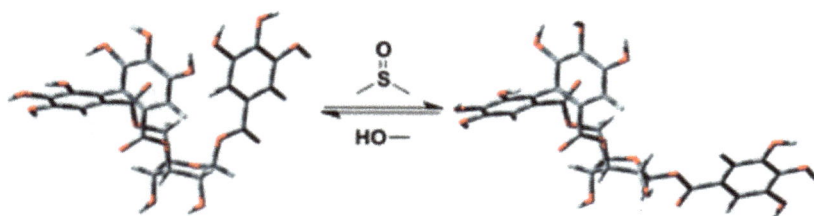

Figure 18.22: Solvent-induced conformational transition of corilagin (Sprenger *et al.* 2016). Copyright ©2016 Royal Society of Chemistry, with permission.

Three new sesquiterpenoids, vetiverianines A–C, and a known eudesmane sesquiterpenoid, (+)-1β,4β,6α-trihydroxyeudesmane, are isolated from the roots of the grass *Vetiveria zizanioides* (Gramineae) by Matsuo *et al.* (2016). The ACs of the sesquiterpenoids are determined by VCD data analysis.

Menthene derivatives, (−)-(3S,4R,5R,6S)-3,5,6-trihydroxy-1-menthene 3-O-β-D-glucopyranoside, and (−)-(3S,4S,6R)-3,6-dihydroxy-1-menthene 3-O-β-D-glucopyranoside, are extracted from the aerial parts of *Ageratina glabrata* (SPRENG.) KING and ROB. (Asteraceae) by Pardo-Novoa *et al.* (2016). The ACs of the series of compounds are determined by comparison of the experimental VCD spectra of the 1,6-acetonide 5-acetate derived from (+)-(1S,4S,5R,6R)-1,5,6-trihydroxy-2-menthene with their DFT-calculated spectra.

Three matrine-type alkaloids, matrine, oxymatrine, and sophoridine, can be extracted from the dry root of *Sophora flavescens*, and have been used in traditional Chinese medicines against cancer and cardiac arrhythmia. Zhang *et al.* (2016) conduct VCD experiments and DFT calculations on the 6-31$^+$G(d,p) level in the PCM of DMSO-d_6 on these compounds and the *anti*-malarial drugs artemisinin and dihydroartemisinin. Zhang *et al.* show that accounting for solvent effects is critical to using IR and VCD spectroscopy to differentiate the potential stereoisomers of the molecules.

Agathisflavone, a dimeric flavonoid isolated from *Schinus terebinthifolius* RADDI (Anacardiaceae), known as Aroeira or the Brazilian pepper tree, is one of the few examples of natural products possessing axial chirality. Covington *et al.* (2016) use ECD, ORD, and VCD

(Figure 18.23) together with the corresponding quantum chemical predictions (DFT, B3LYP/6-311^{++}G(2d,2p)/PCM) to assign the axial chirality of (−)-agathisflavone as (aS).

Reinscheid and Reinscheid (2016) discuss the advantages and limitations of the three chiroptical methods ORD, ECD, and VCD, using (+)-limonene as an example. The authors consider it important to determine the conformer populations using experimental NMR chemical shifts. DFT calculations (MPW1PW91/aug-cc-pvtz, solvent (IEF-PCM) chloroform) of three conformers and the population-weighted mix of them (Figure 18.24) were compared with experimental VCD spectra.

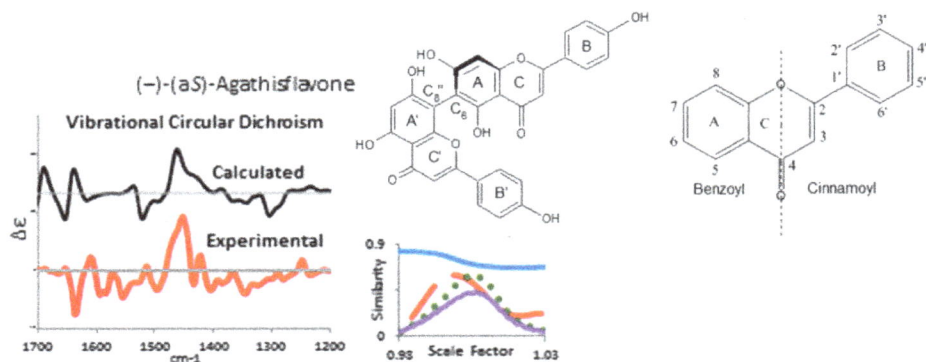

Figure 18.23: Experimental and calculated VCD of agathisflavone (Covington *et al.* 2016). Copyright ©2016 American Chemical Society, with permission.

The known structure of farinosin and its relative stereochemistry are re-evaluated by Ortega *et al.* (2016), using modern methods. Farinosin is isolated from air-dried leaves of *Encelia farinosa* A. GRAY var. *farinosa* (Asteraceae). Conformational search using the B3PW91 functional and the DGDZVP basis set indicates a single low-energy conformer, which is the same as that found in the crystal structure. The VCD spectrum confirms the AC derived by earlier studies.

Zhang *et al.* (2016) study matrine- and artemisinin-type herbal products by IR and VCD spectroscopy (e.g. Figure 18.25). The authors use these spectroscopic techniques for stereochemical characterization of the compounds and study solvent effects as well.

EOs obtained from the aerial parts of *Artemisia herba-alba* ASSO, a species belonging to the Asteraceae family, is studied by Said *et al.* (2016). Using the VCD spectra of the oil's main components (e.g. Figure 18.26), the authors reconstruct the VCD of the natural mixtures to obtain a VCD chiral signature of EOs.

The indole alkaloid (−)-flustramine B and its IR and VCD spectra are studied by Cordero-Rivera *et al.* (2015).

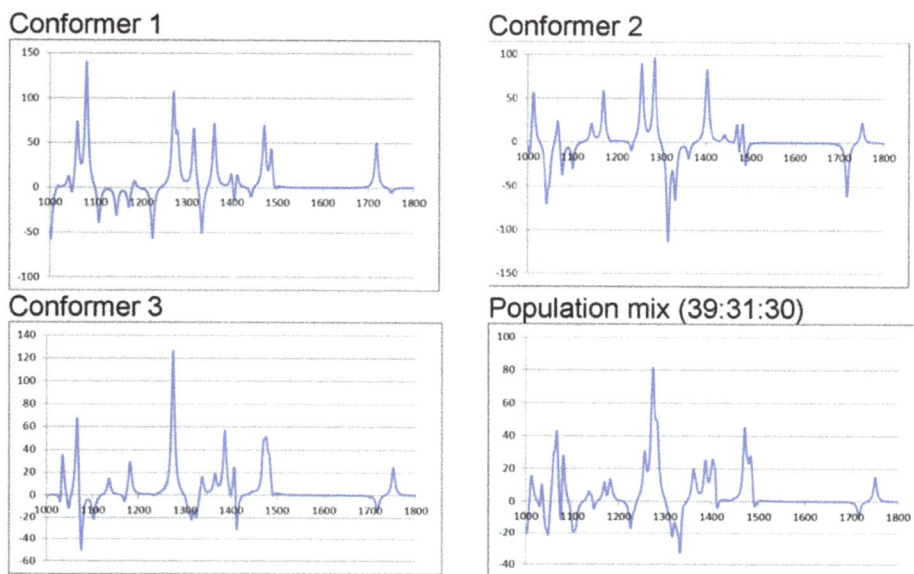

Figure 18.24: Calculated (MPW1PW91/aug-cc-pvtz, solvent (IEF-PCM) chloroform) VCD spectra of (+)-limonene of the three conformers and a population-weighted mix 39:31:30 (*x*-axis: wavenumber in cm^{-1}; *y*-axis: Δε (mol/L cm) (Reinscheid and Reinscheid 2016). Copyright ©2016 Elsevier B.V., with permission.

Esquelane derivatives are found as the main components of *Adesmia boronioides* EO. The ACs of esquel-7-en-9-one, acetylphenol, and cacalol acetate are derived by Cerda-Garcia *et al.* (2015) in a VCD study. The agreement between the experimental and calculated VCD spectra of these bisnorsesquiterpenes reveals that they are (1*S*,4*R*,5*S*)-esquel-6-en-9-one and (1*S*,4*R*,5*R*)-esquel-7-en-9-one.

(+)-Inuloxin A is isolated from the organic extract of the plant *Inula viscosa*, and its relative stereochemistry is deduced to be (7*R*,8*R*,10*S*) by Santoro *et al.* (2015), using a comparison of calculated and experimental VCD. They also assigned the AC (1*S*,2*R*,3a*S*,4*S*,5-*R*,7a*S*) to (−)-seiricardine A from the fungal pathogen for cypress *Seiridium*.

9,12-Cyclomulin-13-ol, or 13-β-hydroxyazorellane, is a diterpene isolated from *Laretia acaulis*, a medicinal plant from the Andes. The compound is subjected to VCD analysis for deducing the previously known AC. Munoz *et al.* (2015) use the B3LYP functional with the DGDZVP basis set for the calculation. The quantitative similarity between experimental and predicted VCD spectra is considered reliable enough to suggest that its AC is the same as that of azorellanol, whose AC is known from X-ray single-crystal diffraction data of its 7-*p*-bromobenzoate derivative and has the normal diterpene AC.

Rhizopine is an opine-like natural product present in nitrogen-fixing nodules of alfalfa infected by rhizobia. For VCD studies, a solution of a synthesized peracetylated sample in CDCl$_3$ is used, and predicted spectra were obtained using B3LYP functional, B3PW91

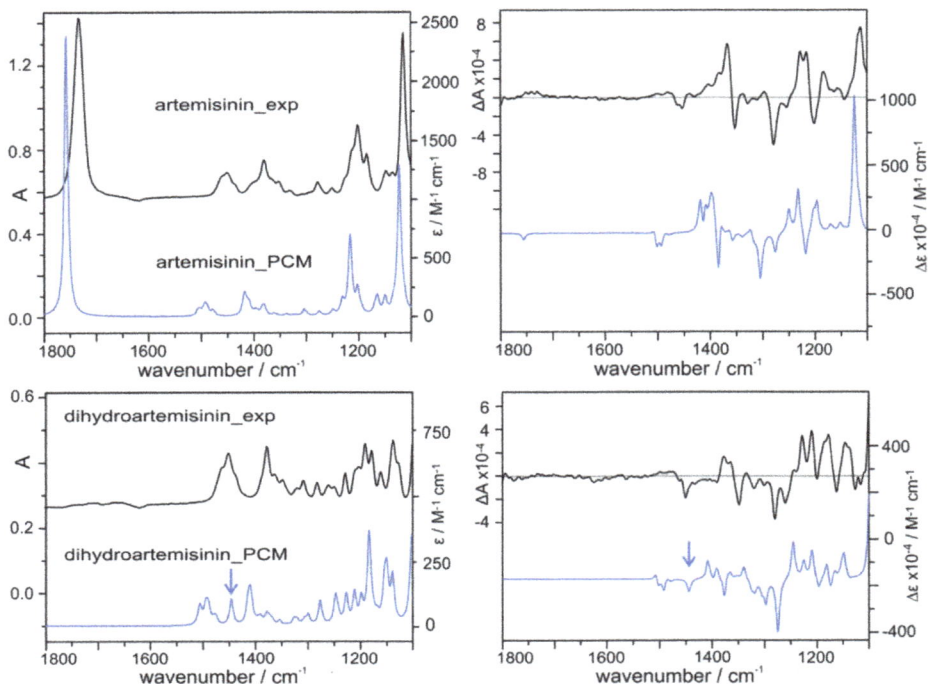

Figure 18.25: Comparison of the experimental and calculated IR and VCD spectra of artemisinin and dihydroartemisinin in DMSO-d$_6$. The calculations were done at the 6-31*G(d,p) level in the PCM of DMSO-d$_6$ (Zhang *et al.* 2016). Copyright ©2016 American Chemical Society, with permission.

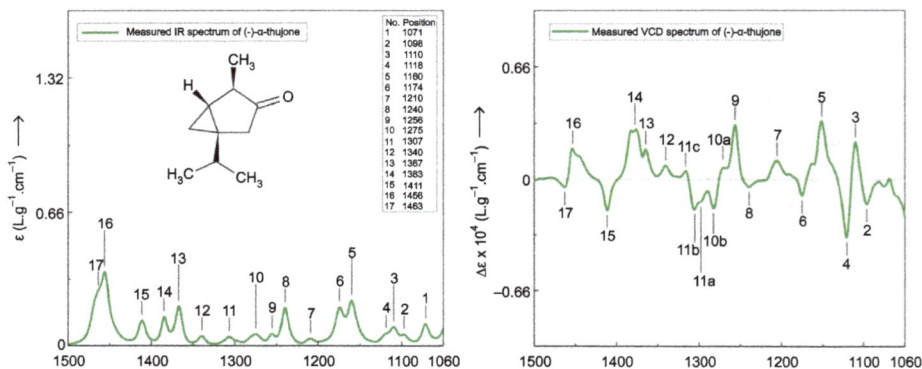

Figure 18.26: (−)-α-Thujone and its IR and VCD spectra (Said *et al.* 2016). Copyright ©2016 Elsevier B.V., with permission.

functional, cc-pVTZ basis set, and PCM for representing the solvent. Krief *et al.* (2015) deduce from the VCD spectrum the stereochemistry of non-derivatized natural rhizopine corresponding to (1*R*,2*S*,3*R*,4*R*,5*S*,6*R*)-4-amino-6-methoxycyclohexane-1,2,3,5-tetraol.

Brevianamide B was originally isolated from the mould *Penicillium brevicompactum*. Bultinck *et al.* (2015) calculate the structure of the dimer of (−)-brevianamide B (Figure 18.27) using (a) B3LYP/6-311^{++}G(d,p) with a continuum solvent field (SCRF) for trifluoroethanol (TFE) and (b) with inclusion of D3 dispersion. The naturally occurring (+)-brevianamide B is enantiomorph of the synthetic product, as shown by comparison of the calculated to the experimental VCD in TFE-d$_3$.

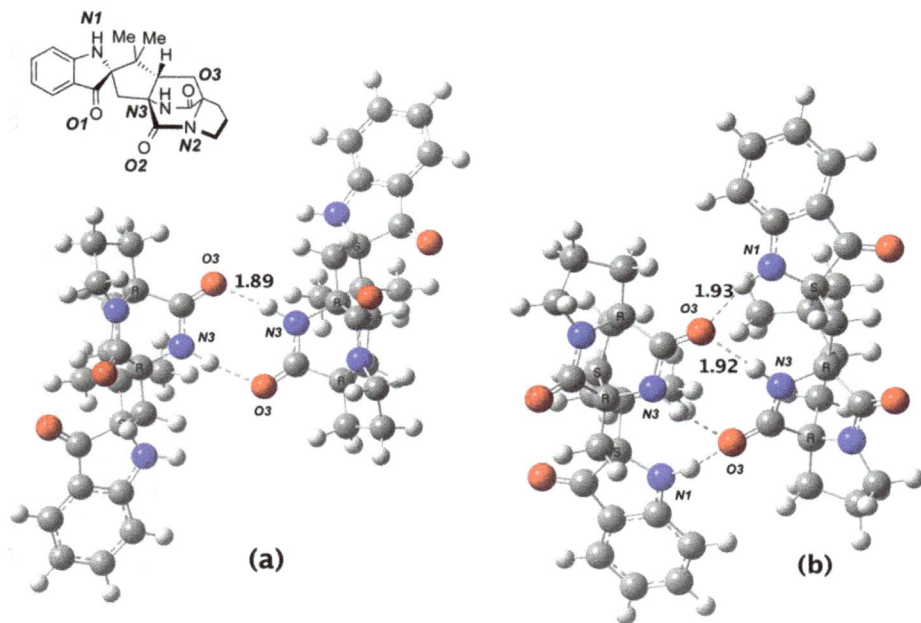

Figure 18.27: Calculated structure of the dimer of (−)-brevianamide B using (a) B3LYP/6-311^{++}G(d,p) with a continuum solvent field (SCRF) for TFE and (b) with inclusion of D3 dispersion (Bultinck *et al.* 2015). Copyright ©2015 American Chemical Society, with permission.

The new prenylated phloroglucinol α-pyrone **1** (3-[5,7-dihydroxy-2,2-dimethyl-8-(2-methylbutanoyl)-2H-chromen-6-ylmethyl]-6-ethyl-4-hydroxy-5-methyl-2H-pyran-2-one) and the new dibenzofuran **4** (Figure 18.28), together with the known 23-methyl-6-*O*-demethylauricepyrone, achyrofuran, and 5,7-dihydroxy-3,8-dimethoxyflavone (gnaphaliin A), are isolated from the aerial parts of the plant *Achyrocline satureioides* (Lam.) DC (Asteraceae), the "marcela blanca" of Argentinean folk medicine. Experimental VCD spectra are measured in CDCl$_3$ solvent by Casero *et al.* (2015) and compared with the calculated spectra. The (*S*)-AC of the α-methylbutyryl chain attached to the phloroglucinol nucleus is thus established.

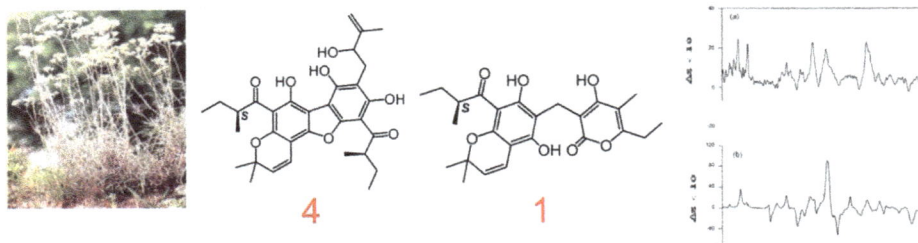

Figure 18.28: *Achyrocline satureioides*, dibenzofuran 4, α-pyrone 1, and their VCD spectra (Casero *et al.* 2015). Copyright ©2015 American Chemical Society, with permission.

Bustos-Brito *et al.* (2015) obtain from the aqueous extract of the leaves of *Ageratina cylindrica* six new *ent*-kaurenoic acid glycosides together with the known diterpenoid paniculoside V, the flavonoid astragalin, chlorogenic acid, and L-chiro-inositol. The structures are elucidated mainly by NMR and MS methods, and the AC is established by VCD spectroscopy.

The relative configuration of (−)-centratherin is known from the literature, but its AC is not. The compound is extracted from the plant *Eremanthus crotonoides*, and its structure and relative configuration is deduced from NMR chemical shifts. Junior *et al.* (2015) deduce from the calculated and experimental VCD the AC of naturally occurring (−)-centratherin to be (6*R*,7*R*,8*S*,10*R*,2′*Z*).

Beta-artemether is a methyl ether derivative of artemisinin (found in *Artemisia annua* plant), and this beta-isomer is a valuable medication for malaria caused by *Plasmodium falciparum*. Wang *et al.* (2015) calculate the optimized molecular structure, Mulliken atomic charges, vibrational spectra (IR, Raman, and VCD), and molecular electrostatic potential by DFT using the B3LYP functional with the 6-311^{++}G(2d,p) basis set. Reliable vibrational assignments for artemether are made on the basis of potential energy distribution (PED). The calculated VCD is compared with the experimental VCD, confirming the AC of beta-artemether (Figure 18.29).

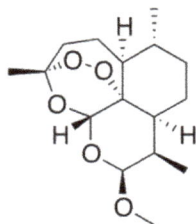

Figure 18.29: Structure of malaria medication beta-artemether (Wikimedia commons).

Ortega *et al.* (2015) study alkaloids from tobacco (*Nicotiana tabacum* L.).

(−)-*S*-Nicotine is the main alkaloid from tobacco. Its conformational landscape in both the gas phase and solution state is investigated by Ortega *et al.* (2015). DFT calculations using B3LYP and B3PW91 functionals in conjunction with the 6-311^{++}G** and aug-cc-pVTZ basis sets and PCM for representing the solvent are used for assessing the conformations. The analysis of VCD spectra of (–)-*S*-nicotine confirms the presence of two main conformers at room temperature.

(–)-*S*-Cotinine is one of the minor alkaloid tobacco constituents, and is the major peripheral oxidative metabolite of (–)-*S*-nicotine in several animal species. The conformational landscape, the rotational isomerism barrier, and solution-state conformational energy profile of (–)-*S*-cotinine are investigated by utilizing experimental and predicted VCD spectra by Ortega *et al.* (2015).

(–)-*S*-Anabasine is one of the minor piperidinic nicotinoids in tobacco leaves. The VCD spectra of (–)-*S*-anabasine are measured as neat liquid as well as in CCl_4 solution. Predicted VCD spectra are obtained using B3PW91 functional and aug-cc-pVTZ basis set by Ortega *et al.* (2015).

The α,β-unsaturated germacranolide, 6-epi-desacetyllaurenobiolide, can be isolated from the plants *Montanoa grandiflora* and *Laurus nobilis*. Its X-ray structure was reported in the literature, but its AC was only assumed. Sanchez *et al.* (2015) prove it by VCD.

Smyrniotopoulos *et al.* (2017) report the first application of VCD spectroscopy for the AC determination of LDs from the brown alga *Bifurcaria bifurcata*. The authors provide experimental and computational procedures, such as a fragmentation approach, for the challenging task of correct assignment of the stereogenic centres of highly flexible LDs.

The absolute stereochemistry of flavour compounds from patchouli (*Pogostemon cablin*) is studied by Yaguchi *et al.* (2015) using the VCD technique. It is at first applied to patchoulol (Figure 18.30), a rigid and fused tricyclic sesquiterpene alcohol. The VCD spectra were calculated by DFT (B3LYP/6-311G(d,p)). The absolute stereochemistry proposed from the observed and calculated VCD data is in agreement with the configuration reported for patchoulol. Moreover, this method is extended to odour-active flavorous furanones having a unique keto-enol tautomeric structure that is prone to racemization. The absolute stereochemistry of flavorous 2- or 5-substituted furanones such as 2,5-dimethyl-4-hydroxy-3(2H)-furanone (DMHF, furaneol), 2,5-dimethyl-4-methoxy-3(2H)-furanone (DMMF, mesifuran), 4-acetoxy-2,5-dimethyl-3(2H)-furanone (ADMF, furaneol acetate), 4,5-dimethyl-3-hydroxy-2(5H)-furanone (sotolon), and 5-ethyl-3-hydroxy-4-methyl-2(5H)-furanone (maple furanone) is investigated.

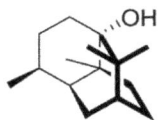

Figure 18.30: Structure of patchoulol (Wikimedia commons).

18.2 Raman optical activity (ROA)

Raman dptical activity (ROA) is a very small effect that can be easily obscured by artefacts. It is most often measured as the dimensionless CID (or better DCID $[I_r - I_l]/[I_r + I_l]$), that is, the difference of right-polarized versus left-polarized Raman intensity divided through the whole Raman intensity. The first ROA spectrometers were derived from scanning Raman spectrometers. The instruments were modified by adding devices to modulate the polarization of the exciting laser beam (Lindner *et al.* 1995).

Lindner *et al.* (1995) use this type of instrument to measure the ROA of enantiomorphic single crystals (Figure 18.31).

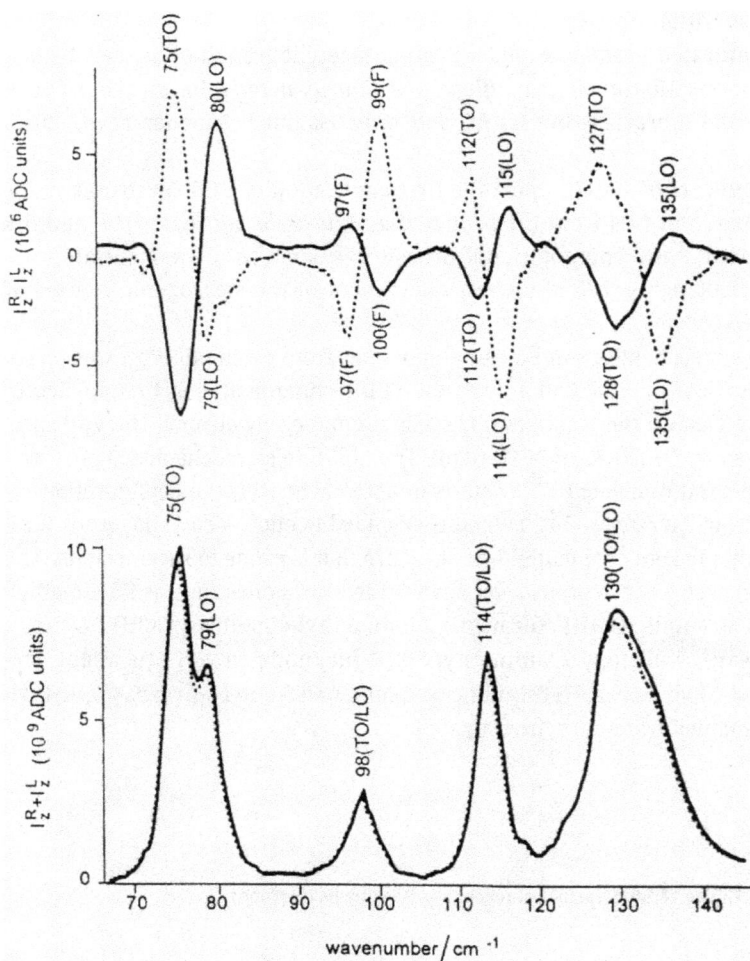

Figure 18.31: Raman optical activity of enantiomorphic single crystals of $NaBrO_4$ (Lindner *et al.* 1995). Copyright ©1995 John Wiley and Sons Ltd., with permission.

As the ROA is very small and the early instruments were not very sensitive, very few spectra were measured before the advent of a new generation of spectrometers. Two main technical advances made the construction of these sensitive instruments possible: first, the development of multichannel detectors culminating in back-thinned CCDs, which are practically ideal light detectors (with quantum efficiencies of about 80% at the wavelength of peak sensitivity), and second, the development of the holographically manufactured line filter. The first improvement shortened the time of measurement to about 20 min for pure samples and allowed the measurement of dilute samples in few hours, and the second allowed the replacement of multiple monochromators by a single polychromator with much higher optical throughput. As these spectra are improved very much, mainly the literature of the last 15 years is covered in this chapter. The interested reader is directed to earlier reviews given in the references for historic data. Nowadays, even commercial instruments are available, vastly increasing the number of papers on ROA. This instrument by Barron *et al.* (2015) is shown in Figure 18.32, and Hug's instrument (SI for Haesler *et al.* 2007) is depicted in Figure 18.33.

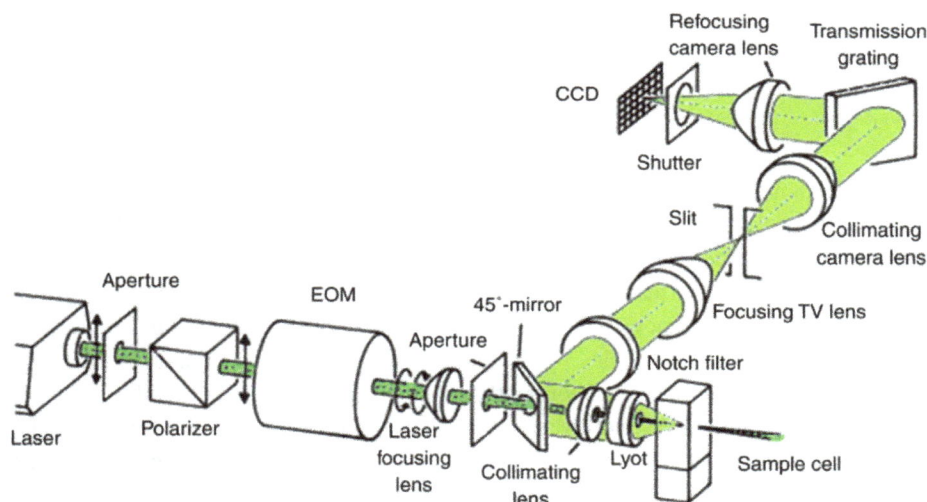

Figure 18.32: The optical design of the final version of the Glasgow backscattering ICP ROA instrument (Barron 2015). Copyright ©2015 John Wiley & Sons Ltd., with permission.

Perhaps the simplest chiral molecule one can imagine is bromochlorofluoromethane. As a liquid, which until now has only been obtained as an enriched but not pure substance, no crystal suitable to deduce the AC could be obtained. It was therefore a great success for ROA together with *ab initio* calculations (with the DZP basis set and electron correlation by the MP2 method), that it was possible for Costante *et al.* (1997) to deduce the absolute conformation of this molecule by comparing experimental and calculated spectrum (Figure 18.34) after more than 100 years.

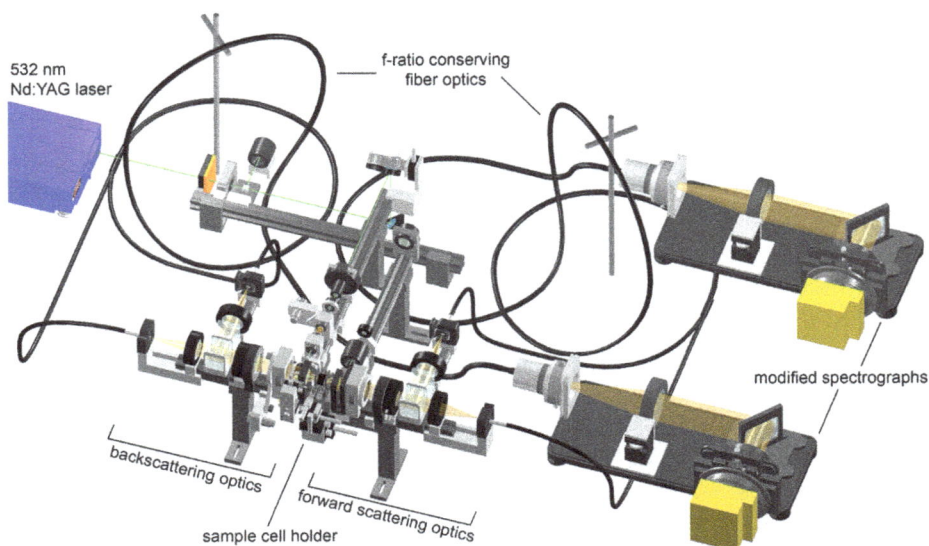

Figure 18.33: Three-dimensional sketch of an ROA instrument, which measures for the first time SCP Raman optical activity in both backward and forward scattering (Haesler *et al.* 2007). Copyright ©2007 Springer Nature, with permission.

Instrumental advances in ROA, combined with quantum chemical computations, made it even possible to determine the AC of (R)-$[^2H_1, \, ^2H_2, \, ^2H_3]$-neopentane (Figure 18.35). This saturated hydrocarbon represents the archetype of all molecules that are chiral as a result of a dissymmetric mass distribution. It is chemically inert and cannot be derivatized to yield molecules that would reveal the AC of the parent compound. Diastereomeric interactions with other molecules, optical rotation, and ECD are, in contrast to the aforementioned case of bromochlorofluoromethane, not expected to be measurable. Vibronic effects in the vacuum ultraviolet circular dichroism might reveal that the molecule is chiral, but the presence of nine rotamers would make it extremely difficult to interpret the spectra, because the spatial arrangement of the rotamers' nuclei resembles that of enantiomers. Although impossible by other techniques, determination of AC of (R)-$[^2H_1, \, ^2H_2, \, ^2H_3]$-neopentane by ROA is done by Haesler *et al.* (2007). The work is extremely difficult, and shows the boundary of the achievable by this spectroscopic technique.

Johannessen *et al.* (2011) examine the glycan structure of a high-mannose glycoprotein with the help of ROA. The protein under study is the external invertase of the yeast *Saccharomyces cerevisiae*, a glycoprotein used as a biocatalyst in the sugar industry. The authors state that the carbohydrate chain imposes a completely disordered state on the protein part.

A VCD study of (R)-(+)-3-methylcyclopentanone is published by He *et al.* (2004). Experimental mid-IR absorption and VCD spectra and theoretical absorption and VCD spectra (DFT, B3LYP, aug-cc-pVDZ, 6-311^{++}G(2d,2p), and aug-cc-pVTZ) are obtained.

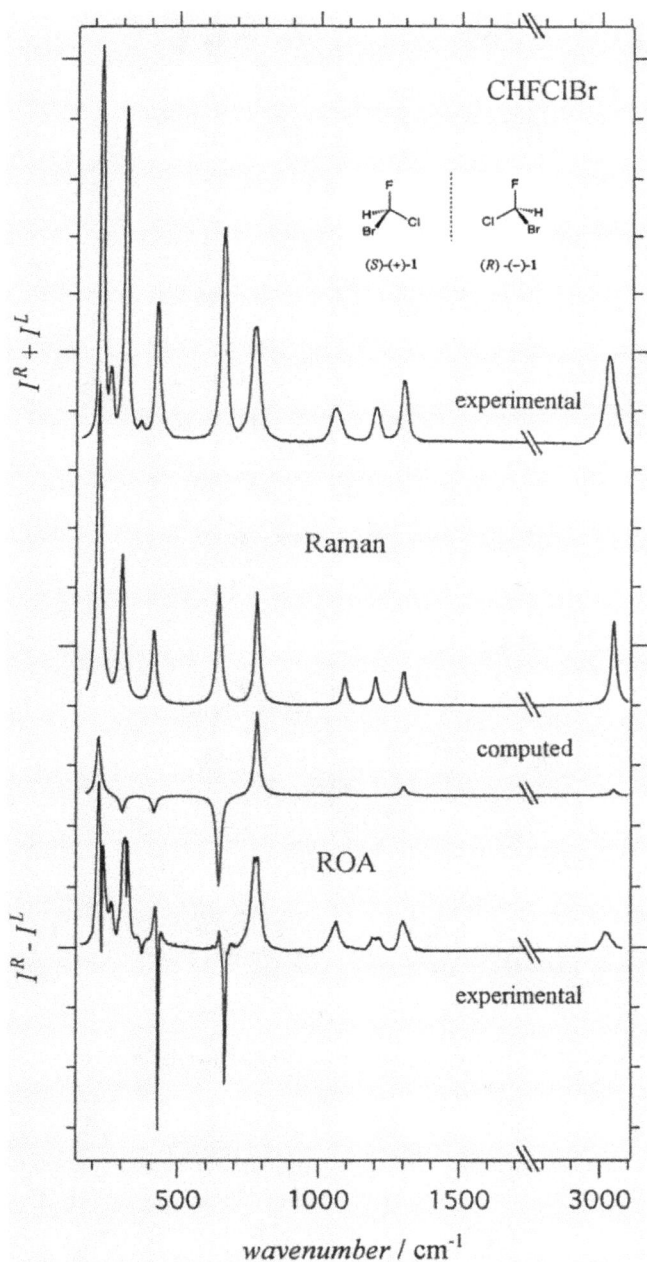

Figure 18.34: Depolarized Raman and ROA spectra of CHFClBr (36% enantiomeric excess) together with the calculated spectrum of the (*R*)-enantiomer. Seven of nine ROA band calculations (MP2, DZP) arrive at the right sign (Costante *et al.* 1997). Copyright ©1997 John Wiley & Sons Ltd., with permission.

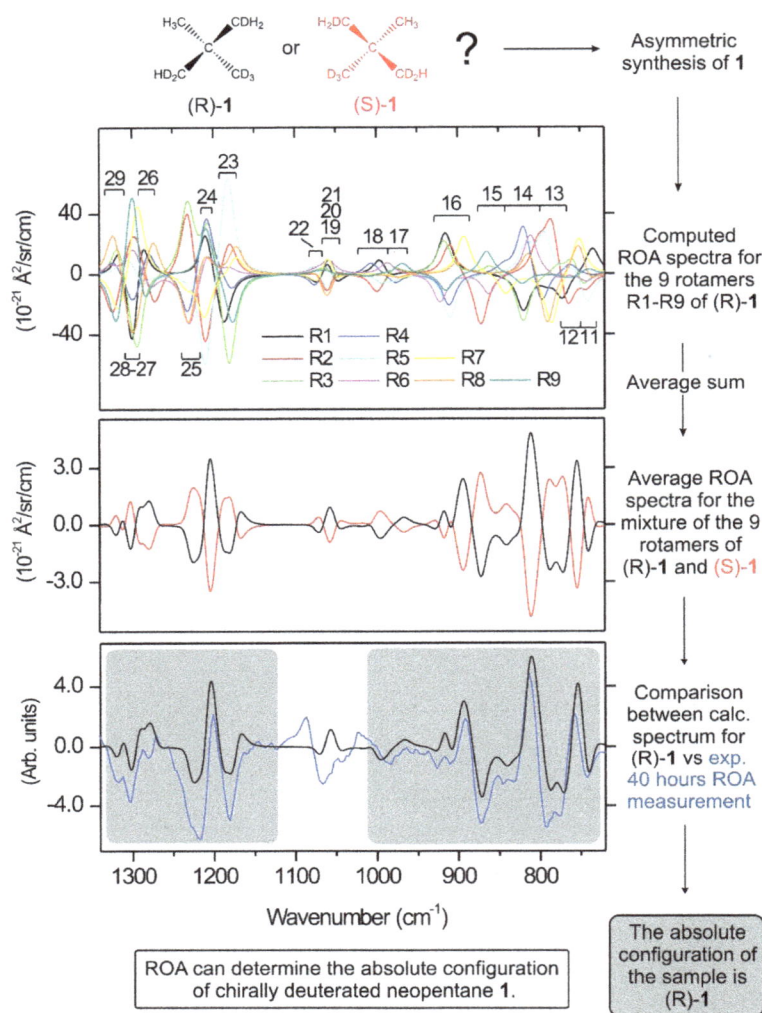

Figure 18.35: History of the assignment of (R)-[^{2}H$_1$, ^{2}H$_2$, ^{2}H$_3$]-neopentane by ROA (Haesler *et al.* 2007). Copyright ©2007 Springer Nature, with permission.

From the temperature-dependent absorption spectra, the ratio of per cent populations of equatorial-methyl:axial-methyl conformers is calculated as 87:13. This ratio is also determined using the experimental and predicted vibrational intensities.

Johannessen *et al.* (2011) report the first combined study of measured and computed ROA of a transition metal complex under non-resonant scattering conditions. The two enantiomers of the dichloro[ethylenebis(4,5,6,7-tetrahydro-1-indenyl)]zirconium(IV) complex yield virtual mirror-image ROA spectra (Figure 18.36). Calculated

ROA spectra, using the B3PW91 functional and the Los Alamos ECP plus DZ basis set (LanL2DZ), match reasonably well.

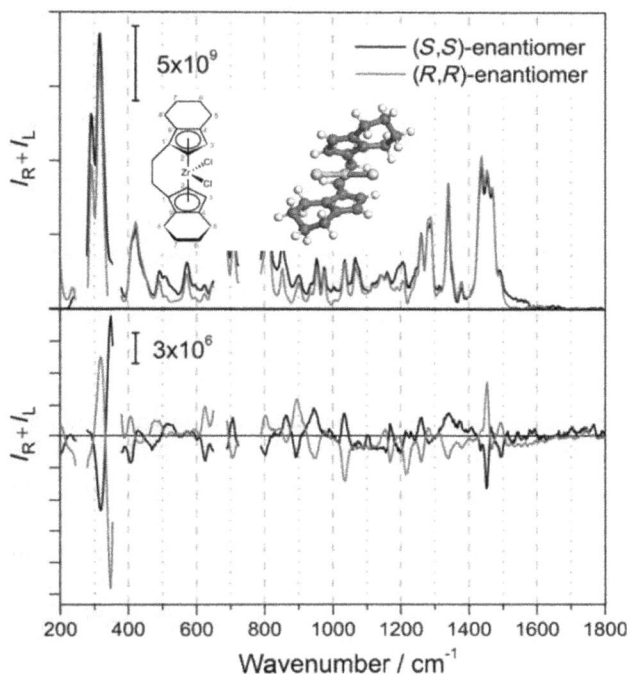

Figure 18.36: Experimental Raman and ROA spectra of both enantiomers of en(thind)$_2$ZrCl$_2$ in chloroform (Johannessen *et al.* 2011). Copyright ©2011 John Wiley & Sons Ltd., with permission.

Barron (2015) reviews the development of biomolecular ROA spectroscopy. Barron describes "the long and tortuous path leading to the first observations of ROA in biomolecules in 1989", and then the road to the large range of biomolecule ROA spectra in aqueous solution available at the time of the review. ROA may be applied to study motif and fold, and secondary structure, of proteins; solution structure of carbohydrates; polypeptide and carbohydrate structure of intact glycoproteins; new insight into structural elements present in unfolded protein sequences; and even protein and nucleic acid structure of intact viruses. The complete three-dimensional structure of biomolecules is available by simulating ROA spectra using quantum chemical calculations, and even conformational dynamics of smaller biomolecules can be obtained by this method.

Pendrill *et al.* (2019) unravel the solution structure of mannobioses by means of ROA. As the analysis of carbohydrate conformations is very complicated, the authors use NMR and ROA as experimental techniques, complemented by MD calculations (to find the major conformations) and DFT. The MD simulations are performed using the CHARMM carbohydrate force field together with the CHARMM and the modified TIP3P parameters for water. Each structure is reoptimized at the B3LYP/6-31G* level, with force field and ROA property tensors subsequently calculated at B3LYP/rDPS level.

Mensch *et al.* (2016) acquire a large database of ROA spectra of peptide model structures. Detailed assignments of experimental ROA patterns to the conformational elements of the peptide in solution are made. The authors validate this method for six polypeptides. Mensch *et al.* also analyse hydrogen/deuterium-exchanged structures and the conformational dependence of the amide modes in Raman spectra using the new database.

Giovannini *et al.* (2016) calculate VCD spectra of chiral systems in aqueous solution. Their method combines classical MD simulations with a fully polarizable quantum mechanical (QM)/molecular mechanics (MM)/PCM Hamiltonian. Polarization effects are included in the MM force field by exploiting an approach based on fluctuating charges. To test their method, the authors calculate VCD spectra of (*R*)-methyloxirane and (L)-alanine in aqueous solution. The calculations give better agreement with experimental spectra than calculations with the PCM.

3,5,4′-Trihydroxystilbene, known as resveratrol, is a stilbenoid phytoalexin produced in many food plants. Chatterjee *et al.* (2018) discover that the non-planar structure of the *cis*-resveratrol conformer possesses certain chiral signals in its simulated VCD and ROA spectra calculated by DFT. The binding of *Z*-resveratrol to the human tyrosyl-tRNA may generate an enantiomeric excess, leading to detectable changes in the vibrational optical activity spectra. The authors identify candidate features at 998, 1,649, and 1,677 cm^{-1} in the ROA and at 1,642 and 3,834 cm^{-1} in the VCD spectra of *Z*-resveratrol that may be useful for this purpose.

The chiroptical vibrational spectra of ladderanoic acids from anammox bacteria are studied by Raghavan *et al.* (2018). Specific rotations as well as ROA and VCD spectra are measured experimentally and compared with QM (DFT, B3LYP/6-311^{++}G (2d,2p)/PCM) calculations. The ACs of the naturally occurring acids and their phenacyl esters (e.g. Figure 18.37) are assigned as *R*. These results are also confirmed by X-ray crystallography.

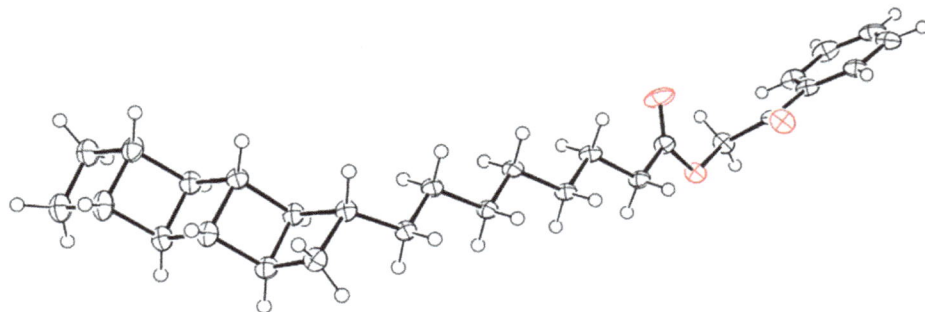

Figure 18.37: ORTEP diagram of the phenacyl ester of [5]-ladderanoic acid (Raghavan *et al.* 2018). Copyright ©2018 American Chemical Society, with permission.

Paclitaxel, also known as taxol (found in the bark of the *Taxus brevifolia* NUTT. tree), is widely used as a natural anticancer agent. Chemical modification of a natural taxoid, baccatin III, is used for the commercial production of paclitaxel. Baccatin III functionalized by 1-*t*-butyloxycarbonyl-3-triethylsilyloxy-4-phenyl-2-azetidinone (Figure 18.38) can lead to four stereoisomers (resulting from two chiral centres), of which the desired one is (3*R*,4*S*), as found out by Profant *et al.* (2017) using ROA (Figure 18.39).

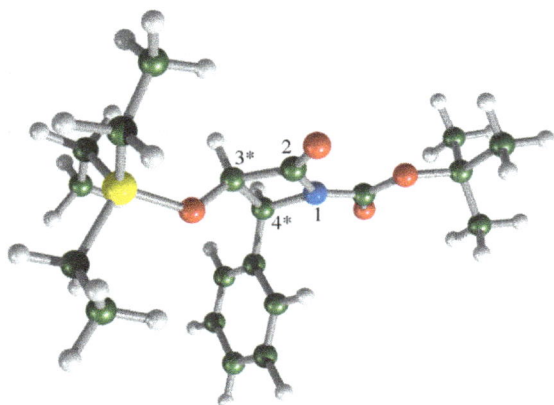

Figure 18.38: Structure of (3*R*,4*S*)-1-*t*-butyloxycarbonyl-3-triethylsilyloxy-4-phenyl-2-azetidinone, a key compound in the synthesis of paclitaxel from baccatin III (Profant *et al.* 2017). Copyright ©2017 American Chemical Society, with permission.

Fagan *et al.* (2017) study cocaine hydrochloride, the tropane alkaloid found in the leaves of *Erythroxylum coca* LAM., by ECD, VCD, and ROA. Calculated IR and VCD spectra of the four lowest energy cocaine conformers are reported, and the experimental spectra as well. Calculated and experimental Raman (I^R+I^L) and ROA (I^R-I^L) spectra of cocaine (Figure 18.40) are shown and discussed. DFT on the B3PW91/6-311^{++}G**/CPCM level is used for the calculations and vibrational assignments are given.

A flavone C-diglycoside may be obtained from the aerial parts of *Peperomia obtusifolia* (Piperaceae). Calisto *et al.* (2017) apply experimental and calculated Raman/ROA data for the stereochemical characterization of the flavone C-diglycoside as isoswertisin-4′-methyl-ether-2″α-L-rhamnoside.

Carotenoids are pigments synthesized by plants, algae, and photosynthetic bacteria. Zeaxanthin, or all-*trans*-(3*R*,3′*R*)-β-carotine-3,3′-diol, a hydroxylated carotenoid, protects human retina against photo-oxidative damage. Dudek *et al.* (2017) analyse two types of chiral zeaxanthin self-assembly, H and J₁, and verify that the aggregation process is not enough to enhance the ROA signal. Essential for the aggregation-induced resonance ROA effect is the fact that the electronic circular absorption band of supramolecular species is in proximity with the ROA excitation line. Dudek *et al.*

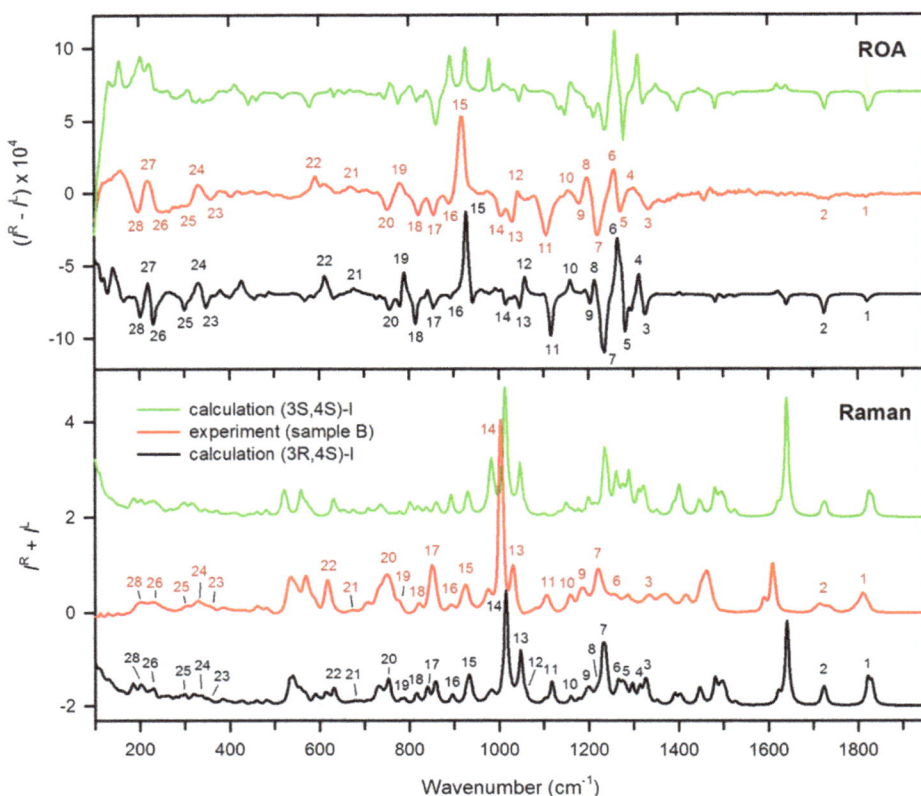

Figure 18.39: Comparison of the experimental (sample B) and simulated Raman and ROA spectra of (3R,4S)- and (3S,4S)-(3R,4S)-1-t-butyloxycarbonyl-3-triethylsilyloxy-4-phenyl-2-azetidinone, a key compound in the synthesis of paclitaxel from baccatin III (Profant *et al.* 2017). Copyright ©2017 American Chemical Society, with permission.

also compare and analyse the spectral differences between J aggregates of astaxanthin and zeaxanthin.

ROA spectra of five different bicyclic terpenes, α-pinene, fenchone, bornyl acetate, 2-carene, and 3-carene were measured by Polavarapu *et al.* (2017). The ACs of these compounds are already known and are consistent with the experimental and theoretical results of the authors.

In the literature, some papers can be found by misidentifying natural products. When the magnitudes of similarity overlap needed to discriminate between incorrect and correct chemical structures are assessed using the experimental and calculated spectra for these compounds, seven natural products, namely elatenyne (derived from algae), aquatolide (plant derived), annuionone A (plant derived), caespitinone (plant derived), palominol (coral derived), sporol (fungal plant pathogen), and klaivanolide (plant derived) are analysed. Based on this information, it is concluded by

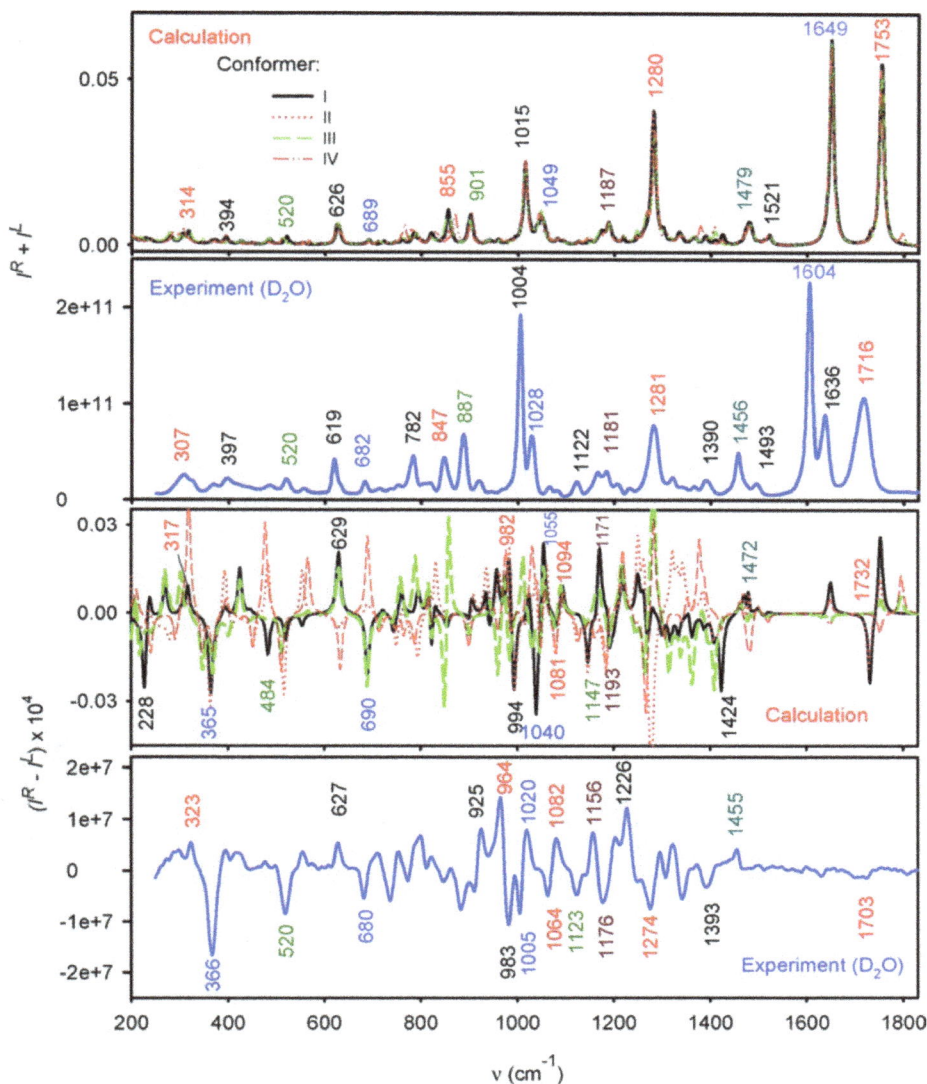

Figure 18.40: Calculated and experimental Raman (I^R+I^L) and ROA (I^R-I^L) spectra of cocaine (Fagan *et al.* 2017). Copyright ©2017 John Wiley & Sons Ltd., with permission.

Polavarapu *et al.* (2017) that if the experimental VCD and ROA spectra for these compounds had been available, the original mis-assignments could have been avoided.

The new diterpenoid (+)-(4R,5S,8R,9S)-18-hydroxy-*ent*-halima-1(10),13-(*E*)-dien-15-oic acid is found by Monteiro *et al.* (2015) in the ethanolic extract of the flowers of *Hymenaea stigonocarpa*. In particular, NOESY correlations allow determining the relative orientations of the substituents at chiral centres, resulting in two possibilities for the AC as

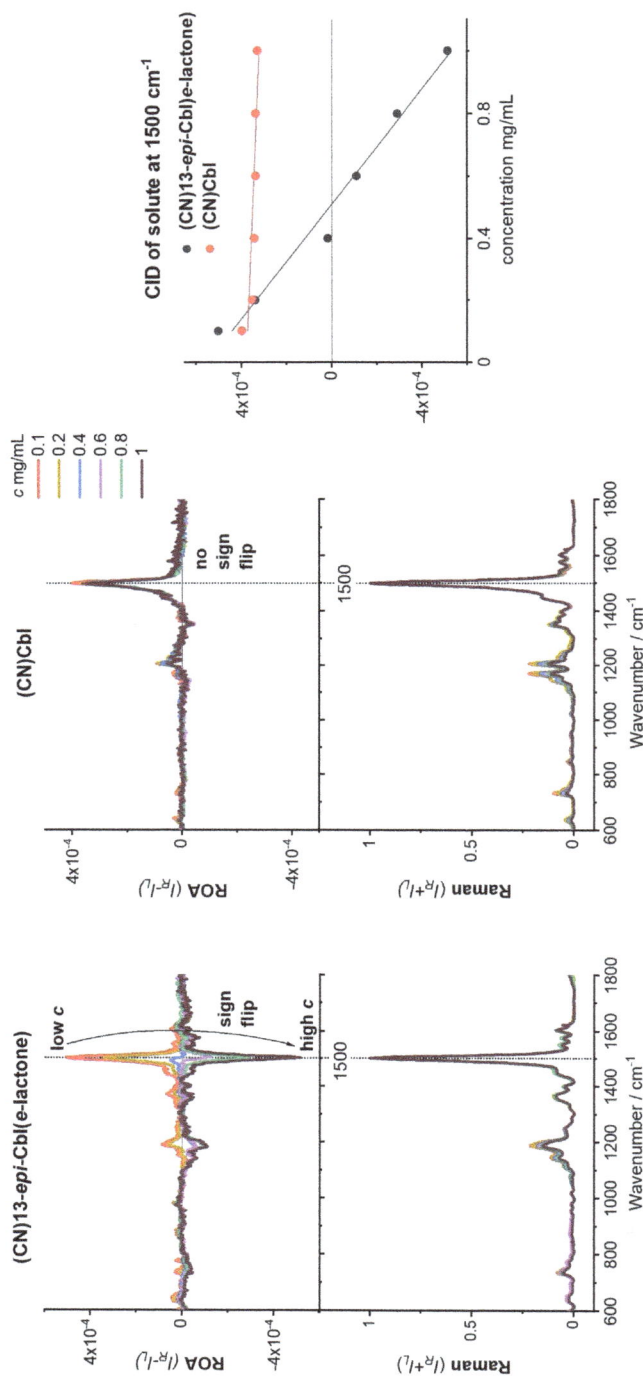

Figure 18.41: ROA and Raman spectra, and CID values measured for various concentrations of (CN)13-*epi*-Cbl(*e*-lactone) and (CN)Cbl in water. The spectra are normalized to the strongest Raman band (Machalska *et al.* 2021). Copyright ©2021 by the authors, licensed by Creative Commons (CC 4.0).

(4R,5S,8R,9S) or its enantiomer. Based on the comparison of Raman and ROA experimental spectra of this new compound with those calculated for the (4R,5S,8R,9S) structure using B3LYP functional, TZVP basis set, and PCM for representing the DMSO solvent, the AC of the new compound is assigned as (4R,5S,8R,9S).

Ribifolin, an orbitide from *Jatropha ribifolia*, is studied by Pinto *et al.* (2015). ROA spectra of the lowest energy conformer (90% at RT) are calculated using the B3LYP functional, 6-311$^+$G(d) basis set, and PCM for representing the water. AC of the compound is established all-L by comparing experimental and calculated spectra.

Machalska *et al.* (2021) discuss the recognition of the true and false resonance ROA. At the example of a series of vitamin B_{12} derivatives, the authors report a strong influence of ECD on the measurement of resonance ROA of coloured chiral compounds (Figure 18.41). DFT calculations on the CAM-B3LYP/GD3/6-31G(d) level of theory are detailed in the supporting information.

Cinchona alkaloids are the subject of Roman *et al.*'s (2015) work. They predict the spectra of quinine (QN), quinidine, cinchonine, and cinchonidine in Figure 18.42, using the B3LYP functional and aug-cc-pVDZ basis set with water solvent represented by PCM (Figure 18.43). ROA allows unequivocal identification of the pseudoenantiomers based on the sign of the characteristic bands from a single measurement. The experiments are supported by the theoretical approach including conformational study followed by wavenumber calculations and PED analysis. For QN, vibrational spectroscopy is additionally used to show its structural changes in aqueous solutions at various pH.

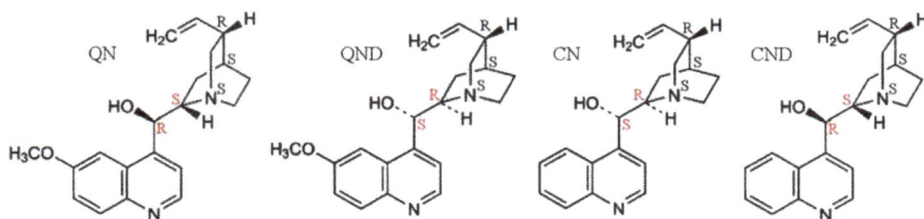

Figure 18.42: Molecular structures of quinine (QN), quinidine (QND), cinchonine (CN), and cinchonidine (CND) (Roman *et al.* 2015). Copyright ©2015 John Wiley & Sons, Ltd., with permission.

The homoisoflavanone 5,7-dihydroxy-6-methoxy-3-(9-hydroxy-phenylmethyl)-chroman-4-one from *Polygonum ferrugineum* (Polygonaceae) is studied by Batista *et al.* (2015). VCD spectral predictions are obtained using the B3PW91 functional and the 6-31$^+$G(d,p) basis set, whereas ROA spectral predictions are obtained using the B3LYP functional and the TZVP basis set. All calculations represent the solvent using PCM. The AC of the compound is determined as 3R,9R.

Kralik *et al.* (2020) succeed in establishing the solution conformers of heroin by ECD, VCD, and DFT (B3LYP/6-311^{++}G(d,p) and other levels of theory) calculations. They also assign the vibrational bands.

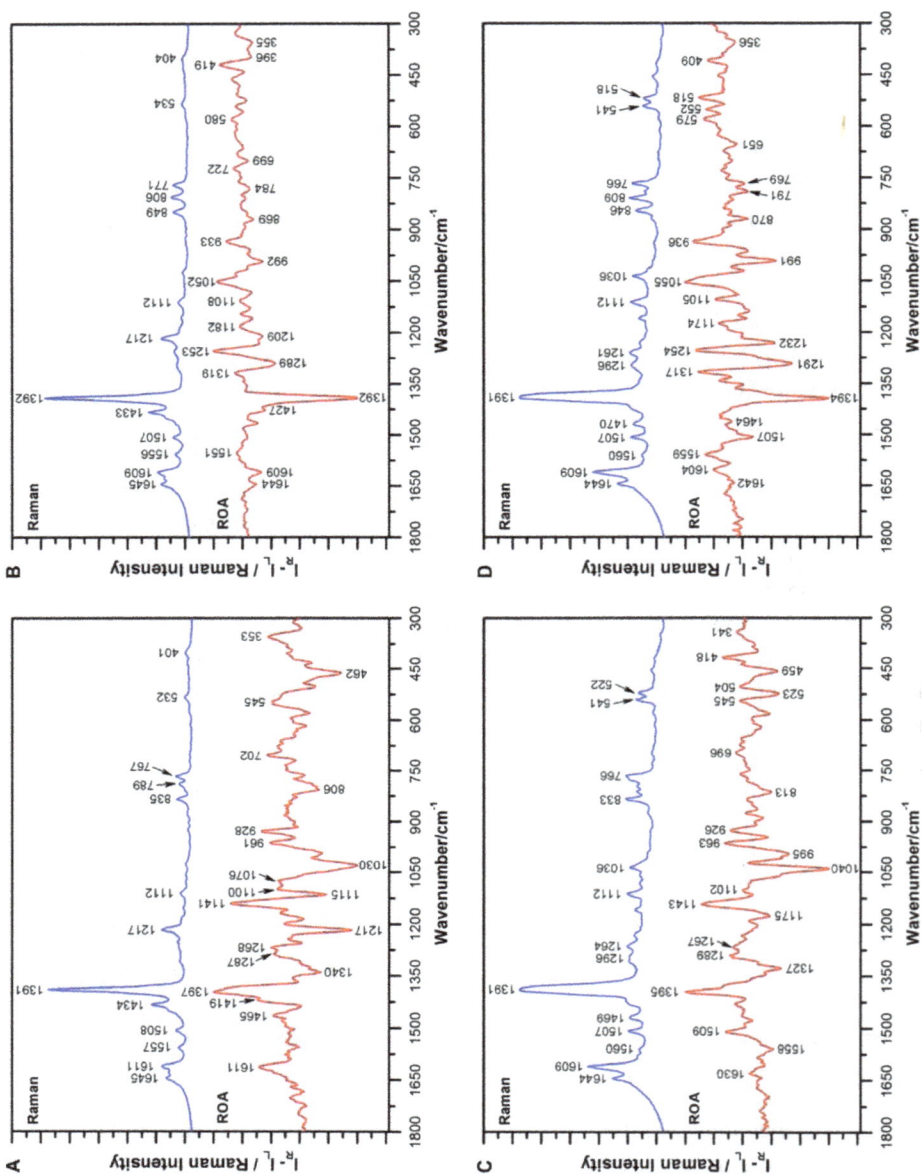

Figure 18.43: Raman and ROA spectra of QN (A), QND (B), CND (C), and CN (D) in aqueous solution at pH = 1. The spectra were measured with an excitation line of 532 nm (Roman *et al.* 2015). Copyright ©2015 John Wiley & Sons, Ltd., with permission.

Abella *et al.* (2020) calculate ROA spectra under off-resonance, near-resonance, and at-resonance conditions for $[Co(en)_3]^{3+}$ and under off-resonance conditions for $[Rh(en)_3]^{3+}$ using complex (damped) time-dependent linear response Kohn–Sham theory. The off-resonance spectra of the two compounds show many similarities. At an incident laser wavelength of 532 nm, used in commercial ROA spectrometers, the spectrum of A is enhanced by near-resonance with the ligand-field transitions of the complex. The near-resonance spectrum exhibits many qualitative differences compared with the off-resonance case, but it remains bi-signate. Even under full resonance with the ligand-field electronic transitions, the ROA spectrum of A remains bi-signate when the electronic transitions are broadened such as to yield absorption line widths that are comparable with those in the experimental UV–vis absorption and ECD spectra.

19 Industrial applications of Raman spectroscopy

Raman spectroscopy finds more and more applications in industry. The often-mentioned advantages of the technique apply here as well: cheap, fast, on-site and remote measurements, no chemicals (that means friendly to the environment), measurement in water is possible. The method is non-destructive, no samples have to be taken, and it can be used *in situ*. It can measure through containers, and with the new hand-held instruments one can measure everywhere.

As a laser-based technique, Raman spectroscopy uses a single laser (e.g. 785 nm) to interrogate a sample, measuring the light that is scattered by the sample of interest to develop the chemical fingerprint. Raman is uniquely able to scan through transparent or translucent containers, eliminating direct contact with the substance and providing significant tactical advantage.

Raman spectroscopy can be used advantageously to differentiate between crystalline polymorphs of the same chemical entity. Wang *et al.* (2000) study the solvent-mediated polymorphic transformation of progesterone using *in situ* Raman spectroscopy. In the two-morphic progesterone system, Form I was found to be thermodynamically more stable than Form II, while Form II was found to be kinetically favoured over Form I.

Fourier-transform Raman spectroscopic online monitoring of the anionic dispersion block copolymerization of styrene and 1,3-butadiene is reported by Bandermann *et al.* (2001). To control the reaction, a light-fibre probe is used to monitor the polymerization progress by Raman spectroscopy and to analyse the concentration of the two monomers and polybutadiene.

Pinzaru *et al.* (2004) discuss identification and characterization of pharmaceuticals using different Raman techniques (dispersive, Fourier transform, resonance Raman, SERS, SERRS, FT-SERS). Several Raman applications such as fundamental structural investigations, quantitative analysis, drug-excipient interaction, formulation, limit of detection, pH-dependent pharmaceutical species, adsorption geometry at a given surface, and functional groups involved in adsorption for several widely used pharmaceutical compounds are presented. The water-soluble vitamins B_1, B_3, B_6, and C, the water-insoluble vitamins A, B_{12}, K_1, and K_3, the alkaloids nicotine, cocaine, and papaverine hydrochloride, and the quinolines were considered. In addition, references are given for the antihypertensive 1,4-dihydrazinophtalazine sulfate, the antimicrobial agent rivanol, chiral β-blockers, antiretrovirals, narcotics, antitumour agents (6-mercaptopurine, 9-phenyl- and 9-aminoacridine, camptothecin and its water-soluble derivative topotecan), and HIV inhibitors (betulinic acid).

Jestel (2005) authors a book chapter about process Raman spectroscopy. She describes for the novice in the field how Raman spectroscopy works, when Raman spectroscopy works well and when it does not, what are the special design issues for

https://doi.org/10.1515/9783110717556-019

process Raman instruments, where Raman spectroscopy is being used, and finally, what is the current state of Raman spectroscopy.

Process measurements by Raman spectroscopy are detailed in a review by Everall *et al.* (2006). The authors explain how compact, sensitive, and robust Raman instruments can be readily interfaced to production processes. Everall *et al.* consider the instrumental and engineering aspects of making Raman measurements under production conditions and discuss some of the areas in which it has been fruitfully employed.

Pharmaceutical applications of Raman spectroscopy are scrutinized in the book by Šašić (2008). With an emphasis on industry issues, the author describes routinely used applications as well as those with promising potential. Šašić covers subjects such as identifying polymorphs, monitoring real-time processes, imaging solid dosage formulations, imaging active pharmaceutical ingredients in cells, and diagnostics.

The review article by Simon *et al.* (2015) about process analytical technology (PAT) aims to bring readers up to date with some of the current trends in the field. The multi-author paper (each chapter written by experts in the field) may act as a comprehensive source of information on PAT topics for the novice reader.

Oligonucleotides are an emergent class of therapeutics for the treatment of a wide variety of diseases. Rydzak *et al.* (2015) report real-time process analytical technology assurance for the flow synthesis of oligonucleotides using mid-infrared and Raman spectroscopy. From the experience with the model oligonucleotide $5'-U_mA_mC_m-G_mC_mU_mA_mG_mA_mU_m-3'$, the authors expect improvements in the production of oligonucleotides with this PAT.

Effective and versatile real-time analysers look for in-quality assurance of drug manufacturing. Raman spectroscopy for PAT of pharmaceutical secondary manufacturing is reviewed by Nagy *et al.* (2018). Results obtained by the real-time application of Raman spectroscopy are discussed, covering the most common secondary process steps of a tablet production line.

Using the example of flufenamic acid (2-[[3-(trifluorometyl)phenyl]amino]-benzoic acid), Hu *et al.* (2005) demonstrate that Raman spectroscopy can measure solute concentration and polymorphic form as a function of time and temperature in batch crystallization operations. Using an immersion probe coupled to a Raman spectrometer, the spectral data contained information about both the solution phase and the crystallized solids.

A review by Esmonde-White *et al.* (2017) features Raman spectroscopy as a process analytical technology for pharmaceutical manufacturing and bioprocessing.

Sixt *et al.* (2018) demonstrate the use of in-line Raman spectroscopy for monitoring the extraction of anethole and fenchone from fennel seed as a typical example. The technique ensures high quality and low costs as well as highly resource-efficient extraction by in-line monitoring and process control.

Widespread microplastic pollution is raising growing concerns as to its detrimental effects upon living organisms. The next few papers descibe the use of Raman spectroscopy in this field.

Identification of microplastics by FTIR and Raman microscopy can be done directly on filters without any visual presorting, provided that the environmental sample was extracted, purified, and filtered before. Because the IR technique is strongly restricted by the limited IR transparency of conventional filter materials, Käppler *et al.* (2015) describe a novel silicon filter substrate that opens the important fingerprint region (1,400–600 cm^{-1}) for FTIR transmission measurements. Therefore, the most common microplastics, polyethylene, and polypropylene, may be identified, even polyesters having quite similar chemical structure, like polyethylene terephthalate and polybutylene terephthalate, can be differentiated.

Latest developments and future prospects in the identification of microplastics using Raman spectroscopy are reviewed by Araujo *et al.* (2018). The authors demand representative data on the abundance, size distribution, and chemical composition of microplastics for risk assessment and call Raman microscopy an "indispensable tool for the analysis of very small microplastics (<20 µm)". Araujo *et al.* also discuss nonconventional Raman techniques such as nonlinear Raman, hyper-spectral imaging, and standoff Raman, permitting real-time Raman detection and imaging of microplastics.

Haefele and Paulus (2018) review confocal Raman microscopy in pharmaceutical development. The authors discuss its use to probe the distribution of components within a formulation, to characterize homogeneity of pharmaceutical samples, to determine the solid state of drug substances and excipients, as well as to characterize contaminations and foreign particulates. Additionally, Haefele and Paulus describe its application as a tool for process analytics, for patent infringements, and counterfeit analysis.

High spatial resolution of confocal Raman microscopy allows for the investigation of small particles. The status of Raman analysis of microplastics from aquatic systems is discussed by Nava *et al.* (2021), and advantages and the drawbacks are highlighted. The authors provide a catalogue with detailed information about peaks of most common plastic polymers and of common additives.

20 Forensic applications of vibrational spectroscopy

Infrared (IR) microscopy, normally done with an FTIR spectrometer attached to an IR microscope, is an indispensable tool for identification of trace evidence in forensics and characterizing small particles in wide varieties of other fields. With the ability to focus directly on micro-sized particles, it can be used for comparative identification of paint and fibres. Spatially separating and measuring mixtures of powdered materials is especially valuable for both drug enforcement and analysis of explosives.

For the identification of prescription-only drugs, Hannah and Pattacini (1973) use IR spectra from 4,000 to 250 cm^{-1}. IR spectra, measured in KBr, of 60 drugs and filling agents are pictured in their paper and a flow chart is given to assist the identification.

Raman spectroscopy is a nearly ideal tool for forensic applications: it can identify substances even if diluted by other chemicals; in the form of confocal Raman spectroscopy, it can detect illicit drugs and explosives in minute amounts on micrometre-sized spots. Chalmers, Edwards, and Hargreaves (2012) dedicate a book to IR and Raman spectroscopy in Forensic Science.

A review of spectroscopy in forensic science is published by Eyring and Martin (2013). The authors compare Raman and IR spectroscopy to other spectroscopic methods used for analysis in the forensic laboratory. Their application to both macroscopic and microscopic samples is described.

20.1 *In situ* crime scene analysis

If an illegal laboratory is found and it is suspected to be in use to prepare methamphetamine, the substances found have to be determined. Iodine, red phosphorus, and lithium can be easily identified by their look. On the contrary white powders and solutions are difficult to identify. Here Raman spectroscopy is very helpful. The white powders can be diagnosed with confidence. Solvents are also easily identified and any dissolved substances as well. This also covers precursors such as ephedrine and pseudoephedrine and the end product methamphetamine.

Small traces of controlled substances can be identified by surface-enhanced Raman spectroscopy. Rana *et al.* (2011) illustrate this technique using the example of morphine, codeine, and hydrocodone.

Bumbrah and Sharma (2016) summarize the use of Raman spectroscopy for the characterization of drugs of abuse: basic principle, instrumentation, and selected applications.

Raman spectroscopy in forensics is compared to other techniques by Pereira De Oliveira *et al.* (2018). They compare forensic devices using the literature of 2013–2017.

https://doi.org/10.1515/9783110717556-020

20.2 Banknotes

Paper banknotes are studied by Frederick *et al.* (2004) by Raman microspectroscopy. Focussing on US currency, they are able to detect single microcrystals of illicit drugs on the dollar bill despite a strong fluorescence background. Frederick *et al.* find out that the fluorescence background from the dark green and grey areas is much weaker than from the white or light green areas.

Imperio *et al.* (2015) report a study of banknotes from the Italian Republic from 1947 to 2001. Three different features of the banknote (serial number, watermark, and security mark) were analysed by micro-Raman and IR spectroscopy.

20.3 Explosives

Hodges and Akhavan (1990) explain the use of Fourier transform Raman spectroscopy in the forensic identification of illicit drugs and explosives. FTIR spectra were recorded of pure and contaminated illicit drug samples, together with some explosive materials. The authors praise the method as simple and superior to others as it needs only negligible sample alignment, and it features reduced sample fluorescence. Shown in the paper are spectra of the alkaloids morphine, heroin, and codeine (Figure 20.1), amphetamine as an example of other illicit drug, the explosives Driftex, TATB, and RDX (Figure 20.2). As an example of the usefulness of the method Hodges and Akhavan identify an unknown sample as Semtex (Figure 20.3).

Ali *et al.* (2008) publish a Raman spectroscopic investigation of cocaine hydrochloride on human nail in a forensic context. The authors are able to detect explosives on human nail too (Ali *et al.* 2009). Raman spectra are obtained from pentaerythritol tetranitrate, trinitrotoluene, ammonium nitrate, and hexamethylenetetramine particles on the surface of the nail with dimensions in the range 5–10 μm.

Zapata *et al.* (2016) use Raman microscopy to detect, identify, and quantify residues in human handmarks of explosives and energetic salts commonly used to manufacture improvised explosive devices including dynamite, ammonium nitrate, single- and double-smokeless gunpowders, and black powder.

Bridoux *et al.* (2016) use micro-Raman in combination with real-time/orbitrap mass spectroscopy in the detection of explosives. Combining these two techniques provides a powerful tool for the screening, comprehensive characterization, and differentiation of particulate explosive samples for forensic sciences and homeland security applications.

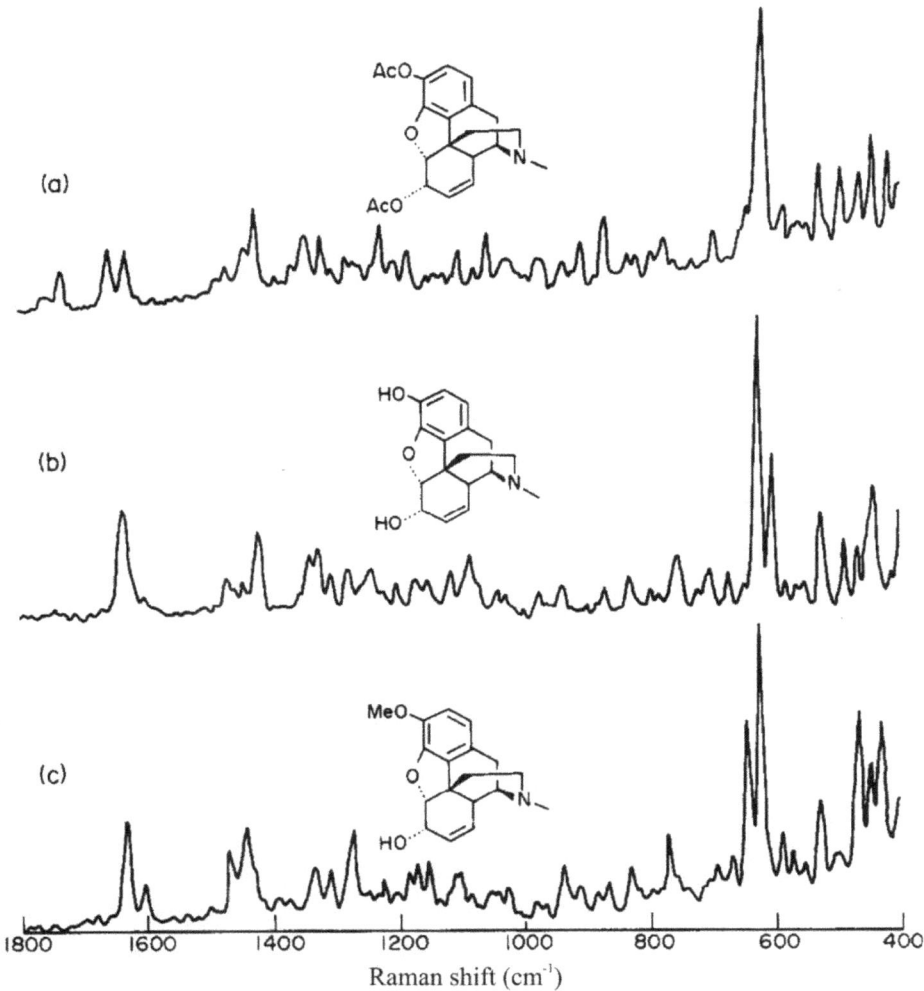

Figure 20.1: FT-Raman spectra of the pure alkaloids: (a) heroin; (b) morphine; and (c) codeine; 50 scans, 6 cm^{-1} resolution, incident laser power 200 mW. Scanning time 3 min. (Hodges and Akhavan 1990). Copyright ©1990 Elsevier B.V., with permission.

20.4 Illicit drugs

Ali *et al.* (2008) use confocal Raman spectroscopy to detect drugs of abuse on different clothing, examining as well drugs on undyed natural or synthetic fibres as drugs on coloured textiles. Despite the presence of spectral bands arising from the textiles, and also in the presence of fluorescence, the drugs could be identified by their characteristic Raman bands.

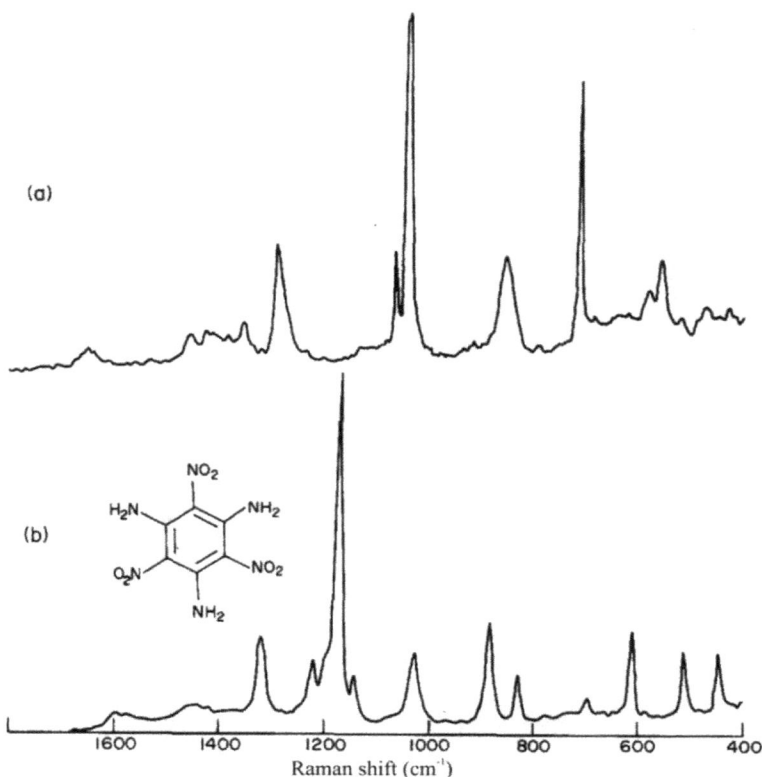

Figure 20.2: FT-Raman spectra of: (a) Driftex – a commercially available nitroglycerine based coalmining explosive and (b) TATB (triaminotrinitrobenzene). Both spectra 100 scans, 6 cm^{-1} resolution, incident laser power 500 mW. Scanning time 6 min (Hodges and Akhavan 1990). Copyright ©1990 Elsevier B.V., with permission.

Brewster *et al.* (2009) test the possibility to detect the date rape drug gamma-hydroxybutyric acid (GHB) and the corresponding gamma-butyrolactone (GBL) in containers and spiked drinks, also in water and 40% ethanol. In all forms GHB and GBL could be verified on levels lower than the common dosage lewel with Raman spectrometer in the laboratory as well as with hand-held spectrometers *in situ*.

Renishaw plc. (2010) publish a recipe to detect 1 fg of the cocaine metabolite benzoylecgonine by SERS. The company use the metabolite on a Klarite substrate and a measurement time of 15 s.

Boyd *et al.* (2011) study Raman spectroscopy of blood samples for forensic applications. In particular parameters such as substrate, sample dilution, individual from which the sample originates, and age of the sample are examined.

Pandey *et al.* (2013) study the vibrational spectra of morphine and heroin with DFT at the B3LYP/6-311G(d,p) level. Equilibrium geometries, FT-Raman, and FT-IR

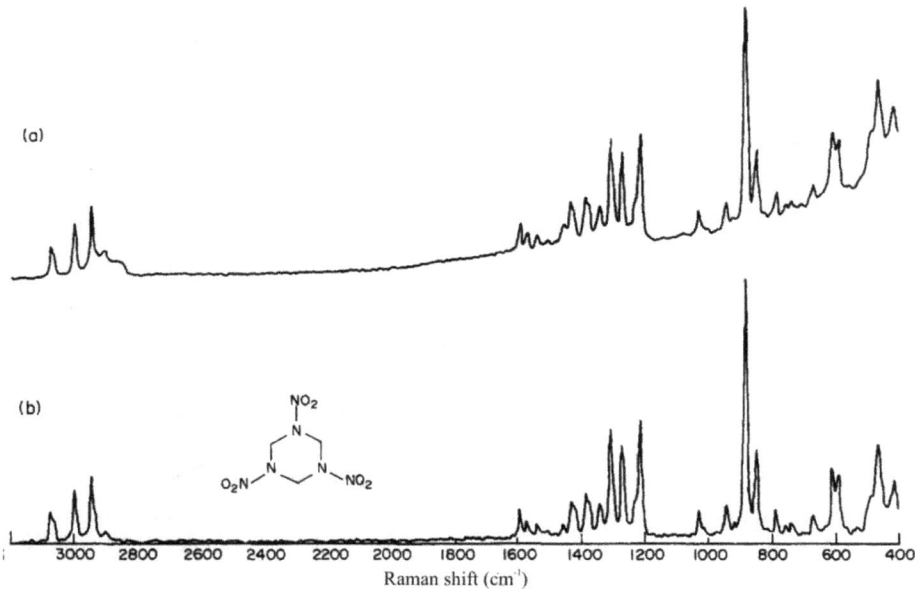

Figure 20.3: FT-Raman spectra of (a) Semtex and (b) the high-explosive 1,3,5-trinitro-1,3,5-triazane (RDX). Both spectra 200 scans, 6 cm^{-1} resolution, incident laser power 700 mW. Scanning time 12 min (Hodges and Akhavan 1990). Copyright ©1990 Elsevier B.V., with permission.

spectra of the two narcotics are shown and the bands assigned using the calculations. The authors advised the readers that the overall structure of the molecule rather than the pharmacophore should be used for studying the receptor binding sites.

Dana *et al.* (2015) report a method for the rapid analysis of cocaine in saliva by surface-enhanced Raman spectroscopy. Applying a solid-phase extraction pretreatment procedure, the authors could measure cocaine in saliva at clinical concentrations as low as 25 ng/mL. Dana *et al.* (2017) state that the method could also be applied to measure not only other basic drugs but also acidic drugs in a saliva matrix.

Pereira *et al.* (2017) directly classify new psychoactive substances (NPS) in seized blotter papers by attenuated total reflection-Fourier transform IR (ATR-FTIR) and partial least square discriminant analysis (PLS-DA). A direct analysis method based on ATR-FTIR and PLS-DA is developed and validated for the rapid screening of NPS in blotter papers. A multivariate model classifies samples of three classes: NBOMe, 2C-H (substituted phenethylamines), and MAL (methallylescaline), with relatively high efficiency, and is able to differentiate them from blank blotter papers. The most discriminant wavenumbers for each class are given in the study. The classification of LSD samples is not possible due to the lower concentration of this drug found in seized blotters and due to the limit of sensitivity of the technique.

A field test on psychoactive drugs is conducted by Gerace *et al.* (2019) in electronic music events. Using a portable Raman spectrometer, drugs in plastic bags held by the owners are sampled. Illicit substances are detected in 304 samples, among them MDMA (106 samples), ketamine (87 samples), cocaine (51 samples), amphetamine (47 samples), methamphetamine (2 samples), heroin (2 samples), and NPS (9 samples).

Using IR spectroscopy, Gomm and Humphreys (1975) describe a method for the comparison and identification of the major excipients encountered in illicit tablets. They either directly use powdered tablet material or a compensating technique using a wedge disc for the resolution of mixtures.

McCord *et al.* (2019) mention in their review on "Forensic DNA analysis" the use of Raman spectroscopy in the analysis of body fluids. Coupled with chemometric modelling, Raman spectroscopy was able to differentiate human blood from various animal bloods. This technique could also differentiate between peripheral blood, saliva, semen, sweat, and vaginal fluids in humans.

Portable IR absorption spectroscopy is used by Ramsay *et al.* (2021) for the analysis of synthetic opioids. Binary mixtures of fentanyl and caffeine and ternary mixtures of fentanyl with caffeine and a sugar alcohol are examined by IR together with a PLS regression model to quantify the fentanyl concentration in the mixtures.

20.5 Counterfeit drugs

As counterfeit pharmaceuticals have become a widespread issue for public health, a powerful tool for the analysis and determination of counterfeit medicines is needed. Raman spectroscopy, combined with chemometric methods, is the technique of choice because it easily provides the information needed and is non-destructive.

Fake artesunate *anti*-malarial tablets with their poor drug quality are a great threat to public health in southeast Asia. Ricci *et al.* (2007) advocate a combination of spatially offset Raman spectroscopy (SORS) and ATR-FTIR imaging for the screening and analysis of fake drugs. They describe the advantages of SORS for chemical analysis through blister and other packages and the spatially resolved analysis by ATR-FTIR imaging, which allows, e.g. for the distribution of active ingredients in the tablet.

For the identity testing of pharmaceutical products through packaging Eliasson *et al.* (2007) use Raman spectroscopy. The authors compare SORS to conventional backscattering Raman spectroscopy and demonstrate the benefits of SORS in samples that emit excessive surface Raman or fluorescence signals from the packaging, capsule shell, tablet coating, or plastic container obscuring the signals from the pharmaceutical.

Neuberger and Neusüß (2015) also advocate Raman spectroscopy for the detection of counterfeit medicines. In addition, they study the small chemical changes that are caused early by inappropriate storage conditions. The authors choose commercial acetylsalicylic acid effervescent tablets stored in five different conditions.

Zhao *et al.* (2015) choose Raman spectroscopy, aided by principal component analysis, as a non-invasive method for the determination of liquid injectables. The authors compare the method to the established analysis by NIR spectroscopy: Raman spectroscopy is not disturbed by water; instead water can be even used as an internal standard. The method is checked against analysis by HPLC using doxofylline liquid injectables, the big glass bottle-containers of levofloxacin lactate, and sodium chloride injections.

Biotherapeutics are also the aim of drug counterfeiting. Paidi *et al.* (2016) study the applicability of Raman spectroscopy to the very challenging task of identity determination of monoclonal antibodies. Testing a set of 22 closely related human and murine antibody drugs in solution, both normal Raman and SERS, enhanced by PLS-DA, give positive results. Using the fingerprint region (950–1,850 cm^{-1}) of the spectra, subtle spectrometric variations from subtle structural variations in the samples are enough to correctly identify the drugs. Raman spectroscopy is also proven to be a powerful method for rapid, on-site biotherapeutic product identification (Neuberger and Neusüß 2015).

20.6 Bioagents

Another important area of vibrational spectroscopy is the detection of bioagents. Stöckel *et al.* (2016) publish a review about the detection of bioagents by vibrational spectroscopy. The authors emphasize the broad information content of the methods on both the chemical composition and the structure of biomolecules within the microorganisms. Slight changes in the chemical composition of microorganisms can be monitored by means of Raman spectroscopy and used to differentiate genera, species, or even strains. Detection of pathogens is possible even from complex matrices such as soil, food, and body fluids. Further, spectroscopic studies of host–pathogen interactions can be addressed as well as the effect of antibiotics on bacteria. As an example, Raman spectra of *Escherichia coli* using different excitation wavelengths are shown (Figure 20.4).

Pahlow *et al.* (2016) report the detection and identification of *Pseudomonas* spp. by Raman spectroscopy. *Pseudomonas* species use the protein pyoverdine to collect iron from the environment. That means that the authors can use ferripyoverdine immobilized on an aluminium chip surface as a capture probe. Raman spectra can then be taken and processed using chemometrics. By this procedure, 85.5% of the Raman spectra are correctly assigned to a *Pseudomonas* species.

Pulmonary tuberculosis worldwide is mainly caused by *Mycobacterium tuberculosis* complex strains. Stöckel *et al.* (2017) focus on the differentiation between pathogenic and commensal non-tuberculous mycobacteria. Raman microspectroscopy together with the authors' chemometric model allow the discrimination between pathogenic and commensal species with an accuracy of 94% and can serve as a basis to further improve Raman microscopy as a first-line diagnostic point-of-care tool for the confirmation of tuberculosis disease.

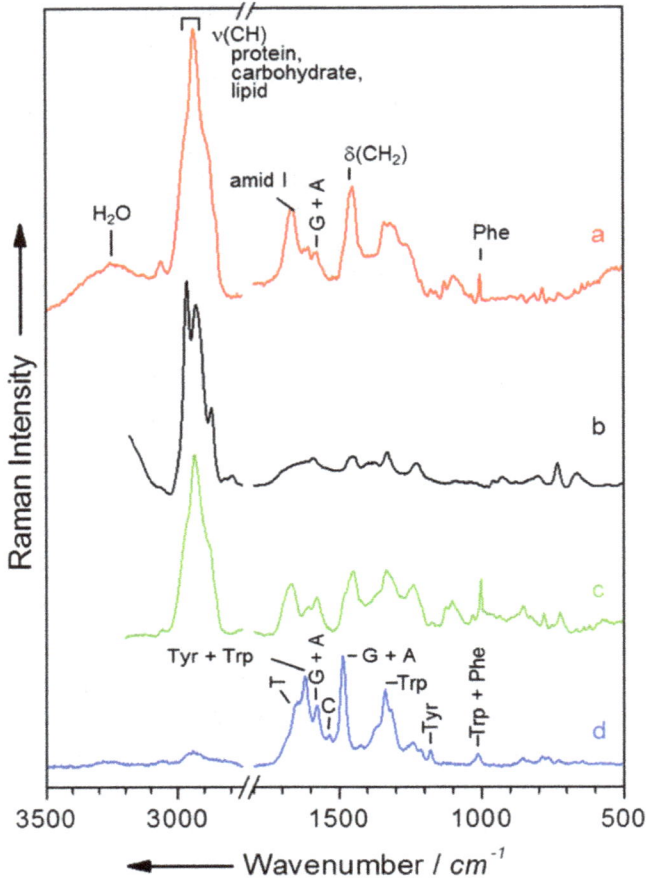

Figure 20.4: Raman spectra of *Escherichia coli* using different excitation wavelengths: (a) Fourier transform Raman spectrum (excited with 1,064 nm); (b) surface-enhanced Raman spectrum (excited with 532 nm in a microfluidic device); (c) visible Raman spectrum (excited with 532 nm); (d) UV resonance Raman spectrum (excited with 244 nm). Spectra a, b, and d belong to bulk samples, and spectrum c is a mean spectrum of single cells (Stöckel, Kirchhoff *et al.* 2016). Copyright ©2016 John Wiley & Sons, Ltd., with permission.

20.7 Art forgeries

Forgeries of valuable stamps are quite common. Imperio *et al.* (2015) study the printing inks of Italian postage stamp issues of 150 years by Raman microscopy and FTIR-ATR spectroscopy. Changes over time were: Prussian blue to copper phthalocyanine (for blue colours), cinnabar (Figure 20.5) to red ochre to red azo pigments (for red colours), a mixture of chrome orange and red ochre to azo pigments (for orange

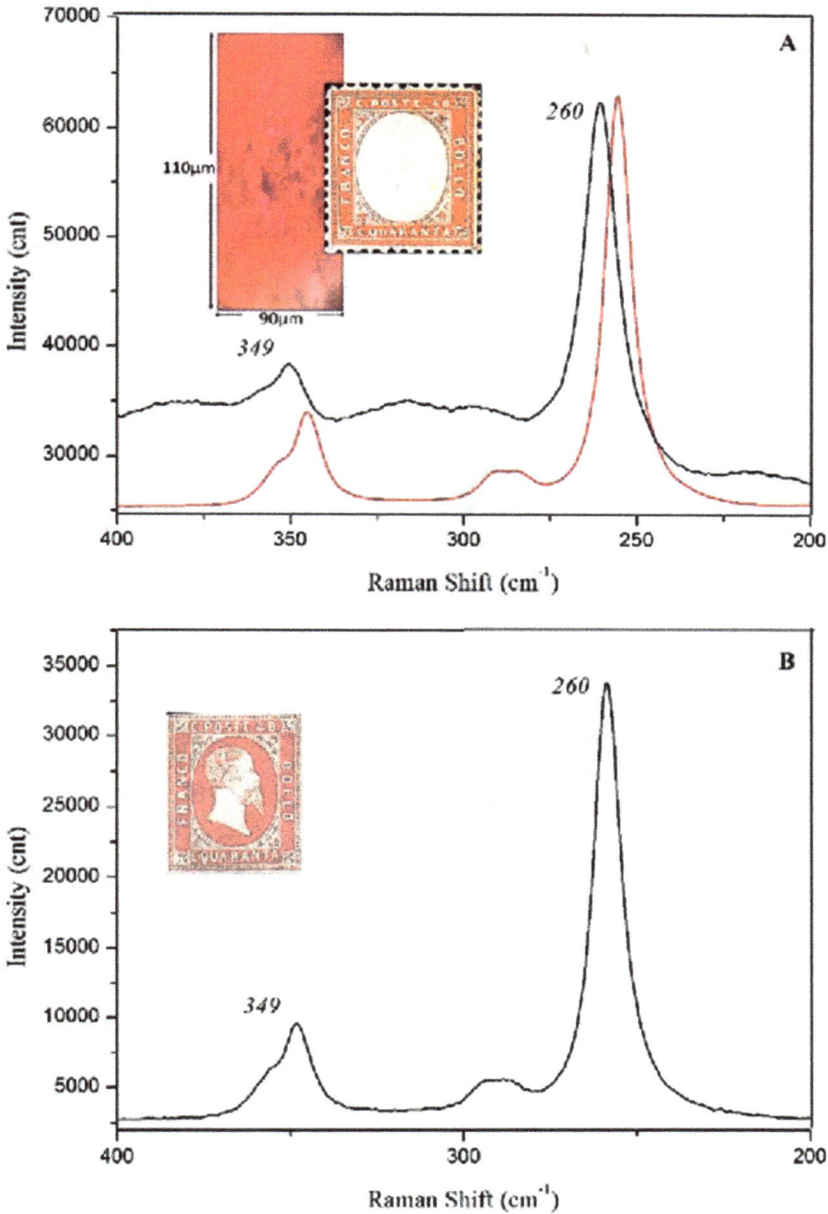

Figure 20.5: Raman spectra of red stamp (denomination of 40 cents) of the first issue of the Kingdom of Italy in 1862 (black) and of a reference sample of Cinnabar powder (red). In the inset, a microphotography of the same sample. (B) Raman spectrum of the red stamp of the Kingdom of Sardinia (denomination of 40 cents) issued in 1851 (Imperio *et al.* 2015). Copyright ©2015 Royal Society of Chemistry, with permission.

colours), and a mixture of chrome orange and Prussian blue to phthalocyanine (for green colours). These findings are of great help in identifying forged stamps.

20.8 Gemstones

Kiefert and Karampelas (2011) review the use of the Raman spectrometer in gemmological laboratories. The authors give examples of routine Raman uses "for the detection of emerald fillers (Figure 20.6), HPHT treatment of diamonds, analysis of the nature of a gemstone, analysis of gemstone inclusions and treatments, and the characterisation of natural or colour enhanced pearls and corals".

Figure 20.6: Filled emerald (left) and cleaned emerald (right) (Kiefert and Karampelas 2011). Copyright ©2011 Elsevier B.V., with permission.

21 Vibrational spectroscopy for the study of works of art and objects of cultural heritage

Vibrational spectroscopy is also used to study other objects of cultural heritage and works of art. As a non-destructive method, it is very valuable for the study of expensive, often priceless, objects.

A review paper by Edwards (2017) describes the study of art works by FT-IR and Raman spectroscopies. The role that FT-IR and FT-Raman in the analysis of artistic samples is illustrated by Edwards by examples from the characterization of genuine and fake artefacts, the strategic restoration of artworks being undertaken by art restorers and museum conservation scientists, the effects of environmental and climatic degradation on exposed artwork, and the identification of pigments.

Candeias and Madariaga (2019) review applications of Raman spectroscopy in art and archaeology. The paper is a report on the RAA 2017 conference in Evora (Portugal), which focused on the characterization of materials, conservation issues affecting cultural heritage (alteration/degradation processes), Raman spectroscopy of organic-based materials, and Raman applications in archaeology and forensics with authenticity research.

Raman microscopy in archaeological science is reviewed by Smith and Clark (2004). The areas covered are rock art and tomb paintings, ceramics and glazes, glass and faience, lithics, metals, ancient technology and pigment synthesis studies, textiles and plant fibres, resins, waxes, and organic residues, and biomaterials (skin, hair, teeth, bones, and ivory). As a conclusion the authors attest the method a very good usefulness for the purpose.

Chinese blue-and-white porcelain or *qinghua* porcelain uses a cobalt-based blue pigment in the underglaze blue decorations. Pinto *et al.* (2019) analyse dendritic crystals present in the dark spot areas of the blue-and-white Ming porcelains. Using Raman spectroscopy, they were able to evidence Mn-rich spinels. Analysing the behaviour of the A_{1g} mode of Mn-rich spinels, phases ranging from jacobsite-like to hausmannite-like compounds could be identified.

Marić-Stojanović *et al.* (2018) apply spectroscopic analysis to the XIV century wall paintings from Patriarchate of Peć Monastery (Church of the Holy Mother of God Hodegetria), Serbia. The painting technique and pigment pallete is examined on micro fragments in thin cross sections by means of several techniques, most helpful of them micro-Raman spectroscopy. The researchers report a palette of findings: the film of paint is multi-layered in most places and 12 pigments were identified. The artists applied mainly natural earth pigments such as red ochre, yellow ochre, and green earth and used a mixture of pigments for attaining desirable optical and aesthetical impressions. Egg white was used as a binder in some places, and only weddelite (calcium oxalate dihydrate) is detected as a decay product of the pigments.

https://doi.org/10.1515/9783110717556-021

New insights into the degradation mechanism of cellulose nitrate in cinematographic films by Raman microscopy are gained by Neves *et al.* (2019). Eventually the new knowledge can be used to find methods to conserve objects of cultural heritage as e.g. old photographic films.

Sodo *et al.* (2019) study the Bosch painting "Saint Wilgefortis Triptych". Raman measurements show the presence of degradation products such as calomel (Hg_2Cl_2) on the red pigment cinnabar (HgS, Figure 21.1), calcium-oxalate, in particular the dihydrate weddellite, localized in a not original external layer, and finally lead soaps in several layers of the investigated samples.

Gomez *et al.* (2019) use linseed oil as a model system for surface-enhanced Raman spectroscopy detection of degradation products in artworks. The authors describe an indirect sampling method, consisting of the extraction of surface products through a silicone strip sampler to the 3D surface-enhanced Raman scattering (SERS) substrate for Raman analysis (Figure 21.2).

Marcaida *et al.* (2019) study red and yellow ochre raw pigments from Pompeii in their original clay pottery bowls. Ochre pigments are mainly based on iron oxides such as hematite (Fe_2O_3) in the case of red ochres, and goethite (FeOOH) in yellow ochre pigments. The researchers are able to discriminate mineral phases of volcanic origin and contaminations on the pigments by Raman microscopy. The mineral phases found are langite [$Cu_4(SO_4)(OH)_6 \cdot 2H_2O$], jarosite [$KFe_3(SO_4)_2(OH)_6$], atacamite [$Cu_2Cl(OH)_3$], and anatase ($\beta$-$TiO_2$), a polymorph of titanium oxide found in igneous rocks. The contamination compounds are Egyptian or Pompeian blue ($CaCuSi_4O_{10}$) and huntite [$Mg_3Ca(CO_3)_4$].

The question: "Is *Ecce Homo* an European or Indian painting?" was put by Antunes *et al.* (2019). Applying µ-Raman spectroscopy in combination with the µ-FT-IR technique to access the chemical composition of a mixture of pigments, the question can be answered for three paintings: the authors consider the paintings from the 17th century as a Goan feature.

Prieto-Taboada *et al.* (2019) investigate the Raman spectra of the Na_2SO_4–K_2SO_4 system, interesting for soluble salts studies in built heritage. The spectrum of aphthitalite ($K_3Na(SO_4)_2$) displays characteristic bands at 1,084 and 1,202 cm^{-1}, while the bands at 1,100, 1,129, and 1,152 cm^{-1} characterize thenardite (Na_2SO_4). Both compounds have their most intense bands at 452 and 993 cm^{-1}. When the secondary bands (above 1,000 cm^{-1}) are not observed or mixtures of both compounds are present, the ratio between the 452 and 993 cm^{-1} band gives correct results.

The decorative polychrome plasterwork in the Royal Baths of Comares (Alhambra monument) is analysed by Arjonilla *et al.* (2019). Using Raman spectroscopy, vermillion, synthetic ultramarine blue, hematite, and carbon black could be identified in red, blue, brown, and black decorations. Raman bands typical of arsenate stretching bands were obtained from green decorations. It is assumed that the green pigments could contain degradation products of copper arsenite.

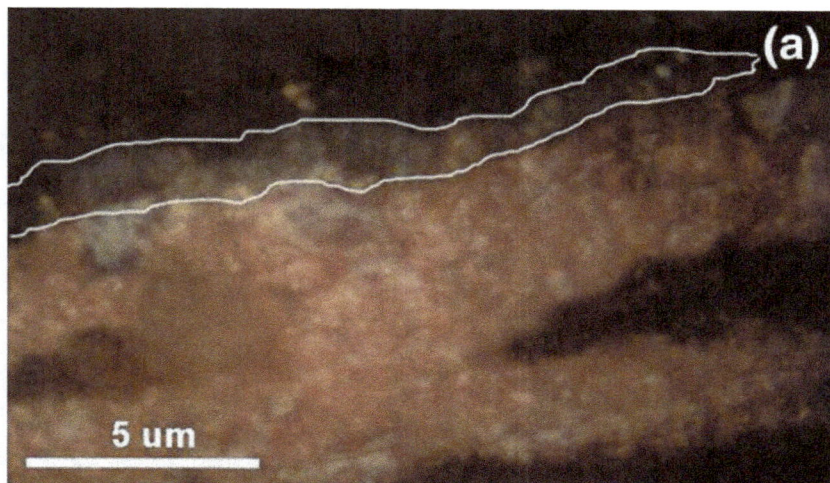

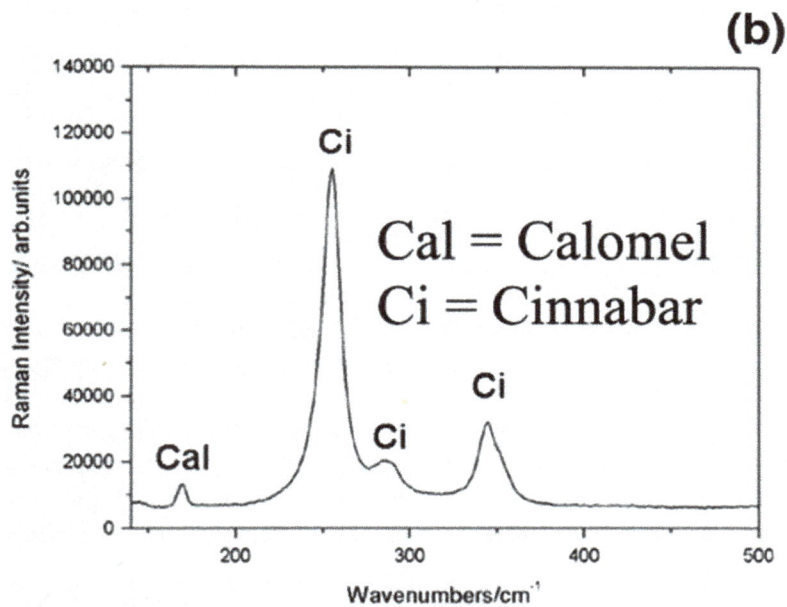

Figure 21.1: Degradation of cinnabar (HgS) to calomel (Hg$_2$Cl$_2$) on the Bosch painting "Saint Wilgefortis Triptych". Shown in (a) is the calomel in the bordered region of the painting's cross section and the Raman spectrum of it in (b) (Sodo *et al.* 2019). Copyright ©2019 Wiley & Sons, Ltd., with permission.

Aramendia *et al.* (2019) combine Raman imaging and LIBS for quantification of original and degradation materials in cultural heritage. To test the method, the authors analyse several dolomitic marble samples. The samples contain some calcite

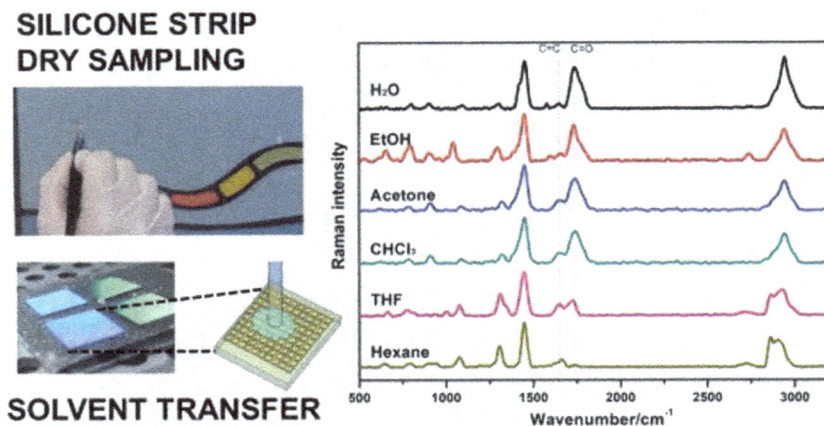

Figure 21.2: Linseed oil as a model system for surface-enhanced Raman spectroscopy detection of degradation products in artworks (Gomez *et al.* 2019). Copyright ©2019 by the authors, licensed by Creative Commons (CC 4.0).

impurities and are partly covered by a calcium oxalate (whewellite, $CaC_2O_4 \cdot H_2O$). The results are promising.

Care and display of resin cast and plastic objects in museum collections is the concern of Klisińska-Kopacz *et al.* (2019). To this purpose they need to know the composition of polymers used and compare the analytical performance of portable and benchtop Raman instruments. The author's conclusion is that portable instruments are suitable for the identification of plastics in museum collections.

From a shipwreck located in Bakio (Basque Country, Northern Spain) an iron anchor and a swivel gun were removed. These pieces of underwater heritage are analysed by Estalayo *et al.* (2019), and cast iron is found as the raw material of the artefacts. Raman spectroscopy and X-ray fluorescence spectroscopy are used to identify the main decaying compound as lepidocrocite (γ-FeO(OH)), a highly reactive iron phase that increases the corrosion rate of the artefacts. Lepidocrocite together with the also-found akaganeite (β-Fe$_2$(OH)$_3$Cl) are probably responsible for the continuous oxidation of the metallic pieces.

Archaeological glass beads are studied by Costa *et al.* (2019), combining Raman spectroscopy and micro-X-ray diffraction. μ-Raman spectroscopy identifies most beads as belonging to the alkaline glass family. The main colorants of the beads are cobalt ions and copper ions for dark blue and turquoise, the iron-sulfur chromophore produced amber or light brown hues, and iron ions are responsible for green, yellow, cream-coloured grey, and black hues. White glass is produced using calcium antimonate phases. Micro-Raman spectroscopy and μ-XRD can also be applied to heavily degraded samples and permit the identification of the opacifying agents and heat treatment to produce opaque glass beads.

Angelin *et al.* (2019) identify pearlescent pigments used to create luster in poly-(methyl methacrylate) artworks by Ângelo de Sousa (1938–2011) by Raman microspectroscopy. Plumbonacrite and bismuth oxychloride are characterized by comparison of their vibrational spectroscopic pattern (infrared and Raman) with reference materials. Using their Raman v_1 stretching mode at around 1,050 cm^{-1}, the two basic pearlescent lead carbonates hydrocerussite $Pb_3(CO_3)_2(OH)_2$ and plumbonacrite $Pb_5(CO_3)_3O(OH)_2$ can be identified.

Applying a hand-held Raman spectrometer, Saviello *et al.* (2019) analyse the dye chemical composition in felt-tip pen drawings. The weak Raman signal of the dye on paper has to be amplified by Ag nanoinks, resulting in SERS. By this technique, the authors reveal the dye content of historical felt tip pens extensively used by the Italian film director Federico Fellini. Hand-held SERS is considered a promising technique for the analysis of ink-based artworks and documents.

22 Medical applications of vibrational spectroscopy

As more and more studies on isolated molecules were published, scientists tried to use their knowledge to understand the complicated world of biology, especially human medicine. They developed and evaluated a plethora of new techniques of vibrational spectroscopy for the analysis of body fluids and tissues both *in vitro* and *in vivo*.

Early reviews of the use of infrared spectroscopy in medicine are published by Mantsch and co-workers (Mantsch 1984; Mantsch and Mcelhaney 1990; Mantsch 1996; Jackson, Sowa *et al.* 1997).

Fendel and Schrader (1998) investigate the skin and skin lesions by NIR-FT-Raman spectroscopy.

Choi *et al.* (2014) review research and application of atomic force microscopy (AFM) and Raman spectroscopy techniques. The basic principles of AFM and Raman spectroscopy are briefly introduced, followed by diagnostic assessments of some selected diseases in biomedical applications using them, including mitochondria isolated from normal and ischemic hearts, hair fibres, individual cells, and human cortical bone. Finally, AFM and Raman spectroscopy applications to investigate the effects of pharmacotherapy, surgery, and medical device therapy in various medicines from cells to soft and hard tissues are discussed. The discussion includes pharmacotherapy: paclitaxel on Ishikawa and HeLa cells, telmisartan on angiotensin II, mitomycin C on strabismus surgery and eye-whitening surgery, and fluoride on primary teeth- and medical device therapy-collagen cross-linking treatment for the management of progressive keratoconus, radiofrequency treatment for skin rejuvenation, physical extracorporeal shockwave therapy for healing of Achilles tendinitis, orthodontic treatment, and toothbrushing time to minimize the loss of teeth after exposure to acidic drinks.

Lipid organization and *stratum corneum* (SC) thickness is studied by Choe *et al.* (2016) using confocal Raman microscopy. The SC thickness can be determined *in vivo* in human skin analysing lipid-keratin peak (2,820–3,030 cm^{-1}, Figure 22.1), where lipids with long-chain carbon backbone (free fatty acids and ceramides) have their polmethylene Raman peak near 2,880 cm^{-1}. In healthy human skin, the sharpness of the peak centred near 2,880 cm^{-1}, which characterizes the polymethylene backbone of long-chain lipids, might be used for determining the SC thickness. The maximum position and broadness of the Gaussian peaks centred at 2,850 cm^{-1} show that near the surface and in the deep-located layers of the SC, the state of the lipids is more fluid and disordered compared to the medium layers (20–40% of the SC thickness).

A new look inside the human body is reviewed by Abramczyk *et al.* (2017): The medical applications of Raman imaging (RI) revolutionize the monitoring of diseases and treatment. The main advantage of RI is that, in contrast to conventional methods, it also provides spatial information about various chemical constituents in defined cellular organelles.

https://doi.org/10.1515/9783110717556-022

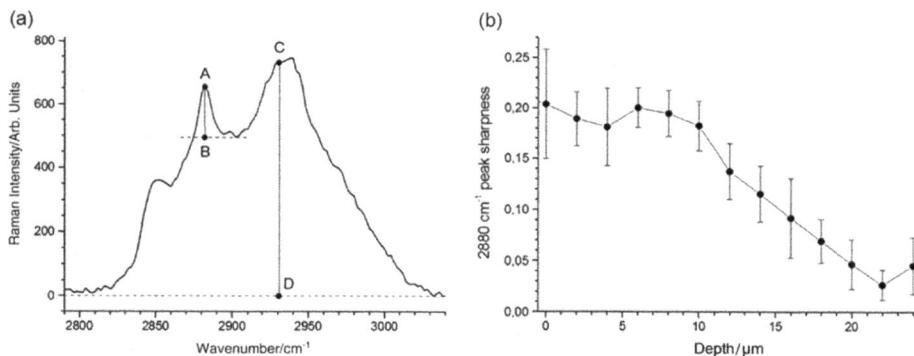

Figure 22.1: (a) shows part of the Raman spectrum of healthy skin. The peak centred near 2,880 cm^{-1}, which characterizes the polymethylene backbone of long-chain lipids, is used to determine the SC thickness by using its sharpness (b) (Choe *et al.* 2016). Copyright ©2016 John Wiley & Sons, Ltd., with permission.

Quite a few papers concentrate on the Raman and IR spectral diagnosis of breast cancer. Some examples are given below.

In a paper on Raman imaging in biochemical and biomedical applications, diagnosis and treatment of breast cancer is reviewed by Abramczyk and Brozek-Pluska (2013).

Raman and infrared spectroscopy is applied by Depciuch *et al.* (2016) in the identification of breast cancer. The authors find changes in the amounts of carotenoids and fats in the spectra collected from breast cancer patients compared to normal breast tissues. Changes in carbohydrate, amino acid, and protein levels are also detected.

Abramczyk and Brozek-Pluska (2016) examine the location and distribution of various biochemical components inside the lumen, epithelial cells of the duct, and the stroma around the human breast duct (Figure 22.2) during cancer development. Raman images of distribution of arachidonic acid, palmitic acid, oleic acid, sphingomyelin, collagen, mammaglobin-A, β-carotene, and stearic acid inside and around the normal human breast duct are shown in Figure 22.3. The authors identify Raman candidates as diagnostic markers for breast cancer prognosis *in situ*: They find four diagnostic markers for breast cancer prognosis: mammaglobin, palmitic acid and sphingomyelin, and carotenoids. Changes in the cancerous duct compared to a normal duct are:

- The composition of the epithelial cells surrounding the lumen of the cancerous duct changes in comparison with a normal duct
- Complete depletion of carotenoids in the cancerous duct
- The cancerous duct contains a smaller amount of monounsaturated fatty acids than the normal duct, which is dominated by oleic acid
- The cancerous duct contains a higher level of saturated lipids and sphingomyelin compared to the normal one

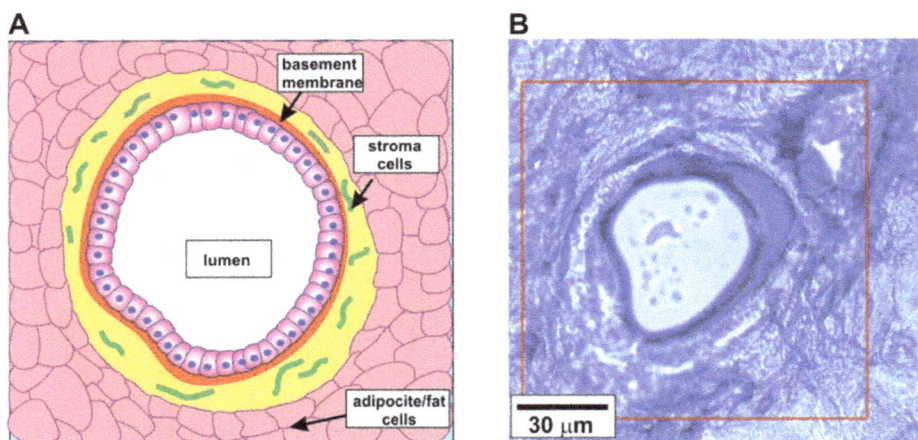

Figure 22.2: Schematic basic structure of epithelial tissue, stromal, and adipocyte cells around the normal duct (A) and microcopy image of normal human breast duct (P123) (B) (Abramczyk and Brozek-Pluska 2016). Copyright ©2016 John Wiley & Sons, Ltd., with permission.

Abramczyk and Brozek-Pluska propose the molecular mechanisms linking onco-genes with lipid programming.

As most histological samples are embedded in paraffine, the question is put by Depciuch *et al.* (2016) if the samples have to be deparaffinized before taking Raman spectra. Comparing paraffined and deparaffinized breast cancer tissue samples after a complete analysis of the peak after Kramers Kronig (KK) transformation, it was found that sample preparation did not affect the result obtained by measuring the reflectance in the mid-infrared range, and it only had a minimal effect relating to the intensity obtained by the measurement of the Raman peak.

Atkins *et al.* (2017) review Raman spectroscopy of blood and blood components. As a non-destructive technique which is not disturbed by water as IR, Raman spectroscopic probing of blood components and of whole blood has proven useful in applications ranging from the understanding of haemoglobin oxygenation, to the discrimination of cancerous cells from healthy lymphocytes, and the forensic investigation of crime scenes.

As thyroid carcinomas are the most common endocrine malignancy, thyroid tis-sues are studied by Medeiros *et al.* (2017). In their study, 30 thyroid samples, including normal thyroid, goitre, and thyroid cancer, are analysed by confocal Raman spectros-copy. Using chemometrics, significant discrimination is observed with a rate of 89.2% for normal thyroid versus cancer, 85.7% for goitre versus cancer, and 80.6% for nor-mal thyroid versus goitre using just the amide III region (1,200–1,400 cm^{-1}).

Takahashi *et al.* (2015) study the NIR Raman spectroscopy of preserved cartilagi-nous tissue. The authors advocate the use of freeze-thaw processes to provide a highly

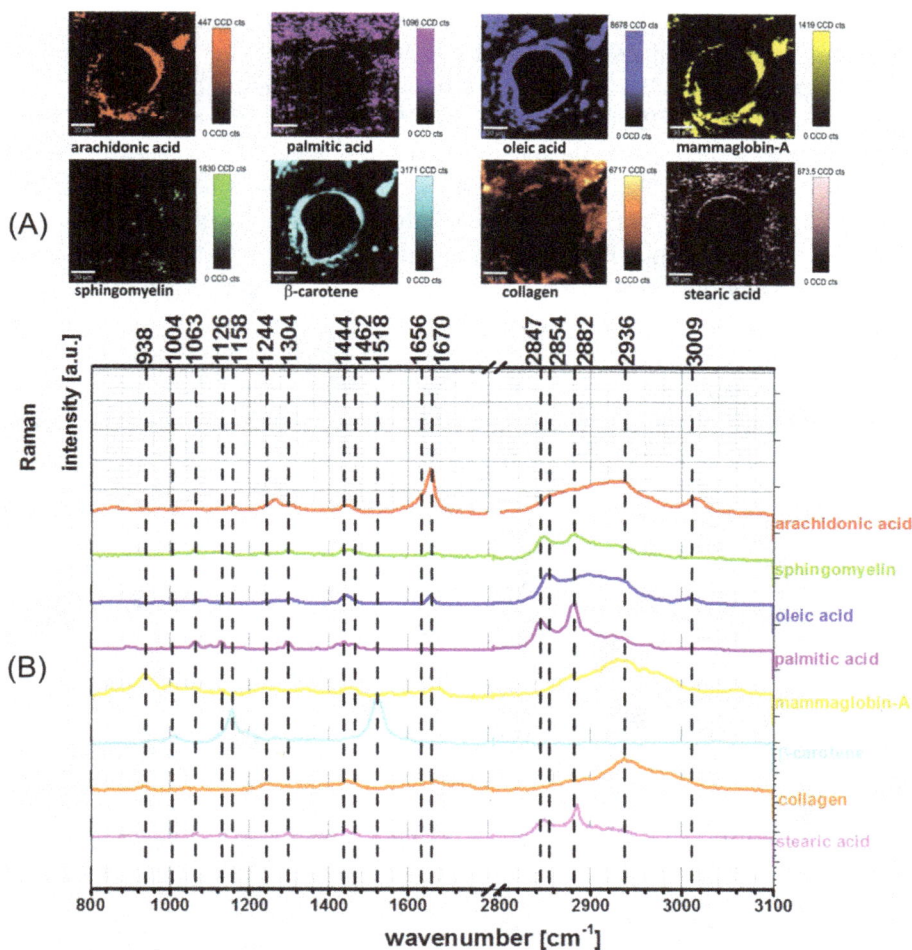

Figure 22.3: Raman images of distribution of pure components inside and around the normal human duct (P123) obtained from the basis analysis with eight components as a reference set: arachidonic acid, palmitic acid, oleic acid, sphingomyelin, collagen, mammaglobin-A, β-carotene, and stearic acid (A) and Raman spectra of the pure components used for the basis analysis (B) (Abramczyk and Brozek-Pluska 2016). Copyright ©2016 John Wiley & Sons, Ltd., with permission.

accurate biochemical evaluation of the extracellular matrix using NIR Raman spectroscopy. In contrast, the common formalin fixation will change the spectra of *ex vivo* samples of human cartilage much more and disturb their analysis.

Biomolecular changes in the cartilage matrix during the early stage of osteoarthritis are the interest of Kumar *et al.* (2015). The authors conclude from their confocal Raman measurements that the technique can classify the different stages of osteoarthritic cartilage and can provide details on biochemical changes. Kumar *et al.* propose a fibre-based miniaturized Raman probe for the development of *in vivo* Raman arthroscopy as a potential diagnostic tool for osteoarthritis.

Kalonakis *et al.* (2016) use cortical and trabecular bovine shinbones in his study of bone composition by Raman spectroscopy. The Raman spectrum of bone is compared to synthetic hydroxyapatite and bovine collagen. Bands associated with the hydroxyapatite at 430, 590, 961, 1,044 and 1,074 cm^{-1} and with the collagen at 1,254, 1,448, 1,668, and 2,939 cm^{-1} can be clearly detected in the Raman spectra of the bones.

The study of pollutant effects on living organisms is the interest of Olmos *et al.* (2018). They study the effect of the pesticide aldrin on human prostate cancer cells (DU145) using Raman spectroscopy and chemometric techniques improved by pre-processing tools. SIMPLISMA-based scaling is used which enhances variables with high relative variation among samples. The most relevant spectral variables related to aldrin effect on DU145 in this study are supposed to be variables related to lipids, proteins, and variations in nucleic acids.

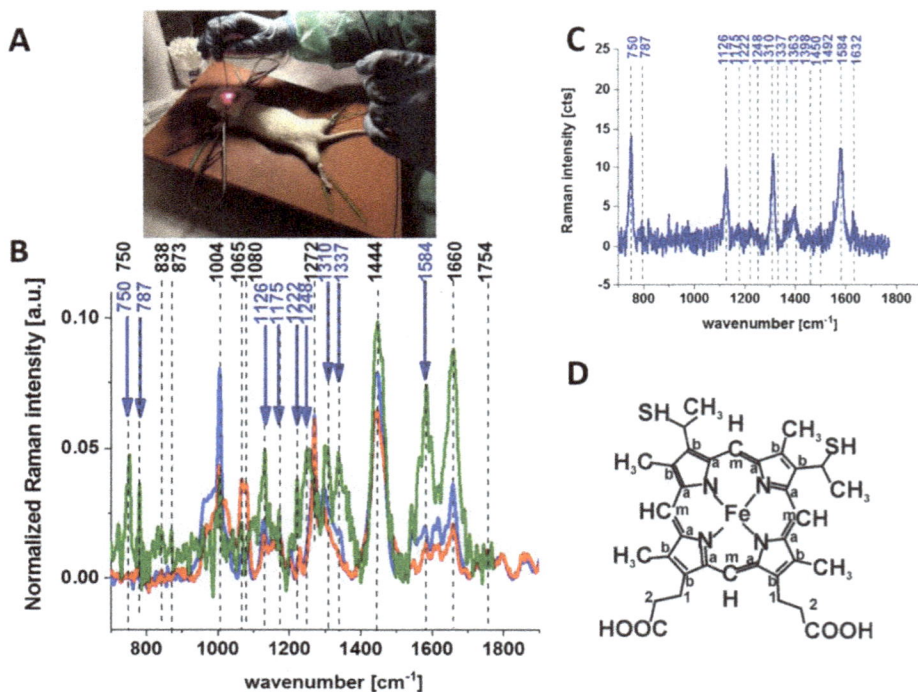

Figure 22.4: Raman- guided *in vivo* animal (rat) brain analysis (A), the average (*n* = 6) Raman spectrum of the *in vivo* brain of animal model (rat) at the excitation 785 nm (blue) and of the ex vivo brain of animal model (rat) at the excitation 532 nm (green) and 785 nm (red) (B), Raman spectrum of cytochrome c (dark blue) (C), and structural formula of heme c in cytochrome c (D) (Abramczyk *et al.* 2021). Copyright ©2021 by the authors, licensed by Creative Commons (CC 4.0).

Abramczyk *et al.* (2021) study brain and breast cancers in rats (Figure 22.4), also medulloblastoma in human brain tissue and breast cancer. Using Raman spectroscopy with various laser excitation lines, they find that the amount of reduced cytochrome becomes abnormally high in human brain tumours and breast cancers. A complete assignment of the resonance Raman spectra of cytochrome c is given by Berezhna *et al.* (2003).

Saleem *et al.* (2020) use Raman spectroscopy and chemometric techniques to diagnose a hepatitis B virus infection in blood plasma. To this purpose, Raman spectra of 24 diseased and 10 healthy samples were taken and processed using chemometrics. The radial basis function-based PCA-SVM model achieved the best performance and yielded an accuracy of 98.82%, a sensitivity of 98.89%, and a specificity of 98.80%.

Martinez *et al.* (2019) use Raman spectroscopy to examine bone composition in infant ribs, and Unal *et al.* (2021) access the composition of bone by Raman spectroscopy. Among other methods, they use the full width at half maximum of the most prominent Raman peak, which is the phosphate band v_1 at 961 cm^{-1} (symmetric stretch).

Scope of this book

The book provides an up-to-date overview of the fast-growing area of vibrational spectroscopy. New techniques are introduced to the reader. Analytic methods using vibrational spectroscopy allow for the chemical analysis of materials, providing even spatial resolution without precedent. Even external perturbations (strain, temperature, and pressure) on molecules and their alignment can be analysed. Vibrational spectroscopy can also provide information about the interactions of components, this again at a high level of spatial resolution. In the forms of nano-IR and tip-enhanced Raman spectroscopy (TERS), the method is a valuable tool for nanotechnology.

The book is intended for researchers or lecturers in chemistry, physics, materials science, and life sciences, who are interested in the composition and properties of their samples. It describes how vibrational spectroscopy will enable them to examine thin layers, surfaces, and interfaces, and also improve their knowledge about the properties of composites. Special chapters introduce VCD, ROA, and TERS.

The book can serve as a short introduction to vibrational spectroscopy too. Students at the first graduate level will benefit from it as well.

https://doi.org/10.1515/9783110717556-023

About the author

Günter Georg Hoffmann received his diploma in chemistry and biochemistry in 1978 in the group of Prof. Dr. Dr.h.c.(H) Günther Snatzke (University of the Ruhr, Bochum, Germany). In 1983, he was promoted to Dr. rer. nat. in the same group, working on organic synthesis and circular dichroism. At the end of 1983, he joined the group of Prof. Dr. Bernhard Schrader (UGE Essen, Germany) as postdoctoral fellow, modifying a commercial FT-IR spectrometer for the measurement of vibrational circular dichroism (VCD). Another postdoc position brought him in the years 1985–1986 to the Max Planck Institute for Radiation Chemistry (Mülheim, Germany), where he studied circularly polarized luminescence in the group of Prof. Dr. Alfred R. Holtzwarth and Prof. Dr. Kurt Schaffner. In 1986, he rejoined the group of Prof. Dr. Bernhard Schrader as assistant professor, investigating VCD and several new applications of Raman spectroscopy. In 1989, he won the Bennigsen-Foerder award of the Minister of Science and Education of North Rhine-Westphalia (Germany). Since 2002, he worked with Prof. Dr. Heinz W. Siesler (University of Duisburg-Essen, Germany) on infrared and near-infrared rheo-optical investigations on the structure of polymers and the Raman spectroscopic determination of temperature in rapidly moving laser spots. In 2004, he was appointed visiting professor in the group of Prof. Dr. Gijsbertus de With (Eindhoven University of Technology, The Netherlands), working on electrically conducting composites, atomic force microscopy (AFM), tip-enhanced Raman spectroscopy (TERS), and tip-enhanced Raman mapping (TERM).

He is the owner and CEO of Hoffmann Datentechnik ꤦꤦꤦ.

His main research topics include:
Vibrational circular dichroism and Raman optical activity of chiral organic molecules
Infrared and near-infrared rheo-optical investigations on the structure of polymers
Ab initio calculation of vibrational spectra
Atomic force microscopy and conductive atomic force microscopy
Tip-enhanced Raman spectroscopy and tip-enhanced Raman mapping
Composites of carbon nanotubes and polymers

GGH authored or co-authored some 70 papers and book chapters as well as 2 books.

https://doi.org/10.1515/9783110717556-024

References

Aamouche, A., F. J. Devlin and P. J. Stephens (2000a). "Conformations of chiral molecules in solution: *ab initio* vibrational absorption and circular dichroism studies of 4,4a,5,6,7,8-hexahydro-4a-methyl-2(3H)-naphthalenone and 3,4,8,8a-tetrahydro-8a-methyl-1,6(2H,7H)-naphthalenedione." Journal of the American Chemical Society **122**(30): 7358–7367.

Aamouche, A., F. J. Devlin and P. J. Stephens (2000b). "Structure, vibrational absorption and circular dichroism spectra, and absolute configuration of Tröger's base." Journal of the American Chemical Society **122**(10): 2346–2354.

Abad, L., D. Bermejo, R. Escribano, V. J. Herrero, J. Santos, I. Tanarro, G. D. Nivellini and L. Ramonat (1994). "Stimulated Raman-spectra of jet-cooled ethylene." Chemical Physics Letters **227**(3): 248–254.

Abella, L., H. D. Ludowieg and J. Autschbach (2020). "Theoretical study of the Raman optical activity spectra of with M = Co, Rh." Chirality **32**(6): 741–752.

Abney, W. and E. R. Festing (1882). Philosophical Transactions **172**: 887.

Abramczyk, H. and B. Brozek-Pluska (2013). "Raman imaging in biochemical and biomedical applications. Diagnosis and treatment of breast cancer." Chemical Reviews **113**(8): 5766–5781.

Abramczyk, H., M. Kopec and M. Jędrzejczyk (2017). Raman Spectroscopy, Medical Applications: A New Look Inside Human Body with Raman Imaging. Encyclopedia of Spectroscopy and Spectrometry (Third Edition). J. C. Lindon, G. E. Tranter and D. W. Koppenaal. Oxford, Academic Press: 915–918.

Abramczyk, H., J. Surmacki and B. Brozek-Pluska (2021). "Redox state changes of mitochondrial cytochromes in brain and breast cancers by Raman spectroscopy and imaging." Journal of Molecular Structure **1252**: 132134.

Adhikari, S. and R. Chowdhury (2011). "Vibration spectra of fullerene family." Physics Letters A **375**(22): 2166–2170.

Agapov, R. L., J. D. Scherger, A. P. Sokolov and M. D. Foster (2015). "Identification of Individual Isotopes in a polymer blend using tip enhanced Raman spectroscopy." Journal of Raman Spectroscopy **46**(5): 447–450.

Alaaeddin, M., S. Sapuan, M. Z. Mohamed Yusoff, E. S. Zainudin and F. Al-Oqla (2019). "Polyvinyl fluoride (PVF); Its Properties, Applications, and Manufacturing Prospects." IOP Conference Series: Materials Science and Engineering **538**: 012010.

Alcolea Palafox, M., J. L. Núñez and M. Gil (2002). "Accurate scaling of the vibrational spectra of aniline and several derivatives." Journal of Molecular Structure: THEOCHEM **593**(1): 101–131.

Ali, E. M. A., H. G. M. Edwards, M. D. Hargreaves and I. J. Scowen (2008). "*In-situ* detection of drugs-of-abuse on clothing using confocal Raman microscopy." Analytica Chimica Acta **615**(1): 63–72.

Ali, E. M. A., H. G. M. Edwards, M. D. Hargreaves and I. J. Scowen (2008). "Raman spectroscopic investigation of cocaine hydrochloride on human nail in a forensic context." Analytical and Bioanalytical Chemistry **390**(4): 1159–1166.

Ali, E. M. A., H. G. M. Edwards, M. D. Hargreaves and I. J. Scowen (2009). "Detection of explosives on human nail using confocal Raman microscopy." Journal of Raman Spectroscopy **40**(2): 144–149.

An, N. T. T., D. Q. Dao, P. C. Nam, B. T. Huy and H. Nhung Tran (2016). "Surface enhanced Raman scattering of melamine on silver substrate: An experimental and DFT study." Spectrochimica Acta Part A: Molecular and Biomolecular Spectroscopy **169**: 230–237.

Ananthakrishnan, R. (1936). "Raman Spectrum of Cyclopropane." Nature **138**(3481): 123.

Andersen, N., P. Klaeboe, C. Nielsen, B. Shumberger and G. Guirgis (2019). "The Raman and infrared spectra, *ab initio* calculations and spectral assignments of 1,1-dicyclopropyl-2,2-dimethylethene (c-C_3H_5)$_2$C =C(CH$_3$)$_2$." Journal of Molecular Structure **1195**: 528–541.

Angelin, E. M., S. Babo, J. L. Ferreira and M. J. Melo (2019). "Raman microscopy for the identification of pearlescent pigments in acrylic works of art." Journal of Raman Spectroscopy **50**(2): 232–241.

https://doi.org/10.1515/9783110717556-025

Antunes, V., V. Serrão, A. Candeias, J. Mirão, A. Cardoso, M. L. Carvalho, N. Fernandes and M. Manso (2019). "Scrutinizing Ecce Homo: European or Indian painting? Assessment by Raman and complementary spectroscopic techniques." Journal of Raman Spectroscopy **50**(2): 161–174.

Aparicio-Cuevas, M. A., I. Rivero-Cruz, M. Sanchez-Castellanos, D. Menendez, H. A. Raja, P. Joseph-Nathan, M. D. Gonzalez and M. Figueroa (2017). "Dioxomorpholines and derivatives from a marine-facultative Aspergillus species." Journal of Natural Products **80**(8): 2311–2318.

Appun, J., H.-U. Siehl, K.-P. Zeller, K. Steinke, S. Berger and D. Sicker (2014). "Colchicin." Chemie in unserer Zeit **48**(1): 36–44.

Aramendia, J., L. Gómez-Nubla, S. Fdez-Ortiz de Vallejuelo, K. Castro, G. Arana and J. M. Madariaga (2019). "The combination of Raman imaging and LIBS for quantification of original and degradation materials in cultural heritage." Journal of Raman Spectroscopy **50**(2): 193–201.

Araujo, C. F., M. M. Nolasco, A. M. P. Ribeiro and P. J. A. Ribeiro-Claro (2018). "Identification of microplastics using Raman spectroscopy: Latest developments and future prospects." Water Research **142**: 426–440.

Arenas, J. F., I. L. Tocon, J. C. Otero and J. I. Marcos (1999). "Vibrational spectra of methylpyridines." Journal of Molecular Structure **476**(1–3): 139–150.

Arjonilla, P., M. J. Ayora-Cañada, R. Rubio Domene, E. Correa Gómez, M. J. de la Torre-López and A. Domínguez-Vidal (2019). "Romantic restorations in the Alhambra monument: Spectroscopic characterization of decorative plasterwork in the Royal Baths of Comares." Journal of Raman Spectroscopy **50**(2): 184–192.

Arjunan, V., R. Anitha, L. Devi, S. Mohan and H. F. Yang (2015). "Comprehensive quantum chemical and spectroscopic (FT-IR, FT-Raman, ^{1}H, ^{13}C NMR) investigations of (1,2-epoxyethyl)benzene and (1,2-epoxy-2-phenyl)propane." Spectrochimica Acta Part A – Molecular and Biomolecular Spectroscopy **135**: 120–136.

Arnett, R. L. and B. L. Crawford (1950). "The vibrational frequencies of ethylene." Journal of Chemical Physics **18**: 118–126.

Atkins, C. G., K. Buckley, M. W. Blades and R. F. B. Turner (2017). "Raman spectroscopy of blood and blood components." Applied Spectroscopy **71**(5): 767–793.

Atkins, P. W. (1983). Molecular Quantum Mechanics. Oxford, Oxford University Press.

Ayala, A. P., H. W. Siesler, R. Boese, G. G. Hoffmann, G. I. Polla and D. R. Vega (2006). "Solid state characterization of olanzapine polymorphs using vibrational spectroscopy." International Journal of Pharmaceutics **326**(1–2): 69–79.

Baiardi, A., C. Latouche, J. Bloino and V. Barone (2014). "Accurate yet feasible computations of resonance Raman spectra for metal complexes in solution: [Ru(bpy)3^{2+}] as a case study." Dalton Transactions **43**(47): 17610–17614.

Bailey, G. F. and R. J. Horvat (1972). "Raman spectroscopic analysis of *cis/trans* isomer composition of edible vegetable oils." Journal of the American Oil Chemists Society **49**(8): 494–498.

Bakker, H., S. R. Meech and E. J. Heilweil (2018). "Time-resolved vibrational spectroscopy." The Journal of Physical Chemistry A **122**(18): 4389–4389.

Balabin, R. M. (2009). "Enthalpy difference between conformations of normal alkanes: Raman spectroscopy study of *n*-pentane and *n*-butane." Journal of Physical Chemistry A **113**(6): 1012–1019.

Balcers, O., U. Miranda and R. Veilande (2022). "Study of ergocalciferol and cholecalciferol (Vitamin D): Modeled optical properties and optical detection using absorption and Raman spectroscopy." Spectrochimica Acta Part A: Molecular and Biomolecular Spectroscopy **269**: 120725.

Balois, M. V., N. Hayazawa, F. C. Catalan, S. Kawata, T. Yano and T. Hayashi (2015). "Tip-enhanced THz Raman spectroscopy for local temperature determination at the nanoscale." Analytical and Bioanalytical Chemistry **407**(27): 8205–8213.

Bandermann, F., I. Tausendfreund, S. Sasic, Y. Ozaki, M. Kleimann, J. A. Westerhuis and H. W. Siesler (2001). "Fourier-transform Raman spectroscopic on-line monitoring of the anionic dispersion block copolymerization of styrene and 1,3-butadiene." Macromolecular Rapid Communications **22**(9): 690–693.

Baraban, J. H., P. B. Changala and J. F. Stanton (2018). "The equilibrium structure of hydrogen peroxide." Journal of Molecular Spectroscopy **343**: 92–95.

Baranska, M. and A. Kaczor (2012). "Morphine studied by vibrational spectroscopy and DFT calculations." Journal of Raman Spectroscopy **43**(1): 102–107.

Baranska, M. and H. Schulz (2009). Chapter 4 Determination of Alkaloids through Infrared and Raman Spectroscopy. The Alkaloids: Chemistry and Biology. G. A. Cordell, Academic Press. **67**: 217–255.

Barletta, R. E., H. H. Claassen and R. L. McBeth (1971). "Raman spectrum of S_2." The Journal of Chemical Physics **55**(11): 5409–5410.

Barron, L. D. (2015). "The development of biomolecular Raman optical activity spectroscopy." Biomedical Spectroscopy and Imaging **4**(3): 223–253.

Barrow, G. M. and S. Searles (1953). "Effect of ring size on the infrared spectra of cyclic ethers – characteristic frequencies of trimethylene oxides and tetrahydrofurans." Journal of the American Chemical Society **75**(5): 1175–1177.

Barth, A. (2007). "Infrared spectroscopy of proteins." Biochimica et Biophysica Acta (BBA) – Bioenergetics **1767**(9): 1073–1101.

Barth, A. and P. Haris (2009). Infrared Spectroscopy – Past and Present. Advances in Biomedical Spectroscopy. P. I. H. Andreas Barth. Amsterdam, IOS Press: 1–52.

Bartlett, N. (1962). "Xenon hexafluoroplatinate(V)." Proceedings of the Chemical Society(June) 197–236.

Barton, B., J. Thomson, E. Lozano Diz and R. Portela (2022). "Chemometrics for Raman Spectroscopy Harmonization." Applied Spectroscopy **76**(9): 1021–1041.

Batista, J. M., B. Wang, M. V. Castelli, E. W. Blanch and S. N. Lopez (2015). "Absolute configuration assignment of an unusual homoisoflavanone from Polygonum ferrugineum using a combination of chiroptical methods." Tetrahedron Letters **56**(44): 6142–6144.

Bava, Y. B., M. V. Cozzarín, C. O. Della Védova, H. Willner and R. M. Romano (2021). "Preparation of FC(S)SF, FC(S)SeF and FC(Se)SeF through matrix photochemical reactions of F_2 with CS_2, SCSe, and CSe_2." Physical Chemistry Chemical Physics **23**(37): 20892–20900.

Beć, K. B., J. Grabska, C. W. Huck, S. Mazurek and M. A. Czarnecki (2021) "Anharmonicity and spectra-structure correlations in MIR and NIR spectra of crystalline menadione (vitamin K3)." Molecules (Basel, Switzerland) **26**(22): 6779.

Becher, H. J. and F. Thévenot (1974). "Infrarotspektroskopische Untersuchung des Borcarbids und seiner isotypen Derivate $B_{12}O_2$, $B_{12}P_2$ und $B_{12}As_2$." Zeitschrift für anorganische und allgemeine Chemie **410**(3): 274–286.

Becke, A. D. (1988). "Density-functional exchange-energy approximation with correct asymptotic behavior." Physical Review A **38**: 3098–3100.

Begley, R. F., A. B. Harvey and R. L. Byer (1974). "Coherent *anti*-Stokes Raman spectroscopy." Applied Physics Letters **25**(7): 387–390.

Bellamy, L. J. (1966). Ultrarotspektrum und chemische Konstitution. Darmstadt, Dr. Dietrich Steinkopff Verlag.

Bellamy, L. J., W. Gerrard, M. F. Lappert and R. L. Williams (1958). "481. Infrared spectra of boron compounds." Journal of the Chemical Society (Resumed) (0): 2412–2415.

Bender, J. S., B. Coasne and J. T. Fourkas (2015). "Assessing polarizability models for the simulation of low-frequency Raman spectra of benzene." Journal of Physical Chemistry B **119**(29): 9345–9358.

Bentley, F. F., L. D. Smithson and A. L. Rozek (1968). Infrared Spectra and Characteristic Frequencies 700 – 300 cm^{-1}. New York, Interscience.

Berard, M. and P. Lallemand (1979). "Etude experimentale de la structure fine du spectre Raman de rotation de l'oxygene." Optics Communications **30**(2): 175–180.

Berezhna, S., H. Wohlrab and P. M. Champion (2003). "Resonance Raman investigations of cytochrome c conformational change upon interaction with the membranes of intact and Ca^{2+}-exposed mitochondria." Biochemistry **42**(20): 6149–6158.

Berg, R. W. (2015). "Investigation of L(+)-ascorbic acid with Raman spectroscopy in visible and UV light." Applied Spectroscopy Reviews **50**(3): 193–239.

Berger, S. and D. Sicker (2009). Classics in Spectroscopy: Isolation and Structure Elucidation of Natural Products. Weinheim, Wiley-VCH Verlag.

Berlizov, A., D. M. L. Ho, A. Nicholl, T. Fanghänel and K. Mayer (2016). "Assessing hand-held Raman spectrometer FirstDefender RM for nuclear safeguards applications." Journal of Radioanalytical and Nuclear Chemistry **307**(1): 285–295.

Bermejo, D., P. Cancio, G. Di Lonardo and L. Fusina (1998). "High resolution Raman spectroscopy from vibrationally excited states populated by a stimulated Raman process: $2v_2–v_2$ of $^{12}C_2H_2$ and $^{13}C_2H_2$." Journal of Chemical Physics **108**(17): 7224–7228.

Bermejo, D., R. Escribano and J. Santos (1997). "The stimulated Raman spectrum of cyanogen." Journal of Molecular Spectroscopy **186**(1): 144–154.

Bhattarai, A. and P. Z. El-Khoury (2017). "Imaging localized electric fields with nanometer precision through tip-enhanced Raman scattering." Chemical Communications **53**(53): 7310–7313.

Bielecki, J. and E. Lipiec (2016). "Basis set dependence using DFT/B3LYP calculations to model the Raman spectrum of thymine." Journal of Bioinformatics and Computational Biology **14**(1): 1650002.

Binnig, G., H. Rohrer, C. Gerber and E. Weibel (1982). "Tunneling through a controllable vacuum gap." Applied Physics Letters **40**: 178.

Bistričić, L., G. Baranović and K. Mlinarić-Majerski (1995). "A vibrational assignment of adamantane and some of its isotopomers. Empirical versus scaled semiempirical force field." Spectrochimica Acta Part A: Molecular and Biomolecular Spectroscopy **51**(10): 1643–1664.

Bloom, S. D., R. C. Harney and F. P. Milanovich (1976). "Stable isotope ratio measurements in nitrogen and oxygen using Raman scattering." Applied Spectroscopy **30**(1): 64–67.

Blumenfeld, S. M., S. P. Reddy and H. L. Welsh (1970). "Vibrational Raman spectra of liquid and solid ethylene." Canadian Journal of Physics **48**(5): 513–520.

Boerio, F. J. and J. L. Koenig (1971). "Vibrational analysis of poly(vinylidene fluoride)." Journal of Polymer Science Part A-2: Polymer Physics **9**(8): 1517–1523.

Bogomolov, A. M. (1960). "Vibrational Spectra of Aromatic Compounds .7. Characteristic Vibrations of Monosubstituted Benzenes." Optika I Spektroskopiya **9**(3): 311–318.

Bogomolov, A. M. (1961). "Vibrational Spectra of Aromatic Compounds .8. Characteristic Vibrations of (1,2)-*ortho*-disubstituted Benzenes." Optika I Spektroskopiya **10**(3): 322–327.

Bogomolov, A. M. (1962a). "Vibrational Spectra of Aromatic Compounds .12. Characteristic Vibrations of *para*-di-Substituted Benzenes." Optika I Spektroskopiya **12**(2): 186–193.

Bogomolov, A. M. (1962b). "Vibrational Spectra of Aromatic Compounds .14. Characteristic Vibrations of *meta*-substituted Benzenes." Optika I Spektroskopiya **13**(2): 159–168.

Bogomolov, A. M. (1962c). "Vibrational Spectra of Aromatic Compounds .16. Characteristic Vibrations of 1,2,3-Substituted Benzenes." Optika I Spektroskopiya **13**(3): 331–334.

Böhme, R., D. Cialla, M. Richter, P. Rösch, J. Popp and V. Deckert (2010). "Biochemical imaging below the diffraction limit – probing cellular membrane related structures by tip-enhanced Raman spectroscopy (TERS)." Journal of Biophotonics **3**(7): 455–461.

Bomem, I. (1987). "Application note DA3-870."

Bonhommeau, S. and S. Lecomte (2018). "Tip-enhanced Raman spectroscopy: a tool for nanoscale chemical and structural characterization of biomolecules." Chemphyschem **19**(1): 8–18.

Borman, S. A. (1982). "Nonlinear Raman spectroscopy." Analytical Chemistry **54**(9): 1021A–1026A.

Bortchagovsky, E., T. Schmid and R. Zenobi (2013). "Internal standard for tip-enhanced Raman spectroscopy." Applied Physics Letters **103**(4): 043111.

Bour, P. and T. A. Keiderling (2004). "Structure, spectra and the effects of twisting of ß-sheet peptides. A density functional theory study." Journal of Molecular Structure – Theochem **675**: 95–105.

Boutilier, G. D., R. M. Irwin, R. R. Antcliff, L. B. Rogers, L. A. Carreira and L. Azarraga (1981). "Computer-controlled absorption, fluorescence, and coherent *anti*-Stokes Raman spectrometry for liquid-chromatographic detection." Applied Spectroscopy **35**(6): 576–581.

Boyd, S., M. F. Bertino and S. J. Seashols (2011). "Raman spectroscopy of blood samples for forensic applications." Forensic Science International **208**(13): 124–128.

Braden, D. A. and B. S. Hudson (2000). "C_6F_6 and *sym*-$C_6F_3H_3$: *Ab initio* and DFT studies of structure, vibrations, and inelastic neutron scattering spectra." Journal of Physical Chemistry A **104**(5): 982–989.

Brandán, S. (2019). "Structural and vibrational studies of equilenin, equilin and estrone steroids." Biointerface Research in Applied Chemistry **9**(6): 4502–4516.

Brandmüller, J., H. Hacker and H. W. Schrötter (1966). "Photoelektrische Registrierung der Raman-Spektren farbiger Substanzen mit He-Ne-6328 Å-Laser-Erregung." Chemische Berichte-Recueil **99**(3): 765–775.

Brecher, C. and R. S. Halford (1961). "Motions of molecules in condensed systems. XI. Infrared spectrum and structure of a single crystal of ethylene." Journal of Chemical Physics **35**: 1109–1117.

Brewster, V. L., H. G. M. Edwards, M. D. Hargreaves and T. Munshi (2009). "Identification of the date-rape drug GHB and its precursor GBL by Raman spectroscopy." Drug Testing and Analysis **1**(1–2): 25–31.

Bridoux, M. C., A. Schwarzenberg, S. Schramm and R. B. Cole (2016). "Combined use of direct analysis in real-time/Orbitrap mass spectrometry and micro-Raman spectroscopy for the comprehensive characterization of real explosive samples." Analytical and Bioanalytical Chemistry **408**(21): 5677–5687.

Brittain, E. F. H., W. O. George and C. H. J. Wells (1970). Introduction to Molecular Spectroscopy. London and New York, Academic Press.

Brodersen, S., D. L. Gray and A. G. Robiette (1977). "Determination of B_0 from v_3 Raman Band of CD_4." Molecular Physics **34**(3): 617–628.

Brown, K. G., S. C. Erfurth, E. W. Small and W. L. Peticolas (1972). "Conformationally dependent low-frequency motions of proteins by laser Raman spectroscopy." Proceedings of the National Academy of Sciences of the United States of America **69**(6): 1467–1469.

Brozek-Pluska, B., G. Waliszewska, M. Jackowicz, S. Kuberski, R. Zarzycki, G. Janowska and H. Abramczyk (2005). "Low temperature Raman study of stable and metastable structures of phenylacetylene in benzene. Vibrational dynamics in undercooled liquid solutions, crystals, and glassy crystals." Journal of Molecular Liquids **121**(2–3): 80–87.

Brunette, J. P. (1974). "Caracterisation des ions $MoOCl_4Br^{2-}$ et $WOCl_4Br^{2-}$." Journal of Inorganic and Nuclear Chemistry **36**: 289–293.

Brunvoll, J., S. J. Cyvin and L. Schäfer (1971). "Normal coordinate analysis for whole molecule of dibenzenechromium." Journal of Organometallic Chemistry **27**(1): 69–71.

Brunvoll, J., S. J. Cyvin and L. Schäfer (1971). "Normal coordinate analysis of total ferrocene complex." Journal of Organometallic Chemistry **27**(1): 107–111.

Bulayev, Y. (2015) "Advances in CMOS image sensors open doors to many applications." Hamamatsu Website.

Bultinck, P., F. L. Cherblanc, M. J. Fuchter, W. A. Herrebout, Y. P. Lo, H. S. Rzepa, G. Siligardi and M. Weimar (2015). "Chiroptical studies on brevianamide B: vibrational and electronic circular dichroism confronted." Journal of Organic Chemistry **80**(7): 3359–3367.

Bumbrah, G. S. and R. M. Sharma (2016). "Raman spectroscopy – Basic principle, instrumentation and selected applications for the characterization of drugs of abuse." Egyptian Journal of Forensic Sciences **6**(3): 209–215.

Burgueno-Tapia, E., K. Chavez-Castellanos, E. Cedillo-Portugal and P. Joseph-Nathan (2017). "Absolute configuration of diterpenoids from Jatropha dioica by vibrational circular dichroism." Tetrahedron-Asymmetry **28**(1): 166–174.

Buric, M. P., J. Falk, S. Woodruff and B. Chorpening (2017). Gas Phase Raman Scattering: Methods and Applications in the Energy Industry. Encyclopedia of Spectroscopy and Spectrometry (Third Edition). J. C. Lindon, G. E. Tranter and D. W. Koppenaal. Oxford, Academic Press: 8–17.

Bustos-Brito, C., M. Sanchez-Castellanos, B. Esquivel, J. S. Calderon, F. Calzada, L. Yepez-Mulia, P. Joseph-Nathan, G. Cuevas and L. Quijano (2015). "ent-Kaurene Glycosides from Ageratina cylindrica." Journal of Natural Products **78**(11): 2580–2587.

Cai, S., A. L. Risinger, C. L. Petersen, T. Grkovic, B. R. O'Keefe, S. L. Mooberry and R. H. Cichewicz (2019). "Anacolosins A-F and Corymbulosins X and Y, clerodane diterpenes from Anacolosa clarkii exhibiting cytotoxicity toward pediatric cancer cell lines." Journal of Natural Products **82**(4): 928–936.

Caldow, G. L. and H. W. Thompson (1958). "Vibrational bands of isothiocyanates, thiocyanates and isocyanates." Spectrochimica Acta **13**(3): 212–216.

Califano, S. (1997). Vibrational States, John Wiley & Sons.

Calisto, I. H., M. Furlan, E. W. Blanch and J. M. Batista (2017). "Raman optical activity of a flavone C-diglycoside: Aqueous solution conformations and absolute configuration." Vibrational Spectroscopy **91**: 136–140.

Candeias, A. and J. M. Madariaga (2019). "Applications of Raman spectroscopy in art and archaeology." Journal of Raman Spectroscopy **50**(2): 137–142.

Canzi, E. F., F. M. dos Santos, E. K. Meneghetti, B. H. L. N. S. Maia and J. M. Batista (2018). "Absolute configuration of a rare dibenzoylmethane derivative from Dahlstedtia glaziovii (Fabaceae)." Tetrahedron Letters **59**(2): 135–137.

Cao, F., Z. H. Meng, X. Mu, Y. F. Yue and H. J. Zhu (2019). "Absolute configuration of bioactive azaphilones from the marine derived fungus Pleosporales sp. CF09-1." Journal of Natural Products **82**(2): 386–392.

Cao, F., C. L. Shao, Y. F. Liu, H. J. Zhu and C. Y. Wang (2017). "Cytotoxic serrulatane-type diterpenoids from the Gorgonian Euplexaura sp and their absolute configurations by vibrational circular dichroism." Scientific Reports **7**: 12548.

Cao, F., T. T. Sun, J. K. Yang, G. Z. Zhao, Q. A. Liu, L. D. Hu, Z. Y. Ma and H. J. Zhu (2019). "The absolute configuration of anti-Vibrio citrinin dimeric derivative by VCD, ECD and NMR methods." Natural Product Research **33**(15): 2192–2199.

Cao, X. and G. Fischer (1999). "Infrared spectral, structural, and conformational studies of zwitterionic l-tryptophan." The Journal of Physical Chemistry A **103**(48): 9995–10003.

Caro, P. E., J. O. Sawyer and L. Eyring (1972). "Infrared-spectra of rare-earth carbonates." Spectrochimica Acta Part A-Molecular Spectroscopy A **28**(6): 1167–1173.

Casacio, C. A., L. S. Madsen, A. Terrasson, M. Waleed, K. Barnscheidt, B. Hage, M. A. Taylor and W. P. Bowen (2021). "Quantum-enhanced nonlinear microscopy." Nature **594**(7862): 201–206.

Casero, C., F. Machín, S. Méndez-Álvarez, M. Demo, Á. G. Ravelo, N. Pérez-Hernández, P. Joseph-Nathan and A. Estévez-Braun (2015). "Structure and antimicrobial activity of phloroglucinol derivatives from Achyrocline satureioides." Journal of Natural Products **78**(1): 93–102.

Castellanos-Gomez, A., V. Leonardo, P. Elsa, O. I. Joshua, K. L. Narasimha-Acharya, I. B. Sofya, J. G. Dirk, B. Michele, A. S. Gary, J. V. Alvarez, W. Z. Henny, J. J. Palacios and S. J. v. d. Z. Herre (2014). "Isolation and characterization of few-layer black phosphorus." 2D Materials **1**(2): 025001.

Cataliotti, R., M. G. Giorgini, G. Paliani and A. Poletti (1975). "The vibrational spectrum of crystalline cyclobutanone." Spectrochimica Acta Part A: Molecular Spectroscopy **31**(12): 1879–1884.

CC 3.0, C. C. "http://creativecommons.org/licenses/by/3.0/."

CC 4.0, C. C. "https://creativecommons.org/licenses/by/4.0/."

Ccana-Ccapatinta, G. V., B. L. Sampaio, F. M. dos Santos, J. M. Batista and F. B. Da Costa (2017). "Absolute configuration assignment of caffeic acid ester derivatives from Tithonia diversifolia by vibrational circular dichroism: the pitfalls of deuteration." Tetrahedron-Asymmetry **28**(12): 1823–1828.

Ceccarelli, M., M. Lutz and M. Marchi (2000). "A Density Functional Normal Mode Calculation of a Bacteriochlorophyll a Derivative." Journal of the American Chemical Society **122**(14): 3532–3533.

Cerda-Garcia-Rojas, C. M., M. A. Bucio, S. B. Gonzalez, H. A. Garcia-Gutierrez and P. Joseph-Nathan (2015). "Absolute configuration of esquelane derivatives from Adesmia boronioides by vibrational circular dichroism." Tetrahedron-Asymmetry **26**(2–3): 136–140.

Chain, F., E. Romano, P. Leyton, C. Paipa, C. A. N. Catalan, M. A. Fortuna and S. A. Brandan (2014). "An experimental study of the structural and vibrational properties of sesquiterpene lactone cnicin using FT-IR, FT-Raman, UV-visible and NMR spectroscopies." Journal of Molecular Structure **1065**: 160–169.

Chalmers, J. M., H. G. M. Edwards and M. D. Hargreaves (2012). Infrared and Raman Spectroscopy in Forensic Science. Chichester, John Wiley & Sons, Ltd.

Chalmers, J. M. and P. R. Griffths (2002). Handbook of vibrational spectroscopy, John Wiley & Sons, Ltd.

Chase, B. (1987). "Fourier-Transform Raman-Spectroscopy." Analytical Chemistry **59**(14): A881–&.

Chatterjee, S., S. Olsen, E. W. Blanch and F. Wang (2018). "Resveratrol's Hidden Hand: A Route to the Optical Detection of Biomolecular Binding." Journal of Physical Chemistry B **122**(11): 2841–2850.

Chen, C., N. Hayazawa and S. Kawata (2014). "A 1.7 nm resolution chemical analysis of carbon nanotubes by tip-enhanced Raman imaging in the ambient." Nature Communications **5**(1): 3312.

Chen, M., H. Zeng, A. Larkum and Z.-L. Cai (2004). "Raman properties of chlorophyll d, the major pigment of Acaryochloris marina: Studies using both Raman spectroscopy and density functional theory." Spectrochimica Acta. Part A, Molecular and Biomolecular Spectroscopy **60**: 527–534.

Chen, X., M. Liu, H. Yuan, S. Huang, J. Tao and J. Zhao (2018). "Analysis of Diethylstilbestrol Residues in Chicken Using Surface-Enhanced Raman Spectroscopy (SERS) Coupled with Multivariate Analysis." Applied Spectroscopy **72**(12): 1798–1806.

Chiang, N. H., X. Chen, G. Goubert, D. V. Chulhai, X. Chen, E. A. Pozzi, N. Jiang, M. C. Hersam, T. Seideman, L. Jensen and R. P. Van Duyne (2016). "Conformational Contrast of Surface-Mediated Molecular Switches Yields Angstrom-Scale Spatial Resolution in Ultrahigh Vacuum Tip-Enhanced Raman Spectroscopy." Nano Letters **16**(12): 7774–7778.

Choe, C., J. Lademann and M. E. Darvin (2016). "Lipid organization and stratum corneum thickness determined *in vivo* in human skin analyzing lipid-keratin peak (2820-3030 cm^{-1}) using confocal Raman microscopy." Journal of Raman Spectroscopy **47**(11): 1327–1331.

Choi, S., G. B. Jung, K. S. Kim, G. J. Lee and H. K. Park (2014). "Medical Applications of Atomic Force Microscopy and Raman Spectroscopy." Journal of Nanoscience and Nanotechnology **14**(1): 71–97.

Christe, K. O. (1971). "Vibrational Spectra of Dimethyl Peroxides." Spectrochimica Acta Part A – Molecular Spectroscopy A 27(3): 463–472.

Cinta, S., C. Morari, E. Vogel, D. Maniu, T. Iliescu, O. Cozar and W. Kiefer (1999). "Vibrational studies of B6 vitamin." Vibrational Spectroscopy **19**(2): 329–334.

Claassen, H., C. L. Chernick and J. G. Malm (1963). "Vibrational Spectra and Structure of Xenon Tetrafluoride." Journal of the American Chemical Society **85**(13): 1927–1928.

Claassen, H. H. and J. L. Huston (1971). "Matrix-Isolation Raman and Infrared Spectra, Molecular Structure, and Force Constants for XeO_3F_2." Journal of Chemical Physics **55**(4): 1505–1507.

Clark, R. J. H., O. H. Ellestad and R. Escribano (1976). "The vapour phase Raman spectra, Raman band contour analyses, Coriolis coupling constants, and force constants for the molecules $F^{12}CH_3$, $F^{13}CH_3$, $H^{12}CCl_3$ and $H^{13}CCl_3$." Molecular Physics **31**(1): 65–81.

Clark, R. J. H. and D. M. Rippon (1972). "The vapor-phase Raman spectra, Raman band contour analyses, Coriolis constants, and force constants of spherical-top molecules MX_4 (M = Group IV element, X = F, Cl, Br, or I)." Journal of Molecular Spectroscopy **44**(3): 479–503.

Clarke, S. J., R. E. Littleford, W. E. Smith and R. Goodacre (2005). "Rapid monitoring of antibiotics using Raman and surface enhanced Raman spectroscopy." Analyst **130**(7): 1019–1026.

Clippard, P. H. and R. C. Taylor (1969). "Raman Spectra and Vibrational Assignments for Trimethylamine." Journal of Chemical Physics **50**(3): 1472–1473.

Cloutis, E., P. Szymanski, D. Applin and D. Goltz (2016). "Identification and discrimination of polycyclic aromatic hydrocarbons using Raman spectroscopy." Icarus **274**: 211–230.

Cole, A. R. H., R. N. Jones and R. C. Lord (1960). "TABLES OF WAVENUMBERS FOR THE CALIBRATION OF INFRARED SPECTROMETERS PARTS III AND IV: 600 – 1 cm^{-1}." Pure and Applied Chemistry **1**(4): 607 –651.

Colle, J. Y., M. Naji, M. Sierig and D. Manara (2017). "A novel technique for Raman analysis of highly radioactive samples using any standard micro-Raman spectrometer." Journal of Visualized Experiments **2017**(122): e54889.

Colthup, N. B., L. H. Daly and S. E. Wiberley (1990). Introduction to Infrared and Raman Spectroscopy. Boston, Academic Press.

Compton, R. and N. Hammer (2014). "Raman under liquid nitrogen (RUN)." Journal of Physics: Conference Series **548**: 012017.

Corbridge, D. E. C. (1969). The Infrared Spectra of Phosphorous Compounds. Topics in Phosphorous Chemistry. M. Grayson and E. J. Griffiths. New York Sydney Toronto, John Wiley and Sons. Vol. **6**: 235–265.

Cordero-Rivera, R. E., M. Melendez-Rodriguez, O. R. Suarez-Castillo, C. I. Bautista-Hernandez, N. Trejo-Carbajal, J. Cruz-Borbolla, L. E. Castelan-Duarte, M. S. Morales-Rios and P. Joseph-Nathan (2015). "Formal synthesis of (-)-flustramine B and its absolute configuration assignment by vibrational circular dichroism exciton chirality." Tetrahedron-Asymmetry **26**(14): 710–720.

Cornell, S. W. and J. L. Koenig (1970). "Raman Spectra of Polyisoprene Rubbers." Rubber Chemistry and Technology **43**(2): 313–321.

Costa, M., A. M. Arruda, L. Dias, R. Barbosa, J. Mirao and P. Vandenabeele (2019). "The combined use of Raman and micro-X-ray diffraction analysis in the study of archaeological glass beads." Journal of Raman Spectroscopy **50**(2): 250–261.

Costante, J., L. Hecht, P. L. Polavarapu, A. Collet and L. D. Barron (1997). "Absolute Configuration of Bromochlorofluoromethane from Experimental and *Ab Initio* Theoretical Vibrational Raman Optical Activity." Angewandte Chemie International Edition in English **36**(8): 885–887.

Cotton, F. A. (1971). Chemical Applications of Group Theory. New York, Wiley-Interscience.

Covington, C. L., F. M. S. Junior, J. H. S. Silva, R. M. Kuster, M. B. de Amorim and P. L. Polavarapu (2016). "Atropoisomerism in Biflavones: The Absolute Configuration of (−)-Agathisflavone via Chiroptical Spectroscopy." Journal of Natural Products **79**(10): 2530–2537.

Cramer, C. J. (2004). Essentials of Computational Chemistry: Theories and Models, John Wiley & Sons, Ltd.

Crocombe, R. A., P. E. Leary and B. W. Kammrath (2021). Portable Spectroscopy and Spectrometry, Volume 1, Technologies and Instrumentation, John Wiley & Sons.

Crocombe, R. A., P. E. Leary and B. W. Kammrath (2021). Portable Spectroscopy and Spectrometry, Volume 2, Applications, John Wiley & Sons.

Cross, A. D. (1964). Introduction to Practical Infrared Spectroscopy. London, Butterworth & Co. Ltd.

Crowder, G. A. (1987). "Vibrational Force-Field for 1-Butyne and Propionitrile." Spectroscopy Letters **20**(4): 343–350.

Culka, A. and J. Jehlička (2019). "A database of Raman spectra of precious gemstones and minerals used as cut gems obtained using portable sequentially shifted excitation Raman spectrometer." Journal of Raman Spectroscopy **50**(2): 262–280.

Cyvin, S. J., J. Brunvoll and L. Schäfer (1971). "Symmetry Coordinates, Harmonic Force Field, and Mean Amplitudes for Light and Heavy Dibenzene Chromium." Journal of Chemical Physics **54**(4): 1517–&.

Czamara, K., K. Majzner, M. Z. Pacia, K. Kochan, A. Kaczor and M. Baranska (2015). "Raman spectroscopy of lipids: a review." Journal of Raman Spectroscopy **46**(1): 4–20.

Dadieu, A. (1931). "Raman effect and constitution problems, II. Announcement: Cyanogen compounds." Berichte der Deutschen Chemischen Gesellschaft **64**: 358–361.

Dadieu, A., F. Jele and K. W. F. Kohlrausch (1931). "Studien zum Ramaneffekt." Monatshefte für Chemie und verwandte Teile anderer Wissenschaften **58**(1): 428–455.

Dahlenborg, H., A. Millqvist-Fureby, B. D. Brandner and B. Bergenstahl (2012). "Study of the porous structure of white chocolate by confocal Raman microscopy." European Journal of Lipid Science and Technology **114**(8): 919–926.

Dana, K., C. Shende, H. Huang and S. Farquharson (2015). "Rapid Analysis of Cocaine in Saliva by Surface-Enhanced Raman Spectroscopy." Journal of analytical & bioanalytical techniques **6**(6): 1–5.

Das, L., G. Dey and A. Chakraborty (2014). "Investigation of the structures, potential energy surface, transition states and vibrational frequencies of a vitamin E precursor-chroman in S0 and S1 states: DFT based computational study." Computational and Theoretical Chemistry **1049**: 115–121.

Das, R. S. and Y. K. Agrawal (2011). "Raman spectroscopy: Recent advancements, techniques and applications." Vibrational Spectroscopy **57**(2): 163–176.

Datta, S. and K. Kumar (2005). "Non-coincidence effect in methyl ethyl ketone: a solvent-dependent Raman study." Journal of Raman Spectroscopy **36**(1): 50–55.

Dazzi, A. and C. B. Prater (2017). "AFM-IR: Technology and Applications in Nanoscale Infrared Spectroscopy and Chemical Imaging." Chemical Reviews **117**(7): 5146–5173.

Dazzi, A., C. B. Prater, Q. Hu, D. B. Chase, J. F. Rabolt and C. Marcott (2012). "AFM-IR: Combining Atomic Force Microscopy and Infrared Spectroscopy for Nanoscale Chemical Characterization." Applied Spectroscopy **66**(12): 1365–1384.

Deckert-Gaudig, T., D. Kurouski, M. A. B. Hedegaard, P. Singh, I. K. Lednev and V. Deckert (2016). "Spatially resolved spectroscopic differentiation of hydrophilic and hydrophobic domains on individual insulin amyloid fibrils." Scientific Reports **6**: 33575.

Deckert-Gaudig, T., A. Taguchi, S. Kawata and V. Deckert (2017). "Tip-enhanced Raman spectroscopy – from early developments to recent advances." Chemical Society Reviews **46**(13): 4077–4110.

del Río, R. E. and P. Joseph-Nathan (2021). "Vibrational Circular Dichroism Absolute Configuration of Natural Products From 2015 to 2019." Natural Product Communications **16**(3): 1–30.

Dellepiane, G. and G. Zerbi (1968). "Normal Coordinate Calculations as a Tool for Vibrational Assignments .1. Fundamental Vibrations of Simple Aliphatic Amines." Journal of Chemical Physics **48**(8): 3573–3583.

Dellepiane, G. and G. Zerbi (1968). "Spectroscopic Quantities from an Overlay Valence Force Field – Trimethylamine." Spectrochimica Acta Part A – Molecular Spectroscopy A **24**(12): 2151–2155.

Dellepiane, G. and G. Zerbi (1968). "Vibrational Mean Amplitudes of Methylamine and Dimethylamine." Zeitschrift für Naturforschung Part A – Astrophysik, Physik und Physikalische Chemie A 23(10): 1661–1664.

Dellepiane, G. and G. Zerbi (1969). "Comment on Molecular Force Field for Trimethylamine." Journal of Chemical Physics **51**(9): 4171–4172.

Demarque, D. P., D. R. Pinho, N. P. Lopes and C. Merten (2018). "Revisiting empirical rules for the determination of the absolute configuration of cascarosides and other (ox-)anthrones." Chirality **30**(4): 432–438.

Denisov, V., B. N. Mavrin and V. Podobedov (1987). "Hyper-Raman scattering by vibrational excitations in crystals, glasses and liquids." Physics Reports **151**: 1–92.

Dennis, T. J., J. P. Hare, H. W. Kroto, R. Taylor, D. R. M. Walton and P. J. Hendra (1991). "The Vibrational Raman-Spectra of C-60 and C-70." Spectrochimica Acta Part A – Molecular and Biomolecular Spectroscopy **47**(9-10): 1289–1292.

Denson, S. C., C. J. S. Pommier and M. B. Denton (2007). "The impact of array detectors on Raman spectroscopy." Journal of Chemical Education **84**(1): 67–74.

Depciuch, J., E. Kaznowska, K. Szmuc, I. Zawlik, M. Cholewa, P. Heraud and J. Cebulski (2016). "Comparing paraffined and deparaffinized breast cancer tissue samples and an analysis of Raman spectroscopy and infrared methods." Infrared Physics & Technology **76**: 217–226.

Depciuch, J., E. Kaznowska, I. Zawlik, R. Wojnarowska, M. Cholewa, P. Heraud and J. Cebulski (2016). "Application of Raman Spectroscopy and Infrared Spectroscopy in the Identification of Breast Cancer." Applied Spectroscopy **70**(2): 251–263.

Dereka, B., Q. Yu, N. H. C. Lewis, W. B. Carpenter, J. M. Bowman and A. Tokmakoff (2021). "Crossover from hydrogen to chemical bonding." Science **371**(6525): 160–164.

Devi, T. G., A. Das and K. Kumar (2004). "Raman anisotropic bandwidth study of C=O stretching vibration of methyl isobutyl ketone: role of van der Waals' volume of the interacting systems." Spectrochimica Acta Part a-Molecular and Biomolecular Spectroscopy **60**(1–2): 211–216.

Dick, L. A., A. J. Haes and R. P. Van Duyne (2000). "Distance and orientation dependence of heterogeneous electron transfer: A surface-enhanced resonance Raman scattering study of cytochrome c bound to carboxylic acid terminated alkanethiols adsorbed on silver electrodes." Journal of Physical Chemistry B **104**(49): 11752–11762.

Diem, M. (2015). Time-Resolved Methods in Vibrational Spectroscopy. Modern Vibrational Spectroscopy and Micro-Spectroscopy: 167–176.

Dieng, S. D. and J. P. M. Schelvis (2010). "Analysis of Measured and Calculated Raman Spectra of Indole, 3-Methylindole, and Tryptophan on the Basis of Observed and Predicted Isotope Shifts." The Journal of Physical Chemistry A **114**(40): 10897–10905.

Ding, S. S., C. C. Zhang, W. S. Shi, M. M. Liang, Q. Yang, H. J. Zhu and Y. Li (2016). "Absolute configuration of two novel compounds from the *Talaromyces aculeatus* using optical rotation, electronic circular dichroism and vibrational circular dichroism." Tetrahedron Letters **57**(1): 75–79.

Djaker, N., P.-F. Lenne, D. Marguet, A. Colonna, C. Hadjur and H. Rigneault (2007). "Coherent *anti*-Stokes Raman scattering microscopy (CARS): Instrumentation and applications." Nuclear Instruments and Methods in Physics Research Section A: Accelerators, Spectrometers, Detectors and Associated Equipment **571**(1–2): 177–181.

Dolenko, T. A., S. A. Burikov, S. A. Dolenko, A. O. Efitorov, I. V. Plastinin, V. I. Yuzhakov and S. V. Patsaeva (2015). "Raman Spectroscopy of Water-Ethanol Solutions: The Estimation of Hydrogen Bonding Energy and the Appearance of Clathrate-like Structures in Solutions." Journal of Physical Chemistry A **119**(44): 10806–10815.

Dollish, F. R., W. G. Fateley and F. F. Bentley (1973). Characteristic Raman Frequencies of Organic Compounds. New York, John Wiley & Sons.

Dong, K., Z. F. Rao, X. Y. Yang, J. C. Lin and P. X. Zhang (2011). "The Study of Several Aldehyde Molecules by Raman Spectroscopy." Spectroscopy and Spectral Analysis **31**(12): 3277–3280.

dos Santos, C. A. A. S. S., J. O. Carvalho, J. G. da Silva Filho, J. L. Rodrigues, R. J. C. Lima, G. S. Pinheiro, P. T. C. Freire and P. F. Façanha Filho (2018). "High-pressure Raman spectra and DFT calculations of l-tyrosine hydrochloride crystal." Physica B: Condensed Matter **531**: 35–44.

Dowling, J. M., P. G. Puranik, A. G. Meister and S. I. Miller (1957). "Vibrational Spectra, Potential Constants, and Calculated Thermodynamic Properties of *cis*-BrHC=CHBr and *trans*-BrHC=CHBr, and *cis*-BrDC=CDBr and *trans*-BrDC=CDBr." Journal of Chemical Physics **26**(2): 233–240.

Dresselhaus, M. S., G. Dresselhaus and M. Hofmann (2007). "The big picture of Raman scattering in carbon nanotubes." Vibrational Spectroscopy **45**(2): 71–81.

Dresselhaus, M. S., G. Dresselhaus, R. Saito and A. Jorio (2005). "Raman spectroscopy of carbon nanotubes." Physics Reports **409**(2): 47–99.

Dresselhaus, M. S., A. Jorio and R. Saito (2010). "Characterizing Graphene, Graphite, and Carbon Nanotubes by Raman Spectroscopy." Annual Review of Condensed Matter Physics **1**: 89–108.

Drosky, P., M. Sander, K. Nakata, H.-U. Siehl, K.-P. Zeller, S. Berger and D. Sicker (2014). "Die optischen Aufheller Fraxin und Aesculin." Chemie in unserer Zeit **48**(6): 450–459.

Drosky, P., H.-U. Siehl, K.-P. Zeller, J. Sieler, S. Berger and D. Sicker (2015). "Scharf, schärfer, Capsaicin!" Chemie in unserer Zeit **49**(2): 114–122.

Dudek, M., G. Zajac, A. Kaczor and M. Baranska (2017). "Resonance Raman optical activity of zeaxanthin aggregates." Journal of Raman Spectroscopy **48**(5): 673–679.

Dudley, R. J., S. F. Mason and R. D. Peacock (1972). "Infrared vibrational circular dichroism." Journal of the Chemical Society, Chemical Communications (19): 1084–1085.

Duncan, J. L., G. K. Speirs and D. C. Mckean (1972). "Infrared-Spectrum of $^{13}CH_3F$ and General Harmonic Force Field of Methyl Fluoride." Molecular Physics **24**(3): 553–565.

Duncan, M. D., J. Reintjes and T. J. Manuccia (1982). "Scanning coherent *anti*-Stokes Raman microscope." Optics Letters **7**(8): 350–352.

Durig, J. R., S. F. Bush and F. G. Baglin (1968). "Infrared and Raman Investigation of Condensed Phases of Methylamine and Its Deuterium Derivatives." Journal of Chemical Physics **49**(5): 2106–2117.

Durig, J. R. and W. H. Green (1968). "Vibrational Spectra and Structure of 4-Membered Ring Molecules 2-Chlorocyclobutanone and 2-Chloro-2 4 4-Trideuterocyclobutanone." Journal of Molecular Spectroscopy **27**(1–4): 95–109.

Durig, J. R., S. Y. Shen, C. Zheng and G. A. Guirgis (2003). "Infrared and Raman spectra, conformational stability, vibrational assignment, and *ab initio* calculations for isopropyl methyl ketone." Canadian Journal of Analytical Sciences and Spectroscopy **48**(1): 106–120.

Durig, J. R. and J. N. Willis (1969). "Vibrational Spectra and Structure of Small Ring Compounds Silacyclopentane." Journal of Molecular Spectroscopy **32**(2): 320–342.

Durig, J. R. and J. N. Willis (1970). "Spectra and Structure of Small Ring Compounds .19. Vibrational Analysis and Barrier to Pseudorotation of Germylcyclopentane." Journal of Chemical Physics **52**(12): 6108–6119.

Durig, J. R., J. N. Willis and W. H. Green (1971). "Spectra and Structure of Small Ring Compounds .20. Fluorocyclobutane." Journal of Chemical Physics **54**(4): 1547–1556.

Durig, J. R., W. Zhao, D. Lewis and T. S. Little (1988). "Conformational stability, barriers to internal rotation, vibrational assignment, and *ab initio* calculations of chloroacetyl fluoride." The Journal of Chemical Physics **89**(3): 1285–1296.

Durocher, G. (1969). "Effect of Matrix and Solute Concentration on Ultraviolet Spectra of Polymethylated Benzene Derivatives at 77 Degrees Kelvin .2. Hexamethylbenzene." Journal de Chimie Physique et de Physico-Chimie Biologique **66**(4): 637–641.

Durocher, G. (1969). "Raman and Infrared Spectra of a Series of Polymethylbenzenes at Frequencies 150–1600 cm^{-1}." Journal De Chimie Physique Et De Physico-Chimie Biologique **66**(5): 988–990.

Eckhardt, K., K.-P. Zeller, H.-U. Siehl, S. Berger and D. Sicker (2016). "Ein Wirkstoff gegen Malaria aus der Natur: Artemisinin." Chemie in unserer Zeit **50**(5): 326–335.

Eckhardt, K., K.-P. Zeller, H.-U. Siehl, S. Berger and D. Sicker (2017). "Berberinchlorid aus Berberitzenrinde." Chemie in unserer Zeit **51**(5): 344–356.

Edwards, H. G. M. (2017). IR and Raman Spectroscopies, The Study of Art Works. Encyclopedia of Spectroscopy and Spectrometry (Third Edition). J. C. Lindon, G. E. Tranter and D. W. Koppenaal. Oxford, Academic Press: 378–393.

Edwards, H. G. M., D. A. Long and S. W. Webb (1988). "Line widths in the pure rotational Raman spectra of nitrogen, oxygen and hydrogen broadened by foreign gases." Journal of Raman Spectroscopy **19**: 457–461.

El-Diasty, F. (2011). "Coherent *anti*-Stokes Raman scattering: Spectroscopy and microscopy." Vibrational Spectroscopy **55**(1): 1–37.

El-Kaaby, E., Z. Al Hattab and A.-A. A (2016). "FT-IR Identification of Capsaicin from callus and seedling of chilli pepper plants Capsicum annuum L. *in vitro*." International Journal of Multidisciplinary and Current Research **4**: 1144–1146.

ElBindary and P. Klaeboe (1991). "Chloroactyl chloride." Acta Chemica Scandinavica **45**: 877–886.

Eliasson, C. and P. Matousek (2007). "Noninvasive authentication of pharmaceutical products through packaging using spatially offset Raman spectroscopy." Analytical Chemistry **79**(4): 1696–1701.

Esmonde-White, K. A., M. Cuellar, C. Uerpmann, B. Lenain and I. R. Lewis (2017). "Raman spectroscopy as a process analytical technology for pharmaceutical manufacturing and bioprocessing." Analytical and Bioanalytical Chemistry **409**(3): 637–649.

Esquivel, B., E. Burgueno-Tapia, C. Bustos-Brito, N. Perez-Hernandez, L. Quijano and P. Joseph-Nathan (2018). "Absolute configuration of the diterpenoids icetexone and conacytone from Salvia ballotaeflora." Chirality **30**(2): 177–188.

Estalayo, E., J. Aramendia, J. M. Matés Luque and J. M. Madariaga (2019). "Chemical study of degradation processes in ancient metallic materials rescued from underwater medium." Journal of Raman Spectroscopy **50**(2): 289–298.

Eyring, M. B. and P. Martin (2013). Spectroscopy in Forensic Science. Reference Module in Chemistry, Molecular Sciences and Chemical Engineering, Elsevier.

Fabrian, J., M. Legrand and P. Poirier (1956). Bulletin de la Société chimique de France **43B**: 364.

Fagan, P., L. Kocourkova, M. Tatarkovic, F. Kralik, M. Kuchar, V. Setnicka and P. Bour (2017). "Cocaine Hydrochloride Structure in Solution Revealed by Three Chiroptical Methods." Chemphyschem **18**(16): 2258–2265.

Fan, Y., D. Jin, X. Wu, H. Fang and X. Yuan (2020). "Facilitating Hotspot Alignment in Tip-Enhanced Raman Spectroscopy via the Silver Photoluminescence of the Probe." Sensors **20**(22): 6687.

Feher, F. and W. Kruse (1958). "Beiträge zur Chemie des Schwefels .53. Über unsymmetrische Sulfanderivate des Typs RSNCl und RSNH." Chemische Berichte-Recueil **91**(11): 2528–2531.

Feldman, T., J. Romanko and H. L. Welsh (1956). "The Raman Spectrum of Ethylene." Canadian Journal of Physics **34**(8): 737–744.

Feldman, T., G. G. Shepherd and H. L. Welsh (1956). "The Raman Spectrum of Acetylene." Canadian Journal of Physics **34**(12): 1425–1430.

Fendel, S. and B. Schrader (1998). "Investigation of skin and skin lesions by NIR-FT-Raman spectroscopy." Fresenius' Journal of Analytical Chemistry **360**(5): 609–613.

Fenlon, P. F., F. F. Cleveland and A. G. Meister (1951). "Substituted Methanes .7. Vibrational Spectra, Force Constants, and Calculated Thermodynamic Properties for CH_3I and CD_3I." Journal of Chemical Physics **19**(12): 1561–1565.

Ferrari, A. C. and D. M. Basko (2013). "Raman spectroscopy as a versatile tool for studying the properties of graphene." Nature Nanotechnology **8**: 235.

Fleischmann, M., P. J. Hendra and A. J. McQuillan (1974). "Raman spectra of pyridine adsorbed at a silver electrode." Chemical Physics Letters **26**(2): 163–166.

Fleming, G. D., I. Golsio, A. Aracena, F. Celis, L. Vera, R. Koch and M. Campos-Vallette (2008a). "Theoretical surface-enhanced Raman spectra study of substituted benzenes I. Density functional theoretical SERS modelling of benzene and benzonitrile." Spectrochimica Acta Part A – Molecular and Biomolecular Spectroscopy **71**(3): 1049–1055.

Fleming, G. D., I. Golsio, A. Aracena, F. Celis, L. Vera, R. Koch and M. Campos-Vallette (2008b). "Theoretical surface-enhanced Raman spectra study of substituted benzenes II. Density functional theoretical SERS modelling of o-, m-, and p-methoxybenzonitrile." Spectrochimica Acta Part A – Molecular and Biomolecular Spectroscopy **71**(3): 1074–1079.

Flett, M. S. C., W. T. Cave, E. E. Vago and H. W. Thompson (1947). "Infra-Red and Raman Spectrum of Cyclo-octatetraene." Nature **159**: 739–739.

Fliegl, H., A. Köhn, C. Hättig and R. Ahlrichs (2003). "*Ab Initio* Calculation of the Vibrational and Electronic Spectra of *trans*- and *cis*-Azobenzene." Journal of the American Chemical Society **125**(32): 9821–9827.

Ford, T. A. (2013). "The vibrational spectra of the boron halides and their molecular complexes. Part 14. *Ab initio* studies of the boron trifluoride-nitrous acid complex: research article." South African Journal of Chemistry **66**(1): 221–230.

Foti, A., F. Barreca, E. Fazio, C. D'Andrea, P. Matteini, O. M. Maragò and P. G. Gucciardi (2018). "Low cost tips for tip-enhanced Raman spectroscopy fabricated by two-step electrochemical etching of 125 μm diameter gold wires." Beilstein Journal of Nanotechnology **9**: 2718–2729.

Foti, A., S. Venkatesan, B. Lebental, G. Zucchi and R. Ossikovski (2022). "Comparing Commercial Metal-Coated AFM Tips and Home-Made Bulk Gold Tips for Tip-Enhanced Raman Spectroscopy of Polymer Functionalized Multiwalled Carbon Nanotubes." Nanomaterials **12**(3): 451.

Frederick, K. A., R. Pertaub and N. W. S. Kam (2004). "Identification of individual drug crystals on paper currency using Raman microspectroscopy." Spectroscopy Letters **37**(3): 301–310.

Freeman, S. K. (1965). Interpretive Spectroscopy. New York, Reinhold Publishing Corporation.

Frei, K. and H. H. Günthard (1961). "Vibrational spectra and normal coordinate treatment of cyclobutanone and α, α, α′, α′ -d₄-cyclobutanone." Journal of Molecular Spectroscopy **5**(1): 218–235.

Freire, P. T. C., F. M. Barboza, J. A. Lima, F. E. A. Melo and J. M. Filho (2017). Raman Spectroscopy of Amino Acid Crystals, Raman Spectroscopy and Applications. Raman Spectroscopy and Applications. K. Maaz, IntechOpen.

Frisch, M. J., G. W. Trucks, H. B. Schlegel, G. E. Scuseria, M. A. Robb, J. R. Cheeseman, G. Scalmani, V. Barone, B. Mennucci, G. A. Petersson, H. Nakatsuji, M. Caricato, X. Li, H. P. Hratchian, A. F. Izmaylov, J. Bloino, G. Zheng, J. L. Sonnenberg, M. Hada, M. Ehara, K. Toyota, R. Fukuda, J. Hasegawa, M. Ishida, T. Nakajima, Y. Honda, O. Kitao, H. Nakai, T. Vreven, J. A. Montgomery Jr., J. E. Peralta, F. Ogliaro, M. J. Bearpark, J. Heyd, E. N. Brothers, K. N. Kudin, V. N. Staroverov, R. Kobayashi, J. Normand, K. Raghavachari, A. P. Rendell, J. C. Burant, S. S. Iyengar, J. Tomasi, M. Cossi, N. Rega, N. J. Millam, M. Klene, J. E. Knox, J. B. Cross, V. Bakken, C. Adamo, J. Jaramillo, R. Gomperts, R. E. Stratmann, O. Yazyev, A. J. Austin, R. Cammi, C. Pomelli, J. W. Ochterski, R. L. Martin, K. Morokuma, V. G. Zakrzewski, G. A. Voth, P. Salvador, J. J. Dannenberg, S. Dapprich, A. D. Daniels, Ö. Farkas, J. B. Foresman, J. V. Ortiz, J. Cioslowski and D. J. Fox (2009). Gaussian 09. Wallingford, CT, USA, Gaussian, Inc.

Frisch, M. J., G. W. Trucks, H. B. Schlegel, G. E. Scuseria, M. A. Robb, J. R. Cheeseman, G. Scalmani, V. Barone, G. A. Petersson, H. Nakatsuji, X. Li, M. Caricato, A. V. Marenich, J. Bloino, B. G. Janesko, R. Gomperts, B. Mennucci, H. P. Hratchian, J. V. Ortiz, A. F. Izmaylov, J. L. Sonnenberg, Williams, F. Ding, F. Lipparini, F. Egidi, J. Goings, B. Peng, A. Petrone, T. Henderson, D. Ranasinghe, V. G. Zakrzewski, J. Gao, N. Rega, G. Zheng, W. Liang, M. Hada, M. Ehara, K. Toyota, R. Fukuda, J. Hasegawa, M. Ishida, T. Nakajima, Y. Honda, O. Kitao, H. Nakai, T. Vreven, K. Throssell, J. A. Montgomery Jr., J. E. Peralta, F. Ogliaro, M. J. Bearpark, J. J. Heyd, E. N. Brothers, K. N. Kudin, V. N. Staroverov, T. A. Keith, R. Kobayashi, J. Normand, K. Raghavachari, A. P. Rendell, J. C. Burant, S. S. Iyengar, J. Tomasi, M. Cossi, J. M. Millam, M. Klene, C. Adamo, R. Cammi, J. W. Ochterski, R. L. Martin, K. Morokuma, O. Farkas, J. B. Foresman and D. J. Fox (2016). Gaussian 16 Rev. C.01. Wallingford, CT.

Frontera, A. (2020). "Noble Gas Bonding Interactions Involving Xenon Oxides and Fluorides." Molecules **25**(15): 3419.

Frunder, H., L. Matziol, H. Finsterholzl, A. Beckmann and H. W. Schrötter (1986). "CARS Spectrometer with CW Intracavity Excitation for High-Resolution Raman-Spectroscopy." Journal of Raman Spectroscopy **17**(1): 143–150.

Fuhrer, H., V. B. Kartha, P. J. Krueger, H. H. Mantsch and R. N. Jones (1972). "Normal modes and group frequencies. Conflict or compromise? In-depth vibrational analysis of cyclohexanone." Chemical Reviews **72**(5): 439–456.

Gall, M. J., P. J. Hendra, C. J. Peacock, M. E. A. Cudby and H. A. Willis (1972). "Laser-Raman spectrum of polyethylene: Part 1. Structure and analysis of the polymer." Polymer **13**(3): 104–108.

Gall, M. J., P. J. Hendra, O. J. Peacock, M. E. A. Cudby and H. A. Willis (1972). "The laser-Raman spectrum of polyethylene: The assignment of the spectrum to fundamental modes of vibration." Spectrochimica Acta Part A: Molecular Spectroscopy **28**(8): 1485–1496.

Gallagher, S. H., R. S. Armstrong, W. A. Clucas, P. A. Lay and C. A. Reed (1997). "Resonance Raman scattering from solutions of C-60." Journal of Physical Chemistry A **101**(16): 2960–2968.

Gao, Y. Y., H. U. Siehl, H. Petzold, D. Sicker, K. P. Zeller and S. Berger (2015). "Remedy for the Pest and mediterranean Spices About Rosmarin and Rosmarinic Acid." Chemie in unserer Zeit **49**(5): 302–311.

Gardiner, D. J., N. J. Lawrence and J. J. Turner (1971). "Some Vibrational Spectroscopic Studies on Oxygen Fluorides." Journal of the Chemical Society A – Inorganic Physical Theoretical **0**(2): 400–404.

Garfinkel, D. and J. T. Edsall (1958). "Raman Spectra of Amino Acids and Related Compounds .10. The Raman Spectra of Certain Peptides and of Lysozyme." Journal of the American Chemical Society **80**(15): 3818–3823.

Gayathri, R. and M. Arivazhagan (2012). "Experimental (FT-IR and FT-Raman) and theoretical (HF and DFT) investigation, NMR, NBO, electronic properties and frequency estimation analyses on 2,4,5-trichlorobenzene sulfonyl chloride." Spectrochimica Acta Part A – Molecular and Biomolecular Spectroscopy **97**: 311–325.

Geiseler, G., H. Kessler and J. Fruwert (1966). "Infrarot- und Ramanspektroskopische Untersuchungen an Homologen und Stellungsisomeren n-Alkanderivaten .5. Über die Schwingungsspektren der sekundären Nitroalkane." Berichte der Bunsen-Gesellschaft für Physikalische Chemie **70**(8): 918–920.

Genin, F., F. Quiles and A. Burneau (2001). "Infrared and Raman spectroscopic study of carboxylic acids in heavy water." Physical Chemistry Chemical Physics **3**(6): 932–942.

Gerace, E., F. Seganti, C. Luciano, T. Lombardo, D. Di Corcia, H. Teifel, M. Vincenti and A. Salomone (2019). "On-site identification of psychoactive drugs by portable Raman spectroscopy during drug-checking service in electronic music events." Drug and Alcohol Review **38**(1): 50–56.

Ghassemi, A. (2017) "Technical note: image sensor product selection." Hamamatsu Website.

Giorgini, M. G., G. Paliani and R. Cataliotti (1977). "Vibrational spectra and assignments for tetrahydrothiophene, tetrahydroselenophene and tetrahydrotellurophene." Spectrochimica Acta Part A: Molecular Spectroscopy **33**(12): 1083–1089.

Giovannini, T., M. Olszowka and C. Cappellit (2016). "Effective Fully Polarizable QM/MM Approach To Model Vibrational Circular Dichroism Spectra of Systems in Aqueous Solution." Journal of Chemical Theory and Computation **12**(11): 5483–5492.

Gocen, T., S. Haman Bayarı and M. Haluk Guven (2017). "Linoleic acid and its potassium and sodium salts: A combined experimental and theoretical study." Journal of Molecular Structure **1150**(Supplement C): 68–81.

Goleb, J. A., H. H. Claassen, M. H. Studier and E. H. Appelman (1972). "Infrared spectrum of matrix-isolated HOF." Spectrochimica Acta Part A: Molecular Spectroscopy **28**(1): 65–67.

Gómez, J. S. (1999). Nonlinear Raman Spectroscopy, Theory* A2 – Lindon, John C. Encyclopedia of Spectroscopy and Spectrometry (Second Edition). Oxford, Academic Press: 1935–1945.

Gomez, M., D. Reggio and M. Lazzari (2019). "Linseed oil as a model system for surface enhanced Raman spectroscopy detection of degradation products in artworks." Journal of Raman Spectroscopy **50**(2): 242–249.

Gomm, P. J. and I. J. Humphreys (1975). "Identification of Major Excipients in Illicit Tablets Using Infrared Spectroscopy." Journal of the Forensic Science Society **15**(4): 293–299.

Gong, N., X. He, M. Zhou, L. Q. He, L. M. Fan, W. Z. Song, C. L. Sun and Z. W. Li (2017). "Effects of Fermi resonance of v_1 and $2v_2$ on the Raman scattering of fundamental mode v_2 from liquid carbon disulfide." Materials Research Bulletin **85**: 104–108.

Goushcha, A. and B. Tabbert (2007). Optical Detectors. Springer Handbook of Lasers and Optics. F. Träger. New York, NY, Springer New York: 503–562.

Grasselli, J. G. and B. J. Bulkin (1991). Analytical Raman Spectroscopy. New York, John Wiley & Sons, Inc.

Gredy, B. (1935). "Application of the Raman spectography to the study of the acetylenic liaison." Annales De Chimie France **4**: 5–82.

Green, J. H. S. (1970). "Vibrational spectra of benzene derivatives – VI: *p*-Disubstituted compounds." Spectrochimica Acta Part A: Molecular Spectroscopy **26**(7): 1503–1513.

Green, J. H. S. (1970). "Vibrational spectra of benzene derivatives – VIII: *m*-Disubstituted compounds." Spectrochimica Acta Part A: Molecular Spectroscopy **26**(7): 1523–1533.

Green, J. H. S. (1970). "Vibrational Spectra of Benzene Derivatives .9. *o*-Disubstituted Compounds." Spectrochimica Acta Part A – Molecular Spectroscopy A 26(9): 1913–1923.

Green, J. H. S. and D. J. Harrison (1971). "Molecular Vibrations and Thermodynamic Functions of 2,2-Dichloropropane." Spectrochimica Acta Part A – Molecular Spectroscopy A **27**(7): 1217–1219.

Green, J. H. S., D. J. Harrison, W. Kynaston and D. W. Scott (1969). "Vibrational Spectrum of 2,2,5,5-Tetramethyl-3,4-Dithiahexane." Spectrochimica Acta Part A – Molecular Spectroscopy A **25**(7): 1313–1314.

Green, W. H., A. B. Harvey and J. A. Greenhouse (1971). "Spectroscopic Determination of Pseudorotation Barrier in Selenacyclopentane." Journal of Chemical Physics **54**(3): 850–856.

Griffiths, P. R. and J. A. de Haseth (2007). Fourier Transform Raman Spectrometry. Fourier Transform Infrared Spectrometry. Hoboken, New Jersey, John Wiley and Sons, Inc.: 375–391.

Gudi, G., A. Krahmer, I. Koudous, J. Strube and H. Schulz (2015). "Infrared and Raman spectroscopic methods for characterization of *Taxus baccata* L. – Improved taxane isolation by accelerated quality control and process surveillance." Talanta **143**: 42–49.

Guirgis, G. A., S. Bell, C. Zheng, P. Groner and J. R. Durig (2005). "Raman, infrared and far infrared spectra, *ab initio* calculations, r_0 structural parameters, and internal rotation of 3-methyl-1-butyne." Journal of Molecular Structure **733**(1): 167–179.

Günzler, H. and H.-U. Gremlich (2003). IR-Spektroskopie. Weinheim, VCH.

Haefele, T. and K. Paulus (2018). Confocal Raman Microscopy in Pharmaceutical Development. Confocal Raman Microscopy. J. Toporski, T. Dieing and O. O. Hollricher: 381–419.

Haesler, J., I. Schindelholz, E. Riguet, C. G. Bochet and W. Hug (2007). "Absolute configuration of chirally deuterated neopentane." Nature **446**(7135): 526–529.

Hagen, G. (1971). "Infrared and Raman Spectra of Chloral, and Normal Coordinate Analyses of Acetaldehyde, Fluoral, Chloral, Bromal." Acta Chemica Scandinavica. **25**: 813.

Hamaguchi, H. and K. Iwata (2017). Time-Resolved Raman Spectroscopy. Encyclopedia of Spectroscopy and Spectrometry (Third Edition). J. C. Lindon, G. E. Tranter and D. W. Koppenaal. Oxford, Academic Press: 463–468.

Hannah, R. W. and S. C. Pattacini (1973). Identifizierung von Drogen mit Hilfe ihrer Infrarot-Spektren. Überlingen, The Perkin-Elmer Corporation. **4**.

Haraguchi, S., M. Hara, T. Shingae, M. Kumauchi, W. D. Hoff and M. Unno (2015). "Experimental Detection of the Intrinsic Difference in Raman Optical Activity of a Photoreceptor Protein under Preresonance and Resonance Conditions." Angewandte Chemie International Edition **54**(39): 11555–11558.

Haress, N. G., F. Al-Omary, A. A. El-Emam, Y. S. Mary, C. Y. Panicker, A. A. Al-Saadi, J. A. War and C. Van Alsenoy (2015). "Spectroscopic investigation (FT-IR and FT-Raman), vibrational assignments, HOMO-LUMO analysis and molecular docking study of 2-(Adamantan-1-yl)-5-(4-nitrophenyl)-1,3,4-oxadiazole." Spectrochimica Acta Part A – Molecular and Biomolecular Spectroscopy **135**: 973–983.

Hargittai, I. (2009). "Neil Bartlett and the first noble-gas compound." Structural Chemistry **20**(6): 953.

Harmony, M. D. (1972). Introduction to Molecular Energies and Spectra, Holt, Rinehart & Winston of Canada.

Harney, R. C. and F. P. Milanovich (1976). "Direct Determination of αe for $^{16}O^{16}O$, $^{14}N^{14}N$, and $^{18}O^{18}O^{*}$." Spectroscopy Letters **9**(8): 513–522.

Harrand, M. and R. Lennuier (1946). "Exaltation of the intensity of certain bands in the Raman spectra emitted from solids with an absorption band near the excitation wavelength." Comptes Rendus de L'Academie Bulgare des Sciences **223**: 356–357.

Harris, W. C. and S. F. Bush (1972). "Vibrational-Spectra and Structure of Acetone Oxime and Acetone Oxime-O-*d*." Journal of Chemical Physics **56**(12): 6147–6155.

Hartmann, C., M. Elsner, R. Niessner and N. Ivleva (2019). "EXPRESS: Nondestructive Chemical Analysis of the Iron-Containing Protein Ferritin Using Raman Microspectroscopy." Applied Spectroscopy **74**(2):193–203.

Hartwig, B. and M. A. Suhm (2021). "Subtle hydrogen bonds: benchmarking with OH stretching fundamentals of vicinal diols in the gas phase." Physical Chemistry Chemical Physics **23**(38): 21623–21640.

Harvey, A. B., J. R. Durig and A. C. Morrissey (1969). "Vibrational Spectra and Structure of 4-Membered Ring Molecules .14. Vibrational Analysis and Ring Puckering Vibration of Trimethylene Selenide and Trimethylene Selenide-d_4." Journal of Chemical Physics **50**(11): 4949–4961.

Haurie, M. and A. Novak (1965). "Spectres de Vibration des Molecules CH_3COOH CH_3COOD CD_3COOH et CD_3COOD .2. Spectres Infrarouges et Raman des Dimeres." Journal de Chimie Physique **62**(2): 146–157.

Haurie, M. and A. Novak (1965). "Spectres de Vibration des Molecules CH_3COOH CH_3COOD CD_3COOH et CD_3COOD .3. Spectres Infrarouges des Cristaux." Spectrochimica Acta **21**(7): 1217–1228.

Haurie, M. and A. Novak (1965). "Spectres de Vibration des Molecules CH_3COOH CH_3COOD CD_3COOH et CD_3COOD .I. Spectres Infrarouges des Monomeres." Journal de Chimie Physique **62**(2): 137–145.

Haurie, M. and A. Novak (1966). "Spectres Infrarouges des Dimeres Mixtes (CH_3COOH – CH_3COOD) et (CD_3COOH – CD_3COOD)." Journal de Chimie Physique **63**(11–1): 1584.

Haurie, M. and A. Novak (1967). "Etude par Spectroscopie Infrarouge et Raman des Complexes de l'Acide Acetique Avec des Accepteurs de Proton." Journal de Chimie Physique et de Physico-Chimie Biologique **64**(4): 679.

Haurie, M. and A. Novak (1967). "Existence d'oligomeres dans l'acide Acetique a l'etat liquide pur – Etude par Spectroscopie Infrarouge et Raman." Comptes Rendus Hebdomadaires des Seances de l'Academie des Sciences Serie B **264**(9): 694.

He, J., A. G. Petrovic and P. L. Polavarapu (2004). "Determining the Conformer Populations of (R)-(+)-3-Methylcyclopentanone Using Vibrational Absorption, Vibrational Circular Dichroism, and Specific Rotation." The Journal of Physical Chemistry B **108**(52): 20451–20457.

Hébert, M., G. Bellavance and L. Barriault (2022). "Total Synthesis of Ginkgolide C and Formal Syntheses of Ginkgolides A and B." Journal of the American Chemical Society **144**(39): 17792–17796.

Hediger, H. J. (1971). Infrarotspektroskopie. Methoden der Chemischen Analyse. Frankfurt am Main, Akademische Verlagsgesellschaft. **11**: 219.

Hendra, P., C. Jones and W. G. (1991). Fourier-Transform Raman Spectroscopy. Chichester, Ellis Horwood.

Hendra, P. J., J. R. Mackenzie and P. Holliday (1969). "The laser-Raman spectrum of polyvinylidene chloride." Spectrochimica Acta Part A: Molecular Spectroscopy **25**(8): 1349–1354.

Hermann, A., S. E. Ulic, C. O. Della Védova, M. Lieb, H. G. Mack and H. Oberhammer (2000). "Vibrational spectra, gas phase structure and conformational properties of perfluorodimethyl trithiocarbonate, $(CF_3S)_2CS$." Journal of Molecular Structure **556**(1): 217–224.

Hermann, R. and M. J. Gordon (2016). "Subdiffraction-limited chemical imaging of patterned phthalocyanine films using tip-enhanced near-field optical microscopy." Journal of Raman Spectroscopy **47**(11): 1287–1292.

Hernandez, B., F. Pflüger, E. Lopez-Tobar, S. G. Kruglik, J. V. Garcia-Ramos, S. Sanchez-Cortes and M. Ghomi (2014). "Disulfide linkage Raman markers: a reconsideration attempt." Journal of Raman Spectroscopy **45**(8): 657–664.

Herschel, W. (1800). "Experiments on the Refrangibility of the Invisible Rays of the Sun." Philosophical Transactions of the Royal Society of London **90**: 284–293.

Herzberg, G. (1945). Infrared and Raman Spectra, Van Nostrand.

Herzfeld, N., C. K. Ingold and H. G. Poole (1946). "Structure of Benzene .21. The Inactive Fundamental Frequencies of Benzene, Hexadeuterobenzene, and the Partly Deuterated Benzenes." Journal of the Chemical Society(Apr): 316–333.

Herzog, K., E. Steger, P. Rosmus, S. Scheithauer and R. Mayer (1969). "Infrared and Raman Spectra of Dithioacetic Acid Methyl Ester and Dimethyltrithiocarbonate." Journal of Molecular Structure **3**(4–5): 339–350.

Heß, G., P. Haiss, D. Wistuba, H.-U. Siehl, S. Berger, D. Sicker and K.-P. Zeller (2016). "Vom Granatapfelbaum zum Cyclooctatetraen." Chemie in unserer Zeit **50**(1): 34–43.

Heyler, R. A., J. T. A. Carriere and F. Havermeyer (2013). "THz-Raman – Accessing molecular structure with Raman spectroscopy for enhanced chemical identification, analysis and monitoring." Proceedings of SPIE **8726**.

Hibben, J. H. (1932). "An investigation of intermediate compound formation by means of the Raman effect." Proceedings of the National Academy of Sciences of the United States of America **18**: 532–538.

Hidalgo, A. (1962). "Estudio de las Vibraciones CN en los Nitrilos." Anales de la Real Sociedad Espanola de Fisica y Quimica Seria A – Fisica A **58**(3–4): 71–73.

Hildebrandt, P., J. Matysik, B. Schrader, B. Scharf and M. Engelhard (1994). "Raman spectroscopic study of the blue copper protein halocyanin from Natronobacterium pharaonis." Biochemistry **33**(38): 11426–11431.

Hiramatsu, K., M. Okuno, H. Kano, P. Leproux, V. Couderc and H. O. Hamaguchi (2012). "Observation of Raman Optical Activity by Heterodyne-Detected Polarization-Resolved Coherent *Anti*-Stokes Raman Scattering." Physical Review Letters **109**(8): 083901.

Hirschfeld, T. and B. Chase (1986). "FT-Raman Spectroscopy – Development and Justification." Applied Spectroscopy **40**(2): 133–137.

Hirschmann, R. P., R. N. Kniseley and V. A. Fassel (1964). "The Infrared Spectra of Alkyl Thiocyanates." Spectrochimica Acta **20**(5): 809–817.

Hodges, C. M. and J. Akhavan (1990). "The Use of Fourier-Transform Raman-Spectroscopy in the Forensic Identification of Illicit Drugs and Explosives." Spectrochimica Acta Part A – Molecular and Biomolecular Spectroscopy **46**(2): 303–307.

Hoffmann, G. G. (1993). "FT-IR and FT-Raman Spectra of (+)-5,6,7,8-Tetrahydro-8-methylindan-1,5-dione." 9th International Conference on Fourier Transform Spectroscopy **2089**: 140–141.

Hoffmann, G. G. (1995). Vibrational Optical Activity (VOA). Infrared and Raman Spectroscopy: Methods and Applications. B. Schrader. Weinheim, VCH: 543–572.

Hoffmann, G. G. (2003). "Infrared, Raman and VCD spectra of (*S*)-(+)-carvone – Comparison of experimental and *ab initio* theoretical results." Journal of Molecular Structure **661**: 525–539.

Hoffmann, G. G. (2010). Raman Optical Activity, Small Molecule Applications A2 – Lindon, John C. Encyclopedia of Spectroscopy and Spectrometry (Second Edition). Oxford, Academic Press: 2378–2386.

Hoffmann, G. G. (2017–2018). DFT Calculational Study of the Vibrational Spectra of the Xylenes.

Hoffmann, G. G. (2019). Raman Spectroscopy, Volume I: Principles and Applications in Chemistry, Physics, Materials Science, and Biology. New York, Momentum Press.

Hoffmann, G. G. (2021). DFT Calculational Study of the Vibrational Spectra of a Small Single-walled Carbon Nanotube of D_2 Symmetry

Hoffmann, G. G. (2022). DFT Calculation of the Vibrational Modes of Benzene.

Hoffmann, G. G. (2022). DFT Calculational Study of the Vibrational Spectra of Small Molecules.

Hoffmann, G. G., O. A. Bârsan, L. G. J. van der Ven and G. de With (2017). "Tip-enhanced Raman mapping of single-walled carbon nanotube networks in conductive composite materials." Journal of Raman Spectroscopy **48**(2): 191–196.

Hoffmann, G. G. and H. J. Hochkamp (1992). "Application of a Digital Signal Processor to Fourier-Transform Infrared-Spectroscopy." Fresenius Journal of Analytical Chemistry **344**(4–5): 227–228.

Hoffmann, G. G. and H. J. Hochkamp (1992). Fourier-transform spectrometer for the measurement of vibrational circular dichroism (VCD) reaches down to the beginning of the far infrared spectral region. 8th International Conference on Fourier Transform Spectroscopy, Lübeck-Travemünde, SPIE **1575**:400.

Hoffmann, G. G., H. U. Menzebach, B. Oelichmann and B. Schrader (1992). "Combined Raman and Fluorescence Spectroscopy with the Same Compact CCD-Based Instrument." Applied Spectroscopy **46**(4): 568–570.

Hoffmann, G. G., B. Schrader and G. Snatzke (1987). "Photoelastic Modulator for the Mid-IR Range down to 33 μm with Inexpensive and Simple Control Electronics." Review of Scientific Instruments **58**(9): 1675–1677.

Hoffmann, G. G., G. Snatzke and B. Schrader (1987). A Dedicated Fourier-Transform Instrument for the Measurement of Vibrational Circular Dichroism. Forty-Second symposium on Molecular Spectroscopy. Ohio, The Ohio State University paper WH2.

Hoffmann, U., F. Pfeifer, S. Okretic, N. Volkl, M. Zahedi and H. W. Siesler (1993). "Rheooptical Fourier-Transform Infrared and Raman-Spectroscopy of Polymers." Applied Spectroscopy **47**(9): 1531–1539.

Hoppe, R., W. Dähne, H. Mattauch and K. Rödder (1962). "Fluorination of Xenon." Angewandte Chemie International Edition in English **1**(11): 599.

Hsu, C. S. (1974). "The Vibrational Spectrum of Allyl Mercaptan." Spectroscopy Letters **7**(9): 439–447.

Hsu, E. C. and G. Holzwarth (1973). "Vibrational circular dichroism observed in crystalline α-NiSO$_4$ · 6H$_2$O and α-ZnSeO$_4$ · 6H$_2$O between 1900 and 5000 cm^{-1}." The Journal of Chemical Physics **59**(9): 4678–4685.

Hu, N. Y., K. Lin, X. G. Zhou and S. L. Liu (2015). "Populations of Ethanol Conformers in Liquid CCl$_4$ and CS$_2$ by Raman Spectra in OH Stretching Region." Chinese Journal of Chemical Physics **28**(3): 245–252.

Hu, Q.-Q., Q. Wang, Y.-K. Luo, Y.-Y. Li, M.-X. Ma, F.-F. Xu, J. Yang and R. Wei (2020). "Measurement of the polarization- and orientation-dependent scalar, vector, and tensor light shifts in ^{87}Rb using the near-resonant, stimulated Raman spectroscopy." Journal of Raman Spectroscopy **51**(4): 660–668.

Hu, Y. R., J. K. Liang, A. S. Myerson and L. S. Taylor (2005). "Crystallization monitoring by Raman spectroscopy: Simultaneous measurement of desupersaturation profile and polymorphic form in flufenamic acid systems." Industrial & Engineering Chemistry Research **44**(5): 1233–1240.

Huang, Y., G. Rui, Q. Li, E. Allahyarov, R. Li, M. Fukuto, G.-J. Zhong, J.-Z. Xu, Z.-M. Li, P. L. Taylor and L. Zhu (2021). "Enhanced piezoelectricity from highly polarizable oriented amorphous fractions in biaxially oriented poly(vinylidene fluoride) with pure β crystals." Nature Communications **12**(1): 675.

Huber, K. P. and G. Herzberg Constants of Diatomic Molecules. New York, Van Nostrand Reinhold.

Huen, J., C. Weikusat, M. Bayer-Giraldi, I. Weikusat, L. Ringer and K. Lösche (2014). "Confocal Raman microscopy of frozen bread dough." Journal of Cereal Science **60**(3): 555–560.

Hummel, D. O. (1974). Polymer Spectroscopy. Weinheim, Verlag Chemie.

Huston, J. L. and H. H. Claassen (1970). "Raman Spectra and Force Constants for OsO$_4$ and XeO$_4$." Journal of Chemical Physics **52**(11): 5646–5648.

Hvoslef, J., P. Klaeboe, B. Pettersson, S. Svensson, J. Koskikallio and S. Kachi (1971). "Vibrational Spectroscopic Studies of L-Ascorbic Acid and Sodium Ascorbate." Acta Chemica Scandinavica **25**: 3043–3053.

Hyams, I. J., E. R. Lippincott and R. T. Bailey (1966). "The Raman and low frequency infrared spectra of C$_6$F$_5$Cl, C$_6$F$_6$Br and C$_6$F$_5$I." Spectrochimica Acta **22**(4): 695–702.

Igarashi, T., M. Hoshi, K. Nakamura, T. Kaharu and K.-i. Murata (2020). "Direct Observation of Bound Water on Cotton Surfaces by Atomic Force Microscopy and Atomic Force Microscopy–Infrared Spectroscopy." The Journal of Physical Chemistry C **124**(7): 4196–4201.

Imperio, E., E. Calo, L. Valli and G. Giancane (2015). "Spectral investigations on 1000 pound banknotes throughout Italian Republic." Vibrational Spectroscopy **79**: 52–58.

Imperio, E., G. Giancane and L. Valli (2015). "Spectral characterization of postage stamp printing inks by means of Raman spectroscopy." Analyst **140**(5): 1702–1710.

Ivanova, M. V., H. P. A. Mercier and G. J. Schrobilgen (2015). "[XeOXeOXe](2+), the Missing Oxide of Xenon (II); Synthesis, Raman Spectrum, and X-ray Crystal Structure of [XeOXeOXe][mu-F(ReO2F3)(2)](2)." Journal of the American Chemical Society **137**(41): 13398–13413.

Jackson, M., M. G. Sowa and H. H. Mantsch (1997). "Infrared spectroscopy: a new frontier in medicine." Biophysical Chemistry **68**(1–3): 109–125.

Jeanmaire, D. L. and R. P. Van Duyne (1977). "Surface raman spectroelectrochemistry: Part I. Heterocyclic, aromatic, and aliphatic amines adsorbed on the anodized silver electrode." Journal of Electroanalytical Chemistry and Interfacial Electrochemistry **84**(1): 1–20.

Jei Lu, F., D. A. Waldman and S. L. Hsu (1984). "A spectroscopic study to interpret the increased piezoelectric effect at high temperature in poly(vinylidene fluoride)." Journal of Polymer Science: Polymer Physics Edition **22**(5): 827–834.

Jenkins, T. E. and J. Lewis (1980). "A Raman study of adamantane ($C_{10}H_{16}$), diamantane ($C_{14}H_{20}$) and triamantane ($C_{18}H_{24}$) between 10 K and room temperatures." Spectrochimica Acta Part A: Molecular Spectroscopy **36**(3): 259–264.

Jensen, J. O. (2003). "Vibrational frequencies and structural determinations of fumaronitrile." Journal of Molecular Structure: THEOCHEM **631**(1–3): 231–240.

Jesson, J. P. and H. W. Thompson (1958). "Vibrational Band Intensities of the CN Group in Aliphatic Nitriles." Spectrochimica Acta **13**(3): 217–222.

Jestel, N. L. (2005). Process Raman Spectroscopy. Process analytical technology. K. A. Bakeev. Oxford, Blackwell Publishing Ltd: 133–163.

Jia, G. Q., S. Qiu, G. N. Li, J. Zhou, Z. C. Feng and C. Li (2009). "Alkali-hydrolysis of D-glucono-delta-lactone studied by chiral Raman and circular dichroism spectroscopies." Science in China Series B – Chemistry **52**(5): 552–558.

Jiang, N., E. T. Foley, J. M. Klingsporn, M. D. Sonntag, N. A. Valley, J. A. Dieringer, T. Seideman, G. C. Schatz, M. C. Hersam and R. P. Van Duyne (2012). "Observation of Multiple Vibrational Modes in Ultrahigh Vacuum Tip-Enhanced Raman Spectroscopy Combined with Molecular-Resolution Scanning Tunneling Microscopy (vol 12, pg 5061, 2012)." Nano Letters **12**(12): 6506–6506.

Jimenez-Romero, C., J. E. Rode and A. D. Rodriguez (2016). "Reassignment of the absolute configuration of plakinidone from the sponge consortium Plakortis halichondrioides-Xestospongia deweerdtae using a combination of synthesis and a chiroptical approach." Tetrahedron-Asymmetry **27**(9–10): 410–419.

Johannessen, C., L. Hecht and C. Merten (2011). "Comparative Study of Measured and Computed Raman Optical Activity of a Chiral Transition Metal Complex." Chemphyschem **12**(8): 1419–1421.

Johannessen, C., R. Pendrill, G. Widmalm, L. Hecht and L. D. Barron (2011). "Glycan Structure of a High-Mannose Glycoprotein from Raman Optical Activity." Angewandte Chemie-International Edition **50**(23): 5349–5351.

Johnson, J. L., D. S. Nair, S. M. Pillai, D. Johnson, Z. Kallingathodii, I. Ibnusaud and P. L. Polavarapu (2019). "Dissymmetry Factor Spectral Analysis Can Provide Useful Diastereomer Discrimination: Chiral Molecular Structure of an Analogue of (-)-Crispine A." Acs Omega **4**(4): 6154–6164.

Jones, R. N., C. L. Angell, T. Ito and R. J. D. Smith (1959). "The Carbonyl Stretching Bands in the Infrared Spectra of Unsaturated Lactones." Canadian Journal of Chemistry – Revue Canadienne de Chimie **37**(12): 2007–2022.

Jones, R. N. and K. Dobriner (1949). Infrared Spectrometry Applied to Steroid Structure and Metabolism. Vitamins & Hormones. R. S. Harris and K. V. Thimann, Academic Press. **7**: 293–363.

Jones, R. N., T. Ito and C. L. Angell (1957). "Die C=O-Valenzschwingung in ungesättigten Lactonen." Angewandte Chemie-International Edition **69**(20): 645–646.

Jones, R. N., V. Z. Williams, M. J. Whalen and K. Dobriner (1948). "Studies in Steroid Metabolism. IV. The Characterization of Carbonyl and Other Functional Groups in Steroids by Infrared Spectrometry1." Journal of the American Chemical Society **70**(6): 2024–2034.

Jones, W. J. (1972). "Use of a Fabry-Perot Interferometer for Study of Raman Spectra of Gases under High-Resolution." Contemporary Physics **13**(5): 419–439.

Jorio, A., R. Saito. (2021). "Raman spectroscopy for carbon nanotube applications." Journal of Applied Physics **129**(2): 021102.

Joyce, L. A., C. C. Nawrat, E. C. Sherer, M. Biba, A. Brunskill, G. E. Martin, R. D. Cohen and I. W. Davies (2018). "Beyond optical rotation: what's left is not always right in total synthesis." Chemical Science **9**(2): 415–424.

Julio, L. F., E. Burgueño-Tapia, C. E. Díaz, N. Pérez-Hernández, A. González-Coloma and P. Joseph-Nathan (2017). "Absolute configuration of the ocimene monoterpenoids from Artemisia absinthium." Chirality **29**(11): 716–725.

Junior, F. M. S., C. L. Covington, A. C. F. de Albuquerque, J. F. R. Lobo, R. M. Borges, M. B. de Amorim and P. L. Polavarapu (2015). "Absolute Configuration of (−)-Centratherin, a Sesquiterpenoid Lactone, Defined by Means of Chiroptical Spectroscopy." Journal of Natural Products **78**(11): 2617–2623.

Kahovec, L. and K. W. F. Kohlrausch (1937). "Study of the Raman effect LXXIV Sulphur bodies 5 (Cyanamide and Derivatives)." Zeitschrift für Physikalische Chemie-Abteilung B – Chemie der Elementarprozesse Aufbau der Materie **37**(5–6): 421–436.

Kalonakis, K., M. Orkoula and C. Kontoyannis (2016). "Analysis of bone Composition with Raman spectroscopy." ACADEMIA 1–6.

Käppler, A., F. Windrich, M. G. J. Löder, M. Malanin, D. Fischer, M. Labrenz, K.-J. Eichhorn and B. Voit (2015). "Identification of microplastics by FTIR and Raman microscopy: a novel silicon filter substrate opens the important spectral range below 1300 cm^{-1} for FTIR transmission measurements." Analytical and Bioanalytical Chemistry **407**(22): 6791–6801.

Karabacak, M., M. Cinar and M. Kurt (2008). "An experimental and theoretical study of molecular structure and vibrational spectra of 2-chloronicotinic acid by density functional theory and *ab initio* Hartree-Fock calculations." Journal of Molecular Structure **885**(1–3): 28–35.

Karabacak, M. and M. Kurt (2008). "Comparison of experimental and density functional study on the molecular structure, infrared and Raman spectra and vibrational assignments of 6-chloronicotinic acid." Spectrochimica Acta Part A – Molecular and Biomolecular Spectroscopy **71**(3): 876–883.

Kartha, V. B., H. H. Mantsch and R. N. Jones (1973). "The Vibrational Analysis of Cyclopentanone." Canadian Journal of Chemistry. **51**: 1749.

Kawano, S., T. Fujiwara and M. Iwamoto (1993). "Nondestructive determination of sugar content in satsuma mandarin using near infrared (NIR) transmittance." Journal of the Japanese Society for Horticultural Science **62**(2): 465–470.

Kawata, S., M. Ohtsu and M. Irie (2001). Near-Field Optics and Surface Plasmon Polaritons. Berlin, Springer.

Kawata, S., M. Ohtsu and M. Irie (2002). Nano-Optics. Berlin, Springer.

Kawata, S. and V. M. Shalaev (2007). Tip Enhancement. Amsterdam, Elsevier Science.

Kellerer, B., J. Brandmüller and H. H. Hacker (1971). "Structure of Azobenzene and Tolane in Solution – Raman Spectra of Azobenzene, Azobenzene-d_{10}, *para-*,*para'*-Azobenzene-d_2, Azobenzene-^{15}N=^{15}N and Tolane." Indian Journal of Pure & Applied Physics **9**(11): 903–909.

Kharintsev, S. S., G. G. Hoffmann, P. S. Dorozhkin and J. Loos (2007). "Atomic force and shear force based tip-enhanced Raman spectroscopy and imaging." Nanotechnology **18**(31): 315502.

Kharintsev, S. S., G. G. Hoffmann, A. I. Fishman and M. K. Salakhov (2013). "Plasmonic optical antenna design for performing tip-enhanced Raman spectroscopy and microscopy." Journal of Physics D: Applied Physics **46**(14): 145501.

Kida, T., Y. Hiejima and K.-h. Nitta (2015). "Rheo-optical Raman study of microscopic deformation in high-density polyethylene under hot drawing." Polymer Testing **44**: 30–36.

Kida, T., Y. Hiejima and K.-h. Nitta (2016). "Raman Spectroscopic Study of High-density Polyethylene during Tensile Deformation." International Journal of Experimental Spectroscopic Techniques **1**: 1.

Kiefer, J. (2017). "Simultaneous Acquisition of the Polarized and Depolarized Raman Signal with a Single Detector." Analytical Chemistry **89**(11): 5726–5729.

Kiefer, J., K. Noack, J. Bartelmess, C. Walter, H. Dörnenburg and A. Leipertz (2010). "Vibrational structure of the polyunsaturated fatty acids eicosapentaenoic acid and arachidonic acid studied by infrared spectroscopy." Journal of Molecular Structure **965**(1): 121–124.

Kiefer, W. (2007). "Recent Advances in linear and nonlinear Raman spectroscopy I." Journal of Raman Spectroscopy **38**(12): 1538–1553.

Kiefert, L. and S. Karampelas (2011). "Use of the Raman spectrometer in gemmological laboratories: Review." Spectrochimica Acta Part A: Molecular and Biomolecular Spectroscopy **80**(1): 119–124.

Kipkemboi, P. K., P. C. Kiprono and J. J. Sanga (2003). "Vibrational spectra of *t*-butyl alcohol, *t*-butylamine and *t*-butyl alcohol plus *t*-butylamine binary liquid mixtures." Bulletin of the Chemical Society of Ethiopia **17**(2): 211–218.

Kirkpatrick, R., T. Masiello, N. Jariyasopit, J. W. Nibler, A. Maki, T. A. Blake and A. Weber (2009). "High-resolution rovibrational study of the Coriolis-coupled v_{12} and v_{15} modes of [1.1.1]propellane." Journal of Molecular Spectroscopy **253**(1): 41–50.

Kirkpatrick, R., T. Masiello, N. Jariyasopit, A. Weber, J. W. Nibler, A. Maki, T. A. Blake and T. Hubler (2008). "High resolution infrared spectroscopy of [1.1.1]propellane." Journal of Molecular Spectroscopy **248**(2): 153–160.

Kirkpatrick, R., T. Masiello, M. Martin, J. W. Nibler, A. Maki, A. Weber and T. A. Blake (2012). "High-resolution infrared studies of the nu(10), nu(11), nu(14), and nu(18) levels of [1.1.1]propellane." Journal of Molecular Spectroscopy **281**: 51–62.

Kitaev, Y. P., B. I. Buzykin and T. V. Troepol'skaya (1970). "The Structure of Hydrazones." Russian Chemical Reviews. **39**: 441.

Klaboe, P. (1970). "The vibrational spectra of 2-chloro, 2-bromo, 2-iodo and 2-cyanopropane." Spectrochimica Acta Part A: Molecular Spectroscopy **26**(1): 87–108.

Klaproth, A., M. Najdanova, M. Minceva, D. Sicker, H.-U. Siehl, K.-P. Zeller and S. Berger (2016). "Chlorophyll." Chemie in unserer Zeit **50**(4): 260–274.

Klimov, E., G. G. Hoffmann, A. Gumenny and H. W. Siesler (2005). "Low-temperature FT-NIR spectroscopy of strain-induced orientation and crystallization in a poly(dimethylsiloxane) network." Macromolecular Rapid Communications **26**(13): 1093–1098.

Klisińska-Kopacz, A., B. Łydźba-Kopczyńska, M. Czarnecka, T. Koźlecki, J. del Hoyo Mélendez, A. Mendys, A. Kłosowska-Klechowska, M. Obarzanowski and P. Frączek (2019). "Raman spectroscopy as a powerful technique for the identification of polymers used in cast sculptures from museum collections." Journal of Raman Spectroscopy **50**(2): 213–221.

Klots, T. D. (1995). "Vibrational spectra of indene .4. Calibration, assignment, and ideal-gas thermodynamics." Spectrochimica Acta Part A – Molecular and Biomolecular Spectroscopy **51**(13): 2307–2324.

Ko, G. B. and J. S. Lee (2015). "Performance characterization of high quantum efficiency metal package photomultiplier tubes for time-of-flight and high-resolution PET applications." Medical Physics **42**(1): 510–520.

Kobayashi, Y., A. Ishigami and H. Ito (2022). "Relating Amorphous Structure to the Tear Strength of Polylactic Acid Films." Polymers **14**(10): 1965.

Koenig, J. L. (1972). Raman Spectroscopy of Biological Molecules: A Review. Journal of Polymer Science – Part D John Wiley & Sons. Inc. **6**: 59–177.

Koenig, J. L. and B. G. Frushour (1972). "Raman-Scattering of Chymotrypsinogen-a, Ribonuclease, and Ovalbumin in Aqueous-Solution and Solid-State." Biopolymers **11**(12): 2505–2520.

Kohlrausch, K. W. F. (1931). Der Smekal-Raman-Effekt. Berlin, Julius Springer.

Kohlrausch, K. W. F. (1938). Der Smekal-Raman-Effekt – Ergänzungsband. Berlin, Julius Springer.

Kohlrausch, K. W. F. (1943). Ramanspektren. Leipzig, Akad.Verl.Ges. Becker & Erler.

Kohlrausch, K. W. F., F. Köppl and A. Pongratz (1933). "Studies on the Raman effect. Announcement XXVI. The Raman spectrum of methyl and ethyl esters of mono-basic fatty acids." Zeitschrift für Physikalische Chemie-Abteilung B – Chemie der Elementarprozesse: Aufbau der Materie **22**(5–6): 359–372.

Kohlrausch, K. W. F. and A. W. Reitz (1940). "Studies on the Raman effect. Announcement III: Saturated heterocyclic compounds." Zeitschrift für Physikalische Chemie-Abteilung B – Chemie der Elementarprozesse: Aufbau der Materie **45**(4): 249–271.

Kohn, W. and L. J. Sham (1965). "Self-Consistent Equations Including Exchange and Correlation Effects." Physical Review **140**(4A): 1133–1138.

Kokaislová, A. and P. Matějka (2012). "Surface-enhanced vibrational spectroscopy of B vitamins: what is the effect of SERS-active metals used?" Analytical and Bioanalytical Chemistry **403**(4): 985–993.

Kovner, M. A. and A. M. Bogomolov (1958). "Vibrational Spectra of Aromatic Compounds .5. Calculation and Interpretation of the Vibrational Spectra of *meta*-Xylene." Optika I Spektroskopiya **4**(3): 301–308.

Kovner, M. A. and A. M. Bogomolov (1959). "Vibrational Spectra of Aromatic Compounds .6. Calculation and Interpretation of Vibrational Spectra of *ortho*-Xylene." Optika I Spektroskopiya **7**(6): 751–755.

Krafft, C., C. Cervellati, C. Paetz, B. Schneider and J. Popp (2012). "Distribution of Amygdalin in Apricot (Prunus armeniaca) Seeds Studied by Raman Microscopic Imaging." Applied Spectroscopy **66**(6): 644–649.

Krafft, C., L. Neudert, T. Simat and R. Salzer (2005). "Near infrared Raman spectra of human brain lipids." Spectrochimica Acta Part A: Molecular and Biomolecular Spectroscopy **61**(7): 1529–1535.

Krafft, C., E. Pigorsch, B. Weber, F. Ott, S. Brennecke, G. E. Krammer and R. Salzer (2007). "Determination of configurational isomers in cyclic polysulfides by Raman spectroscopy." Vibrational Spectroscopy **43**(1): 49–52.

Králík, F., P. Fagan, M. Kuchař and V. Setnička (2020). "Structure of heroin in a solution revealed by chiroptical spectroscopy." Chirality **32**(6): 854–865.

Krayev, A., S. Bashkirov, V. Gavrilyuk, V. Zhizhimontov, M. Chaigneau, M. Shanmugasundaram and A. E. Robinson (2016). "TERS at work: 2D materials, from graphene to 2D semiconductors." SPIE Proceedings **9925**, Nanoimaging and Nanospectroscopy IV: 99250A.

Krief, A., M. Dunkle, M. Bahar, P. Bultinck, W. Herrebout and P. Sandra (2015). "Elucidation of the absolute configuration of rhizopine by chiral supercritical fluid chromatography and vibrational circular dichroism." Journal of Separation Science **38**(14): 2545–2550.

Kriegsmann, H. (1955). "Ramanspektroskopische Untersuchungen zur Isomerie und Tautomerie der Organischen Derivate der Schwefligen Säure." Angewandte Chemie-International Edition **67**(17–8): 530.

Krishnakumar, V. and R. Mathammal (2009). "A joint FTIR, FT-Raman and scaled quantum mechanical study of 1,3-dibromo-2,4,5,6-tetrafluoro benzene (DTB) and 1,2,3,4,5-pentafluoro benzene(PB)." Journal of Raman Spectroscopy **40**(9): 1104–1109.

Křížová, J., P. Matějka, G. Budínová and K. Volka (1999). "Fourier-transform Raman spectroscopic study of surface of Norway spruce needles." Journal of Molecular Structure **480–481**: 547–550.

Kübler, R., W. Lüttke and S. Weckherlin (1960). "Infrarotspektroskopische Untersuchungen an Isotopen Stickstoffverbindungen .1. Die Lokalisierung der Valenzfrequenz der N=N-Doppelbindung." Zeitschrift für Elektrochemie **64**(5): 650–658.

Kumar, R., K. M. Gronhaug, N. K. Afseth, V. Isaksen, C. D. Davies, J. O. Drogset and M. B. Lilledahl (2015). "Optical investigation of osteoarthritic human cartilage (ICRS grade) by confocal Raman spectroscopy: a pilot study." Analytical and Bioanalytical Chemistry **407**(26): 8067–8077.

Kumar, V. R. and S. Umapathy (2016). "Solvent effects on the structure of the triplet excited state of xanthone: a time-resolved resonance Raman study." Journal of Raman Spectroscopy **47**(10): 1220–1230.

Kurouski, D., A. Dazzi, R. Zenobi and A. Centrone (2020). "Infrared and Raman chemical imaging and spectroscopy at the nanoscale." Chemical Society Reviews **49**(11): 3315–3347.

Kurouski, D., T. Deckert-Gaudig, V. Deckert and Igor K. Lednev (2014). "Surface Characterization of Insulin Protofilaments and Fibril Polymorphs Using Tip-Enhanced Raman Spectroscopy (TERS)." Biophysical Journal **106**(1): 263–271.

Kuzmany, H. (2009). Solid-State Spectroscopy. Berlin Heidelberg, Springer Verlag.

Laane, J. (1970). "Vibrational Spectra and Normal-Coordinate Analyses of Silacyclobutanes." Spectrochimica Acta Part A – Molecular Spectroscopy A **26**(3): 517–540.

Lancaster, J. E., R. F. Stamm and N. B. Colthup (1961). "The Vibrational Spectra of s-Triazine and s-Triazine-d_3." Spectrochimica Acta **17**(2): 155–165.

Landsberg, G. S. and L. I. Mandelstam (1928). "Eine neue Erscheinung bei der Lichtzerstreuung in Krystallen." Naturwissenschaften **16**: 557.

Landsberg, G. S. and L. I. Mandelstam (1928). "New phenomenon in scattering of light (preliminary report)." Journal of the Russian Physico-Chemical Society **60**: 335.

Landsberg, G. S. and L. I. Mandelstam (1928). "Über die Lichtzerstreuung in Kristallen." Zeitschrift für Physik **50**: 769.

Larkin, P. J., M. Dabros, B. Sarsfield, E. Chan, J. T. Carriere and B. C. Smith (2014). "Polymorph Characterization of Active Pharmaceutical Ingredients (APIs) Using Low-Frequency Raman Spectroscopy." Applied Spectroscopy **68**(7): 758–776.

Larkin, P. J., J. Wasylyk and M. Raglione (2015). "Application of Low- and Mid-Frequency Raman Spectroscopy to Characterize the Amorphous-Crystalline Transformation of Indomethacin." Applied Spectroscopy **69**(11): 1217–1228.

Leadbeater, R. (1950). "Sur les Spectres Raman et Infrarouges des Peroxydes Organiques – Recherches de la Frequence Caracteristique O-O." Comptes Rendus Hebdomadaires des Seances de l'Academie des Sciences **230**(9): 829–831.

Lecomte, J. and J. P. Mathieu (1941). "Raman spectra and infrared of some alkyl nitrates – Structure and modes of vibration of these compounds." Comptes Rendus Hebdomadaires Des Seances De L Academie Des Sciences **213**: 721–723.

Lee, C., W. Young and R. G. Parr (1988). "Development of the Colic-Salvetti correlation-energy formula into a functional of the electron density." Physical Review B **37**(2): 785.

Lee, J., K. T. Crampton, N. Tallarida and V. A. Apkarian (2019). "Visualizing vibrational normal modes of a single molecule with atomically confined light." Nature **568**(7750): 78–82.

Legge, E. J., K. R. Paton, M. Wywijas, G. McMahon, R. Pemberton, N. Kumar, A. P. A. Raju, C. P. Dawson, A. J. Strudwick, J. W. Bradley, V. Stolojan, S. R. P. Silva, S. A. Hodge, B. Brennan and A. J. Pollard (2020). "Determining the Level and Location of Functional Groups on Few-Layer Graphene and Their Effect on the Mechanical Properties of Nanocomposites." Acs Applied Materials & Interfaces **12**(11): 13481–13493.

Lehmann, K. K., J. Smolarek, O. S. Khalil and L. Goodman (1979). "Vibrational assignments for the Raman and the phosphorescence spectra of 9,10-anthraquinone and 9,10-anthraquinone-d8." The Journal of Physical Chemistry **83**(9): 1200–1205.

Leutzsch, M., A. Roth, K. Steinke, D. Sicker, H. U. Siehl, M. Schmid, K. P. Zeller and S. Berger (2013). "Eine Boswelliasäure aus Weihrauch." Chemie in unserer Zeit (Chiuz) **47**(6): 344–354.

Lewis, I. R. and H. G. M. Edwards (2001). Handbook of Raman Spectroscopy – From the Research Laboratory to the Process Line. New York, Basel, Marcel Dekker.

Lieber, E., C. N. R. Rao and J. Ramachandran (1959). "The Infrared Spectra of Organic Thiocyanates and Isothiocyanates." Spectrochimica Acta **13**(4): 296–299.

Lin-Vien, D., N. B. Colthup, W. G. Fateley and J. G. Grasselli (1991). Handbook of Infrared and Raman Characteristic Frequencies of Organic Molecules. New York, Academic Press, Inc.

Lindner, M., B. Schrader and L. Hecht (1995). "Raman optical activity of enantiomorphic single crystals." Journal of Raman Spectroscopy **26**(8–9): 877–882.

Lipiec, E., R. Sekine, J. Bielecki, W. M. Kwiatek and B. R. Wood (2014). "Molecular Characterization of DNA Double Strand Breaks with Tip-Enhanced Raman Scattering." Angewandte Chemie-International Edition **53**(1): 169–172.

Lippert, E. and H. Prigge (1963). "Der Einfluss der Ringgrösse auf die Schwingungsspektren cyclischer Äther, Thioäther und Imine." Berichte der Bunsen-Gesellschaft für Physikalische Chemie **67**(6): 554–560.

Lippincott, E. R., R. C. Lord and R. S. Mcdonald (1948). "The Infraed and Raman Spectra of Heavy Cyclooctatetraene." Journal of Chemical Physics **16**: 548–549.

Lippincott, E. R., R. C. Lord and R. S. Mcdonald (1950). "Structure of Cyclo-Octatetraene." Nature **166**(4214): 227–227.

Liu, H. B., G. Q. Chen, J. Gu, C. H. Zhu, X. S. Song and X. H. Zhang (2017). "Raman spectroscopic analysis and fast identification of several saturated monohydroxy alcohols." Spectroscopy Letters **50**(6): 347–351.

Liu, K. and F. Yu (2013). "Accurate wavelength calibration method using system parameters for grating spectrometers." Optical Engineering **52**(1): 013603.

Liu, P., X. Chen, H. Ye and L. Jensen (2019). "Resolving Molecular Structures with High-Resolution Tip-Enhanced Raman Scattering Images." ACS Nano **13**(8): 9342–9351.

Liu, Z., X. Wang, K. Dai, S. Jin, Z.-C. Zeng, M.-D. Zhuang, Z.-L. Yang, D.-Y. Wu, B. Ren and Z.-Q. Tian (2009). "Tip-enhanced Raman spectroscopy for investigating adsorbed nonresonant molecules on single-crystal surfaces: tip regeneration, probe molecule, and enhancement effect." Journal of Raman Spectroscopy **40**(10): 1400–1406.

Lodder, R. A., G. M. Hieftje, W. Moorehead, S. P. Robertson and P. Rand (1989). "Assessment of the feasibility of determination of cholesterol and other blood constituents by near-infrared reflectance analysis." Talanta **36**(1): 193–198.

Long, D. A. (2002). The Raman Effect. A unified treatment of the theory of Raman scattering by molecules. Chichester, Wiley.

Long, D. A., A. H. S. Matterson and L. A. Woodward (1954). "Raman Intensities of the Totally Symmetric Vibrations of Neopentane." Proceedings of the Royal Society of London Series A – Mathematical and Physical Sciences **224**(1156): 33–43.

López-Ramírez, M. R., J. F. Arenas, J. C. Otero and J. L. Castro (2004). "Surface-enhanced Raman scattering of D-penicillamine on silver colloids." Journal of Raman Spectroscopy **35**(5): 390–394.

López-Tobar, E., B. Hernandez, A. Chenal, Y. M. Coic, J. G. Santos, E. Mejia-Ospino, J. V. Garcia-Ramos, M. Ghomi and S. Sanchez-Cortes (2017). "Large size citrate-reduced gold colloids appear as optimal SERS substrates for cationic peptides." Journal of Raman Spectroscopy **48**(1): 30–37.

Lord, R. C. and J. Ocampo (1951). "The Raman Spectrum of Allene-d_4." Journal of Chemical Physics **19**(2): 260–261.

Lu, S., A. E. Russell and P. J. Hendra (1998). "The Raman spectra of high modulus polyethylene fibres by Raman microscopy." Journal of Materials Science **33**(19): 4721–4725.

Lucazeau, G. and A. Novak (1969). "Spectres de vibration des aldéhydes CCl_3CHO, CCl_3CDO et CBr_3CHO." Spectrochimica Acta Part A: Molecular Spectroscopy **25**(9): 1615–1629.

Lucazeau, G. and A. Novak (1970). "Spectres de vibration et structure du dichloroacétaldéhyde." Journal of Molecular Structure **5**(1): 85–99.

Lupinetti, C. and K. M. Gough (2002). "*Ab initio* analysis of Raman trace scattering intensities in alkenes and silanes." Journal of Raman Spectroscopy **33**(3): 147–154.

Luther, H., F. Lampe, J. Goubeau and B. W. Rodewald (1950). "Die Qualitative und Quantitative Analyse der Isomeren Hexachlorcyclohexane mit Hilfe der Raman-Spektroskopie." Zeitschrift für Naturforschung Section A – Journal of Physical Sciences **5**(1): 34–40.

Lutz, M. and W. Mäntele (1991). Vibrational spectroscopy of chlorophylls. Chlorophylls. H. Scheer. Boca Raton, CRC Press: 855–902.

Machalska, E., G. Zajac, A. J. Wierzba, J. Kapitán, T. Andruniów, M. Spiegel, D. Gryko, P. Bouř and M. Baranska (2021). "Recognition of the True and False Resonance Raman Optical Activity." Angewandte Chemie International Edition **60**(39): 21205–21210.

Mahadevan, D., S. Periandy, M. Karabacak and S. Ramalingam (2011). "FT-IR and FT-Raman, UV spectroscopic investigation of 1-bromo-3-fluorobenzene using DFT (B3LYP, B3PW91 and MPW91PW91) calculations." Spectrochimica Acta Part A – Molecular and Biomolecular Spectroscopy **82**(1): 481–492.

Mahadevan, D., S. Periandy and S. Ramalingam (2011). "Comparative vibrational analysis of 1,2-Dinitro benzene and 1-Fluoro-3-nitro benzene: A combined experimental (FT-IR and FT-Raman) and theoretical study (DFT/B3LYP/B3PW91)." Spectrochimica Acta Part A-Molecular and Biomolecular Spectroscopy **84**(1): 86–98.

Maiman, T. H. (1960). "Stimulated Optical Radiation in Ruby." Nature **187**(4736): 493–494.

Maki, A., A. Weber, J. W. Nibler, T. Masiello, T. A. Blake and R. Kirkpatrick (2010). "High resolution infrared spectroscopy of [111]propellane The region of the v(9) band." Journal of Molecular Spectroscopy **264**(1): 26–36.

Malard, L. M., L. Lafeta, R. S. Cunha, R. Nadas, A. Gadelha, L. G. Cançado and A. Jorio (2021). "Studying 2D materials with advanced Raman spectroscopy: CARS, SRS and TERS." Physical Chemistry Chemical Physics **23**(41): 23428–23444.

Mammone, J. F., S. K. Sharma and M. Nicol (1980). "Raman spectra of methanol and ethanol at pressures up to 100 kbar." The Journal of Physical Chemistry **84**(23): 3130–3134.

Mantsch, H. H. (1984). "Biological Applications of Fourier-Transform Infrared-Spectroscopy – a Study of Phase-Transitions in Biomembranes." Journal of Molecular Structure **113**(Mar): 201–212.

Mantsch, H. H. (1996). "Infrared spectroscopy: A new frontier in medicine." Progress in Biophysics & Molecular Biology **65**: Sh501-Sh501.

Mantsch, H. H. and R. N. Mcelhaney (1990). "Applications of Infrared-Spectroscopy to Biology and Medicine." Journal of Molecular Structure **217**: 347–362.

Marcaida, I., M. Maguregui, H. Morillas, M. Veneranda, S. Prieto-Taboada, S. Fdez-Ortiz de Vallejuelo and J. M. Madariaga (2019). "Raman microscopy as a tool to discriminate mineral phases of volcanic origin and contaminations on red and yellow ochre raw pigments from Pompeii." Journal of Raman Spectroscopy **50**(2): 143–149.

Marić-Stojanović, M., D. Bajuk-Bogdanovic, S. Uskokovic-Markovic and I. Holclajtner-Antunovic (2018). "Spectroscopic analysis of XIV century wall paintings from Patriarchate of Pec Monastery, Serbia." Spectrochimica Acta Part A – Molecular and Biomolecular Spectroscopy **191**: 469–477.

Marshall, P., G. N. Srinivas and M. Schwartz (2005). "A computational study of the thermochemistry of bromine- and iodine-containing methanes and methyl radicals." Journal of Physical Chemistry A **109**(28): 6371–6379.

Martin, D. H. and E. Puplett (1970). "Polarised interferometric spectrometry for the millimetre and submillimetre spectrum." Infrared Physics **10**(2): 105–109.

Martin, M. A., A. Perry, T. Masiello, K. D. Schwartz, J. W. Nibler, A. Weber, A. Maki and T. A. Blake (2010). "High-resolution infrared spectra of bicyclo[1.1.1]pentane." Journal of Molecular Spectroscopy **262**(1): 42–48.

Martinez, M. E. S., C. M. Crowder and X. H. Bi (2019). "The Use of Raman Spectroscopy to Examine Bone Composition in Infant Ribs." American Journal of Physical Anthropology **168**: 234–234.

Mashiko, Y. (1959). "Vibrational Spectra of Methyl- and Ethyl Ethers." Journal of the Chemical Society of Japan, Pure Chemistry Section **80**: 593.

Mashiko, Y., S. Saeki, K. Nukada, Y. Kanazawa and T. Suzuki (1962). International Symposium on Molecular Spectroscopy. Tokyo **A220**.

Masi, M., S. Meyer, M. Gorecki, G. Pescitelli, S. Clement, A. Cimmino and A. Evidente (2018). "Phytotoxic Activity of Metabolites Isolated from Rutstroemia sp.n., the Causal Agent of Bleach Blonde Syndrome on Cheatgrass (Bromus tectorum)." Molecules **23**(7): 1734.

Masiello, T., N. Vulpanovici, J. Barber, E. t. H. Chrysostom, J. W. Nibler, A. Maki, T. A. Blake, R. L. Sams and A. Weber (2004). "Analysis of high-resolution infrared and CARS spectra of $^{32}S_{18}O_3$." Journal of Molecular Spectroscopy **227**(1): 50–59.

Mathurin, J., E. Pancani, A. Deniset-Besseau, K. Kjoller, C. B. Prater, R. Gref and A. Dazzi (2018). "How to unravel the chemical structure and component localization of individual drug-loaded polymeric nanoparticles by using tapping AFM-IR." Analyst **143**(24): 5940–5949.

Matousek, P., I. P. Clark, E. R. Draper, M. D. Morris, A. E. Goodship, N. Everall, M. Towrie, W. F. Finney and A. W. Parker (2005). "Subsurface probing in diffusely scattering media using spatially offset Raman spectroscopy." Applied Spectroscopy **59**(4): 393–400.

Matousek, P., C. Conti, M. Realini and C. Colombo (2016). "Micro-scale spatially offset Raman spectroscopy for non-invasive subsurface analysis of turbid materials." Analyst **141**(3): 731–739.

Matsuo, Y., S. Maeda, C. Ohba, H. Fukaya and Y. Mimaki (2016). "Vetiverianines A, B, and C: Sesquiterpenoids from Vetiveria zizanioides Roots." Journal of Natural Products **79**(9): 2175–2180.

Matysik, J., P. Hildebrandt, K. Smit, F. Mark, W. Gartner, S. E. Braslavsky, K. Schaffner and B. Schrader (1997). "Raman spectroscopic analysis of isomers of biliverdin dimethyl ester." Journal of Pharmaceutical and Biomedical Analysis **15**(9–10): 1319–1324.

May, P., S. Ashworth, C. Pickard, M. Ashfold, T. Peakman and J. Steeds (1998). "Interactive Raman spectra of adamantane, diamantane and diamond, and their relevance to diamond film deposition." Physchemcomm **1**(4): 35–44.

Mazzeo, G., A. Cimmino, M. Masi, G. Longhi, L. Maddau, M. Memo, A. Evidente and S. Abbate (2017). "Importance and Difficulties in the Use of Chiroptical Methods to Assign the Absolute Configuration of Natural Products: The Case of Phytotoxic Pyrones and Furanones Produced by Diplodia corticola." Journal of Natural Products **80**(9): 2406–2415.

McCord, B. R., Q. Gauthier, S. Cho, M. N. Roig, G. C. Gibson-Daw, B. Young, F. Taglia, S. C. Zapico, R. F. Mariot, S. B. Lee and G. Duncan (2019). "Forensic DNA Analysis." Analytical Chemistry **91**(1): 673–688.

Mdluli, P. S., N. M. Sosibo, N. Revaprasadu, P. Karamanis and J. Leszczynski (2009). "Surface enhanced Raman spectroscopy (SERS) and density functional theory (DFT) study for understanding the regioselective adsorption of pyrrolidinone on the surface of silver and gold colloids." Journal of Molecular Structure **935**(1–3): 32–38.

Mecke, R., R. Mecke and A. Lüttringhaus (1957). "Spektroskopische Untersuchungen an Carbonylverbindungen und Thiocarbonylverbindungen." Chemische Berichte-Recueil **90**(6): 975–986.

Medeiros Neto, L. P., L. F. das Chagas e Silva de Carvalho, L. D. Santos, C. A. Tellez Soto, R. de Azevedo Canevari, A. B. de Oliveira Santos, E. S. Mello, M. A. Pereira, C. R. Cernea, L. G. Brandão and A. A. Martin (2017). "Micro-Raman spectroscopic study of thyroid tissues." Photodiagnosis and Photodynamic Therapy **17**: 164–172.

Meiklejohn, R. A., R. J. Meyer, S. M. Aronovic, H. A. Schuette and V. W. Meloche (1957). "Characterization of Long-Chain Fatty Acids by Infrared Spectroscopy." Analytical Chemistry **29**(3): 329–334.

Meister, A. G., S. E. Rosser and F. F. Cleveland (1950). "Substituted Methanes .1. Raman and Infra-Red Spectral Data, Assignments, and Force Constants for Some Tribromomethanes." Journal of Chemical Physics **18**(3): 346–354.

Melchor-Martinez, E. M., D. A. Silva-Mares, E. Torres-Lopez, N. Waksman-Minsky, G. F. Pauli, S. N. Chen, M. Niemitz, M. Sanchez-Castellanos, A. Toscana, G. Cuevas and V. M. Rivas-Galindo (2017). "Stereochemistry of a Second Riolozane and Other Diterpenoids from Jatropha dioica." Journal of Natural Products **80**(8): 2252–2262.

Melveger, A. J., L. R. Anderson, W. B. Fox and C. T. Ratcliffe (1972). "Identification of O-O Stretching Frequency in Some Fluoroperoxides by Raman Spectroscopy." Applied Spectroscopy **26**(3): 381–384.

Mendelovici, E., R. L. Frost and T. Kloprogge (2000). "Cryogenic Raman spectroscopy of glycerol." Journal of Raman Spectroscopy **31**(12): 1121–1126.

Menezes, D. B., A. Reyer, A. Marletta and M. Musso (2017). "Glass transition of polystyrene (PS) studied by Raman spectroscopic investigation of its phenyl functional groups." Materials Research Express **4**(1): 015303.

Menges, F., H. Yang, S. Berweger, A. Roy, T. Jiang and M. B. Raschke (2021). "Substrate-enhanced photothermal nano-imaging of surface polaritons in monolayer graphene." APL Photonics **6**(4): 041301.

Mensch, C., L. D. Barron and C. Johannessen (2016). "Ramachandran mapping of peptide conformation using a large database of computed Raman and Raman optical activity spectra." Physical Chemistry Chemical Physics **18**(46): 31757–31768.

Menzies, A. C. and R. Whiddington (1939). "Fine structure of the Raman lines of carbon tetrachloride." Proceedings of the Royal Society of London. Series A. Mathematical and Physical Sciences **172**(948): 89–94.

Merrick, J. P., D. Moran and L. Radom (2007). "An evaluation of harmonic vibrational frequency scale factors." Journal of Physical Chemistry A **111**(45): 11683–11700.

Merten, C., M. Dirkmann and F. Schulz (2017). "Stereochemical assignment of fusiccocadiene from NMR shielding constants and vibrational circular dichroism spectroscopy." Chirality **29**(8): 409–414.

Merten, C., V. Smyrniotopoulos and D. Tasdemir (2015). "Assignment of absolute configurations of highly flexible linear diterpenes from the brown alga Bifurcaria bifurcata by VCD spectroscopy." Chemical Communications **51**(90): 16217–16220.

Meyer, C., S. Hühn, M. Jungbauer, S. Merten, B. Damaschke, K. Samwer and V. Moshnyaga (2017). "Tip-enhanced Raman spectroscopy (TERS) on double perovskite La_2CoMnO_6 thin films: field enhancement and depolarization effects." Journal of Raman Spectroscopy **48**(1): 46–52.

Minaeva, V., B. Minaev and D. Hovorun (2008). "Vibrational spectra of the steroid hormones, estradiol and estriol, calculated by density functional theory. The role of low-frequency vibrations." Ukrainskiĭ biokhimicheskiĭ zhurnal **80**: 82–95.

Mircescu, N. E., M. Oltean, V. Chiş and N. Leopold (2012). "FTIR, FT-Raman, SERS and DFT study on melamine." Vibrational Spectroscopy **62**: 165–171.

Miyaoka, R., M. Hosokawa, M. Ando, T. Mori, H.-o. Hamaguchi and H. Takeyama (2014). "*In Situ* Detection of Antibiotic Amphotericin B Produced in Streptomyces nodosus Using Raman Microspectroscopy." Marine Drugs **12**(5): 2827–2839.

Miyazawa, T., T. Shimanouchi and S. I. Mizushima (1956). "Characteristic Infrared Bands of Monosubstituted Amides." Journal of Chemical Physics **24**(2): 408–418.

Monteiro, A. F., J. M. Batista, M. A. Machado, R. P. Severino, E. W. Blanch, V. S. Bolzani, P. C. Vieira and V. G. P. Severino (2015). "Structure and Absolute Configuration of Diterpenoids from Hymenaea stigonocarpa." Journal of Natural Products **78**(6): 1451–1455.

Moreno, J. R. A., F. P. Urena and J. J. L. Gonzalez (2013). "Hydrogen bonding network in a chiral alcohol: (1R,2S,5R)-(-)-menthol. Conformational preference studied by IR-Raman-VCD spectroscopies and quantum chemical calculations." Structural Chemistry **24**(2): 671–680.

Moretti, M., R. P. Zaccaria, E. Descrovi, G. Das, M. Leoncini, C. Liberale, F. De Angelis and E. Di Fabrizio (2013). "Reflection-mode TERS on Insulin Amyloid Fibrils with Top-Visual AFM Probes." Plasmonics **8**(1): 25–33.

Motoyama, M., M. Ando, K. Sasaki, I. Nakajima, K. Chikuni, K. Aikawa and H.-o. Hamaguchi (2016). "Simultaneous imaging of fat crystallinity and crystal polymorphic types by Raman microspectroscopy." Food Chemistry **196**: 411–417.

Mrđenović, D., Z.-F. Cai, Y. Pandey, G. L. Bartolomeo, R. Zenobi and N. Kumar (2023). "Nanoscale chemical analysis of 2D molecular materials using tip-enhanced Raman spectroscopy." Nanoscale **15**(3): 963–974.

Müller, M. and A. Zumbusch (2007). "Coherent *anti*-Stokes Raman scattering microscopy." Chemphyschem **8**(15): 2156–2170.

Munoz, M. A., O. Munoz and P. Joseph-Nathan (2006). "Absolute Configuration of Natural Diastereoisomers of 6ß-Hydroxyhyoscyamine by Vibrational Circular Dichroism." Journal of Natural Products **69**(9): 1335–1340.

Munoz, M. A., A. San-Martin and P. Joseph-Nathan (2015). "Vibrational Circular Dichroism Absolute Configuration of 9,12-Cyclomulin-13-ol, a Diterpene from Azorella and Laretia Species." Natural Product Communications **10**(8): 1343–1344.

Murray, M. J. and F. F. Cleveland (1938). "Raman spectra of acetylenes .I. Derivatives of phenylacetylene, $C_6H_5C\equiv CR$." Journal of the American Chemical Society **60**: 2664–2666.

Nabiev, S. S., V. B. Sokolov and B. B. Chaivanov (2014). "Molecular and crystal structures of noble gas compounds." Russian Chemical Reviews **83**(12): 1135–1180.

Nafie, L. A. (2011). Vibrational Optical Activity – Principles and Applications. Chichester, John Wiley & Sons, Ltd.

Nafie, L. A. (2015). "Recent advances in linear and non-linear Raman spectroscopy. Part IX." Journal of Raman Spectroscopy **46**(12): 1173–1190.

Nafie, L. A., J. C. Cheng and P. J. Stephens (1975). "Vibrational circular dichroism of 2,2,2-trifluoro-1-phenylethanol." Journal of the American Chemical Society **97**(13): 3842–3843.

Nafie, L. A., M. Diem and D. W. Vidrine (1979). "Fourier transform infrared vibrational circular dichroism." Journal of the American Chemical Society **101**(2): 496–498.

Nagy, B., A. Farkas, E. Borbás, P. Vass, Z. K. Nagy and G. Marosi (2018). "Raman Spectroscopy for Process Analytical Technologies of Pharmaceutical Secondary Manufacturing." Aaps Pharmscitech **20**(1): 1.

Naik, J. L., B. V. Reddy and N. Prabavathi (2015). "Experimental (FTIR and FT-Raman) and theoretical investigation of some pyridine-dicarboxylic acids." Journal of Molecular Structure **1100**: 43–58.

Najjar, S., D. Talaga, L. Schue, Y. Coffinier, S. Szunerits, R. Boukherroub, L. Servant, V. Rodriguez and S. Bonhommeau (2014). "Tip-Enhanced Raman Spectroscopy of Combed Double-Stranded DNA Bundles." Journal of Physical Chemistry C **118**(2): 1174–1181.

Nakahashi, A., A. K. C. Siddegowda, M. A. S. Hammam, S. G. B. Gowda, Y. Murai and K. Monde (2016). "Stereochemical Study of Sphingosine by Vibrational Circular Dichroism." Organic Letters **18**(10): 2327–2330.

Nakamoto, K. (1997). Infrared and Raman Spectra of Inorganic and Coordination Compounds. New York, John Wiley & Sons, Inc.

Narayanan, U., A. Annamalai and T. A. Keiderling (1988). "Vibrational spectra of 1,3-dideuterioallene and normal mode calculations for allene." Spectrochimica Acta Part A: Molecular Spectroscopy **44**(8): 785–791.

Narayanan, V. A., N. A. Stump, G. D. Del Cul and T. Vo-Dinh (1999). "Vibrational spectrum of strychnine: Detection at the nanogram level using a Raman microscope." Journal of Raman Spectroscopy **30**(6): 435–439.

Nava, V., M. L. Frezzotti and B. Leoni (2021). "Raman Spectroscopy for the Analysis of Microplastics in Aquatic Systems." Applied Spectroscopy **75**(11): 1341–1357.

Nemecek, D., J. Stepanek and G. J. Thomas. (2013). "Raman Spectroscopy of Proteins and Nucleoproteins." Current Protocols in Protein Science **71**(1): 17.18.11–17.18.52.

Neuberger, S. and C. Neusüß (2015). "Determination of counterfeit medicines by Raman spectroscopy: Systematic study based on a large set of model tablets." Journal of Pharmaceutical and Biomedical Analysis **112**: 70–78.

Neves, A., E. M. Angelin, É. Roldão and M. J. Melo (2019). "New insights into the degradation mechanism of cellulose nitrate in cinematographic films by Raman microscopy." Journal of Raman Spectroscopy **50**(2): 202–212.

Newton, H., L. L. Walkup, N. Whiting, L. West, J. Carriere, F. Havermeyer, L. Ho, P. Morris, B. M. Goodson and M. J. Barlow (2014). "Comparative study of *in situ* N_2 rotational Raman spectroscopy methods for probing energy thermalisation processes during spin-exchange optical pumping." Applied Physics B-Lasers and Optics **115**(2): 167–172.

Ngai, L. H., F. E. Stafford and L. Schaefer (1969). "Symmetry of gaseous dibenzenechromium." Journal of the American Chemical Society **91**(1): 48–49.

Nielsen, J. R., H. H. Claassen and D. C. Smith (1950a). "Infra-Red and Raman Spectra of Fluorinated Ethanes .2. 1,1,1-Trifluoroethane." Journal of Chemical Physics **18**(11): 1471–1476.

Nielsen, J. R., H. H. Claassen and D. C. Smith (1950b). "Infra-Red and Raman Spectra of Fluorinated Ethylenes .2. 1,1-Difluoro-2,2-Dichloroethylene." Journal of Chemical Physics **18**(4): 485–489.

Nielsen, J. R., H. H. Claassen and D. C. Smith (1950c). "Infra-Red and Raman Spectra of Fluorinated Ethylenes .3. Tetrafluoroethylene." Journal of Chemical Physics **18**(6): 812–817.

Nishinari, K., R. K. Cho and M. Iwamoto (1989). "Near Infra-red Monitoring of Enzymatic Hydrolysis of Starch." Starch – Stärke **41**(3): 110–112.

Nomoto, T., H. Hosoi, T. Fujino, T. Tahara and H. O. Hamaguchi (2007). "Excited-state structure and dynamics of 1,3,5-tris(phenylethynyl)benzene as studied by Raman and time-resolved fluorescence spectroscopy." Journal of Physical Chemistry A **111**(15): 2907–2912.

Novotny, L. and B. Hecht (2012). Principles of Nano-Optics. Cambridge, Cambridge University Press.

Nyquist, R. A. (1986). "Infrared Group Frequency Data for 1,2-Epoxyalkanes." Applied Spectroscopy **40**(2): 275–278.

Nyquist, R. A., Y. S. Lo and J. C. Evans (1964). "The Vibrational Spectra and Vibrational Assignments of Halopropadienes." Spectrochimica Acta **20**(4): 619–627.

Nyquist, R. A., T. L. Reder, F. F. Stec and G. J. Kallos (1971). "Vibrational Spectra of 3-Bromopropyne-1-*d* and 1-Bromopropadiene-1-*d*." Spectrochimica Acta Part A – Molecular Spectroscopy A **27**(6): 897–903.

Nyquist, R. A., T. L. Reder, G. R. Ward and G. J. Kallos (1971). "Vibrational Spectrum of 3-Chloropropyne-1-*d*." Spectrochimica Acta Part A-Molecular Spectroscopy A **27**(4): 541.

Ojha, J. K., B. Venkatram Reddy and G. Ramana Rao (2012). "Vibrational analysis and valence force field for nitrotoluenes, dimethylanilines and some substituted methylbenzenes." Spectrochimica Acta Part A: Molecular and Biomolecular Spectroscopy **96**: 632–643.

Okajima, H. and H. Hamaguchi (2015). "Accurate intensity calibration for low wavenumber (-150 to 150 cm^{-1}) Raman spectroscopy using the pure rotational spectrum of N_2." Journal of Raman Spectroscopy **46**(11): 1140–1144.

Okuno, M. (2021). "Hyper-Raman spectroscopy of alcohols excited at 532 nm: Methanol, ethanol, 1-propanol, and 2-propanol." Journal of Raman Spectroscopy **52**(4): 849–856.

Okuno, Y., S. Vantasin, I. S. Yang, J. Son, J. Hong, Y. Y. Tanaka, Y. Nakata, Y. Ozaki and N. Naka (2016). "Side-illuminated tip-enhanced Raman study of edge phonon in graphene at the electrical breakdown limit." Applied Physics Letters **108**(16): 163110.

Olmos, V., C. Bedia, R. Tauler and A. d. Juan (2018). "Preprocessing Tools Applied to Improve the Assessment of Aldrin Effects on Prostate Cancer Cells Using Raman Spectroscopy." Applied Spectroscopy **72**(3): 489–500.

Olschewski, K., E. Kammer, S. Stöckel, T. Bocklitz, T. Deckert-Gaudig, R. Zell, D. Cialla-May, K. Weber, V. Deckert and J. Popp (2015). "A manual and an automatic TERS based virus discrimination." Nanoscale **7**(10): 4545–4552.

Onchoke, K. K. (2020). "Vibrational Spectroscopic Studies of Nitrated Polycyclic Aromatic Hydrocarbons (NPAHs): A Review (1960-2019)." Vibrational Spectroscopy **109**: 103072.

Ortega, A. R., M. Sanchez-Castellanos, N. Perez-Hernandez, R. E. Robles-Zepeda, P. Joseph-Nathan and L. Quijano (2016). "Relative Stereochemistry and Absolute Configuration of Farinosin, a Eudesmanolide From Encelia farinosa." Chirality **28**(5): 415–419.

Ortega, H. E., J. M. Batista, W. G. P. Melo, J. Clardy and M. T. Pupo (2017). "Absolute configurations of griseorhodins A and C." Tetrahedron Letters **58**(50): 4721–4723.

Ortega, P. G. R., M. Montejo and J. J. L. Gonzalez (2015). "Vibrational Circular Dichroism and Theoretical Study of the Conformational Equilibrium in (-)-S-Nicotine." Chemphyschem **16**(2): 342–352.

Ortega, P. G. R., M. Montejo, F. Marquez and J. J. L. Gonzalez (2015). "Conformational properties of chiral tobacco alkaloids by DFT calculations and vibrational circular dichroism: (-)-S-anabasine." Journal of Molecular Graphics & Modelling **60**: 169–179.

Ortega, P. G. R., M. Montejo, F. Marquez and J. J. L. Gonzalez (2015). "DFT-Aided Vibrational Circular Dichroism Spectroscopy Study of (-)-S-cotinine." Chemphyschem **16**(7): 1416–1427.

Ortlieb, N., K. Bretzel, A. Kulik, J. Haas, S. Ludeke, N. Keilhofer, S. D. Schrey, H. Gross and T. H. J. Niedermeyer (2018). "Xanthocidin Derivatives from the Endophytic Streptomyces sp AcE210 Provide Insight into Xanthocidin Biosynthesis." Chembiochem **19**(23): 2472–2480.

Otting, W. (1956). "Infrarotspektren reaktionsfähiger N-Acyl-Verbindungen." Chemische Berichte **89**(8): 1940–1945.

Otvos, J. W. and J. T. Edsall (1939). "Raman spectra of deuterium substituted guanidine and urea." Journal of Chemical Physics **7**(8): 632–632.

Ozaki, Y., C. Huck, S. Tsuchikawa and S. B. Engelsen (2021). Near-Infrared Spectroscopy: Theory, Spectral Analysis, Instrumentation, and Applications.

Pachler, K. G. R., F. Matlok and H.-U. Gremlich (1988). Merck FT-IR Atlas -Eine Sammlung von FT-IR Spektren. Weinheim, VCH Verlagsgesellschaft mbH.

Paddubskaya, A., A. Dementjev, A. Devizis, R. Karpicz, S. Maksimenko and G. Valusis (2018). "Coherent *anti*-Stokes Raman scattering as an effective tool for visualization of single-wall carbon nanotubes." Optics Express **26**(8): 10527–10534.

Pagliai, M., I. Osticioli, A. Nevin, S. Siano, G. Cardini and V. Schettino (2018). "DFT calculations of the IR and Raman spectra of anthraquinone dyes and lakes." Journal of Raman Spectroscopy **49**(4): 668–683.

Pahlow, S., S. Stöckel, S. Pollok, D. Cialla-May, P. Rösch, K. Weber and J. Popp (2016). "Rapid Identification of *Pseudomonas spp*. via Raman Spectroscopy Using Pyoverdine as Capture Probe." Analytical Chemistry **88**(3): 1570–1577.

Paidi, S. K., S. Siddhanta, R. Strouse, J. B. McGivney, C. Larkin and I. Barman (2016). "Rapid Identification of Biotherapeutics with Label-Free Raman Spectroscopy." Analytical Chemistry **88**(8): 4361–4368.

Palafox, M. A. (2018). "DFT computations on vibrational spectra: Scaling procedures to improve the wavenumbers." Physical Sciences Reviews **3**(6): 184.

Palma, F. E., F. F. Cleveland, E. A. Piotrowski and S. Sundaram (1964). "Substituted Methanes .34. Raman + Infrared Spectral Data + Calculated Thermodynamic Properties for CH_2Cl_2 $CHDCl_2$ + CD_2Cl_2." Journal of Molecular Spectroscopy **13**(1): 119–131.

Pandey, A. K., A. Dwivedi and N. Misra (2013). "Quantum Mechanical Study on the Structure and Vibrational Spectra of Cyclobutanone and 1,2-Cyclobutanedione." Journal of Spectroscopy **2013**: 1–11.

Pandey, A. K., A. Dwivedi, S. A. Siddiqui and N. Misra (2013). "Vibrational Spectra of Two Narcotics-A DFT Study." Chinese Journal of Physics **51**(3): 473–499.

Paoloni, L., G. Mazzeo, G. Longhi, S. Abbate, M. Fusè, J. Bloino and V. Barone (2020). "Toward Fully Unsupervised Anharmonic Computations Complementing Experiment for Robust and Reliable

Assignment and Interpretation of IR and VCD Spectra from Mid-IR to NIR: The Case of 2,3-Butanediol and trans-1,2-Cyclohexanediol." The Journal of Physical Chemistry A **124**(5): 1011–1024.

Pardo-Novoa, J. C., H. M. Arreaga-Gonzalez, M. A. Gomez-Hurtado, G. Rodriguez-Garcia, C. M. Cerda-Garcia -Rojas, P. Joseph-Nathan and R. E. del Rio (2016). "Absolute Configuration of Menthene Derivatives by Vibrational Circular Dichroism." Journal of Natural Products **79**(10): 2570–2579.

Parker, S. F., J. Tomkinson, D. A. Braden and B. S. Hudson (2000). "Experimental test of the validity of the use of the n-alkanes as model compounds for polyethylene." Chemical Communications(2): 165–166.

Parrott, A. J., P. Dallin, J. Andrews, P. M. Richardson, O. Semenova, M. E. Halse, S. B. Duckett and A. Nordon (2019). "Quantitative *In Situ* Monitoring of Parahydrogen Fraction Using Raman Spectroscopy." Applied Spectroscopy **73**(0): 88–97.

Pasquini, C. (2003). "Near Infrared Spectroscopy: fundamentals, practical aspects and analytical applications." The Journal of the Brazilian Chemical Society **14**(2): 198–219.

Patel, H. (2017). "Near Infrared Spectroscopy: Basic principles and use in tablet evaluation." International Journal of Chemical and Life sciences **6**(2): 2006–2015.

Paulite, M., C. Blum, T. Schmid, L. Opilik, K. Eyer, G. C. Walker and R. Zenobi (2013). "Full Spectroscopic Tip-Enhanced Raman Imaging of Single Nanotapes Formed from beta-Amyloid(1-40) Peptide Fragments." ACS Nano **7**(2): 911–920.

Pendrill, R., S. T. Mutter, C. Mensch, L. D. Barron, E. W. Blanch, P. L. A. Popelier, G. Widmalm and C. Johannessen (2019). "Solution Structure of Mannobioses Unravelled by Means of Raman Optical Activity." Chemphyschem **20**(5): 695–705.

Peng, Y.-C., S.-L. Chou, J.-I. Lo, M.-Y. Lin, H.-C. Lu, B.-M. Cheng and J. F. Ogilvie (2016). "Infrared and Ultraviolet Spectra of Diborane: B_2H_6 and B_2D_6." The Journal of Physical Chemistry A **120**(28): 5562–5572.

Peng, Y. J., S. Sun, Y. F. Song and Y. Q. Yang (2018). "Coherent *anti*-Stokes Raman scattering spectrum of vibrational properties of liquid nitromethane molecules." Acta Physica Sinica **67**(2): 024208.

Pennington, R. E., D. W. Scott, H. L. Finke, J. P. Mccullough, J. F. Messerly, I. A. Hossenlopp and G. Waddington (1956). "The Chemical Thermodynamic Properties and Rotational Tautomerism of 1-Propanethiol." Journal of the American Chemical Society **78**(14): 3266–3272.

Pereira De Oliveira, L., D. P. Rocha, W. Reis De Araujo, R. A. Abarza Muñoz, T. R. Longo Cesar Paixão and M. Oliveira Salles (2018). "Forensics in hand: New trends in forensic devices (2013–2017)." Analytical Methods **10**(43): 5135–5163.

Pereira, L. S. A., F. L. C. Lisboa, J. C. Neto, F. N. Valladao and M. M. Sena (2017). "Direct classification of new psychoactive substances in seized blotter papers by ATR-FTIR and multivariate discriminant analysis." Microchemical Journal **133**: 96–103.

Perrier-Datin, A. and J.-M. Lebas (1969). "Etude vibrationelle comparee des molecules $(C_6H_5)_2C=NH$, $(C_6H_5)_2C=ND)$ et $(C_6H_5)_2C=O$ en solution." Spektrochimica Acta A **25A**: 169–185.

Perry, A., M. A. Martin, J. W. Nibler, A. Maki, A. Weber and T. A. Blake (2012). "Coriolis analysis of several high-resolution infrared bands of bicyclo[111]pentane-d_0 and -d_1." Journal of Molecular Spectroscopy **276–277**: 22–32.

Pettinger, B., G. Picardi, R. Schuster and G. Ertl (2000). "Surface enhanced Raman spectroscopy: Towards single molecular spectroscopy." Electrochemistry **68**(12): 942–949.

Pettinger, B., G. Picardi, R. Schuster and G. Ertl (2003). "Surface-enhanced and STM tip-enhanced Raman spectroscopy of CN- ions at gold surfaces." Journal of Electroanalytical Chemistry **554**: 293–299.

Pinan, J. P., R. Ouillon, P. Ranson, M. Becucci and S. Califano (1998). "High resolution Raman study of phonon and vibron bandwidths in isotopically pure and natural benzene crystal." Journal of Chemical Physics **109**(13): 5469–5480.

Pinto, A., P. Sciau, T. Zhu, B. Zhao and J. Groenen (2019). "Raman study of Ming porcelain dark spots: Probing Mn-rich spinels." Journal of Raman Spectroscopy **50**(5): 711–719.

Pinto, M. E. F., J. M. Batista, J. Koehbach, P. Gaur, A. Sharma, M. Nakabashi, E. M. Cilli, G. M. Giesel, H. Verli, C. W. Gruber, E. W. Blanch, J. F. Tavares, M. S. da Silva, C. R. S. Garcia and V. S. Bolzani (2015). "Ribifolin, an Orbitide from Jatropha ribifolia, and Its Potential Antimalarial Activity." Journal of Natural Products **78**(3): 374–380.

Pinzaru, S. C., N. Leopold and W. Kiefer (2002). "Vibrational spectroscopy of betulinic acid HIV inhibitor and of its birch bark natural source." Talanta **57**(4): 625–631.

Placzek, G. (1934). Rayleigh-Streuung und Raman-Effekt. Handbuch der Radiologie. Leipzig, Pt2. **6**.

Placzek, G. (1934). The Rayleigh and Raman Scattering. Handbuch der Radiologie. **VI**: 209 –374.

Polavarapu, P. L., C. L. Covington, K. Chruszcz-Lipska, G. Zajac and M. Baranska (2017). "Vibrational Raman optical activity of bicyclic terpenes: comparison between experimental and calculated vibrational Raman, Raman optical activity, and dimensionless circular intensity difference spectra and their similarity analysis (vol 48, pg 305, 2017)." Journal of Raman Spectroscopy **48**(5): 777–777.

Polavarapu, P. L., C. L. Covington and V. Raghavan (2017). "To Avoid Chasing Incorrect Chemical Structures of Chiral Compounds: Raman Optical Activity and Vibrational Circular Dichroism Spectroscopies." Chemphyschem **18**(18): 2459–2465.

Polavarapu, P. L. and E. Santoro (2020). "Vibrational optical activity for structural characterization of natural products." Natural Product Reports **37**(12): 1661–1699.

Poliani, E., M. R. Wagner, A. Vierck, F. Herziger, C. Nenstiel, F. Gannott, M. Schweiger, S. Fritze, A. Dadgar, J. Zaumsei, A. Krost, A. Hoffmann and J. Maultzsch (2017). "Breakdown of Far-Field Raman Selection Rules by Light-Plasmon Coupling Demonstrated by Tip-Enhanced Raman Scattering." Journal of Physical Chemistry Letters **8**(22): 5462–5471.

Porto, S. P. S., P. A. Fleury and T. C. Damen (1967). "Raman Spectra of TiO$_2$, MgF$_2$, ZnF$_2$, FeF$_2$, and MnF$_2$." Physical Review **154**(2): 522–526.

Prabhu, T., S. Periandy and S. Mohan (2011). "Spectroscopic (FTIR and FT Raman) analysis and vibrational study on 2,3-dimethyl naphthalene using *ab-initio* HF and DFT calculations." Spectrochimica Acta Part A – Molecular and Biomolecular Spectroscopy **78**(2): 566–574.

Prabhu, T., S. Periandy and S. Ramalingam (2011). "FT-IR and FT-Raman investigation, computed vibrational intensity analysis and computed vibrational frequency analysis on m-Xylol using *ab-initio* HF and DFT calculations." Spectrochimica Acta Part A: Molecular and Biomolecular Spectroscopy **79**(5): 948–955.

Prasad, R. and N. Dube (1987). "Infrared and Raman-Spectra of Isomeric Pyridine Carboxylic-Acids." Indian Journal of Pure & Applied Physics **25**(4): 178–179.

Prasath, M., M. Govindammal and B. Sathya (2017). "Spectroscopic investigations (FT-IR & FT-Raman) and molecular docking analysis of 6-[1-methyl-4-nitro-1H-imidazol-5-yl) sulfonyl]-7H-purine." Journal of Molecular Structure **1146**: 292–300.

Prieto-Taboada, N., S. Fdez-Ortiz de Vallejuelo, M. Veneranda, E. Lama, K. Castro, G. Arana, A. Larrañaga and J. M. Madariaga (2019). "The Raman spectra of the Na$_2$SO$_4$-K$_2$SO$_4$ system: Applicability to soluble salts studies in built heritage." Journal of Raman Spectroscopy **50**(2): 175–183.

Primpke, S., M. Wirth, C. Lorenz and G. Gerdts (2018). "Reference database design for the automated analysis of microplastic samples based on Fourier transform infrared (FTIR) spectroscopy." Analytical and Bioanalytical Chemistry **410**(21): 5131–5141.

Profant, V., A. Jegorov, P. Bour and V. Baumruk (2017). "Absolute Configuration Determination of a Taxol Precursor Based on Raman Optical Activity Spectra." Journal of Physical Chemistry B **121**(7): 1544–1551.

Prokhorov, K., D. Aleksandrova, E. Sagitova, G. Nikolaeva, T. Vlasova, P. Pashinin, C. Jones and S. Shilton (2016). "Raman Spectroscopy Evaluation of Polyvinylchloride Structure." Journal of Physics: Conference Series **691**: 012001.

Pudney, P. D. A. and T. M. Hancewicz (2010). The Role of Confocal Raman Spectroscopy in Food Science. Handbook of Vibrational Spectroscopy.

Quick, M., M. A. Kasper, C. Richter, R. Mahrwald, A. L. Dobryakov, S. A. Kovalenko and N. P. Ernsting (2015). "beta-Carotene Revisited by Transient Absorption and Stimulated Raman Spectroscopy." Chemphyschem **16**(18): 3824–3835.

Rabinovich, D. (2016). Raman's Gift to the Art World. Chemistry International. **38**: 17.

Radu, A. I., M. Kuellmer, B. Giese, U. Huebner, K. Weber, D. Cialla-May and J. Popp (2016). "Surface-enhanced Raman spectroscopy (SERS) in food analytics: Detection of vitamins B-2 and B-12 in cereals." Talanta **160**: 289–297.

Raghavan, V., J. L. Johnson, D. F. Stec, B. Song, G. Zajac, M. Baranska, C. M. Harris, N. D. Schley, P. L. Polavarapu and T. M. Harris (2018). "Absolute Configurations of Naturally Occurring [5]- and [3]-Ladderanoic Acids: Isolation, Chiroptical Spectroscopy, and Crystallography." Journal of Natural Products **81**(12): 2654–2666.

Ragunathan, N., N. S. Lee, T. B. Freedman, L. A. Nafie, C. Tripp and H. Buijs (1990). "Measurement of Vibrational Circular Dichroism Using a Polarizing Michelson Interferometer." Applied Spectroscopy **44**(1): 5–7.

Ramalingam, S., S. Periandy, M. Govindarajan and S. Mohan (2010a). "FT-IR and FT-Raman vibrational spectra and molecular structure investigation of nicotinamide: A combined experimental and theoretical study." Spectrochimica Acta Part A – Molecular and Biomolecular Spectroscopy **75**(5): 1552–1558.

Ramalingam, S., S. Periandy, M. Govindarajan and S. Mohan (2010b). "FTIR and FTRaman spectra, assignments, *ab initio* HF and DFT analysis of 4-nitrotoluene." Spectrochimica Acta Part A – Molecular and Biomolecular Spectroscopy **75**(4): 1308–1314.

Raman, C. V. and K. S. Krishnan (1928). "A New Type of Secondary Radiation." Nature **121**: 501.

Ramanauskaite, L. and V. Snitka (2015). "Surface enhanced Raman spectroscopy of L-alanyl-L-tryptophan dipeptide adsorbed on Si substrate decorated with triangular silver nanoplates." Chemical Physics Letters **623**: 46–50.

Ramanauskaite, L., H. Xu, E. Griskonis, D. Batiuskaite and V. Snitka (2018). "Comparison and Evaluation of Silver Probe Preparation Techniques for Tip-Enhanced Raman Spectroscopy." Plasmonics **13**(6): 1907–1919.

Ramos, G. Q., S. S. Moreira De Almeida, A. C. F. Amaral, M. Maia Da Costa and H. D. Da Fonseca (2019). "FT-Raman and GC/MS of Extract and Epicuticular Waxes of *Anacardium occidentale* L. Leaves from Amazon." Periodico Tche Quimica **16**(31): 704–710.

Ramsay, M., L. Gozdzialski, A. Larnder, B. Wallace and D. Hore (2021). "Fentanyl quantification using portable infrared absorption spectroscopyA framework for community drug checking.". Vibrational Spectroscopy **114**: 103243.

Rana, V., M. V. Canamares, T. Kubic, M. Leona and J. R. Lombardi (2011). "Surface-enhanced Raman Spectroscopy for Trace Identification of Controlled Substances: Morphine, Codeine, and Hydrocodone." Journal of Forensic Sciences **56**(1): 200–207.

Rasmussen, A. and V. Deckert (2006). "Surface- and tip-enhanced Raman scattering of DNA components." Journal of Raman Spectroscopy **37**(1–3): 311–317.

Rauscher, B. J., N. Boehm, S. Cagiano, G. S. Delo, R. Foltz, M. A. Greenhouse, M. Hickey, R. J. Hill, E. Kan, D. Lindler, D. B. Mott, A. Waczynski and Y. Wen (2014). "New and Better Detectors for the JWST Near-Infrared Spectrograph." Publications of the Astronomical Society of the Pacific **126**(942): 739–749.

Reader, J. and C. H. Corliss (2017). Line spectra of the elements. CRC Handbook of Chemistry and Physics. J. R. Rumble. Boca Raton London New York, Taylor & Francis Ltd.: 10–11 – 10–92.

Redlich, O. and W. Stricks (1936). "Streuspektrum des Deuterobromoforms." Monatshefte für Chemie und verwandte Teile anderer Wissenschaften **67**(1): 328–331.

Reinscheid, F. and U. M. Reinscheid (2016). "Stereochemical analysis of (+)-limonene using theoretical and experimental NMR and chiroptical data." Journal of Molecular Structure **1106**: 141–153.

Reinscheid, F., M. Schmidt, H. Abromeit, S. Liening, G. K. E. Scriba and U. M. Reinscheid (2016). "Structural and chiroptical analysis of naturally occurring (-)-strychnine." Journal of Molecular Structure **1106**: 200–209.

Reitz, A. W. and R. Sabathy (1937a). "Studien zum Ramaneffekt." Monatshefte für Chemie und verwandte Teile anderer Wissenschaften **71**(1): 131–143.

Reitz, A. W. and R. Sabathy (1937b). "Studien zum Ramaneffekt." Monatshefte für Chemie und verwandte Teile anderer Wissenschaften **71**(1): 100–108.

Reitz, A. W. and R. Skrabal (1937). "Studien zum Ramaneffekt." Monatshefte für Chemie und verwandte Teile anderer Wissenschaften **70**(1): 398–404.

Ren, J., D. Zhao, S. J. Wu, J. Wang, Y. J. Jia, W. X. Li, H. J. Zhu, F. Cao, W. Li, C. U. Pittman and X. J. He (2019). "Reassigning the stereochemistry of bioactive cepharanthine using calculated versus experimental chiroptical spectroscopies." Tetrahedron **75**(9): 1194–1202.

Renuga Devi, T. S. and S. Gayathri (2010). "FTIR AND FT-RAMAN SPECTRAL ANALYSIS OF PACLITAXEL DRUGS." International Journal of Pharmaceutical Sciences Review and Research **2**(2): 106–110.

Reparaz, J. S., N. Peica, R. Kirste, A. R. Goñi, M. R. Wagner, G. Callsen, M. I. Alonso, M. Garriga, I. C. Marcus, A. Ronda, I. Berbezier, J. Maultzsch, C. Thomsen and A. Hoffmann (2013). "Probing local strain and composition in Ge nanowires by means of tip-enhanced Raman scattering." Nanotechnology **24**(18): 185704.

Reuß, F., K.-P. Zeller, H.-U. Siehl, S. Berger and D. Sicker (2016). "Steviosid aus Süßkraut." Chemie in unserer Zeit **50**(3): 198–208.

Rey, M., I. S. Chizhmakova, A. V. Nikitin and V. G. Tyuterev (2021). "Towards a complete elucidation of the ro-vibrational band structure in the SF_6 infrared spectrum from full quantum-mechanical calculations." Physical Chemistry Chemical Physics **23**(21): 12115–12126.

Ricci, C., C. Eliasson, N. A. Macleod, P. N. Newton, P. Matousek and S. G. Kazarian (2007). "Characterization of genuine and fake artesunate *anti*-malarial tablets using Fourier transform infrared imaging and spatially offset Raman spectroscopy through blister packs." Analytical and Bioanalytical Chemistry **389**(5): 1525–1532.

Ricci, M., C. Lofrumento, M. Becucci and E. M. Castellucci (2018). "The Raman and SERS spectra of indigo and indigo-Ag_2 complex: DFT calculation and comparison with experiment." Spectrochimica Acta Part A: Molecular and Biomolecular Spectroscopy **188**: 141–148.

Ricciardi, A., G. Piuri, M. D. Porta, S. Mazzucchelli, A. Bonizzi, M. Truffi, M. Sevieri, R. Allevi, F. Corsi, R. Cazzola and C. Morasso (2020). "Raman spectroscopy characterization of the major classes of plasma lipoproteins ". Vibrational Spectroscopy **109**: 103073.

Rieke, G. H. (2007). "Infrared Detector Arrays for Astronomy." Annual Review of Astronomy and Astrophysics **45**(1): 77–115.

Robert, B. (2004). Resonance Raman Studies in Photosynthesis – Chlorophyll and Carotenoid Molecules. **3**: 161–176.

Roberts, J. D. and C. W. Sauer (1949). "Small-Ring Compounds. III. Synthesis of Cyclobutanone, Cyclobutanol, Cyclobutene and Cyclobutane." Journal of the American Chemical Society **71**(12): 3925–3929.

Rodgers, E. G. and D. P. Strommen (1981). "A Multifunctional Spinning Cell for Obtaining Raman Spectra of Micro Samples in the Backscattering Geometry." Applied Spectroscopy **35**(2): 215–217.

Rodrigues Silva, D., L. de Azevedo Santos, T. A. Hamlin, F. M. Bickelhaupt, M. P. Freitas and C. Fonseca Guerra (2021). "Dipolar repulsion in α-halocarbonyl compounds revisited." Physical Chemistry Chemical Physics **23**(37): 20883–20891.

Rodríguez-Cabello, J. C., J. C. Merino, J. M. Pastor, U. Hoffmann, S. Okretic and H. W. Siesler (1995). "Rheo-optical FT-Raman study of uniaxially stretched poly(vinylidene fluoride)." Macromolecular Chemistry and Physics **196**(3): 815–824.

Rogalski, A. (2012). "History of infrared detectors." Opto-Electronics Review **20**(3): 279–308.

Roman, M., K. Chruszcz-Lipska and M. Baranska (2015). "Vibrational analysis of cinchona alkaloids in the solid state and aqueous solutions." Journal of Raman Spectroscopy **46**(11): 1041–1052.

Rossi, D., K. M. Ahmed, R. Gaggeri, S. Della Volpe, L. Maggi, G. Mazzeo, G. Longhi, S. Abbate, F. Corana, E. Martino, M. Machado, R. Varandas, M. D. Sousa and S. Collina (2017). "(R)-(-)-Aloesaponol III 8-Methyl Ether from Eremurus persicus: A Novel Compound against Leishmaniosis." Molecules **22**(4): 519.

Rud Nielsen, J. and H. D. Brandt (1965). "Vibrational spectra of 1,3,5-trifluoro-2,4,6-trichlorobenzene." Journal of Molecular Spectroscopy **17**(2): 334–340.

Rudo, A., H.-U. Siehl, K.-P. Zeller, S. Berger and D. Sicker (2015). "Diosgenin aus Yams als Hormonvorstufe." Chemie in unserer Zeit **49**(6): 372–384.

Rudo, A., H. U. Siehl, K. P. Zeller, S. Berger and D. Sicker (2013). "Refuted as a Potency Remedy, but . . . Cantharidin." Chemie in unserer Zeit **47**(5): 310–316.

Ruggeri, F. S., B. Mannini, R. Schmid, M. Vendruscolo and T. P. J. Knowles (2020). "Single molecule secondary structure determination of proteins through infrared absorption nanospectroscopy." Nature Communications **11**(1): 2945.

Rydzak, J. W., D. E. White, C. Y. Airiau, J. T. Sterbenz, B. D. York, D. J. Clancy and Q. Y. Dai (2015). "Real-Time Process Analytical Technology Assurance for Flow Synthesis of Oligonucleotides." Organic Process Research & Development **19**(1): 203–214.

Rygula, A., K. Majzner, K. M. Marzec, A. Kaczor, M. Pilarczyk and M. Baranska (2013). "Raman spectroscopy of proteins: a review." Journal of Raman Spectroscopy **44**(8): 1061–1076.

Ryu, S. M., H. M. Lee, E. G. Song, Y. H. Seo, J. Lee, Y. Guo, B. S. Kim, J.-J. Kim, J. S. Hong, K. H. Ryu and D. Lee (2017). "Antiviral Activities of Trichothecenes Isolated from Trichoderma albolutescens against Pepper Mottle Virus." Journal of Agricultural and Food Chemistry **65**(21): 4273–4279.

Rzeźnicka, I. I., H. Horino, N. Kikkawa, S. Sakaguchi, A. Morita, S. Takahashi, T. Komeda, H. Fukumura, T. Yamada and M. Kawai (2013). "Tip-enhanced Raman spectroscopy of 4,4'-bipyridine and 4,4'-bipyridine N,N'-dioxide adsorbed on gold thin films." Surface Science **617**: 1–9.

Sackett, P. H., R. W. Hannah and W. Slavin (1978). "Polystyrene fractionation by liquid chromatography and identification of the compounds by spectroscopy." Chromatographia **11**(11): 634–639.

Said, M. E. A., I. Bombarda, J. V. Naubron, P. Vanloot, M. Jean, A. Cheriti, N. Dupuy and C. Roussel (2017). "Isolation of the major chiral compounds from Bubonium graveolens essential oil by HPLC and absolute configuration determination by VCD." Chirality **29**(2): 70–79.

Said, M. E. A., P. Vanloot, I. Bombarda, J. V. Naubron, E. Dahmane, A. Aamouche, M. Jean, N. Vanthuyne, N. Dupuy and C. Roussel (2016). "Analysis of the major chiral compounds of Artemisia herba-alba essential oils (EOs) using reconstructed vibrational circular dichroism (VCD) spectra: En route to a VCD chiral signature of EOs." Analytica Chimica Acta **903**: 121–130.

Sajan, D., G. D. Sockalingum, M. Manfait, I. Hubert Joe and V. S. Jayakumar (2008). "NIR-FT Raman, FT-IR and surface-enhanced Raman scattering spectra, with theoretical simulations on chloramphenicol." Journal of Raman Spectroscopy **39**(12): 1772–1783.

Sakhaee, N., S. A. Jalili and F. Darvish (2016). "Spherical conformational landscape shed new lights on fluxional nature of cyclopentane and its derivatives, confirmed by their Raman spectra." Computational and Theoretical Chemistry **1090**: 193–202.

Saleem, M., S. Ali, M. B. Khan, A. Amin, M. Bilal, H. Nawaz and M. Hassan (2020). "Optical diagnosis of hepatitis B virus infection in blood plasma using Raman spectroscopy and chemometric techniques." Journal of Raman Spectroscopy **51**(7): 1067–1077.

Sanchez-Castellanos, M., M. A. Bucio, A. Hernandez-Barragan, P. Joseph-Nathan, G. Cuevas and L. Quijano (2015). "Vibrational Circular Dichroism (VCD), VCD Exciton Coupling, and Xray Determination of the Absolute Configuration of an alpha,beta-Unsaturated Germacranolide." Chirality **27**(3): 247–252.

Santoro, E., G. Mazzeo, A. G. Petrovic, A. Cimmino, J. Koshoubu, A. Evidente, N. Berova and S. Superchi (2015). "Absolute configurations of phytotoxins seiricardine A and inuloxin A obtained by chiroptical studies." Phytochemistry **116**: 359–366.

Satoh, T., T. Tsuji, H. Matsuda and S. Sudoh (2007). "DFT Calculations and IR Studies on 2-Hydroxy1,4-naphthoquinone and Its 3-Substituted Derivatives." Bulletin of The Chemical Society of Japan **80**: 321–323.

Saviello, D., A. Di Gioia, P.-I. Turenne, M. Trabace, R. Giorgi, A. Mirabile, P. Baglioni and D. Iacopino (2019). "Handheld surface-enhanced Raman scattering identification of dye chemical composition in felt-tip pen drawings." Journal of Raman Spectroscopy **50**(2): 222–231.

Schachtschneider, J. H. and R. G. Snyder (1963). "Vibrational analysis of the n-paraffins – II: Normal co-ordinate calculations." Spectrochimica Acta **19**(1): 117–168.

Scherer, J. R. and J. C. Evans (1963). "Vibrational Spectra and Assignments for 16 Chlorobenzenes." Spectrochimica Acta **19**(11): 1739–1775.

Scherer, J. R., J. C. Evans and W. W. Muelder (1962). "Vibrational Spectra and Urey-Bradley Force Constants of sym-Trifluoro and Tribromo Benzene." Spectrochimica Acta **18**(12): 1579–1592.

Scherger, J. D., E. A. Evans, J. A. Dura and M. D. Foster (2016). "Extending nanoscale spectroscopy with titanium nitride probes." Journal of Raman Spectroscopy **47**(11): 1332–1336.

Scherger, J. D. and M. D. Foster (2017). "Tunable, Liquid Resistant Tip Enhanced Raman Spectroscopy Probes: Toward Label-Free Nano-Resolved Imaging of Biological Systems." Langmuir **33**(31): 7818–7825.

Schlücker, S., A. Szeghalmi, M. Schmitt, J. Popp and W. Kiefer (2003). "Density functional and vibrational spectroscopic analysis of β-carotene." Journal of Raman Spectroscopy **34**(6): 413–419.

Schlücker, S. H. (2010). Surface Enhanced Raman Spectroscopy. Weinheim, Wiley-VCH.

Schrader, B. (1987). US Patent 4714345; UK patent GB 2 162961; German patent DE 3 424108 C 2.

Schrader, B. (1991). "Raman-Spectroscopy of Mineral-Oil Products .1. Nir Ft-Raman Spectra of Polycyclic Aromatic-Hydrocarbons." Applied Spectroscopy **45**(8): 1230–1232.

Schrader, B. (1995). Infrared and Raman Spectroscopy: Methods and Applications. Weinheim, VCH.

Schrader, B. (1996). Raman/Infrared Atlas of Organic Compounds, 2nd Ed., John Wiley & Sons Inc.

Schrader, B., G. Baranovic, A. Epding, G. G. Hoffmann, P. J. M. Van Kan, S. Keller, P. Hildebrandt, C. Lehner and J. Sawatzki (1993). "Time-Resolved and 2-Dimensional NIR FT-Raman Spectroscopy." Applied Spectroscopy **47**(9): 1452–1456.

Schrader, B. and E. H. Korte (1972). "Infrarot-Rotationsdispersion (IRD)." Angewandte Chemie. **84**: 218.

Schrader, B. and W. Meier (1972). "Raman Spectroscopy and Molecular Structure .5. Scheme for Determination of Substitution Pattern of Benzene Derivatives." Fresenius Zeitschrift für Analytische Chemie **260**(3): 248.

Schrader, B. and E. Steigner (1970). "[Raman spectroscopy and molecular structure. I. Raman and infrared spectra of steroids]." Justus Liebigs Annalen der Chemie **735**: 6–14.

Schrobilgen, G. J., D. Martin-Rovet, P. Charpin and M. Lance (1980). "XeOF$_5^-$ and [(XeOF$_4$)$_3$F]$^-$ anions." Journal of the Chemical Society, Chemical Communications(19): 894–897.

Schulze, F., J. Titus, P. Mettke, S. Berger, H. U. Siehl, K. P. Zeller and D. Sicker (2013). "Carminic Acid: The Red of Cochineal Lice." Chemie in unserer Zeit **47**(4): 222–228.

Schweitzer-Stenner, R. (2001). "Visible and UV-resonance Raman spectroscopy of model peptides." Journal of Raman Spectroscopy **32**(9): 711–732.

Schweitzer-Stenner, R., F. Eker, Q. Huang, K. Griebenow, P. A. Mroz and P. M. Kozlowski (2002). "Structure Analysis of Dipeptides in Water by Exploring and Utilizing the Structural Sensitivity of Amide III by Polarized Visible Raman, FTIt Spectroscopy and DFT Based Normal Coordinate Analysis." Journal of Physical Chemistry B **106**: 4294–4304.

Schweitzer-Stenner, R., J. B. Soffer, S. Toal and D. Verbaro (2012). Structural Analysis of Unfolded Peptides by Raman Spectroscopy. Intrinsically Disordered Protein Analysis: Volume 1, Methods and Experimental Tools. V. N. Uversky and A. K. Dunker. Totowa, NJ, Humana Press: 315–346.

Scott, D. W. and M. Z. Elsabban (1969). "A Valence Force Field for Aliphatic Sulfur Compounds – Alkanethiols and Thioalkanes." Journal of Molecular Spectroscopy **30**(2): 317–337.

Scott, D. W. and M. Z. Elsabban (1969). "A Valence Force Field for Aliphatic Sulfur Compounds – Dithiaalkanes." Journal of Molecular Spectroscopy **31**(3): 362–367.

Seiler, P. and J. D. Dunitz (1979). "New Interpretation of the Disordered Crystal-Structure of Ferrocene." Acta Crystallographica Section B – Structural Science **35**(May): 1068–1074.

Seiler, P. and J. D. Dunitz (1979). "Structure of Triclinic Ferrocene at 101, 123 and 148 K." Acta Crystallographica Section B – Structural Science **35**(Sep): 2020–2032.

Seupel, R., A. Roth, K. Steinke, D. Sicker, H. U. Siehl, K. P. Zeller and S. Berger (2015). "Non screwed up Friedelin from Cork." Chemie in unserer Zeit **49**(1): 60–72.

Shalaev, V. M. and S. Kawata (2007). Nanophotonics with Surface Plasmons. Amsterdam, Elsevier Science.

Sharma, B., R. R. Frontiera, A.-I. Henry, E. Ringe and R. P. Van Duyne (2012). "SERS: Materials, applications, and the future." Materials Today **15**(1–2): 16–25.

Sheinker, Y. N. and Y. K. Syrkin (1959). Isv. Akad. Nauk. SSSR, Ser. Fiz. **14**: 478.

Sheppard, N. (1949). "The Infra-Red Spectrum, and the Assignment of the Fundamental Modes of Vibration of Thioacetic Acid." Transactions of the Faraday Society **45**(7): 693–697.

Sheremet, E., R. D. Rodriguez, A. L. Agapov, A. P. Sokolov, M. Hietschold and D. R. T. Zahn (2016). "Nanoscale imaging and identification of a four-component carbon sample." Carbon **96**: 588–593.

Shigorin, D. N. and Y. K. Syrkin (1949). Zh. Fiz. Chim. **23**.

Shimanouchi, T. and I. Suzuki (1962). "Infrared Spectra and Force Constants of CH_2Cl_2, $CHCl_2$, and CD_2Cl_2." Journal of Molecular Spectroscopy **8**(3): 222.

Shirk, J. S. and H. H. Claassen (1971). "Raman Spectra of Matrix-Isolated Molecules." The Journal of Chemical Physics **54**(7): 3237–3238.

Sicker, D., K.-P. Zeller, H. U. Siehl and S. Berger (2019). Natural Products: Isolation, Structure Elucidation, History. Weinheim, Wiley-VCH.

Siddiqui, S., A. Dwivedi, A. Pandey, P. Singh, T. Hasan, S. Jain and N. Misra (2009). "Molecular Structure, Vibrational Spectra and Potential Energy Distribution of Colchicine Using *ab Initio* and Density Functional Theory." Journal of Computer Chemistry, Japan **8**: 59–72.

Siesler, H. W., Y. Ozaki, S. Kawata and H. M. Heise (2002). Near-Infrared Spectroscopy – Principles, Instruments, Applications. Weinheim, Wiley-VCH Verlag GmbH.

Simeral, M. L. and J. H. Hafner (2022). "The Raman Active Vibrational Modes of Anthraquinones." Astrobiology **22**(10): 1165–1175.

Simon, L. L., H. Pataki, G. Marosi, F. Meemken, K. Hungerbuhler, A. Baiker, S. Tummala, B. Glennon, M. Kuentz, G. Steele, H. J. M. Kramer, J. W. Rydzak, Z. P. Chen, J. Morris, F. Kjell, R. Singh, R. Gani, K. V. Gernaey, M. Louhi-Kultanen, J. O'Reilly, N. Sandler, O. Antikainen, J. Yliruusi, P. Frohberg, J. Ulrich, R. D. Braatz, T. Leyssens, M. von Stosch, R. Oliveira, R. B. H. Tan, H. Q. Wu, M. Khan, D. O'Grady, A. Pandey, R. Westra, E. Delle-Case, D. Pape, D. Angelosante, Y. Maret, O. Steiger, M. Lenner, K. Abbou-Oucherif, Z. K. Nagy, J. D. Litster, V. K. Kamaraju and M. S. Chiu (2015). "Assessment of Recent Process Analytical Technology (PAT) Trends: A Multiauthor Review." Organic Process Research & Development **19**(1): 3–62.

Sixt, M., G. Gudi, H. Schulz and J. Strube (2018). "In-line Raman spectroscopy and advanced process control for the extraction of anethole and fenchone from fennel (*Foeniculum vulgare* L. MILL.)." Comptes Rendus Chimie **21**(2): 97–103.

Sloane, H. J. and R. Bramston-Cook (1973). "Raman Spectroscopy of Some Polymers and Copolymers of Styrene, Butadiene, and Methylmethacrylate." Applied Spectroscopy **27**(3): 217–225.

Smekal, A. (1923). "Zur Quantentheorie der Dispersion." Naturwissenschaften **11**: 873–875.

Smith, B. C. (2021). "The Infrared Spectra of Polymers II: Polyethylene." Spectroscopy **36**(9): 24–29.

Smith, B. C. (2021). "The Infrared Spectra of Polymers III: Hydrocarbon Polymers." Spectroscopy **36**(11): 22–25.

Smith, B. C. (2021). "The Infrared Spectra of Polymers, Part I: Introduction." Spectroscopy **36**(7): 17 –22.

Smith, B. C. (2022). "The Infrared Spectra of Polymers IV: Rubbers." Spectroscopy **37**(1): 8–12.

Smith, B. C. (2022). "The Infrared Spectra of Polymers V: Epoxies." Spectroscopy **37**(3): 17–19.

Smith, B. C. (2022). "The Infrared Spectra of Polymers, VI: Polymers With C-O Bonds." Spectroscopy **37**(5): 15–19, 27.

Smith, D. C., J. R. Nielsen and H. H. Claassen (1950). "Infra-Red and Raman Spectra of Fluorinated Ethylenes .1. 1,1-Difluoroethylene." Journal of Chemical Physics **18**(3): 326–331.

Smith, D. C., C. Y. Pan and J. R. Nielsen (1950). "Vibrational Spectra of the 4 Lowest Nitroparaffins." Journal of Chemical Physics **18**(5): 706–712.

Smith, D. F. (1963). "Xenon Difluoride." Journal of Chemical Physics **38**(1): 270–271.

Smith, G. D. and R. J. H. Clark (2004). "Raman microscopy in archaeological science." Journal of Archaeological Science **31**(8): 1137–1160.

Smith, G. P. S., S. E. Holroyd, D. C. W. Reid and K. C. Gordon (2017). "Raman imaging processed cheese and its components." Journal of Raman Spectroscopy **48**(3): 374–383.

Smyrniotopoulos, V., C. Merten, M. Kaiser and D. Tasdemir (2017). "Bifurcatriol, a New Antiprotozoal Acyclic Diterpene from the Brown Alga Bifurcaria bifurcata." Marine Drugs **15**(8): 245.

Socrates, G. (2001). Infrared and Raman Characteristic Group Frequencies. Chichester, John Wiley & Sons.

Sodo, A., L. Tortora, P. Biocca, A. Casanova Municchia, E. Fiorin and M. A. Ricci (2019). "Raman and time of flight secondary ion mass spectrometry investigation answers specific conservation questions on Bosch painting Saint Wilgefortis Triptych." Journal of Raman Spectroscopy **50**(2): 150–160.

Soliman, U. A., A. M. Hassan and T. A. Mohamed (2007). "Conformational stability, vibrational assignmenents, barriers to internal rotations and *ab initio* calculations of 2-aminophenol (d$_0$ and d$_3$)." Spectrochimica Acta Part A: Molecular and Biomolecular Spectroscopy **68**(3): 688–700.

Spectroscopy, C. o. M. S. a. (1960). "TABLES OF WAVENUMBERS FOR THE CALIBRATION OF INFRA-RED SPECTROMETERS." Pure and Applied Chemistry **1**(4): 537–699.

Spiekermann, M., D. Bougeard and B. Schrader (1982). "Coupled Calculations of Vibrational Frequencies and Intensities .3. IR and Raman-Spectra of Ethylene Oxide and Ethylene Sulfide." Journal of Computational Chemistry **3**(3): 354–362.

Spiller, S., P. Mettke, H. U. Siehl, K. P. Zeller, D. Sicker and S. Berger (2014). "Thymoquinone-the Yellow of the Oil." Chemie in unserer Zeit **48**(2): 114–122.

Sprenger, R. F., S. S. Thomasi, A. G. Ferreira, Q. B. Cass and J. M. Batista (2016). "Solution-state conformations of natural products from chiroptical spectroscopy: the case of isocorilagin." Organic & Biomolecular Chemistry **14**(13): 3369–3375.

Srivastava, A., P. Tandon, A. P. Ayala and S. Jain (2011). "Solid state characterization of an antioxidant alkaloid boldine using vibrational spectroscopy and quantum chemical calculations." Vibrational Spectroscopy **56**(1): 82–88.

Srivastava, M., P. Rani, N. P. Singh and R. A. Yadav (2014). "Experimental and theoretical studies of vibrational spectrum and molecular structure and related properties of pyridoxine (vitamin B6)." Spectrochimica Acta Part A: Molecular and Biomolecular Spectroscopy **120**: 274–286.

Srivastava, M., N. P. Singh and R. A. Yadav (2014). "Experimental Raman and IR spectral and theoretical studies of vibrational spectrum and molecular structure of Pantothenic acid (vitamin B-5)." Spectrochimica Acta Part a-Molecular and Biomolecular Spectroscopy **129**: 131–142.

Srivastava, R., H. V. Lauer, L. L. Chase and W. E. Bron (1971). "Raman Frequencies of Fluorite Crystals." Physics Letters A **36**(4): 333–334.

Stagi, L., D. Chiriu, M. Scholz, C. M. Carbonaro, R. Corpino, A. Porcheddu, S. Rajamaki, G. Cappellini, R. Cardia and P. C. Ricci (2017). "Vibrational and optical characterization of s-triazine derivatives." Spectrochimica Acta Part A: Molecular and Biomolecular Spectroscopy **183**: 348–355.

Stahl, E. and W. Schild (1981). Isolierung und Charakterisierung von Naturstoffen. Stuttgart, Gustav Fischer Verlag.

Stammreich, H. and R. Forneris (1956). "Raman Spectrum of Iodoform." Spectrochimica Acta **8**(1): 52–53.

Stammreich, H., Y. Tavares and D. Bassi (1961). "The Vibrational Spectrum and Force Constants of Carbon Tetraiodide." Spectrochimica Acta **17**(6): 661–664.

Stani, C., C. Invernizzi, G. Birarda, P. Davit, L. Vaccari, M. Malagodi, M. Gulmini and G. Fiocco (2022). "A Nanofocused Light on Stradivari Violins: Infrared s-SNOM Reveals New Clues Behind Craftsmanship Mastery." Analytical Chemistry **94**(43): 14815–14819.

Steele, D. (1962). "The Vibrational Spectra of Tetra-Fluoro-Benzenes." Spectrochimica Acta **18**(7): 915–925.

Steele, D. and D. H. Whiffen (1959). "The Vibrational Frequencies of Hexafluorobenzene." Transactions of the Faraday Society **55**(3): 369–376.

Steele, D. and D. H. Whiffen (1960). "The Force Field of Hexafluorobenzene." Transactions of the Faraday Society **56**(1): 5–7.

Steele, D. and D. H. Whiffen (1960). "The Vibration Frequencies of Pentafluorobenzene." Spectrochimica Acta **16**(3): 368–375.

Steigner, E. and B. Schrader (1970). "[Raman spectroscopy and molecular structure. II. Elucidation of structure of steroids by raman spectroscopy]." Justus Liebigs Annalen der Chemie **735**: 15–22.

Steinfeld, J. I., I. Burak, D. G. Sutton and A. V. Nowak (1970). "Infrared Double Resonance in Sulfur Hexafluoride." The Journal of Chemical Physics **52**(10): 5421–5434.

Steinke, K., E. Jose, D. Sicker, H. U. Siehl, K. P. Zeller and S. Berger (2013). "A Flavone Sinensetin." Chemie in unserer Zeit **47**(3): 158–163.

Steinke, K., E. Jose, H.-U. Siehl, K.-P. Zeller and S. Berger (2013). "Campher." Chemie in unserer Zeit **47**(2): 102–107.

Steinke, K., E. Jose, H.-U. Siehl, K.-P. Zeller and S. Berger (2013). "Kokain." Chemie in unserer Zeit **47**(1): 56–60.

Steudel, R. (1970). "S=O Stretching Frequency in Thionyl Compounds." Zeitschrift für Naturforschung Part B – Chemie, Biochemie, Biophysik, Biologie und verwandte Gebiete B 25(2): 156–165.

Stewart, J. E. (1957). "Infrared Absorption Spectra of Urea, Thiourea, and Some Thiourea-Alkali Halide Complexes." Journal of Chemical Physics **26**(2): 248–254.

Stewart, J. E. and J. C. Richmond (1957). "Infrared Emission Spectrum of Silicon Carbide Heating Elements." Journal of Research of the National Bureau of Standards **59**(6): 405–409.

Stöckel, S., J. Kirchhoff, U. Neugebauer, P. Rösch and J. Popp (2016). "The application of Raman spectroscopy for the detection and identification of microorganisms." Journal of Raman Spectroscopy **47**(1): 89–109.

Stöckel, S., S. Meisel, B. Lorenz, S. Kloss, S. Henk, S. Dees, E. Richter, S. Andres, M. Merker, I. Labugger, P. Rösch and J. Popp (2017). "Raman spectroscopic identification of *Mycobacterium tuberculosis*." Journal of Biophotonics **10**(5): 727–734.

Stöckle, R. M., Y. D. Suh, V. Deckert and R. Zenobi (2000). "Nanoscale chemical analysis by tip-enhanced Raman spectroscopy." Chemical Physics Letters **318**(1-3): 131–136.

Su, W., N. Kumar, A. Krayev and M. Chaigneau (2018). "*In situ* topographical chemical and electrical imaging of carboxyl graphene oxide at the nanoscale." Nature Communications **9**(1): 2891.

Su, W. T., A. Esfandiar, O. Lancry, J. Q. Shao, N. Kumar and M. Chaigneau (2021). "Visualising structural modification of patterned graphene nanoribbons using tip-enhanced Raman spectroscopy." Chemical Communications **57**(56): 6895–6898.

Sugahara, K., M. Yoshida, T. Sugahara and K. Ohgaki (2004). "High-pressure phase behavior and cage occupancy for the CF_4 hydrate system." Journal of Chemical and Engineering Data **49**(2): 326–329.

Sundaraganesan, N., S. Kalaichelvan, C. Meganathan, B. D. Joshua and J. Cornard (2008). "FT-IR, FT-Raman spectra and *ab initio* HF and DFT calculations of 4-N,N'-dimethylamino pyridine." Spectrochimica Acta Part A – Molecular and Biomolecular Spectroscopy **71**(3): 898–906.

Superchi, S., P. Scafato, M. Gorecki and G. Pescitelli (2018). "Absolute Configuration Determination by Quantum Mechanical Calculation of Chiroptical Spectra: Basics and Applications to Fungal Metabolites." Current Medicinal Chemistry **25**(2): 287–320.

Suzuki, I. (1960). "Infrared Spectra and Normal Vibrations of Formamide – $HCONH_2$, $HCOND_2$, $DCONH_2$ and $DCOND_2$." Bulletin of the Chemical Society of Japan **33**(10): 1359–1365.

Suzuki, I. (1962). "Infrared Spectra and Normal Vibrations of Acetamide and Its Deuterated Analogues." Bulletin of the Chemical Society of Japan **35**(8): 1279–1286.

Suzuki, M., T. Yokoyama and M. Ito (1968). "Polarized Raman Spectra of Naphthalene and Anthracene Single Crystals." Spectrochimica Acta Part A – Molecular Spectroscopy A **24**(8): 1091–1107.

Suzuki, S., T. Shimanouchi and M. Tsuboi (1963). "Normal Vibrations of Glycine and Deuterated Glycine Molecules." Spectrochimica Acta **19**(7): 1195–1208.

Synge, E. H. (1928). "A suggested method for extending microscopic resolution into the ultra-microscopic region." The London, Edinburgh and Dublin Philosophical Magazine and Journal of Science Series 7, **6**(Series 7): 356–362.

Szasz, G., A. Kovacs, I. Hargittai, I. Jeon and G. P. Miller (1998). "Molecular structure and molecular vibrations of 1,3,5,7-tetramethyl-2,4,6,8,9,10-hexathiaadamantane." Journal of Physical Chemistry A **102**(2): 484–489.

Takahashi, H., T. Shimanouchi, K. Fukushima and T. Miyazawa (1964). "Infrared spectrum and normal vibrations of cyclohexane." Journal of Molecular Spectroscopy **13**(1): 43–56.

Takahashi, Y., T. Shishido, K. Yamamoto, Y. Sawaji, J. Nishida and G. Pezzotti (2015). "Do formalin fixation and freeze-thaw affect near-infrared Raman spectroscopy of cartilaginous tissue? A preliminary ex vivo analysis of native human articular cartilage." Journal of Raman Spectroscopy **46**(11): 1166–1172.

Talari, A. C. S., Z. Movasaghi, S. Rehman and I. U. Rehman (2015). "Raman Spectroscopy of Biological Tissues." Applied Spectroscopy Reviews **50**(1): 46–111.

Tan, C. L. and H. Mohseni (2018). "Emerging technologies for high performance infrared detectors." Nanophotonics **7**(1): 169–197.

Tanaka, N., H. Kitano and N. Ise (1990). "Raman-Spectroscopic Study of Hydrogen-Bonding in Aqueous Carboxylic-Acid Solutions." Journal of Physical Chemistry **94**(16): 6290–6292.

Tanaka, N., H. Kitano and N. Ise (1991a). "Raman-Spectroscopic Study of Hydrogen-Bonding in Aqueous Carboxylic-Acid Solutions .2. Deuterio Analogs in Heavy-Water." Journal of Physical Chemistry **95**(3): 1503–1507.

Tanaka, N., H. Kitano and N. Ise (1991b). "Raman-Spectroscopic Study of Hydrogen-Bonding in Aqueous Carboxylic-Acid Solutions .3. Polyacrylic-Acid." Macromolecules **24**(10): 3017–3019.

Tarakeshwar, P. and S. Manogaran (1994). "Ground-State Vibrations of Citric-Acid and the Citrate Trianion – an *Ab-Initio* Study." Spectrochimica Acta Part A – Molecular and Biomolecular Spectroscopy **50**(14): 2327–2343.

Tarantili, P. A., A. G. Andreopoulos and C. Galiotis (1998). "Real-time micro-Raman measurements on stressed polyethylene fibers. 1. Strain rate effects and molecular stress redistribution." Macromolecules **31**(20): 6964–6976.

Tasumi, M., I. Harada, T. Takamatsu and S. Takahashi (1982). "Raman studies of L-histidine and related compounds in aqueous solutions." Journal of Raman Spectroscopy **12**(2): 149–151.

Tasumi, M., S. Ikeda, H. Tanaka and T. Shimanouchi (1964). "Stereoregulated Polydideuteroethylene .2. Infrared Spectra + Normal Vibration Analysis." Journal of Polymer Science Part a-General Papers **2**(4pa): 1607–1631.

Tavčar, G. and B. Žemva (2013). "[Li(XeF2)n](AF6) (A = P, As, Ru, Ir), the first xenon(II) compounds of lithium. Synthesis, Raman spectrum, and crystal structure of [Li(XeF$_2$)$_3$](AsF$_6$)." Inorganic chemistry **52** 8: 4319–4323.

Terhune, R. (1963). "Coherent *anti*-Stokes Raman spectroscopy." Bulletin of the American Physical Society **8**: 359.

Tewari, B., G. Beaulieu-Houle, A. Larsen, R. Kengne-Momo, K. Auclair and I. Butler (2011). "An Overview of Molecular Spectroscopic Studies on Theobromine and Related Alkaloids." Applied Spectroscopy Reviews **47**(3): 163–179.

Thomas, H. C. (1974). Interpretation of the Infrared Spectra of Organophosphorous Compounds. London, Heyden and Son.

Thomas, N. C. (1991). "The Early History of Spectroscopy." Journal of Chemical Education **68**(8): 631–634.

Timón, V., B. Maté, V. J. Herrero and I. Tanarro (2021). "Infrared spectra of amorphous and crystalline urea ices." Physical Chemistry Chemical Physics **23**(39): 22344–22351.

Tobin, M. C. (1968). "Raman Spectra of Crystalline Lysozyme, Pepsin, and Alpha Chymotrypsin." Science **161**(3836): 68–69.

Trachta, G., B. Schwarze, G. Brehm, S. Schneider, M. Hennemann and T. Clark (2004). "Near-infrared Fourier transform surface-enhanced Raman scattering spectroscopy of 1,4-benzodiazepine drugs employing gold films over nanospheres." Journal of Raman Spectroscopy **35**(5): 368–383.

Träger, F. (2007). Springer Handbook of Lasers and Optics. Dordrecht Heidelberg London New York, Springer.

Tuma, R. (2005). "Raman spectroscopy of proteins: from peptides to large assemblies." Journal of Raman Spectroscopy **36**(4): 307–319.

Tuschel, D. (2016). "Selecting an Excitation Wavelength for Raman Spectroscopy." Spectroscopy **31**(3): 14–23.

Tuschel, D. D., A. V. Mikhonin, B. E. Lemoff and S. A. Asher (2010). "Deep Ultraviolet Resonance Raman Excitation Enables Explosives Detection." Applied Spectroscopy **64**(4): 425–432.

Tuteja, M., M. Kang, C. Leal and A. Centrone (2018). "Nanoscale partitioning of paclitaxel in hybrid lipid–polymer membranes." Analyst **143**(16): 3808–3813.

Uchida, T., R. Ohmura and A. Hori (2010). "Raman Peak Frequencies of Fluoromethane Molecules Measured in Clathrate Hydrate Crystals: Experimental Investigations and Density Functional Theory Calculations." Journal of Physical Chemistry A **114**(1): 317–323.

Unal, M., R. Ahmed, A. Mahadevan-Jansen and J. S. Nyman (2021). "Compositional assessment of bone by Raman spectroscopy." Analyst **146**(24): 7464–7490.

Urena, F. P., M. F. Gomez, J. J. L. Gonzalez and E. M. Torres (2003). "A new insight into the vibrational analysis of pyridine." Spectrochimica Acta Part A – Molecular and Biomolecular Spectroscopy **59**(12): 2815–2839.

Urlaub, E., J. Popp, W. Kiefer, G. Bringmann, D. Koppler, H. Schneider, U. Zimmerman and B. Schrader (1998). "FT-Raman investigation of alkaloids in the liana Ancistrocladus heyneanus." Biospectroscopy **4**(2): 113–120.

Vagnini, M., F. Gabrieli, A. Daveri and D. Sali (2017). "Handheld new technology Raman and portable FT-IR spectrometers as complementary tools for the *in situ* identification of organic materials in modern art." Spectrochimica Acta – Part A: Molecular and Biomolecular Spectroscopy **176**: 174–182.

van Dalen, G., E. J. J. van Velzen, P. C. M. Heussen, M. Sovago, K. F. van Malssen and J. P. M. van Duynhoven (2017). "Raman hyperspectral imaging and analysis of fat spreads." Journal of Raman Spectroscopy **48**(8): 1075–1084.

van den Akker, C. C., T. Deckert-Gaudig, M. Schleeger, K. P. Velikov, V. Deckert, M. Bonn and G. H. Koenderink (2015). "Nanoscale Heterogeneity of the Molecular Structure of Individual hIAPP Amyloid Fibrils Revealed with Tip-Enhanced Raman Spectroscopy." Small **11**(33): 4131–4139.

Vandenabeele, P. (2013). Practical Raman Spectroscopy – an Introduction. Chichester, John Wiley & Sons.

Varsányi, G. (1969). Vibrational Spectra of Benzene Derivatives. New York & London, Academic Press.

Varsanyi, G., G. Horvath, L. Imre, J. Schawartz, P. Sohar and F. Soti (1977). "Infrared-Spectra of 1,2,3,5-Tetrasubstituted Benzene-Derivatives." Acta Chimica Academiae Scientarium Hungaricae **93**(3–4): 315–355.

Vašková, H. and V. Křesálek (2011). "Quasi real-time monitoring of epoxy resin crosslinking via Raman microscopy." International Journal of Mathematical Models and Methods in Applied Sciences **5**: 1197–1204.

Vedad, J., E.-R. E. Mojica and R. Z. B. Desamero (2018). "Raman spectroscopic discrimination of estrogens." Vibrational Spectroscopy **96**: 93–100.

Velázquez-Jiménez, R., J. M. Torres-Valencia, A. Valdez-Calderón, J. G. Alvarado-Rodríguez, J. D. Hernández-Hernández, L. U. Román-Marín, C. M. Cerda-García-Rojas and P. Joseph-Nathan (2016). "Absolute configuration of stegane lignans by vibrational circular dichroism." Tetrahedron: Asymmetry **27**(4): 193–200.

Venkateswarlu, K. and S. Mariam (1962). "Raman Spectra of Alcohols – Intensity Measurements." Zeitschrift für Physik **168**(2): 195–198.

Vergura, S., E. Santoro, M. Masi, A. Evidente, P. Scafato, S. Superchi, G. Mazzeo, G. Longhi and S. Abbate (2018). "Absolute configuration assignment to anticancer Amaryllidaceae alkaloid jonquailine." Fitoterapia **129**: 78–84.

Verma, S. P. and D. F. H. Wallach (1977). "Raman-Spectra of Some Saturated, Unsaturated and Deuterated C18 Fatty-Acids in HCH-Deformation and CH-Stretching Regions." Biochimica et Biophysica Acta **486**(2): 217–227.

Vincent, T., M. Hamer, I. Grigorieva, V. Antonov, A. Tzalenchuk and O. Kazakova (2019). Strongly Absorbing Nanoscale Infrared Domains within Graphene Bubbles.

Vinod, K. S., S. Periandy and M. Govindarajan (2016). "Spectroscopic [FT-IR and FT-Raman] and molecular modeling (MM) study of benzene sulfonamide molecule using quantum chemical calculations." Journal of Molecular Structure **1116**: 226–235.

Vogel, C., G. G. Hoffmann and H. W. Siesler (2009). "Rheo-optical FT-IR spectroscopy of poly(3-hydroxybutyrate)/poly(lactic acid) blend films." Vibrational Spectroscopy **49**(2): 284–287.

Volkmann, H. (1972). Handbuch der Infrarot-Spektroskopie. Weinheim, Verlag Chemie GmbH.

Volkov, S. Y., D. N. Kozlov, P. V. Nikles, A. M. Prokhorov, V. V. Smirnov and S. M. Chuksin (1981). "Infrared–CARS spectrometer with 0.001 cm^{-1} resolution in the 1900–5000 cm^{-1} range." Soviet Journal of Quantum Electronics **11**(1): 135.

Volochanskyi, O., M. Svecova and V. Prokopec (2019). "Detection and identification of medically important alkaloids using the surface-enhanced Raman scattering spectroscopy." Spectrochimica Acta Part A-Molecular and Biomolecular Spectroscopy **207**: 143–149.

Wagner, J. (1938a). "Studies on the Raman-effect. LXXXIV. Methyl derivatives." Zeitschrift für Physikalische Chemie – Abteilung B – Chemie der Elementarprozesse: Aufbau der Materie **40**(1–2): 36–50.

Wagner, J. (1938b). "Studies on the RAMAN-Effect. LXXXVI. Ethyl derivatives." Zeitschrift für Physikalische Chemie – Abteilung B – Chemie der Elementarprozesse: Aufbau der Materie **40**(6): 439–449.

Wang, D. K., K. Mittauer and N. Reynolds (2009). "Raman scattering of carbon disulfide: The temperature effect." American Journal of Physics **77**(12): 1130–1134.

Wang, F., N. Mohammadi, S. P. Best, D. Appadoo and C. T. Chantler (2021). "Dominance of eclipsed ferrocene conformer in solutions revealed by the IR spectra between 400 and 500 cm^{-1}." Radiation Physics and Chemistry **188**: 109590.

Wang, F., J. A. Wachter, F. J. Antosz and K. A. Berglund (2000). "An Investigation of Solvent-Mediated Polymorphic Transformation of Progesterone Using *in situ* Raman Spectroscopy." Organic Process Research & Development **4**(5): 391–395.

Wang, K., Y. X. Wang, R. F. Liang, J. Q. Wang and P. Qiu (2016). "Contributed Review: A new synchronized source solution for coherent Raman scattering microscopy." Review of Scientific Instruments **87**(7): 071501.

Wang, X., S. C. Huang, T. X. Huang, H. S. Su, J. H. Zhong, Z. C. Zeng, M. H. Li and B. Ren (2017). "Tip-enhanced Raman Spectroscopy for Surfaces and Interfaces." Chemical Society Reviews **46**(13): 4020–4041.

Wang, Z. Q., J. C. Chen, L. W. Li, Z. X. Zhou, Y. D. Geng and T. M. Sun (2015). "Detailed structural study of beta-artemether: Density functional theory (DFT) calculations of Infrared, Raman spectroscopy, and vibrational circular dichroism." Journal of Molecular Structure **1097**: 61–68.

Wang, Z. Q., C. J. Wu, Z. H. Wang, Z. Huang, J. Huang, J. H. Wang and T. M. Sun (2017). "Determination of absolute configuration of an isopimarane-type diterpenoid by experimental and theoretical electronic circular dichroism and vibrational circular dichroism." Journal of Molecular Structure **1146**: 484–489.

Wawer, I. (1990). "Hindered rotation in amidines" .Journal of Molecular Structure **218**: 165–167.

Wei, L., Z. X. Chen, L. X. Shi, R. Long, A. V. Anzalone, L. Y. Zhang, F. H. Hu, R. Yuste, V. W. Cornish and W. Min (2017). "Super-multiplex vibrational imaging." Nature **544**(7651): 465–470.

Weidlein, J., U. Müller and K. Dehnicke (1981). Schwingungsfrequenzen I – Hauptgruppenelemente. Stuttgart, New York, Georg Thieme Verlag.

Weidlein, J., U. Müller and K. Dehnicke (1982). Schwingungsspektroskopie. Stuttgart – New York, Georg Thieme Verlag.

Weidlein, J., U. Müller and K. Dehnicke (1986). Schwingungsfrequenzen II – Nebengruppenelemente. Stuttgart, New York, Georg Thieme Verlag.

Weisman, R. B. and J. Kono (2019). Introduction to Optical Spectroscopy of Single-Wall Carbon Nanotubes Handbook of Carbon nanomaterials R. B. Weisman and J. Kono, World Scientific Publishing Company.

Welsh, H. L., M. F. Crawford, T. R. Thomas and G. R. Love (1952). "Raman Spectroscopy of Low Pressure Gases and Vapors." Canadian Journal of Physics **30**(5): 577–596.

Wiercigroch, E., E. Szafraniec, K. Czamara, M. Z. Pacia, K. Majzner, K. Kochan, A. Kaczor, M. Baranska and K. Malek (2017). "Raman and infrared spectroscopy of carbohydrates: A review." Spectrochimica Acta Part A – Molecular and Biomolecular Spectroscopy **185**: 317–335.

Wieser, H., W. G. Laidlaw, P. J. Krueger and H. Fuhrer (1968). "Vibrational Spectra and a Valence Force Field for Conformers of Diethyl Ether and Deuterated Analogues." Spectrochimica Acta Part A – Molecular Spectroscopy A **24**(8): 1055–1089.

Wilson, E. B., J. C. Decius and P. C. Cross (1955). Molecular Vibrations: The Theory of Infrared and Raman Vibrational Spectra, McGraw-Hill.

Winkler, M., K. Steinke, R. Oehme, S. Berger, D. Sicker, H. U. Siehl and K. P. Zeller (2014). "Raffinose plus Raffinase = Refined sugar: Refined!" Chemie in unserer Zeit **48**(3): 190–199.

Wittek, H. (1942). "Studies on the Raman-effect – Announcement 136 – Nitrogen bodies XXIV – Alkyl nitrates." Zeitschrift für Physikalische Chemie – Abteilung B – Chemie der Elementarprozesse: Aufbau der Materie **52**(3): 153–166.

Wood, R. W. and D. H. Rank (1935). "The Raman spectrum of heavy chloroform." Physical Review **48**(1): 63–65.

Woodward, L. A. (1972). Introduction to the Theory of Molecular Vibrations and Vibrational Spectroscopy. Oxford, Oxford University Press.

Workman, J. J. (2000). The Handbook of Organic Compounds, Three-Volume Set: NIR, IR, R, and UV-Vis Spectra Featuring Polymers and Surfactants. New York, Academic Press.

Workman, J. J. and L. Weyer (2012). Practical Guide to Interpretive Near-Infrared Spectroscopy (English Edition). Boca Raton, CRC Press, Taylor & Francis Group.

Wu, N. N., S. L. Ouyang, Z. W. Li, J. Y. Liu and S. Q. Gao (2011). "Using Raman Spectroscopy and *ab initio* Calculations to Investigate Intermolecular Hydrogen Bonds in Binary Mixture (Tetrahydrofuran plus Water)." Chemical Research in Chinese Universities **27**(4): 693–696.

Wu, Y. H., M. Onomichi, S. Sasaki and H. Shimizu (1993). "High-Pressure Raman-Study of Liquid and Crystalline CHF_2Cl and CHF_3." Journal of Raman Spectroscopy **24**(12): 845–849.

Wu, Y. H., M. Onomichi, S. Sasaki and H. Shimizu (1994). "High-Pressure Raman-Study of Liquid and Crystalline CH_2F_2 up to 29-GPa." Journal of the Physical Society of Japan **63**(3): 934–940.

Wu, Y. H., S. Sasaki and H. Shimizu (1995). "High-Pressure Raman-Study of Dense Methane – CH_4 and CD_4." Journal of Raman Spectroscopy **26**(10): 963–967.

Wu, Y. H. and H. Shimizu (1995). "High-Pressure Raman-Study of Liquid and Crystalline CH_3F up to 12 Gpa." Journal of Chemical Physics **102**(3): 1157–1163.

Xiong, H. and W. Min (2022). Chapter 13 – Stimulated Raman excited fluorescence (SREF) microscopy: Combining the best of two worlds. Stimulated Raman Scattering Microscopy. J.-X.Cheng, W. Min, Y. Ozeki and D. Polli, Elsevier: 179–188.

Xiong, H., N. Qian, Y. Miao, Z. Zhao, C. Chen and W. Min (2021). "Super-resolution vibrational microscopy by stimulated Raman excited fluorescence." Light: Science & Applications **10**(1): 87.

Xiong, H., N. Qian, Y. Miao, Z. Zhao and W. Min (2019). "Stimulated Raman Excited Fluorescence Spectroscopy of Visible Dyes." The Journal of Physical Chemistry Letters **10**: 3563–3570.

Xiong, H., L. Shi, L. Wei, Y. Shen, R. Long, Z. Zhao and W. Min (2019). "Stimulated Raman Excited Fluorescence Spectroscopy and Imaging." Nature photonics **13**(6): 412–417.

Xue, L. J., W. Z. Li, G. G. Hoffmann, J. G. P. Goossens, J. Loos and G. de With (2011). "High-Resolution Chemical Identification of Polymer Blend Thin Films Using Tip-Enhanced Raman Mapping." Macromolecules **44**(8): 2852–2858.

Yaguchi, Y., A. Nakahashi, N. Miura, T. Taniguchi, D. Sugimoto, M. Emura, K. Zaizen, Y. Kusano and K. Monde (2015). Vibrational CD (VCD) Spectroscopy as a Powerful Tool for Chiral Analysis of Flavor Compounds. Importance of Chirality to Flavor Compounds, American Chemical Society. **1212**: 35–56.

Yamaguchi, T., M. Ohkubo, N. Ikeda and T. Nomura (1999). "Measurement of Semiconductor Surface Temperature Using Raman Spectroscopy." Furukawa Review **18**: 73–77.

Yamakita, Y., Y. Isogai and K. Ohno (2006). "Large Raman-scattering activities for the low-frequency modes of substituted benzenes: Induced polarizability and stereo-specific ring-substituent interactions." Journal of Chemical Physics **124**(10): 104301.

Yamamoto, S., H. Watabe, K. Kato and J. Hatazawa (2012). "Performance comparison of high quantum efficiency and normal quantum efficiency photomultiplier tubes and position sensitive photomultiplier tubes for high resolution PET and SPECT detectors." Medical Physics **39**(11): 6900–6907.

Yang, J., Z. E. Krix, S. Kim, J. Tang, M. Mayyas, Y. Wang, K. Watanabe, T. Taniguchi, L. H. Li, A. R. Hamilton, I. Aharonovich, O. P. Sushkov and K. Kalantar-Zadeh (2021). "Near-Field Excited Archimedean-like Tiling Patterns in Phonon-Polaritonic Crystals." ACS Nano **15**(5): 9134–9142.

Yang, Q., M. M. Liang, H. J. Wang, Q. Q. Zhao, H. J. Zhu, L. Liu and C. U. Pittman (2017). "Investigating cyclic sotolon, maple furanone and their dimers in solution using optical rotation, electronic circular dichroism and vibrational circular dichroism." Tetrahedron **73**(17): 2432–2438.

Yang, Y., Z. Y. Li, M. Nogami, M. Tanemura and Z. R. Huang (2014). "The controlled fabrication of "Tip-On-Tip" TERS probes." Rsc Advances **4**(9): 4718–4722.

Yang, Y., T. Xu, S. R. Kirk and S. Jenkins (2020). "Bond flexing, twisting, anharmonicity and responsivity for the infrared-active modes of benzene." International Journal of Quantum Chemistry **121**(8): e26584.

Yates, J. T. M., Theodore E. (1987). Vibrational Spectroscopy of Molecules on Surfaces. New York and London, Plenum Press.

You, T. T., X. F. Lang, A. P. Huang and P. G. Yin (2018). "A DFT study on surface-enhanced Raman spectroscopy of aromatic dithiol derivatives adsorbed on gold nanojunctions." Spectrochimica Acta Part A – Molecular and Biomolecular Spectroscopy **188**: 222–229.

You, T. T., X. Liang, Y. K. Gao, P. G. Yin, L. Guo and S. H. Yang (2016). "A computational study on surface-enhanced Raman spectroscopy of para-substituted Benzenethiol derivatives adsorbed on gold nanoclusters." Spectrochimica Acta Part a-Molecular and Biomolecular Spectroscopy **152**: 278–287.

Yu, A., D. Zuo and X. Wang (2019). "Optimization of parabolic cell for gas Raman analysis." Journal of Raman Spectroscopy **50**: 731–740.

Yu, Y. Q., K. Lin, X. G. Zhou, H. Wang, S. L. Liu and X. X. Ma (2007). "New C-H stretching vibrational spectral features in the Raman spectra of gaseous and liquid ethanol." Journal of Physical Chemistry C **111**(25): 8971–8978.

Yurdakul, S., N. C. Yasayan and S. Badoglu (2014). "FT-IR, FT-Raman Spectra and Quantum Mechanical Study of Piperidine-3-carboxylic Acid and Its Tautomers, Isomers." Optics and Spectroscopy **116**(6): 906–918.

Yurtseven, H. and E. C. Iseri (2013). "Raman Study of Benzene near the Melting Point." International Journal of Modern Physics B **27**(14): 1350065.

Yurtseven, H. and H. Özdemir (2017). "Raman frequencies calculated as functions of temperature and pressure using volume data for solid phase I of benzene." Optik **144**: 224–231.

Zaitsu, S.-i. and T. Imasaka (2014). "Intracavity Phase-matched Coherent *Anti*-Stokes Raman Spectroscopy for Trace Gas Detection." Analytical Sciences **30**(1): 75–79.

Zając, G. and P. Bouř (2022). "Measurement and Theory of Resonance Raman Optical Activity for Gases, Liquids, and Aggregates. What It Tells about Molecules." The Journal of Physical Chemistry B **126**(2): 355–367.

Zapata, F., M. A. F. de la Ossa, E. Gilchrist, L. Barron and C. Garcia-Ruiz (2016). "Progressing the analysis of Improvised Explosive Devices: Comparative study for trace detection of explosive residues in handprints by Raman spectroscopy and liquid chromatography." Talanta **161**: 219–227.

Zeller, M. V. and M. P. Juszli (1975). Infrarot-Vergleichsspektren von Mineralien. Überlingen, The Perkin-Elmer Corporation: 37.

Zeng, Y. B., X. L. Liu, Y. Zhang, C. J. Li, D. M. Zhang, Y. Z. Peng, X. Zhou, H. F. Du, C. B. Tan, Y. Y. Zhang and D. J. Yang (2016). "Cantharimide and Its Derivatives from the Blister Beetle *Mylabris phalerata* Palla." Journal of Natural Products **79**(8): 2032–2038.

Zeng, Z. C., S. C. Huang, D. Y. Wu, L. Y. Meng, M. H. Li, T. X. Huang, J. H. Zhong, X. Wang, Z. L. Yang and B. Ren (2015). "Electrochemical Tip-Enhanced Raman Spectroscopy." Journal of the American Chemical Society **137**(37): 11928–11931.

Zerbi, G. and P. J. Hendra (1968). "Laser-excited Raman spectra of polymers: Hexagonal and orthorhombic polyoxymethylene." Journal of Molecular Spectroscopy **27**(1): 17–26.

Zhan, G. Q., J. F. Zhou, R. Liu, T. T. Liu, G. L. Guo, J. P. Wang, M. Xiang, Y. B. Xue, Z. W. Luo, Y. H. Zhang and G. M. Yao (2016). "Galanthamine, Plicamine, and Secoplicamine Alkaloids from Zephyranthes candida and Their *Anti*-acetylcholinesterase and *Anti*-inflammatory Activities." Journal of Natural Products **79**(4): 760–766.

Zhang, K., B. Guo, P. Colarusso and P. F. Bernath (1996). "Far-Infrared Emission Spectra of Selected Gas-Phase PAHs: Spectroscopic Fingerprints." Science **274**(5287): 582–583.

Zhang, M., J. Wang and Q. Tian (2014). "Tip-enhanced Raman spectroscopy mapping with strong longitudinal field excitation." Optics Communications **315**: 164–167.

Zhang, R., Y. Zhang, Z. C. Dong, S. Jiang, C. Zhang, L. G. Chen, L. Zhang, Y. Liao, J. Aizpurua, Y. Luo, J. L. Yang and J. G. Hou (2013). "Chemical mapping of a single molecule by plasmon-enhanced Raman scattering." Nature **498**(7452): 82–86.

Zhang, Y., Y.-R. Zhen, O. Neumann, J. K. Day, P. Nordlander and N. J. Halas (2014). "Coherent *anti*-Stokes Raman scattering with single-molecule sensitivity using a plasmonic Fano resonance." Nature Communications **5**: 4424.

Zhang, Y. F., M. R. Poopari, X. L. Cai, A. Savin, Z. Dezhahang, J. Cheramy and Y. J. Xu (2016). "IR and Vibrational Circular Dichroism Spectroscopy of Matrine- and Artemisinin-Type Herbal Products: Stereochemical Characterization and Solvent Effects." Journal of Natural Products **79**(4): 1012–1023.

Zhang, Z. L., L. Chen, S. X. Sheng, M. T. Sun, H. R. Zheng, K. Q. Chen and H. X. Xu (2014). "High-vacuum tip enhanced Raman spectroscopy." Frontiers of Physics **9**(1): 17–24.

Zhao, Y., N. Ji, L. H. Yin and J. Wang (2015). "A Non-invasive Method for the Determination of Liquid Injectables by Raman Spectroscopy." AAPS PharmSciTech **16**(4): 914–921.

Zhao, Z. L., Y. H. Shen, F. H. Hu and W. Min (2017). "Applications of vibrational tags in biological imaging by Raman microscopy." Analyst **142**(21): 4018–4029.

Zhou, R., L. Chen, T. Kong, H. Chen, Z. Zhang, H. Shen and H. Zheng (2021). "The Vector Beam Assisted "hot-spot" optimization in tip-enhanced Raman spectroscopy." Journal of Raman Spectroscopy **52**(10): 1698–1704.

Zhuravlev, K. K., K. Traikov, Z. H. Dong, S. T. Xie, Y. Song and Z. X. Liu (2010). "Raman and Infrared Spectroscopy of Pyridine under High Pressure." Physical Review B **82**(6): 064116.

Ziegler, L. D. (1990). "Hyper-Raman spectroscopy." Journal of Raman Spectroscopy **21**(12): 769–779.

Ziegler, L. D. (2001). Instrumentation for Raman Spectroscopy – Hyper-Raman Spectroscopy. Handbook of Vibrational Spectroscopy: 1 –12.

Zietlow, J. P., F. F. Cleveland and A. G. Meister (1950a). "Raman Spectrum and Force Constants for Fluorotrichloromethane." Physical Review **77**(5): 739–740.

Zietlow, J. P., F. F. Cleveland and A. G. Meister (1950b). "Substituted Methanes .3. Raman Spectra, Assignments, and Force Constants for Some Trichloromethanes." Journal of Chemical Physics **18**(8): 1076–1080.

Zou, Q., X. Y. Du, C. Zhang, X. C. Li and Y. Li (2013). "Vibrational Assignment Analysis of Raman Spectra of Fatty Alcohols." Spectroscopy and Spectral Analysis **33**(1): 106–110.

Author index

https://doi.org/10.1515/9783110717556-026

Compound index

https://doi.org/10.1515/9783110717556-027

Formula index

https://doi.org/10.1515/9783110717556-028

Subject index

https://doi.org/10.1515/9783110717556-029

www.ingramcontent.com/pod-product-compliance
Lightning Source LLC
Chambersburg PA
CBHW080135220326
41598CB00032B/5077